# Student Study Guide

## to accompany

# Chemistry

## Tenth Edition

**Raymond Chang**
*Williams College*

**Prepared by**
**Kim Woodrum**
*University of Kentucky*

 **Higher Education**

Boston   Burr Ridge, IL   Dubuque, IA   New York   San Francisco   St. Louis
Bangkok   Bogotá   Caracas   Kuala Lumpur   Lisbon   London   Madrid   Mexico City
Milan   Montreal   New Delhi   Santiago   Seoul   Singapore   Sydney   Taipei   Toronto

Student Study Guide to accompany
CHEMISTRY, TENTH EDITION
RAYMOND CHANG

Published by McGraw-Hill Higher Education, an imprint of The McGraw-Hill Companies, Inc., 1221 Avenue of the Americas, New York, NY 10020. Copyright © 2010, 2007, 2005, 2002, and 1998 by The McGraw-Hill Companies, Inc. All rights reserved.

This book is printed on recycled, acid-free paper containing 10% post consumer waste.

3 4 5 6 7 8 9 0 QDB/QDB 11

ISBN: 978–0–07–322676–7
MHID: 0–07–322676–9

www.mhhe.com

# Table of Contents

# Preface

General Chemistry is a course required by many different majors. Most students who are required to take this course are not chemistry majors. At most institutions, chemistry majors make up only a small percentage of the total enrollment. Chemistry is sometimes referred to as the "Central Science" because it is used in so many other areas. And while many students understand that the chemistry they learn will serve them in their chosen major, as I stand in front of my students, I know that many wonder why they are there. The civil engineer, for example, could be sitting there thinking, "This course is a waste of time. I'll never use this for my degree." To put energy into an endeavor, it is important to understand why. Why bother putting the 5-10 hours a week into a class? Even if you don't see the application of the material, right now, you need to know that the skills you will develop in this course will serve you in all your courses and beyond, even if the chemistry is never used again. The courses you are required to take were not put together lightly. Much thought and debate goes into deciding what courses a student must take. But what you develop in this course is much more than chemistry knowledge. It is a road (and for many students, their first road) on a journey to become a "problem solver". I find that some freshmen have entered college being pretty good memorizers and mimics. They can quote back definitions, they can follow a set of procedures, but if given a new challenge (a new problem to solve) which uses the same skills, they are lost.

Your time in General Chemistry as not only a time to learn theories and laws on the subject but also to develop problem solving skills. Let's compare being a good problem solver to being a good carpenter. A good carpenter learns how to use certain tools; a hammer, a saw, a sander, a planer. He knows what the tools do and when to use them. When the master carpenter builds a beautiful piece of furniture, he is not going to need someone to say, "Pick up a board. Now pick up the saw and cut 3-inches off one end...." The carpenter will use the tools and skills as he decides; based on the design he has planned. You, as a problem solver will need to do the same. In this course you must 1) learn how to use the tools and then 2) learn when these tools are useful so you can decide when to use them. You must get away from the notion that you must see an example just like an assigned problem in order to work the problem. You must also get away from thinking that a test should only consist of problems just like the ones you have already worked. Your instructor wants to test you on both the use of tools and whether you can pick them up and use them as necessary. They don't want to know if you merely memorized some facts and can mimic how someone else solves problems.

The Student Study Guide will summarize and emphasize points as discussed in Chang's Chemistry 9th Edition. In it, you will learn to use the "tools" of chemistry.

The icon to the left will point out the tools that must be mastered. As you master the use of the tools, you will then be able to develop problem solving skills. Often the tools needed are mathematical tools. There is no way around the fact that much of your success in General Chemistry is dependent on being able to use the pre-algebra and algebra skills you have already developed.

Each chapter contains:

- A summary of the material covered in the text, with tools indicated where appropriate. Some students may want to start here, reading through the summary to gain a different perspective of topics covered in the textbook. The summary content is built upon the work of Sharon Neal (8th Edition) and Ken Watkis (7th Edition.)

- A glossary of new terms used in the chapter, organized by association. This is the same list which is present in your textbook.
- A list of equations presented in the chapter.
- Worked Examples, with references to matched "Exercises & Problems."
- Exercises & Problems section. Some students may want to start here, and refer to the summary and Worked Examples, when needed.
- Practice test. To give you a good sense of how well you know the material, try the following:
    - Detach Practice Test pages which apply to the exam you will be taking.
    - Cut the questions apart along the dashed lines and randomly pick several from each chapter.
    - Find a quiet place, without notes, and work through the questions.
    - Grade it. The back of each question gives the chapter number. This will aide you in finding the answer in the appendix.
    - If you didn't do well, repeat after going over the ones you missed. *Seek out help, when needed.*

Thank you to Raymond Chang, and the McGraw-Hill team, for granting me this opportunity. Thanks to Kevin (my husband) and Logan, Jenni and Megan (my kids) for their patience with me as I spent too much time in front of the computer!

# Chapter 1.  Chemistry:  The Study of Change

**Introduction (Sections 1.2 – 1.3)**
**Matter and Its Properties (Sections 1.4 – 1.6)**
**Recording Measurements (Sections 1.7 – 1.8)**
**Calculations with Observed Quantities (Section 1.9)**


## SUMMARY

### Introduction (Sections 1.2 – 1.3)

***Chemistry & the Scientific Method.*** Chemistry is the study of matter and the changes matter undergoes.  Matter is defined as anything that has mass and occupies space.  The substances that make up matter have unique properties that chemists use to identify them.  Chemists are concerned with developing tools to study matter and concepts to describe the properties of matter.  They direct their efforts toward purposefully changing given forms of matter into new and different substances, and to the discovery of the properties and uses of these new materials.  Chemists usually observe matter and the changes it undergoes using macroscopic (large enough to be seen using the naked eye) amounts of material.  However, they interpret the properties and changes in matter in terms of atoms and molecules, microscopic entities that are too small to be seen.

Chemists use the scientific method to guide their search for new knowledge and to deduce facts about the microscopic world of atoms and molecules from macroscopic observations of reactions and properties.  The scientific method can be broken down into four stages:  observation and experiment, hypothesis formation, law development, and theory adoption.  The first stage involves defining the problem clearly and suggesting a systematic, logical approach towards a possible solution.  The investigator must decide what s/he is trying to find out and what approach to take.  During experiments, observations and measurements are made and information collected about the phenomenon being studied.  After significant amounts of data related to the phenomenon have been collected, the relationships indicated by the data are collected into a simple verbal or mathematical statement, often called a law.  At this point, a hypothesis may be formulated to explain the facts summarized by the law.  A hypothesis is only temporary and is meant to serve as a working model that is adjusted as new information about the phenomenon is discovered.  As a hypothesis develops and survives many experimental tests, it becomes a theory.  A theory is a unifying principle that explains a large body of observations and laws.  Theories are tested using carefully devised experiments that confirm or disprove their predictions.  A theory is important when it leads to predictions across a wide range of observations.  The theory of gravitational motion is a famous example of this process.  Holy men, stargazers and sailors mapped the position of the stars and planets for centuries.  In the late seventeenth century, Newton summarized these observations using a single inverse squared law.  The validity of the law was established by the use of Newton's theories to locate unknown planets.  Newton's theory stood uncontested for more than one hundred years until Einstein's theory of general relativity explained its limitations around very dense objects.

Examples
1.1, 1.2

Exercises
1, 2

## Matter and Its Properties (Sections 1.4 – 1.6)

***Elements and Compounds.*** A pure substance is a form of matter that has a fixed composition and distinct properties. Examples are water, table salt, and aluminum. Just as each individual person has a set of unique characteristics, such as fingerprints, voiceprint and retinal pattern, each pure substance has characteristic properties, such as density, color, and hardness. There are two types of pure substances: elements and compounds. An element is a pure substance that cannot be decomposed (broken down) into simpler substances by chemical reaction. (One element may be converted to another by nuclear reactions, more about this in Chapter 23.) Elements are the building blocks of which all compounds are composed. Oxygen (O), mercury (Hg) and iron (Fe) are examples of elements. Table 1.1 below, lists common elements and their symbols. Memorize these elements soon unless your instructor has other plans. Compounds are pure substances that are composed of two or more elements in fixed proportions. Water is a compound that contains twice as much hydrogen as oxygen ($H_2O$). Compounds can be broken down into simpler substances by chemical reactions: electrolysis (a reaction induced by passing current) of water produces hydrogen and oxygen gas; decomposition (a reaction induced by heating) of chalk produces calcium oxide and carbon dioxide.

### Table 1.1  Common Elements & Symbols

| Symbol | Element | Symbol | Element | Symbol | Element |
|--------|---------|--------|---------|--------|---------|
| Na | sodium | C | carbon | O | oxygen |
| Al | aluminum | N | nitrogen | Cl | chlorine |
| Fe | iron | S | sulfur | I | iodine |
| Cu | copper | Si | silicon | F | fluorine |
| Au | gold | H | hydrogen | He | helium |
| Ag | silver | K | potassium | Ne | neon |

***Mixtures.*** When pure substances are combined, they form mixtures. Unlike elements and compounds, mixtures do not always consist of specific proportions of substances. For example, we can buy milk that contains >3.5% (whole), 2% (reduced fat) or 0% (skim) fat. All three of these are mixtures of water, lactose (sugar), protein and fat. Mixtures can be homogeneous or heterogeneous depending on the extent of intermingling of the pure components. Sugar water is a homogeneous mixture; a mixture which is uniform throughout. The sugar crystals dissolve in the water and disperse evenly throughout the water. Homogeneous mixtures are called solutions, especially when the mixture is fluid. Granite is a heterogeneous mixture of rocks, namely quartz, feldspar and mica. The component rocks are not evenly dispersed; the characteristics of a heterogeneous mixture. The pieces of the components in the mixture are large enough for us to see them by eye. A mixture can be separated into its components by physical methods, such as filtration.

***Properties of Matter.*** Chemistry is the study of matter and the changes it undergoes. Changes in matter are detected by observing changes in matter properties. Physical properties are those that can be observed and measured without changing the identity or chemical composition of the substance. Physical properties include color, hardness, density, melting point and boiling point. Heating ice (solid) above its melting point produces water (liquid). Additional heating above the boiling point produces steam (gas), but all three forms are the same substance: $H_2O$. Solid, liquid, and gas are three states of matter. In a solid, particles of matter are close and move very little. (The particles will be described in Chapter 2.) The distance between particles <u>and</u> extent of particle motion increases a little in liquids, but increases substantially in gases. Observation of chemical properties can only occur as a substance is converted to another species (element or compound) in a chemical reaction. Chemical properties include reactivity to

| |
|---|
| Examples<br>1.3 - 1.4 |
| Exercises<br>3. 4 |

acids, bases, hydrogen, oxygen or water.  Potassium, a metal, reacts violently (bubbling and fire!) with water to produce potassium hydroxide, KOH, but gold, also a metal, is stable in water indefinitely.  All properties of matter are either extensive or intensive.  Extensive properties vary with the amount of matter being considered.  For example, the more material there is, the larger the mass or volume. Intensive properties are independent of the amount of material.  For example, the color of a small piece of pure gold is the same as the color of a large piece of gold.

## Recording Measurements (Sections 1.7 – 1.8)

*The International System of Units.*  Scientists and engineers have adopted a uniform, international system of units for labeling observations.  SI (abbreviation for Le Systeme International) is a metric system built on a foundation of base units for fundamental phenomena such as mass, length (distance) and time.  A list of the base SI units used frequently in this text is given in Table 1.2.  Units for other phenomena can be derived (calculated) from the base units.

### Table 1.2  Base SI Units

| Quantity | Name of Unit | Symbol |
|---|---|---|
| mass | kilogram | kg |
| length | meter | m |
| time | second | s |
| temperature | Kelvin | K |
| amount | mole | mol |

*Base and Derived Units.*  The mass of an object is a measure of its resistance to external forces.  The greater the mass, the less external forces will affect an object's motion.  A tractor-trailer requires brakes that exert more friction (force) on the wheel than a compact car does, because it has a much larger mass.  The SI base unit for mass is the kilogram (kg).  It would make more sense if the base unit was the gram (no prefix), but grams are small relative to the size of people and most of our possessions, so the kilogram wins out.  The terms mass and weight are often used interchangeably, although they actually refer to distinct properties.  The mass is a measure of the amount of matter in an object and is the same everywhere in the universe.  The weight is the force exerted on the object by gravity.  The weight of an object on the moon is 1/6$^{th}$ its weight on the earth because the moon's gravity is six times smaller than the earth's.  If you weigh 150 lbs on earth, you would weigh only 25 lbs on the moon.

The volume of an object is a measure of the space it occupies.  The larger the volume, the more space the object requires.  The SI unit for volume is a derived unit.  The volume of a cube or rectangular solid is the product of the length, width and height, all measured in meters.

$$V = \text{length} \times \text{width} \times \text{height}$$
$$= \text{m} \times \text{m} \times \text{m} = \text{m}^3$$

Of course, the units most scientists use for volume are based on the liter (L).  The liter is defined as the volume of a cube that has 10 cm (= 0.1 m) edges.

$$V = 0.1\,\text{m} \times 0.1\,\text{m} \times 0.1\,\text{m} = 0.001\,\text{m}^3 = 1\,\text{L}$$

A convenient relationship to remember is 1 mL = 1 cm$^3$.

Example
1.5

Exercises
5, 6

The density of an object is the ratio of the mass to the volume, or the mass per unit volume, so it also has a derived unit. The ratio of the fundamental units for mass and volume is $kg/m^3$, but this unit is rarely used. Densities of liquids and solids are usually reported in g/mL or $g/cm^3$. Densities of gases are usually reported in g/L. Look back at the definition of intensive vs. extensive properties. Is density intensive or extensive? If you consider two pieces of gold, one twice as big as the other, will they have different densities? The mass of the two pieces doubles but so does the volume. Density = mass/volume so the value for density does not change. Density is intensive.

| Conversion between | Tool |
|---|---|
| g $\leftrightarrow$ mL (or $cm^3$) | density |

**Unit Prefixes.** Rather than report observations using very small or very large numbers, SI units are scaled (multiplied) by decimal prefixes. See Table 1.3. To make sense of this table, let us focus on the prefix kilo. Look at the table: What does kilo mean? Kilo means 1000. Therefore if something has as mass of 5 kg, it has a mass of 5000 g. If the prefix means something smaller than one, you can think of it in two different ways. For example, look at the table again: what does centi mean? Centi means $10^{-2}$ (one hundredth). If something is 1 cm long, it is 0.01 m, or it takes 100 cm to make 1 m. You need to memorize the prefix and the definition of each. The ones in boldface type are used most frequently in general chemistry.

**Table 1.3  Commonly Used SI Prefixes**

| Prefix | Symbol | Definitions | |
|---|---|---|---|
| tera | T | 1,000,000,000,000 | $10^{12}$ |
| giga | G | 1,000,000,000 | $10^{9}$ |
| mega | M | 1,000,000 | $10^{6}$ |
| **kilo** | **k** | **1,000** | $10^{3}$ |
| deci | d | 0.1 | $10^{-1}$ |
| **centi** | **c** | **0.01** | $10^{-2}$ |
| **milli** | **m** | **0.001** | $10^{-3}$ |
| micro | $\mu$ | 0.000001 | $10^{-6}$ |
| **nano** | **n** | **0.000000001** | $10^{-9}$ |
| pico | p | 0.000000000001 | $10^{-12}$ |

| Conversion between | Tool |
|---|---|
| g $\leftrightarrow$ kg (for example) | Prefix Meanings (Table 1.3 above) |

**Significant Figures.** It is not enough to label all measurements and calculations with appropriate units, the numerical values must be reported in a way that does not mislead the reader about the precision of the measurement. If a scientist reports that a sample weighs 0.12546 g, the number implies that the balance used could distinguish 0.1254 g from 0.1255 g. If that balance is only reliable to within 1 milligram, the value is being reported using too many digits, called significant figures. Scientists have adopted the convention of using all the certain digits and the first uncertain digit to report a measurement. Therefore, the scientist should report the weight as 0.1255 g, even if the

Examples
1.6 – 1.10

Exercises
8 – 13

display on the balance provides more digits. This notation informs the reader that the measurement is really 0.1255 ± 0.0001 g and the last digit is uncertain.

Consider the diagram below. The ruler is marked off in mm (that is, 1/10 of a cm). To measure the length of the pencil, we see that it is on or close to the mark "4.7 cm." We can then estimate one more place, by imagining 10 subdivisions between the 4.7 and 4.8. I'll estimate the pencil as just past the 4.7 mark and call it 4.71 cm. You might say, "It looks right on the line." If so, you would call it 4.70 cm.

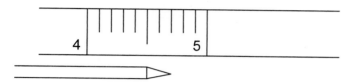

We know two digits with certainty, but the last digit is uncertain but significant. It lets the reader of the measurement 4.71 cm know that the pencil is somewhere between 4.70 and 4.72 cm, that is, 4.71 ± 0.01 cm.

***Algorithm for Identifying Significant Figures.*** To determine the number of significant digits in a written number use the following rules:
1. All nonzero digits are significant (this includes the 1 uncertain digit).
2. Zeros between nonzero digits are significant.
3. Leading zeros, to the left of the first nonzero digit, are **not** significant. These zeros only locate the decimal point and reveal nothing about certainty. Thus, 0.6 g and 0.0006 g both have 1 significant figure. Watch out. This is probably the most frequently forgotten rule.
4. Trailing zeros, to the right of the last nonzero digit, are definitely significant if the number contains a decimal point. Therefore, 1.0 mg has two significant figures, 0.500 mg has three significant figures and 600.00 g has five significant figures.
5. Trailing zeros are ambiguous if they appear in numbers without a decimal point. The number 450 g has 2 significant figures if it represents 450 ± 10 g and three if it represents 450 ± 1 g. To avoid this ambiguity, use scientific notation (see Math review in Appendix I). The number 450 g should be written as $4.5 \times 10^2$ g if it has two significant figures, and as $4.50 \times 10^2$, if it has three.
6. Numbers obtained from definitions are exact. For example, acceleration due to gravity (here on earth) is exactly 9.8 $m/s^2$. In a calculation, this value would be treated as if it had an infinite number of significant figures rather than two.

***Algorithms for Calculating with Significant Figures.*** In many cases, the results we seek are calculated from the measurements we make. The precision (number of significant figures) is limited by the least precise measurement. For example, when we want to know the density of an object, we measure its mass and volume and then calculate their ratio. The number of significant figures depends on the accuracy of the balance and glassware we used to measure the mass and volume, not the number of digits our calculator displays when we calculate the ratio. Say an investigator needs to determine if an object is metal quickly and cheaply, its density is a good indicator. The mass of the object is 2.1572 g (5 significant figures) and the volume is measured as 1.66 mL (3 significant figures). The density of the object is 1.30 g/mL (3 significant figures) even though the calculator reads 1.29939759.

To determine the number of significant figures in a calculation use the following rules:
1.  The number of significant figures in a product (x) or quotient (÷) is the number of significant figures in the least accurate factor
2.  The number of significant figures in a sum (+) or difference (-) depends on the number of digits to the right of the decimal point (decimal digits) rather than the number of significant figures in the components. In general chemistry calculations we multiply and divide more frequently so students remember rule one and forget about this rule.

| Addition | Subtraction |
|---|---|
| 102.226 g | |
| 2.51  g | 102.25 g |
| + 96    g | - 99.3  g |
| Sum =  200.736 g | Difference =  2.95 g |
| Correctly reported result =   201    g | Correctly reported result =   3.0 g |

If adding two numbers which are in scientific notation but have different exponents, the numbers must be converted to a common exponent. For example,

| | |
|---|---|
| $(4.5 \times 10^{-5}) + (1.5 \times 10^{-6})$ | Convert one of the numbers to match the others exponent. We will convert the -6 to -5 |
| $(4.5 \times 10^{-5}) + (0.15 \times 10^{-5})$ | Now, add the coefficients (i.e., the numbers in front) |
| $(4.5 + 0.15) \times 10^{-5} = 4.65 \times 10^{-5}$ | Use your rounding rules (least number of decimal places) |
| $4.7 \times 10^{-5}$ | This is YOUR ANSWER to the correct significant figures. |

The significant figure convention was adopted to insure that measurement values reflect the precision of the measurement; the uncertainty associated with precise measurements is small. The precision says nothing about the accuracy of a measurement. Accurate measurements are close to the true value of the quantity under observation. Precise measurements can be inaccurate, as when a scale consistently reads 5 lbs too low. Accurate measurements can be imprecise, but this is less probable. The careful work that leads to accurate measurements, generally leads to small measurement uncertainty too.

## Calculations with Observed Quantities (Section 1.9)

*Dimensional Analysis.* The key to dimensional analysis (also called the factor-label method) is to consider the units as an essential part of any quantity. We should not even think of a quantity without its units. The mass of a sample is never 5.00; it is 5.00 g. In everyday speech, it is acceptable to say that a man is "six-three" and leave the units implied. In scientific speech, he is always 6 feet 3 inches tall. The great benefit of this is that when quantities labeled with the correct units are subjected to calculations, the units are correctly computed as well.

Dimensional analysis uses two math facts.
1.  A number divided by itself is one.
2.  A number multiplied by 1 gives that number.

Let's use donuts as an example. A dozen donuts are equivalent to 12 donuts so;

$$\frac{1 \text{ dozen donuts}}{12 \text{ donuts}} = 1 \qquad \text{or} \qquad \frac{12 \text{ donuts}}{1 \text{ dozen donuts}} = 1 \qquad \text{(Fact 1)}$$

If you had 4 ½ dozen donuts and wanted to use dimensional analysis to determine the actual number of donuts, the conversion would look like this:

$$4.5 \text{ dozen donuts} \times \frac{12 \text{ donuts}}{1 \text{ dozen donuts}} = 54 \text{ donuts} \qquad \text{(Fact 2)}$$

Note that the units "dozen donuts" will now cancel out leaving you only the units "donuts."

In general, conversion factors are simple ratios. Any time units of a given quantity are a ratio (as in m/s, g/mL, etc.) a corresponding conversion factor can be written. For example if the density of a substance is given as 1.78 g/mL, we can write the density as the following conversion factor:

$$\frac{1.78 \text{ g}}{1 \text{ mL}}$$

Conversion factors can be used according to the following general form

$$\text{desired value} = \text{given value} \times \frac{\# \text{ desired unit}}{\# \text{ given unit}} .$$

Some conversions, for example temperature conversions, are complicated by differences in the reference value. The conversion factor for donuts to dozen is a simple ratio because 0 donuts = 0 dozen. Freezing water is 32° on the Fahrenheit scale but 0° on the Celsius scale, so the conversion from °F to °C consists of a subtraction to adjust for the difference in the reference temperatures in addition to the ratio between units. The generalized formula becomes

$$\text{desired value} = (\text{given value} - \text{given reference}) \times \frac{\# \text{ desired unit}}{\# \text{ given unit}}$$

or

$$\text{desired value} = \text{given value} \times \frac{\# \text{ desired unit}}{\# \text{ given unit}} + \text{desired reference}$$

The ratio of "desired units" to "given units" is called a conversion factor. The temperature of boiling water is 212 °F. The conversion factor is 5 °C for every 9 °F, or 5 °C per 9 °F. On the Celsius scale 212 °F is

$$\left( 212 \text{ °F} - 32 \text{ °F} \right) \times \frac{5 \text{ °C}}{9 \text{ °F}} = 100 \text{ °C} .$$

Examples
1.11–1.15

Exercises
14 – 18

Example
1.16

Exercises
19, 20

The formula for conversions from Celsius to Fahrenheit can be demonstrated as follows: to convert 100 °C to °F:

$$100 \; °C \times \frac{9 \; °F}{5 \; °C} + 32 \; °F = 212 \; °F$$

The formula for conversions from Celsius to Kelvin is

$$(100 \; °C + 273.15 \; °C) \times \frac{1 \; K}{1 \; °C} = 373.15 \; K$$

## GLOSSARY LIST

| | | | |
|---|---|---|---|
| chemistry | substance | extensive properties | density |
| scientific method | element | intensive properties | Kelvin |
| qualitative | compound | macroscopic properties | significant |
| quantitative | mixture | microscopic properties | figures |
| hypothesis | homogeneous mixture | international system of units | accuracy |
| law | heterogeneous mixture | weight | precision |
| theory | physical properties | mass | |
| matter | chemical properties | volume | |

## EQUATIONS

| Algebraic Equation | English Translation |
|---|---|
| $d = \dfrac{m}{V}$ | The density is the mass per unit volume. Usually in g/mL or g/cm³. |
| $y \; °C = (x \; °F - 32 \; °F) \times \dfrac{5 \; °C}{9 \; °F}$ | Used to convert from °F to °C. Formula simplified: °C=(°F-32)5/9 |
| $y \; °F = x \; °C \times \dfrac{9 \; °F}{5 \; °C} + 32 \; °F$ | Used to convert from °F to °C. Formula simplified: °F=(9/5)°C + 32 |
| $y \; K = (x \; °C + 273.15 \; °C) \times \dfrac{1 \; K}{1 \; °C}$ | Used to convert from °C to K. Formula simplified: K=°C + 273.15 |
| $y \; °C = (x \; K - 273.15 \; K) \times \dfrac{1 \; °C}{1 \; K}$ | Used to convert from K to °C. Formula simplified: °C=K − 273.15 |

## WORKED EXAMPLES

---

**EXAMPLE 1.1 Element, Compound or Mixture**

Classify each of the following as an element, compound or mixture.
a.     Aluminum foil                                  c.       Kool-aid
b.     Table salt

**• Solution**

At the early stages of your chemical education, you may not have the body of knowledge to address these topics properly. If it appears on the periodic table, it will be an element. To decide between a compound and a mixture is sometimes a bit more difficult.
a.   Aluminum foil is simply made of aluminum flattened into sheets. This is an element (atomic number 13).
b.   Table salt is a compound it consists of sodium and chlorine chemically combined with the formula NaCl.
c.   If you think of how you make Kool-aid, you will decide it is a mixture. Sugar, water and flavoring are mixed together. Each substance retains its identity. No reaction occurs.

**Work EXERCISES & PROBLEMS: 1**

---

**EXAMPLE 1.2  Homogeneous or Heterogeneous Mixture**

Classify each of the following as a homogeneous or a heterogeneous mixture.
a.     The beverage tea                              c.       Cow's milk
b.     Oil and water                                    d.       Wine

**• Solution**

Recall that homogeneous mixtures are uniform throughout, while heterogeneous mixtures have components that can be observed to be physically separate.
a.   The mixture called tea is uniform throughout and no particles of tea can be observed (if there are no tea leaves). It is a homogeneous mixture.
b.   Since oil floats on water, the oil and water components can be observed to be separate. Samples from the top part of the mixture have different properties from samples taken from the bottom. It is a heterogeneous mixture.
c.   Cow's milk contains fats and solids suspended in water.  On standing, the cream (fat) will rise to the top and the solids will settle. It is a heterogeneous mixture.
d.   Wine contains ethyl alcohol, water, flavor components, color components, and other substances.  The composition is uniform throughout.  No solid particles are visible in the liquid; it is clear. It is a homogeneous mixture.

**Work EXERCISES & PROBLEMS: 2**

---

**EXAMPLE 1.3  Chemical and Physical Changes**

Classify the following changes as physical or chemical.
a.  Solid shortening turns to liquid when heated
b.  Gasoline burning in air

• **Solution**

a.  Solid fats melt at a fairly low temperature. Melting is a physical change.  Melting can be reversed by lowering the temperature.  Removing the source of heat in this case will cause the fat to solidify.
b.  Gasoline undergoes combustion with oxygen (burning) to produce carbon dioxide, water and heat.  Removing the heat does not regenerate gasoline, so burning is a chemical change.

**Work EXERCISES & PROBLEMS: 3**

---

**EXAMPLE 1.4  Chemical and Physical Properties**

The following are properties of the element silicon; classify them as physical or chemical properties.
a.  Melting point, 1410 °C
b.  Reacts with fluorine to form silicon tetrafluoride
c.  Gray color
d.  Not affected by most acids

• **Solution**

Physical properties can be observed without a change in composition, while chemical properties describe reactions with other substances.
a.  Melting involves a change in physical state but no chemical change. The melting point is a physical property.
b.  This statement describes the change of silicon into another substance on reaction with fluorine. The reaction is a chemical property.
c.  The color of a substance is a physical property. No change in composition occurs while observing the color.
d.  The lack of reactivity with another substance or class of substances such as acids is a chemical property.

**Work EXERCISES & PROBLEMS: 4**

---

**EXAMPLE 1.5  Density**

A flask filled to the 25.0 mL mark contained 29.97 g of a concentrated salt–water solution. What is the density of the solution?

• **Solution**

The density (d) of an object or a solution is defined as the ratio of its mass (m) to its volume (V). Substituting the given quantities:

$$d = \frac{m}{V} = \frac{29.97 \text{ g}}{25.0 \text{ mL}} = 1.20 \text{ g/mL (rounded to 3 significant figures)}$$

**Work EXERCISES & PROBLEMS: 5, 6**

**EXAMPLE 1.6  Scientific Notation**

Write the following numbers using scientific (exponential) notation.
a.  7,620,000                    b.       0.000495

• **Solution**

a.  7,620,000—Count the number of places that the decimal point must be moved to the *left* to give a number between 1 and 10, in this case, 7.62.  Since this requires six places, the exponent of 10 must be 6. The value is $7.62 \times 10^6$.

b.  0.000495—Fractional numbers like this have negative exponents in scientific notation. Count the number of places that the decimal point must be moved to the *right* to give 4.95. Since this requires a move of four places, the exponent must be –4. The value is $4.95 \times 10^{-4}$.

**Work EXERCISES & PROBLEMS: 7, 8**

**EXAMPLE 1.7  Math Operations with Scientific Notation**

Multiply $4.0 \times 10^4$ by $3.5 \times 10^{-6}$.

• **Solution**

$$(4.0 \times 10^4)(3.5 \times 10^{-6})$$

Regroup so that the coefficients are separated from the exponentials, and add the exponents.

$$(4.0 \times 3.5) \times 10^{4+(-6)} = 14 \times 10^{-2} = 1.4 \times 10^{-1}$$

• **Calculator Calculation**

Most people use an electronic calculator for these types of calculations.  On most, but not all, calculators, you would carry out the above multiplication problem in the following way:

$$(4.0 \times 10^4)(3.5 \times 10^{-6})$$

When entering the exponential terms, the exponents (4) and (–6) are entered as [EXP] then 4, and [EXP] then [+/–] 6.  Therefore, $4.0 \times 10^4$ is entered as 4.0 [EXP] 4, and $3.5 \times 10^{-6}$ is entered as 3.5 [EXP] [+/–] 6. Note that the × sign within the exponential must not be entered. Some calculators have an [EE] key for exponents.

The calculation can be carried out with the following sequence of key strokes:

4.0 [EXP] 4 [×] 3.5 [EXP] [+/–] 6 [=]

The calculator will display the answer as 0.14, or in exponential form as $1.4^{-01}$, or as 1.4 –01 depending on the brand of calculator.  Write your answer in exponential notation.

**EXAMPLE 1.8  Math Operations with Scientific Notation**

Divide $4.2 \times 10^{-7}$ by $5.0 \times 10^{-5}$.

• **Solution**

$$\frac{4.2 \times 10^{-7}}{5.0 \times 10^{-5}}$$

Divide the coefficients, and subtract the exponent in the denominator from the exponent in the numerator.

$$= 0.84 \times 10^{-7-(-5)}$$
$$= 0.84 \times 10^{-7+5}$$
$$= 0.84 \times 10^{-2}$$

Because 0.84 is less than 1, the usual practice is to move the decimal point to give a coefficient between 1 and 10. Move the decimal point one place to the right, and add (−1) to the exponent.

$$= 8.4 \times 10^{-3}$$

• **Calculator Calculation**

On most calculators, you carry out the division problem by entering the numerator first, then pushing the divide key [÷], and then entering the denominator followed by the [=] key. As before, the exponential key is used to enter exponents.

If written out, the calculation would look like this:

4.2 [EXP] [+/−] 7 [÷] 5.0 [EXP] [+/−] 5 [=]

The calculator will display the answer as:  $8.4^{-03}$  or  8.4 −03; both stand for $8.4 \times 10^{-3}$.

**EXAMPLE 1.9  Significant Figures**

Determine the number of significant figures in each of the following numbers.
a.  6.02
b.  0.012
c.  $1.23 \times 10^7$
d.  1.5400

• **Solution**

a.  3 significant figures. Recall that zeros between nonzero digits are significant.
b.  2 significant figures. Zeros to the left of the first nonzero digit are not significant.
c.  3 significant figures. Scientific notation implies that significant figures are shown.
d.  5 significant figures. When a number contains a decimal point and has trailing zeros, these zeros are significant.

**Work EXERCISES & PROBLEMS: 9, 10**

## EXAMPLE 1.10  Significant Figures

Carry out the following operations, rounding off the answer to the correct number of significant digits.
a.   $287.12 - 95.333 =$

b.   $7.25 \times 10^{12} \div 92 =$

c.   $\dfrac{(6.0 + 5.21)}{4781} =$

### • Solution

a.   First, subtract the numbers giving: 191.787.  Only two digits to the right of the decimal are significant, so the answer is 191.79.
b.   The exponent plays no part in significant figures when multiplying or dividing. Look at total significant figures in your original numbers, your answer must have the same as the one with the least (i.e., two significant figures). Answer: $7.9 \times 10^{10}$.
c.   When 6.0 and 5.21 are added, the answer is 11.21. However in this addition, you only keep one place to the right, giving 11.2. The number 11.2 has 3 significant figures and it is divided by 4781 which has 4 significant figures. This quotient can only have three significant figures. Answer: $2.34 \times 10^{-3}$

**Work EXERCISES & PROBLEMS: 11, 12**

## EXAMPLE 1.11  Conversions with SI Prefixes

Express the following amounts using prefixes from the SI.
a.   Convert 827 m to kilometers.
b.   Convert 257 mg to micrograms.

### • Solution

a.   Conversions within the SI are based on knowing the SI prefixes. First, state the problem in the format:

desired quantity = given quantity × conversion factor.

? km = 827 m

The road map is:  m  →  km

There are two possible conversion factors between m and km:

$$\frac{1 \times 10^3 \, m}{1 \, km} \quad \text{and} \quad \frac{1 \, km}{1 \times 10^3 \, m}$$

Choose the one that allows cancellation of the unit m.

$$? \, km = 827 \, m \times \frac{1 \, km}{1 \times 10^3 \, m}$$

? km = 0.827 km

b.   First, state the problem:

? μg = 275 mg

Since we know how many mg are in a gram and how many μg are in a gram, we will convert mg to grams and then grams to μg.

Road map:   mg → g → μg

$$? \ \mu m = 275 \ mg \times \frac{1 \times 10^{-3} \ g}{1 \ mg} \times \frac{1 \ \mu g}{1 \times 10^{-6} \ g} = 275 \times 10^{3} \ \mu g$$

? μm = $2.75 \times 10^{5}$ μg

**Work EXERCISES & PROBLEMS: 13, 14**

---

**EXAMPLE 1.12  Using Density**

What is the mass in grams of a lead (Pb) brick that measures 20.5 cm × 9.7 cm × 62 mm? The density of Pb is 11.4 g/cm$^3$.

• **Solution**

Density is defined by the formula:

$$d = \frac{m}{V}$$

If you know any two of the variables in the equation, you can solve for the third. However, let's see how to do this using dimensional analysis. Remember, we can use density as a conversion factor:

$$\frac{11.4 \ g \ Pb}{1 \ cm^{3} \ Pb}$$

This conversion factor can convert between grams and cubic centimeters.  We will first need to determine the volume of the lead, For a regular solid the volume is:

V = length × width × height

We note that the dimensions are all in cm except the 62 mm. Convert this to centimeters.

$$62 \ mm \times \frac{1 \ cm}{10 \ mm} = 6.2 \ cm$$

The volume is

V = 20.5 cm × 9.7 cm × 6.2 cm = 1233 cm$^3$

V = $1.2 \times 10^{3}$ cm$^3$   (two significant figures)

Road map: $V \rightarrow g$

$$\text{mass} = 1.2 \times 10^3 \text{ cm}^3 \times \frac{11.4 \text{ g}}{1 \text{ cm}^3} = 1.4 \times 10^4 \text{ g}$$

The last step in solving the problem is checking the answer. Is it reasonable in size and units? One way to check an answer is to estimate it in your head with rounded numbers. If the density of lead is about 10 and the volume was 1200 cm³, then the answer should be around 12,000 g, which it is. Also, the units are correct for mass.

**Work EXERCISES & PROBLEMS: 15**

---

### EXAMPLE 1.13 Dimensional Analysis

The Voyager 2 mission to the outer planets of the solar system transmitted spectacular photographs of Neptune to Earth by radio. Radio waves, like light waves, travel at 3.00 × 10⁸ m/s. If Neptune was 2.75 billion miles from Earth during these transmissions, how many hours were required for radio signals to reach Earth from Neptune?

**• Solution**

First let's determine what conversion factor is given in the problem:

$$\text{speed of radio waves gives} \quad \frac{3.00 \times 10^8 \text{ m}}{1 \text{ s}}$$

Any time units are a ratio (as in m/s, g/mL, etc.) a corresponding conversion factor can be written.
Use conversion factors to convert the distance and speed to the units requested.

Road Map: miles → km → m → s → min → h

Insert one conversion factor for each arrow in the road map.

$$\text{hours} = 2.75 \times 10^9 \text{ mi} \times \frac{1.61 \text{ km}}{1.0 \text{ mi}} \times \frac{10^3 \text{ m}}{1 \text{ km}} \times \frac{1 \text{ s}}{3.00 \times 10^8 \text{ m}} \times \frac{1 \text{ min}}{60 \text{ sec}} \times \frac{1 \text{ h}}{60 \text{ min}} = 4.10 \text{ h}$$

**Work EXERCISES & PROBLEMS: 16**

---

### EXAMPLE 1.14 Using Percent

How many grams of sucrose ($C_{12}H_{22}O_{11}$) are in 2.0 kg of a 5.5% (by mass) solution of sucrose?

**•  Solution**

Whenever you are given percentage, you are given a conversion factor. Percent means the number of parts in 100 parts. Therefore, if the solution is 5.5% (by mass), there is 5.5 g sucrose in 100 g of solution or:

$$\frac{5.5 \text{ g sucrose}}{100 \text{ g solution}}$$

Road map: kg sol'n → g sol'n → g sucrose

$$2.0 \, kg \, sol'n \times \frac{1000 \, g \, sol'n}{1 kg \, sol'n} \times \frac{5.5 \, g \, sucrose}{100 \, g \, sol'n} = 110 \, g \, sucrose$$

**Work EXERCISES & PROBLEMS: 17**

---

**EXAMPLE 1.15  A Conversion of Volume Units**

How many cubic centimeters are equal to 5 m$^3$?

**• Solution**

The desired unit is cm$^3$.

Road map:    m$^3$ → cm$^3$

The relationship of meters to centimeters is:  1 cm = 10$^{-2}$ m
Writing in the conversion factor

$$? \, cm^3 = 5 \, m^3 \times \frac{1 \, cm}{1 \times 10^{-2} \, m}$$

We can see that the "m" unit does not entirely cancel the "m$^3$" unit.

The conversion factor must also be cubed.

$$\left( \frac{1 \, cm}{10^{-2} \, m} \right)^3 = \frac{1 \, cm}{10^{-2} \, m} \times \frac{1 \, cm}{10^{-2} \, m} \times \frac{1 \, cm}{10^{-2} \, m} = \frac{1 \, cm^3}{10^{-6} \, m^3}$$

Therefore, the solution is:

$$? \, cm^3 = 5 \, m^3 \times \frac{1 \, cm^3}{10^{-6} \, m^3} = 5 \times 10^6 \, m^3$$

**Work EXERCISES & PROBLEMS: 18**

---

**EXAMPLE 1.16  Temperature Scale Conversion**

The freezing point of a 50-50 mixture of antifreeze (ethylene glycol) and water is –36.5 °C. Convert this temperature to degrees Fahrenheit.

**• Solution**

First, write the equation for temperature conversions:

$$°C = (°F - 32 \, °F) \times \frac{5 \, °C}{9 \, °F}$$

Next, substitute –36.5 °C and solve for degrees Fahrenheit:

$$-36.5\ °C = (°F - 32\ °F) \times \frac{5\ °C}{9\ °F}$$

$$-36.5\ °C \times \frac{9\ °F}{5\ °C} = (°F - 32\ °F)$$

$$°F = -65.7\ °F + 32\ °F = -33.7\ °F$$

**Work EXERCISES & PROBLEMS: 19, 20**

## EXERCISES & PROBLEMS

1. Classify each of the following as an element, compound or mixture
   a. Dry ice
   b. Iodine
   c. Beer
   d. Margarine
   e. Steel

2. Consider the following and decide whether they are homogeneous or heterogeneous mixtures:
   a. Wood
   b. Wine
   c. Cranberry juice
   d. Milk
   e. Italian dressing

3. Identify the following as chemical or physical properties
   a. Melting of Iron
   b. Density of wood
   c. Reaction of wood with oxygen (burning)
   d. Reaction of iron with oxygen

4. Identify the following as chemical or physical changes
   a. Frost formation on windshield
   b. Steak charring on grill
   c. Boiling water to steam
   d. Fading jeans with bleach

5. Calculate the volume of a brick that is 34 cm long by 7.0 cm wide by 14 cm high.

6. A certain metal ingot has a mass of 3951 g, and measures 10.2 cm by 8.2 cm by 4.2 cm. Calculate the density of the metal.

7. Express the following numbers in scientific notation:
   a. 24,000
   b. 0.00014
   c. 740,000,000
   d. 0.0906

8. Write the following exponential numbers in standard decimal form by moving the decimal point in the proper direction:
   a. $5.2 \times 10^{-3}$
   b. $1.4 \times 10^{3}$
   c. $7.5 \times 10^{6}$
   d. $7.06 \times 10^{-5}$

9. Determine the number of significant figures expressed in the following numbers:
    a. 0.609
    b. $1.0 \times 10^3$
    c. 0.000222
    d. 238.0
    e. $1.030 \times 10^{-2}$

10. Round off the following numbers to the number of significant figures requested:
    a. 0.60945 to three significant figures
    b. $1.012 \times 10^3$ to two significant figures
    c. 0.00022174 to three significant figures
    d. 237.95 to four significant figures
    e. 1.303 to two significant figures

11. Carry out the following operations and express the result to the correct number of significant figures:
    a. $12 \times 2143.1$
    b. $3.09 \div 7$
    c. $(2.2 \times 10^{-3})(1.40 \times 10^6)$
    d. $12.70 + 1.222$
    e. $595.2 \times (24.33 - 16.271)$

12. Carry out the following operations and express the result to the correct number of significant figures:
    a. $125 \text{ g} + 64 \text{ g} + 10.837 \text{ g} =$
    b. $11.2 \text{ cm} + 0.0093 \text{ cm} + 0.80 \text{ cm} =$
    c. $\dfrac{15.01 \text{ g}}{(7.13 \text{ mL} - 6.2 \text{ mL})} =$

13. Write the following amounts in scientific notation in terms of the base SI unit.
    a. 7 µg
    b. 8.0 nm
    c. 0.14 ML
    d. 1.0 ks

14. Make the following conversions of metric lengths.
    a. 12.5 cm = _____ m
    b. $8.0 \times 10^{-8}$ m = _____ nm
    c. 445 cm = _____ km
    d. 32.5 mm = _____ µm
    e. $5.73 \times 10^3$ nm = _____ mm

15. Calculate the density of mercury given that a spherical droplet of mercury with a radius of 0.328 cm has a mass of 2.00 g. The volume of a sphere is $4/3\pi r^3$.

16. The world record for the 100 m dash is 9.78 s. What is the runner's average speed in miles per hour?

17. The density of a 26% salt solution is 1.199 g/mL. What is the volume, in mL, occupied by 20.0 g of this solution?

18. The displacement volume of a certain automobile engine is 350 in$^3$. How many liters is this?

19. The melting point of cesium metal is 28.4 °C. What is the melting point in degrees Fahrenheit?

20. On a sunny summer day the temperature inside a closed automobile can reach 128 °F. Convert this temperature to degrees Celsius.

## PRACTICE TEST QUESTIONS
See notes on taking practice test in the Preface

1. Which of the following describe physical properties and which describe chemical properties?
   a. Color
   b. Decomposition upon heating
   c. Boiling point
   d. Hardness
   e. Density
   f. A change of color on exposure to air

2. Which of the following statements describe physical changes and which describe chemical changes?
   a. Fresh cut apples turning brown
   b. Baking bread
   c. Cutting a soft metal with a knife
   d. Drying clothes
   e. Fermentation of sugar

3. Classify each of the following as a pure substance, a heterogeneous mixture, or a homogeneous mixture.
   a. Raw milk
   b. Tap water
   c. Gold
   d. Table sugar
   e. Concrete

4. Classify each of the following substances as an element, a compound, or a mixture.
   a. Hot tea
   b. Dry ice
   c. Sulfur
   d. Pure aspirin
   e. Paint
   f. Beer

5. Convert:
   a. 9.8 cm to millimeters
   b. 170 mm to centimeters
   c. 325 mL to liters
   d. 50 mi/h to m/s
   e. 20 GJ to kilojoules

6. Carry out the following calculation and report the answer to the correct number of significant figures.

   $$\frac{82.0715}{(32.107 - 28.224)} =$$

7. To determine the density of a piece of metal, a student finds its mass to be 13.62 g. The metal is placed into a flask that can hold exactly 25.00 mL. The student finds that 19.26 g of water is required to fill the flask. Given that the density of water is 0.9970 g/mL, what is the density of the metal?

8. The density of lead is 11.4 g/cm$^3$. What is the mass in kilograms of a lead brick with dimensions 2.0 in by 2.0 cm by 80.0 mm?

9. What volume does 1.00 lb of gold occupy at 20 °C? The density of gold is 19.3 g/cm$^3$ at 20 °C.

10. If gasoline costs $3.59 per gallon, compare the expense of driving exactly 10,000 miles in an SUV that gets 12 miles per gallon to that of a compact car that gets 12 km/L.

11. Calculate the density of water in units of kilograms per cubic meter. Density = $1.00 \text{ g/cm}^3$

12. The radius of an aluminum (Al) atom is 0.125 nm. How many Al atoms would have to be lined up in a row to form a line 1 cm in length?

13. Convert atmospheric pressure from 14.7 pounds per square inch to an equivalent number of megagrams per square centimeter.

14. Soft solder contains 70.0% tin and 30.0% lead by mass. If you had 10.0 g of tin, how many grams of solder could you make?

15. Copper (Cu) is a trace element that is essential for nutrition. Newborn infants require 80 µg of Cu per kilogram of body mass per day. The Cu content of a popular baby formula is 0.48 µg of Cu per milliliter. How many milliliters of formula should a 7.0 lb baby consume per day to obtain the minimum daily Cu requirement?

16. It takes 31,000 J of energy to make 1.0 g of aluminum (Al) metal from the ore bauxite. Burning 10.0 g of gasoline releases $4.20 \times 10^5$ J. If burning gasoline is used to supply the energy to produce aluminum from bauxite, then how many grams of gasoline are needed to make enough aluminum for one beverage can containing approximately 15 g Al?

17. What mass of carbon monoxide (CO), in kilograms, is present in the lower 500 m of air above a section of a city that measures 10 km × 10 km if the concentration of CO is 10 mg/m$^3$?

18. A horse can run a 10 furlong race in 2 min and 5.5 s. Calculate the average speed in miles per hour. Given: 1 furlong = 1/8 mile

19. The melting point of iron is 1540 °C. What is this temperature in degrees Fahrenheit?

20. A typical sunny summer day in Kentucky is 90 °F. What is the temperature in Kelvins?

# Chapter 2.   Atoms, Molecules & Ions

**Atomic Theory (Section 2.1)**
**The Structure of the Atom (Sections 2.2 – 2.3)**
**The Periodic Table (Section 2.4)**
**Ions and Molecules (Section 2.5)**
**Representation and Nomenclature of Elements & Compounds (Sec. 2.6- 2.7)**
**Introduction to Organic Compounds (Section 2.8)**

## SUMMARY

Note: The tools presented in this chapter are not unit conversion tools, as in most of the other chapters. Most of the tools point out the rules you must learn to name compounds.

## Atomic Theory (Section 2.1)

***Atoms.*** According to Dalton's atomic theory (proposed and accepted in 1808), elements are composed of extremely small, indivisible particles called *atoms*. Dalton assumed that atoms were the smallest units of elements that can combine to form compounds. All atoms of the same element have the same mass; atoms of different elements have different masses. Atomic theory describes diamond, the hardest solid, as a tightly packed collection of carbon particles (atoms). The atomic theory consists of a set of postulates summarized below.

---

**Dalton's Atomic Theory**
1. All matter is made of extremely small, indivisible particles called atoms. An *atom* is the smallest unit of an element that can enter into chemical combination. Atoms of the same element are identical in size, mass, and chemical properties. Atoms of one element are different from atoms of any other element.
2. Compounds are formed when atoms of two or more elements combine. Atoms combine in small whole-number ratios. For a given compound, the ratio of atoms of one element that combine per atom of the other element is fixed. The smallest particle that has the properties of the compound is called a molecule.
3. Atoms are the smallest units of chemical change. Chemical reactions involve the combination, separation, or rearrangement of atoms. Atoms are not destroyed or created in chemical processes.

---

These postulates explain the law of definite proportions, the law of multiple proportions, and law of conservation of mass. The law of definite proportions follows from postulate #2. It states that all purified samples of a compound contain its constituent elements in the same proportions by mass. For example, all pure samples of the compound carbon monoxide, no matter their source, contain 43% carbon and 57% oxygen by mass. Dalton's proposal was that the smallest particle of carbon monoxide was a molecule consisting of one carbon atom and one oxygen atom. Assuming all carbon atoms have the same mass, and all oxygen atoms have a mass about 1.33 times greater than carbon atoms, the composition of carbon monoxide is fixed at the percentages given above.

While working on the atomic theory, Dalton recognized the law of multiple proportions, also a consequence of postulate #2. It states that when two elements form more than one compound, the masses of the elements in the compounds are related by small whole-number ratios. Consider two compounds of nitrogen and oxygen: nitrogen monoxide, and nitrogen dioxide. The mass of oxygen combining with nitrogen in nitrogen dioxide was found to be twice as great as the mass of oxygen combining with nitrogen in nitrogen monoxide. Dalton proposed that the two compounds of N and O were the result of atoms of the two elements combining in different ratios to form two different molecules. When one atom of nitrogen combined with one atom of oxygen, nitrogen monoxide was formed. When one atom of nitrogen combined with two atoms of oxygen, then nitrogen dioxide was the compound formed. The ratio of the mass of oxygen in nitrogen dioxide to that in nitrogen monoxide is 2 to 1, a small whole-number ratio.

Postulate #3 is the basis of the law of conservation of mass. This observation has been made many times:  there is no detectable gain or loss of mass during a chemical reaction. Atoms are neither created nor destroyed in a chemical reaction; rather they are rearranged to form new combinations of atoms, which we would recognize as new compounds. Mass is conserved in chemical reactions because atoms are conserved.

## The Structure of the Atom (Sections 2.2 – 2.3)

*Subatomic Particles.* In the late 1800's, evidence began to mount that atoms were not indivisible, but were comprised of smaller particles. In a remarkable series of experiments, scientists learned that all atoms are constructed from the same three subatomic particles: the electron, proton, and neutron. The Table 2.1 summarizes these experiments and the conclusions drawn regarding the particles that make up atoms. Concurrent investigations into the nature of radioactivity play an important role in the discovery of subatomic particles. Radioactive substances decay (break down) into three types of particles: positively charged alpha particles, negatively charged beta particles (electrons), and gamma rays (high energy 'light' particles), called photons. You will learn more about photons in chapter 7.

Table 2.1

| Year | Investigator | Experiment | Conclusions |
|------|-------------|-----------|-------------|
| 1897 | J.J. Thompson | Observed negative particle beam in cathode ray tube (CRT) | atom contains electrons with single unit of negative charge, charge/mass ratio is $-1.76 \times 10^8$ C/g |
| 1909 | R.A. Millikan | Calculated charge on electron from the rate of descent charged oil droplets in an electric field | charge of electron is $-1.6 \times 10^{-18}$ C mass of electron is $9.11 \times 10^{-28}$ g, 1/1840 mass of hydrogen atom |
| 1911 | E. Rutherford | Observed deflection of alpha particle by gold foil | atom contains very small positively charged nucleus surrounded by negative electrons |
| 1920 | E. Rutherford | Observes protons ejected from elements bombarded by alpha particles | nucleus contains proton with single unit of positive charge, mass of proton = $1.67262 \times 10^{-24}$ g |
| 1932 | J. Chadwick | Observed neutral particles heavy enough to eject protons from wax | nucleus contains neutrons (no charge) with mass similar to proton, mass = $1.67493 \times 10^{-24}$ g |

***Atomic Number and Mass Number.*** Atoms are distinguished from atoms of other elements by the number of protons in the nucleus. The number of protons in the atomic nucleus is called the atomic number, Z, of an element. For example, the atomic number of oxygen is 8; therefore, all oxygen atoms contain eight protons and also eight electrons.

The total mass of an atom is determined almost entirely by the number of protons and neutrons. In most cases, the mass of the electrons can be neglected because it is so much smaller. The mass number, A, is the total number of neutrons and protons in the nucleus of an atom. The number of neutrons in an atom is $A - Z$.

***Isotopes.*** Atoms of a given element that differ in the number of neutrons and consequently in mass, are called isotopes. For example, there are two isotopes of the element lithium, one with a mass number of 6 and another with a mass number of 7. Isotopes are identified by mass numbers. The lithium isotopes are called lithium-6 (pronounced lithium-six) and lithium-7. The different mass numbers are the result of different numbers of neutrons per atom. An atom of lithium-6 contains three protons, three electrons, and three neutrons, whereas an atom of lithium-7 contains three protons, three electrons and four neutrons. In nature, elements are found as mixtures of isotopes.

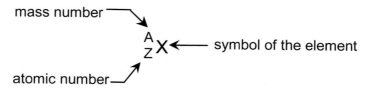

Chemical symbols are labeled with mass and atomic numbers to designate specific isotopes. The general symbol for an isotope is:

The symbols for the lithium isotopes are $^{6}_{3}Li$ and $^{7}_{3}Li$.

| Examples 2.1, 2.2 |
| Exercises 1 – 3 |

## The Periodic Table (Section 2.4)

***Element Classification.*** Chemists realized that there was an underlying periodic pattern to the chemical behavior of the elements as early as the 1830's. For example, it was noted that chlorine, bromine, and iodine undergo very similar chemical reactions. The essential nature of the pattern was discovered in 1861 by a professor writing a chemistry textbook -- Dmitri Mendeleev. In the modern table, elements are arranged by atomic number (Mendeleev's table was based on atomic masses) in vertical columns called groups and horizontal rows call periods.

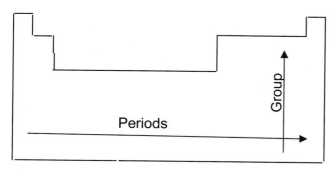

Elements in a group have similar chemical and physical properties. In fact, the element at the top of the group is a prototype for the group. Knowing the chemical and physical properties of lithium (Group 1A, Column 1) takes us a long way to understanding the properties of sodium, potassium, rubidium, cesium and francium. Several element groups have special names that are listed with the periodic table on below. The elements can be divided into three categories: metals to the left of the table, nonmetals to the right and metalloids between them (shaded in grey.) The metals tend to lose electrons in chemical reactions, whereas the nonmetals tend to gain them. The noble gases (Group 8A) are known to be quite inert, and only react under extreme conditions, with some noble gases not reacting at all.

| 1 1A | 2 2A | 3 3B | 4 4B | 5 5B | 6 6B | 7 7B | 8 | 9 8B | 10 | 11 1B | 12 2B | 13 3A | 14 4A | 15 5A | 16 6A | 17 7A | 18 8A |
|---|---|---|---|---|---|---|---|---|---|---|---|---|---|---|---|---|---|
| 1 **H** 1.008 | | | | | | | | | | | | | | | | | 2 **He** 4.003 |
| 3 **Li** 6.941 | 4 **Be** 9.012 | | | | | | | | | | | 5 **B** 10.81 | 6 **C** 12.01 | 7 **N** 14.01 | 8 **O** 16.00 | 9 **F** 19.00 | 10 **Ne** 20.18 |
| 11 **Na** 22.99 | 12 **Mg** 24.31 | | | | | | | | | | | 13 **Al** 26.98 | 14 **Si** 28.09 | 15 **P** 30.97 | 16 **S** 32.07 | 17 **Cl** 35.45 | 18 **Ar** 39.95 |
| 19 **K** 39.10 | 20 **Ca** 40.08 | 21 **Sc** 44.96 | 22 **Ti** 47.88 | 23 **V** 50.94 | 24 **Cr** 52.00 | 25 **Mn** 54.94 | 26 **Fe** 55.85 | 27 **Co** 58.93 | 28 **Ni** 58.69 | 29 **Cu** 63.55 | 30 **Zn** 65.39 | 31 **Ga** 69.72 | 32 **Ge** 75.59 | 33 **As** 74.92 | 34 **Se** 78.96 | 35 **Br** 79.90 | 36 **Kr** 83.80 |
| 37 **Rb** 85.47 | 38 **Sr** 87.62 | 39 **Y** 88.91 | 40 **Zr** 91.22 | 41 **Nb** 92.91 | 42 **Mo** 95.94 | 43 **Tc** (98) | 44 **Ru** 101.1 | 45 **Rh** 102.9 | 46 **Pd** 106.4 | 47 **Ag** 107.9 | 48 **Cd** 112.4 | 49 **In** 114.8 | 50 **Sn** 118.7 | 51 **Sb** 121.8 | 52 **Te** 127.6 | 53 **I** 126.9 | 54 **Xe** 131.3 |
| 55 **Cs** 132.9 | 56 **Ba** 137.3 | 57 **La** 138.9 | 72 **Hf** 178.5 | 73 **Ta** 180.9 | 74 **W** 183.9 | 75 **Re** 186.2 | 76 **Os** 190.2 | 77 **Ir** 192.2 | 78 **Pt** 195.1 | 79 **Au** 197.0 | 80 **Hg** 200.6 | 81 **Tl** 204.4 | 82 **Pb** 207.2 | 83 **Bi** 209.0 | 84 **Po** (210) | 85 **At** (210) | 86 **Rn** (222) |
| 87 **Fr** (223) | 88 **Ra** (226) | 89 **Ac** (227) | 104 **Rf** (257) | 105 **Db** (260) | 106 **Sg** (263) | 107 **Bh** (262) | 108 **Hs** (265) | 109 **Mt** (266) | 110 **Ds** (281) | 111 **Rg** (272) | 112 | 113 | 114 | 115 | 116 | 117 | 118 |

Alkali Metals
Alkaline Earth Metals
Noble Gases
Halogens

| Lanthanide series | 58 **Ce** 140.1 | 59 **Pr** 140.9 | 60 **Nb** 144.2 | 61 **Pm** (147) | 62 **Sm** 150.4 | 63 **Eu** 152.0 | 64 **Gd** 157.3 | 65 **Tb** 158.9 | 66 **Dy** 162.5 | 67 **Ho** 164.9 | 68 **Er** 167.3 | 69 **Tm** 168.9 | 70 **Yb** 173.0 | 71 **Lu** 175.0 |
|---|---|---|---|---|---|---|---|---|---|---|---|---|---|---|---|
| Actinide series | 90 **Th** 232.0 | 91 **Pa** (231) | 92 **U** 238.0 | 93 **Np** (237) | 94 **Pu** (242) | 95 **Am** (243) | 96 **Cm** (247) | 97 **Bk** (247) | 98 **Cf** (249) | 99 **Es** (254) | 100 **Fr** (253) | 101 **Md** (256) | 102 **No** (254) | 103 **Lr** (257) |

Book-mark this page. You will need to refer to a periodic table often.

## Ions and Molecules (Section 2.5)

*Ions.* Atoms are electrically neutral because they have equal numbers of protons and electrons. When an atom acquires a charge by gaining or losing electrons, it becomes an ion. These processes do not change the identity of the atom because the atomic number (the number of protons) is unaffected. Metals tend to lose electrons, forming positive ions called cations. Nonmetals tend to gain electrons forming negative ions called anions. Consider magnesium that has 12 protons in the nucleus and 12 electrons outside the nucleus. It can lose two of its electrons, producing a magnesium ion (or cation), which has 12 protons and 10 electrons. The symbol for the magnesium ion is $Mg^{2+}$. The superscript 2+ represents the net charge on the ion: the number of positive charges outnumbers that of the negative charges by two. On the other hand, negative ions are formed when atoms gain extra electrons. For example, a neutral oxygen atom contains 8 protons and 8 electrons but can gain two electrons. The result is an oxygen anion that has 8 protons and 10 electrons. The symbol for the oxygen ion is $O^{2-}$ because there are two more negative charges than positive charges in the ion.

Example 2.3

Exercises 4, 5

When a group of atoms acquires a net charge, it is called a polyatomic ion. The poison cyanide is a common example. Cyanide is an anion of carbon and nitrogen, CN⁻ (the 1 is not written in the formula of ions with a net charge of ±1). Ionic compounds are combinations of anions and cations held together by electrostatic forces. Sodium chloride (table salt, NaCl) is an example of an ionic compound.

***Molecules.*** A molecule is a discrete particle composed of atoms that are bonded together in a fixed arrangement. A molecule can consist of atoms of the same element, such as $O_2$ or $S_8$. If a molecule consists of atoms of different elements, the molecule is the smallest particle that has the properties of that compound. Diatomic molecules contain two atoms, while polyatomic molecules contain more than two atoms. (The prefix "poly" is Greek for many.)

## Representation and Nomenclature of Elements & Compounds (Sec. 2.6 - 2.7)

One of the biggest challenges for new chemistry students is mastering the rules for constructing molecular formulas and chemical names. Chemists use chemical formulas to represent elements and compounds in terms of their chemical symbols. The chemical formula reflects the number of the different types of atoms in the substance. Chemists use molecular models and structural formulas to represent elements and compounds pictorially. The structural formula depicts the spatial arrangement of the atoms in the substance. Chemists use chemical names to represent elements and compounds verbally. Since these are all different ways of representing the same microscopic reality, this guide will present the three representations for prototype elements and compounds together.

***Representing Elements.*** Elements can exist as atoms, e.g., helium gas (He), or molecules, e.g., hydrogen gas ($H_2$), sulfur ($S_8$) and aggregates of atoms of varying size, e.g., sodium (Na). The nomenclature rules for elements are straightforward. The element name refers to the element as it occurs in nature. Oxygen refers to the diatomic molecule; oxygen atoms are represented as O.

**Table 2.2**

| Class | Chemical Formula | Structural Model | Chemical Name | Nomenclature Rule |
|---|---|---|---|---|
| Elements | He | (He) | helium | element name |
| | $H_2$ | H——H | hydrogen | element name |
| | $O_2$ | O——O | oxygen | element name |
| | $S_8$ | | sulfur | element name |
| | Na | | sodium | element name |

***Naming Compounds (Nomenclature).*** The rules for naming compounds depend on the classification of the compound. Compounds can be ionic or molecular.
- Ionic compounds are composed of cations and anions bound in a three-dimensional network (often containing a metal and a nonmetal).
- Molecular compounds are composed of discrete combinations of atoms (typically atoms are all nonmetals).

***Naming Ionic Compounds.*** The formula of an ionic compound is generally written as the empirical formula (formula in which elements have smallest whole-number ratio) because these compounds do not exist as discrete molecular units. These compounds exist as three-dimensional networks of ions in the solid phase or ions surrounded by water in the aqueous phase. The principal feature of these compounds is that the negative and positive charges must balance. Consequently, ionic compounds of equally charged ions, such as LiF and MgS, have equal numbers of cations and anions in their formulas. Ionic compounds composed of cations and anions of different charge have ratios that equalize the positive and negative charge. Some metals form more than one ion and there are two ways to name them. The modern method uses the Stock convention in which the charge on the ion is denoted using Roman numerals. Metals which do not need the Roman numeral are the Alkali Metals (1A), the Alkaline Earth Metals (2A), as well as Silver (Ag), Cadmium (Cd), Zinc (Zn) and Aluminum (Al) because these metals always have but one charge in an ionic compound. Their charge will be the same as their group number. Table 2.3 lists common (unsystematic) names of several metal ions in the left hand column. The "-ic" ending indicates the larger positive charge. The Stock system for naming the ionic compounds does not use the common name. The right hand column of Table 2.3 lists names of various anions. For a more complete list, see and memorize Table 2.3 *in your textbook*. Some ionic compounds have specific numbers of water molecules attached to them, even in the solid state. These compounds are called "hydrates." The number of water molecules is denoted by Greek numerical prefixes. The prefixes are listed in Table 2.4. Rules for naming ionic compounds and their hydrates are given in Table 2.5.

| Example 2.4 |
|---|
| Exercises 6, 7 |

### Table 2.3 Common Names of Cations and Anions

| | |
|---|---|
| ammonium ($NH_4^+$) | cyanide ($CN^-$) |
| chromic ($Cr^{3+}$) | chlorate ($ClO_3^-$) |
| cuprous ($Cu^+$), cupric ($Cu^{2+}$) | dihydrogen phosphate ($H_2PO_4^-$) |
| ferrous ($Fe^{2+}$), ferric ($Fe^{3+}$) | bicarbonate or hydrogen carbonate ($HCO_3^-$) |
| plumbous ($Pb^{2+}$) | hydrogen phosphate ($HPO_4^{2-}$) |
| manganous ($Mn^{2+}$) | bisulfate or hydrogen sulfate ($HSO_4^-$) |
| mercurous ($Hg_2^{2+}$), mercuric ($Hg^{2+}$) | thiocyanate ($SCN^-$) |
| stannous ($Sn^{2+}$), stannic ($Sn^{4+}$) | nitrite ($NO_2^-$) |

### Table 2.4 Numerical Prefixes

| 1 | mono- | 4 | tetra- | 7 | hepta- | 10 | deca- |
|---|---|---|---|---|---|---|---|
| 2 | di- | 5 | penta- | 8 | octa- | | |
| 3 | tri- | 6 | hexa- | 9 | nona- | | |

## Table 2.5  Naming Ionic Compounds using the Stock System

| Class | General Formula | Nomenclature Rules | Example |
|---|---|---|---|
| Ionic Compounds | MX | Cation name (which is typically the name of the metal) + Roman numeral if needed + anion name | LiF ($1Li^+$ + $1F^-$) lithium fluoride |
| | | | $MgCl_2$ ($1Mg^{2+}$ +$2Cl^-$) magnesium chloride |
| | | | $FeCl_3$  ($1Fe^{3+}$ +$3Cl^-$) iron(III) chloride |
| | | | $KClO_3$ potassium chlorate |
| | | | $Mg(NO_2)_2$ magnesium nitrite |
| Hydrates | $MX{\cdot}nH_2O$ | Ionic compound name + # prefix + "hydrate" | $BaCl_2{\cdot}2H_2O$ barium chloride dihydrate |
| | | | $CuSO_4{\cdot}6H_2O$ copper(II) sulfate hexahydrate |

**Naming Molecular Compounds.** For our discussion, there are two types of molecular compounds: inorganic compounds and organic compounds. Molecular inorganic compounds are usually combinations of nonmetal elements. The first element is the one that is furthest to the left of the periodic table. Compounds containing different numbers of the same elements are distinguished using the numerical prefixes in Table 2.4.
Study table 2.6 to learn to name molecular inorganic compounds.

## Table 2.6  Naming Molecular Inorganic Compounds

| Class | General Formula | Nomenclature Rules | Example | Structural Model |
|---|---|---|---|---|
| Molecular Compounds | $X_pY_q$ | p# prefix + element 1 name + q# prefix + root of element 2 + 'ide' | $CO_2$ carbon dioxide | O=C=O |
| | | | $N_2O_4$ dinitrogen tetraoxide | |

Some molecular compounds are named with their common, not systematic names. Table 2.7 lists a few to keep in mind.

## Table 2.7  Common Names & Formulas of Molecular Inorganic Compounds

| | | | |
|---|---|---|---|
| $H_2O$ | water | $H_2S$ | hydrogen sulfide |
| $NH_3$ | ammonia | $PH_3$ | phosphine |

Examples
2.5, 2.6

Exercises
8-13

Compounds can also be classified as Acids, Bases and Salts.
- Acids are compounds that can "dissolve" reactive metals (by consuming loose electrons). Acids are also described as compounds which when dissolved in water produce $H^+$ ions.
- Bases are compounds that neutralize (counteract) acids. When bases dissolve in water, they produce hydroxide ions.
- Salts are compounds that form when acids are mixed with bases and water is removed. All salts are ionic compounds and for now, these terms are used interchangeably.

***Naming Acids.*** The rules are covered in Table 2.8.

### Table 2.8  Naming Acids

| Class | | General Formula | Nomenclature Rules | Example | Structural Model |
|---|---|---|---|---|---|
| Acids | Binary Acid | HX | root of hydrogen + root of element X +"-ic" acid | HF hydrofluoric acid | H——F |
| | Oxo-acids | $H_aXO_n$* | root of element X + "-ic" acid | $HClO_3$ chloric acid |  |
| | | $H_aXO_{n-1}$ | root of element X + "-ous" acid | $HClO_2$ chlorous acid | |
| | | $H_aXO_{n-2}$ | hypo- + root of element X + "-ous" acid | HClO hypochlorous acid | |
| | | $H_aXO_{n+1}$ | per- + root of element X + "-ic" acid | $HClO_4$ perchloric acid | |

\* *The value n depends on the element; it is 3 for chlorine, 3 for nitrogen and 4 for sulfur.*

***Naming Bases.*** Most bases are metal hydroxides that dissociate into metal cations and hydroxide ions in water. Metal hydroxides are a form of ionic compounds. The rules for naming these bases are the same as that of naming salts. Ammonia and compounds like it are also bases because they also generate hydroxide ions in water. Many of these bases are organic compounds whose names must be memorized at this point.  Study Table 2.9 to familiarize yourself with the rules for naming bases.

### Table 2.9  Naming Bases

| Class | | General Formula | Nomenclature Rules | Example | Structural Model |
|---|---|---|---|---|---|
| Ionic Compounds: Bases | Metallic Base | $M(OH)_n$ | metal name + hydroxide | LiOH lithium hydroxide | $Li^+ OH^-$ |
| | | | | $Mg(OH)_2$ Magnesium hydroxide | $Mg^{2+} 2OH^-$ |
| | Amines | $NR_3$ | Common names | $NH_3$ ammonia |  |

## Introduction to Organic Compounds (Section 2.8)

Hydrocarbons are molecular compounds which contain only carbon and hydrogen. They are the simplest type of organic compounds. Throughout the book, organic compounds are used to demonstrate principles of chemistry. Look at Table 2.10 and familiarize yourself with how the condensed structural formula of the hydrocarbon gives information as to how the atoms are connected. Table 2.8 in your textbook gives a more exhaustive list and shows the three dimensional ball and stick model of molecules. Note how the H's in the condensed structural formulas are directly attached to the carbon which they follow. Table 2.11 lists organic compounds you will see often within your textbook. Become familiar with the formula and names of these compounds as well as those listed in Table 2.8 of your textbook.

Let's extend the concept to other organic compounds (containing other elements besides C and H). The compound $CH_3NH_2$ has three hydrogen atoms connected to the carbon which is then connected to the nitrogen:

| | Example 2.7 |
|---|---|
| | Exercise 14 |

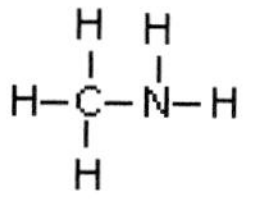

Keep in mind that hydrogen atoms can only be connected to **one** atom and will not be between 2 atoms:

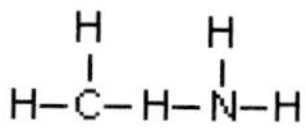

**NOT THIS!**

### Table 2.10  Formula and Name of Hydrocarbons

| Molecular Formula | Condensed Structural Formula | Structural Formula | Name |
|---|---|---|---|
| $CH_4$ | $CH_4$ | 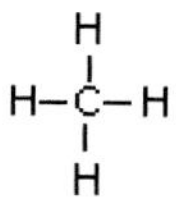 | methane |
| $C_2H_6$ | $CH_3CH_3$ | 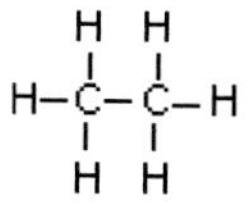 | ethane |
| $C_3H_8$ | $CH_3CH_2CH_3$ | 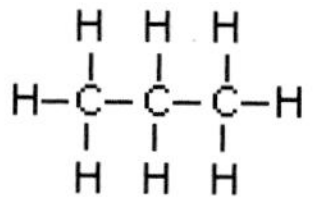 | propane |
| $C_4H_{10}$ | $CH_3CH_2CH_2CH_3$ | 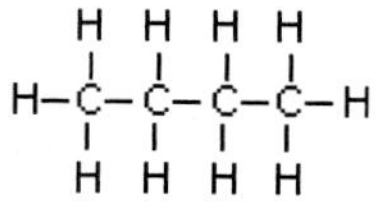 | butane |

### Table 2.11  Names of Common Organic Compounds

| Formula | Name | Formula | Name |
|---|---|---|---|
| $CH_2O$ | formaldehyde | $CH_3OH$ | methanol |
| $CH_3CH_2OH$ | ethanol | $CH_3COOH$ | acetic acid |
| $C_6H_6$ | benzene | $C_6H_{12}O_6$ | glucose |

## GLOSSARY LIST

| | | | |
|---|---|---|---|
| atom | atomic number | periodic table | nomenclature |
| definite proportions | mass number | periods | binary compounds |
| multiple proportions | isotope | groups | ternary compounds |
| conservation of mass | | families | inorganic compounds |
| | ion | metals | organic compounds |
| electron | anion | nonmetals | molecular compounds |
| nucleus | cation | metalloids | acid |
| proton | monatomic ion | | base |
| neutron | polyatomic ion | alkali metals | salt |
| | | alkaline earth metals | hydrate |
| radioactivity | molecule | halogens | oxoacid |
| radiation | allotrope | noble gases | oxoanion |
| alpha particle | diatomic molecule | | |
|   alpha ray | polyatomic molecule | chemical formula | |
| beta particle | | molecular formula | |
|   beta ray | | structural formula | |
| gamma ray | | empirical formula | |

## WORKED EXAMPLES

### EXAMPLE 2.1  Isotopic Symbols

The three isotopes of oxygen found in nature are oxygen-16, oxygen-17, and oxygen-18. Write their isotopic symbols.

**• Solution**

The general form of the symbols is $_Z^A O$, where $A$ is the mass number and $Z$ is the atomic number. The atomic number of oxygen is 8, and so all oxygen atoms contain eight protons. The mass numbers are 16, 17, and 18, respectively. The isotopic symbols are:

$$_8^{16}O \qquad\qquad _8^{17}O \qquad\qquad _8^{18}O$$

### EXAMPLE 2.2  Isotopic Symbols

How many neutrons are present in the nucleus of each of the oxygen isotopes in the previous example?

**• Solution**

The number of neutrons is given by $A - Z$, the mass number minus the proton number.

> For $^{16}O$, there are $16 - 8 = 8$ neutrons.
> For $^{17}O$, there are $17 - 8 = 9$ neutrons.
> For $^{18}O$, there are $18 - 8 = 10$ neutrons.

**Work EXERCISES & PROBLEMS: 1 – 3**

**EXAMPLE 2.3  Electrons and Protons in Ions**

Determine the number of protons and electrons in the ions represented by $K^+$, $Fe^{3+}$, and $I^-$.

**• Solution**

The net positive charge of the $K^+$ ion is due to the loss of an electron from a K atom.
A K atom has 19 $p^+$ and 19 $e^-$. Therefore, the $K^+$ ion has 19 $p^+$ and 18 $e^-$.

The net positive charge of the $Fe^{3+}$ ion is due to the loss of three electrons from an Fe atom.
An Fe atom has 26 $p^+$ and 26 $e^-$. Therefore, the $Fe^{3+}$ ion has 26 $p^+$ and 23 $e^-$.

The net negative charge of the $I^-$ ion is due to a gain of one electron by an I atom.
An I atom has 53 $p^+$ and 53 $e^-$. Therefore, the $I^-$ ion has 53 $p^+$ and 54 $e^-$.

**Work EXERCISES & PROBLEMS: 4, 5**

**EXAMPLE 2.4  Empirical Formulas**

What are the empirical formulas of the following compounds?
a. $B_2H_6$    b. $P_4O_{10}$

**• Solution**

To find the smallest whole-number ratio of atoms, reduce the numbers as if reducing a fraction, i.e., divide both by the largest number which give whole-numbers.
a. The simplest ratio of B to H is 2 to 6 which reduces to 1 to 3. The empirical formula is $BH_3$.
b. The simplest ratio of P to O is 4 to 10 which reduces to 2 to 5. Notice that you must keep a whole-number ratio. An atom ratio of 2.5 to 1 would not be physically reasonable. The empirical formula is $P_2O_5$.

**Work EXERCISES & PROBLEMS: 6, 7**

**EXAMPLE 2.5  Formulas of Ionic Compounds**

Write the formulas of the following compounds:
a. magnesium chloride    b. magnesium oxide    c. magnesium phosphate

**• Solution**

The formula must represent an electrically neutral grouping of ions. Identify the ions first.
a. The ions in magnesium chloride are $Mg^{2+}$ and $Cl^-$ ions.  Two chloride ions are needed for each magnesium ion. 1 cation(+2) + 2 anions(−1) = 0    The formula is $MgCl_2$.

b. The ions in magnesium oxide are $Mg^{2+}$ and $O^{2-}$. In this case just one cation and one anion will give electrical neutrality.  (+2) + (−2) = 0    The formula is MgO.

c. The ions in magnesium phosphate are $Mg^{2+}$ and $PO_4^{3-}$.  In this case 2 anions have a −6 charge and it will take 3 cations to have the needed +6 charge.    3 cations (+2) + 2 anions(−3) = 0    The formula is $Mg_3(PO_4)_2$.

---

**EXAMPLE 2.6  Naming Ionic Compounds**

Name the following compounds according to the Stock system:
a. CuBr
b. $CuSO_4$

**• Solution**

In the Stock system the name of the metallic element in a compound is followed by a Roman numeral derived from the charge of the cation.

a. Assuming bromide to be –1 as in the $Br^-$ ion, Cu must be a +1 ion in order to have a neutral compound. Therefore, CuBr is named copper(I) bromide.

b. Since the sulfate ion has a -2 charge, copper must be a +2 ion and the compound is copper(II) sulfate.

**Work EXERCISES & PROBLEMS: 8 - 13**

---

**EXAMPLE 2.7 Organic Compounds**

How would the following organic compound be represented using molecular, empirical and condensed structural formulas?

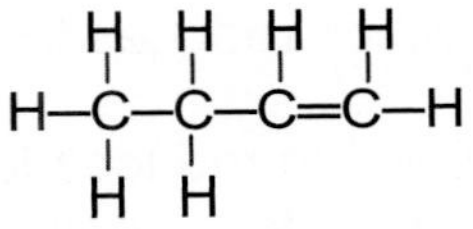

**• Solution**

Molecular formula gives the element symbols with the number of atoms in the molecule as subscripts:

$$C_4H_8$$

The empirical formula gives the lowest whole-number ratio:

$$CH_2$$

Follow each carbon in the condensed structure with the hydrogen atoms which are directly connected to the carbon:

$$CH_3CH_2CHCH_2 \text{ or } CH_3CH_2CH=CH_2$$

**Work EXERCISES & PROBLEMS: 14**
**Work the rest of the EXERCISES & PROBLEMS.**

---

## EXERCISES & PROBLEMS

1. What is the mass number of a sodium atom that has 13 neutrons?

2. How many neutrons are in an atom of $^{109}Ag$?

3. Write the symbol for each of the following isotopes.
   a. An atom with Z = 30 and 37 neutrons.
   b. An atom with Z = 51 and 69 neutrons.

4. What are the three fundamental particles from which atoms are made? What are their electric charges. Which particles are in the atomic nucleus?

5. Give the number of protons and electrons in each of the following ions.
   a. $Li^+$   b. $Sr^{2+}$   c. $Fe^{3+}$   d. $N^{3-}$   e. $Se^{2-}$   f. $Cl^-$

6. Styrene has the molecular formula $C_8H_8$. What is its empirical formula?

7. What are the empirical formulas of the following compounds?
   a. $N_2O_4$   b. $C_4H_8$   c. $AlCl_3$   d. $Fe_2O_3$   e. $S_2F_{10}$

8. Which of the following compounds are likely to be ionic?
   a. KCl   b. $CH_4$   c. $AlCl_3$   d. $SO_2$   e. MgO   f. $CCl_4$

9. Write formulas of the following binary compounds.
   a. barium chloride
   b. magnesium nitride
   c. iron(III) oxide
   d. iron(II) fluoride

10. Write the formulas for the following compounds.
   a. ammonium chloride
   b. sodium phosphate
   c. potassium sulfate
   d. calcium carbonate
   e. potassium hydrogen carbonate
   f. magnesium nitrite
   g. sodium nitrate
   h. ammonium perchlorate
   i. strontium hydroxide
   j. copper(II) cyanide

11. Name the following compounds:
   a. $K_3N$        b. $Ag_2CO_3$    c. $Mg(OH)_2$   d. NaCN    e. $NH_4I$
   f. $Fe(NO_3)_2$   g. $CaSO_4 \cdot 2H_2O$

12. Name the following compounds:
   a. $PCl_5$    b. $SO_3$    c. $P_4O_{10}$    d. $N_2O$    e. $NO_2$

13. Name the following acids:
   a. $HNO_3$    b. $HNO_2$    c. HBr    d. HCN    e. $HClO_2$

14. Show the connectivity of atoms in the following organic compounds.
   a. $CH_3CH_3$                b. $CH_3CH_2CH_2CH_2CF_3$

15. Summarize the four postulates of Dalton's atomic theory in your own words.

16. What evidence did Rutherford find that supported his theory of the atomic nucleus?

17. What is the difference between an atom and a molecule?

18. Give an example of:  a. a monatomic cation.  b. a polyatomic anion

19. Identify the following as elements or compounds.
    a. $ClF_3$   b. HCl   c. $O_3$   d. $I_2$   e. Se   f. NaI

20. Explain the difference in the meaning of the symbols $O_3$ and 3O.

21. Why are the chemical formulas of ionic compounds the same as the empirical formulas?

## PRACTICE TEST QUESTIONS
See notes on taking practice test in the Preface

1. Complete the following isotope table:

| Name | Symbol | Number of Protons | Number of Electrons | Number of Neutrons | Mass Number |
|---|---|---|---|---|---|
| Sodium | $^{23}Na$ | 11 | _____ | 12 | 23 |
| _____ | $^{40}Ar$ | _____ | _____ | 22 | _____ |
| Arsenic | $^{75}As$ | _____ | _____ | _____ | _____ |
| Lead | _____ | _____ | _____ | 126 | _____ |
| _____ | _____ | 19 | _____ | 20 | _____ |

2.
a. What is the total number of protons, neutrons, and electrons in an atom of $^{56}_{26}Fe$ ?

b. What is the mass number of a copper atom that has 35 neutrons?

3. List the number of protons, neutrons, and electrons in atoms of the following.

a. $^{17}_{8}O^{2-}$   b. $^{107}_{47}Ag^{+}$   c. $^{222}_{86}Rn$

4. Which of the following are isotopes of element X?

$^{46}_{20}X, \quad ^{20}_{46}X, \quad ^{43}_{20}X, \quad ^{46}_{43}X$

5. Take the mass of an atom to be the sum of the masses of its protons, neutrons, and electrons. Determine the percentages by mass of each subatomic particle in a $^{12}_{6}C$ atom. See table 2.1 for masses of protons, neutrons and electrons.

6. Which of the following molecules are forms of a pure element?

$P_4, \quad He, \quad N_2, \quad O_3, \quad N_2O_3$

7. What is the empirical formula of each of the following compounds?
a. $C_6H_8O_6$
b. $C_2H_2$
c. $Hg_2Cl_2$
d. $H_2O_2$
e. $C_2H_2O_4$
f. $MgCl_2$

8. Write the formulas for the following compounds.
a. calcium hypochlorite
b. mercury(II) sulfate
c. barium sulfite
d. zinc oxide
e. dinitrogen oxide
f. sodium carbonate
g. copper(II) sulfide
h. lead(IV) oxide

9. Name the following compounds:
a. $Na_2HPO_4$
b. HI (gas)
c. $P_4O_6$
d. $LiNO_3$
e. HI (solution)
f. $Sr(NO_2)_2$
g. $NaHCO_3$
h. $K_2SO_3$
i. $Na_3PO_4$
j. $Al(OH)_3$

2

2

2

2

2

2

2

2

2

10. Name the following acids:
   a. $H_2SO_3$
   b. $HClO$
   c. $HClO_4$
   d. $H_3PO_4$
   e. $HCN$

2

# Chapter 3.  Mass Relationships in Chemical Reactions

**Atomic & Molecular Masses (Sections 3.1, 3.3)**
**Avogadro's Number, the Mole and Molar Mass (Section 3.2)**
**The Mass Spectrometer (Section 3.4)**
**Percent Composition & Chemical Formulas (Sections 3.5 – 3.6)**
**Chemical Reactions and Chemical Equations (Section 3.7)**
**Amounts of Reactants and Products (Section 3.8)**
**Limiting Reagents & Reaction Yield (Sections 3.9 - 3.10)**


## SUMMARY

Note: Many of the tools you learn in this chapter will be very important and will be used through your general chemistry course.

### Atomic & Molecular Masses (Sections 3.1, 3.3)

*The Atomic Mass Scale.* The masses of individual atoms are too small to be measured with a balance; but the *relative* masses of the atoms of different elements can be measured. For instance, it is possible to determine that an atom of $^4_2He$ is very close to 1/3 the mass of an atom of $^{12}_6C$.  This is the basis of the atomic mass scale. By international agreement, an atom of carbon-12 is assigned a mass of exactly 12 atomic mass units (amu), making carbon-12 the standard (or reference) for the amu scale. On this scale, a helium-4 atom has a mass of 4.00 amu (1/3 of 12). Other measurements have shown that oxygen-16 atoms are 1.33 times heavier than carbon-12 atoms, making the mass of an oxygen-16 atom 16.00 amu. In this way, the masses of atoms of all the elements have been established.

*Average Atomic Mass.* The atomic masses that appear in the modern periodic table reflect the fact that elements occur in nature as combinations of isotopes. The atomic mass reported in the table is a weighted average of the atomic masses of the isotopes that make up the element. The mass of each isotope is weighted (or scaled) by its percent abundance in nature.

For example, the element lithium has two isotopes that occur in nature: $^6_3Li$ with 7.5 percent abundance, and $^7_3Li$ with 92.5 percent abundance. The atomic mass of lithium-6 is 6.01513 amu, and that of lithium-7 is 7.01601 amu. The average mass of such a mixture of Li atoms is given by:

average atomic mass = (fraction of isotope X)(mass of isotope X)
                    + (fraction of isotope Y)(mass of isotope Y)

$$= (0.075)(6.01513 \text{ amu}) + (0.925)(7.0161 \text{ amu})$$
$$= 0.45 \text{ amu} + 6.49 \text{ amu}$$
$$= 6.94 \text{ amu}$$

> Example
> 3.1
>
> Exercises
> 1, 2

Note that neither $^6_3Li$ nor $^7_3Li$ has an atomic mass of 6.94 amu. This value is the weighted average of the masses of the two Li isotopes and the value is closer to the mass of the lithium-7 isotope which reflects the fact that most of the mixture of isotopes is lithium-7.

***Molecular Masses.*** Molecules are composed of a number of atoms bonded together in a fixed arrangement. By the law of conservation of mass, the molecular mass is the sum of the atomic masses of the atoms in the molecular formula. For example, the molecular masses of two nitrogen oxides $NO_2$ and $N_2O_5$ are as follows:

$$\text{molecular mass of } NO_2 = \text{atomic mass of N} + 2(\text{atomic mass of O})$$
$$= 14.01 \text{ amu} + 2(16.00 \text{ amu})$$
$$= 46.01 \text{ amu}$$

$$\text{atomic mass of } N_2O_5 = 2(\text{ atomic mass of N}) + 5(\text{atomic mass of O})$$
$$= 2(14.01 \text{ amu}) + 5(16.00 \text{ amu})$$
$$= 108.02 \text{ amu}$$

## Avogadro's Number, the Mole and Molar Mass (Section 3.2)

***The Mole.*** The atomic mass scale is useful for small numbers of atoms or molecules, but is much too small for the quantities encountered in the laboratory, pharmacy or manufacturing plant. Macroscopic amounts of elements and compounds are too large to be measured conveniently on the atomic mass scale. For example, a vitamin C tablet would have a mass over $10^{21}$ amu. Macroscopic amounts of elements and compounds are measured in grams, but to measure out equal *numbers* of atoms of two elements, say carbon and oxygen, we cannot simply weigh out equal masses of the two elements. (Remember that oxygen atoms are 1.33 times heavier than carbon atoms.)

| Example 3.2 |
| Exercises 3,4 |

> If you wanted the same number of grapes and apples you couldn't weigh out 2 pounds of each and expect to have the same number.

Instead, we must measure a gram ratio of the two that is the same as the mass ratio of one C atom to one O atom. The atomic masses of C and O are 12.01 amu and 16.00 amu, respectively, so any amounts of C and O that have a mass ratio of 1.0 : 1.33, will contain equal numbers of C and O atoms. Therefore, 12.01 g C contains the same number of atoms as 16.00 g O. The number of C atoms in 12.01 g C is $6.022 \times 10^{23}$, called Avogadro's number. The quantity of a substance that contains Avogadro's number of atoms or other entities is called a mole (abbreviated mol).

The concept of a mole is a simple concept but often difficult for students to grasp. "Mole" is a word which represents a number, the number being a very large number: $6.022 \times 10^{23}$. Having a word represent a number isn't new to you. A dozen is a word that means 12. If someone said they had a ½ dozen donuts for breakfast, you wouldn't panic. You would know precisely what that meant. Similarly, if you find that you have ½ mole of carbon atoms, it too is just a number, but a really big number.

TOOL

| Conversion between | Tool |
|---|---|
| mole ←→ actual number (of anything) | Avogadro's number ($6.022 \times 10^{23}$) |

> How many atoms are 0.5 moles of carbon?
>
> $$0.5 \text{ mol C} \times \frac{6.022 \times 10^{23} \text{ atoms C}}{1 \text{ mol C}} = 3 \times 10^{23} \text{ atoms C}$$

The molar mass of an element or compound is the mass of one mole of its atoms or molecules. The molar mass of an element or molecule is numerically equal to its atomic or molecular mass expressed in grams rather than atomic mass units. For example, the atomic mass of Na is 22.99 amu, or we can say that there are 22.99 grams of Na in one mole, that is, 22.99 grams/mol. This gives us another conversion tool:

| Conversion between | Tool |
|---|---|
| gram A ←→ mol A | molar mass |

---

How many moles are in 15.0 g Na?

$$15.0 \ g \, Na \times \frac{1 \, mol \, Na}{22.99 \, g \, Na} = 0.653 \ mol \ Na$$

---

With the two tools we just learned, we can put them together and if you know any one of the three: moles, gram and actual number, you can solve for any of the other two:

| grams A | ←→ | moles A | ←→ | actual number A |

Example 3.4-3.6

Exercises 5-9

The term *mole* can be used in relation to any kind of particle, such as atoms, ions, or molecules. For clarity, the particle must always be specified. We say 1 mole of $O_3$ (ozone), or 1 mole of $O_2$ (diatomic oxygen), or 1 mole of $Na^+$ (sodium ions).  Other examples of molar amounts are:

1 mole $Na^+$ ions = $6.022 \times 10^{23}$ $Na^+$ ions = 29.00 g $Na^+$
1 mole $O_2$ molecules = $6.022 \times 10^{23}$ $O_2$ molecules = 32.00 g $O_2$
0.5 mole $O_3$ molecules = $3.011 \times 10^{23}$ $O_3$ molecules = 24.00 g $O_3$

Another conversion we can accomplish is between moles of molecules and moles of atoms within the molecule. The subscript of the formula determines this conversion. Consider $H_2O$.

How many hydrogen atoms are in each water molecule?
 • Two hydrogen atoms.
If there are two dozen water molecules, how many dozen hydrogen atoms are there?
 • Four dozen hydrogen atoms.
If there are 8 moles of water molecules, how many moles of hydrogen atoms are there?
 • 16 moles.

To use dimensional analysis to solve this last problem we do the following:

$$8 \, mol \, H_2O \times \frac{2 \, mol \, H}{1 \, mol \, H_2O} = 16 \, mol \, H$$

| Conversion between | Tool |
|---|---|
| mol compound ←→ mol atoms in compound | subscripts |
| no. molecules of compound ←→ no. atoms in compound | subscripts |

## The Mass Spectrometer (Section 3.4)

The mass spectrometer is an electronic instrument for measuring the mass of ionized, gas phase compounds. It is the most accurate method available for determining atomic and molecular masses. There are many different types of mass spectrometers. They all consist of the same basic 4 parts:  an ionization chamber to convert neutral samples to ions; ion optics to direct the ions into the mass analyzer; a mass analyzer to sort the ions by mass-to-charge ratio; and a detector to count the different types of ions sorted by the analyzer. The type of ionizer and mass analyzer determine the type of sample the mass spectrometer is best suited to analyze.

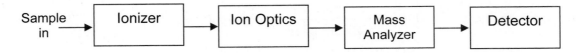

The instrument depicted in Figure 3.3 of the text is called a magnetic sector spectrometer because ions of different mass-to-charge ratios (m/Z) are deflected into different paths by the magnet. This occurs because magnetic fields change the motion of charged particles. The arrival of ions at the detector produces an electrical signal (current) that is proportional to the number of ions, so we can determine the abundance of each type of ion in the sample as well as the mass.

## Percent Composition & Chemical Formulas (Sections 3.5 – 3.6)

*Percent Composition.*  The percent composition of a compound is the percentage by mass of each element in the compound. It measures the relative mass of an element in a compound.  The formula for the percent composition of an element is:

$$\% \text{ composition of element} = \frac{n \times \text{element molar mass}}{\text{compound molar mass}} \times 100\%$$

> Keep in mind that percent is always calculated as *the part* over *the whole* times 100. In this case, the part is the grams of the individual element; the whole is the grams of the entire compound.

Consider sodium chloride (NaCl) as an example.  The molar mass, 58.44 g, is the sum of the mass of 1 mole of Na, 22.99 g, and the mass of 1 mole of Cl, 35.45 g. The percentage of Na by mass is

$$\%\text{Na} = \frac{1 \times 22.99 \text{ gNa}}{58.44 \text{ gNaCl}} \times 100\% = 39.33\%$$

The percentage of Cl by mass is

$$\%\text{Cl} = \frac{1 \times 35.45 \text{ gCl}}{58.44 \text{ gNaCl}} \times 100\% = 60.66\%$$

The percentages sum to 99.99% rather than 100% because the atomic masses and molar mass of NaCl were rounded to two decimal places.

Example
3.7, 3.8

Exercises
10, 11

*Percent Composition & Chemical Formulas.* Percent composition can be calculated directly from the formula of the compound (as above) or can be determined by chemical analysis if the formula is not known. You might expect, then, that given the percent composition you could calculate the molecular formula. This is almost true. Remember, the percent composition is a measure of the relative contribution of an element to a compound. So we can calculate the *empirical formula*, reflecting the simplest whole-number ratio of the different kinds of atoms in the compound, from the percent composition.

General procedure to determine the empirical formula from percent composition:
- Assume a 100 g sample of the compound and treat the percents as grams. (If the percent composition of a substance is 30.43% N and 60.56% O, and there are 100.00 grams of the sample, 30.43 grams is nitrogen and 60.56 grams is oxygen)
- Convert the mass to moles. Remember, an empirical formula is a ratio of the number of atoms so we must leave the "mass world" and enter the "number world," and moles is a number.
- Divide by the smallest number of moles to get the mole ratio.
- If this gives fractions (like ½ or ¾) multiply by whatever number is necessary to remove the fraction. Empirical formulas have the smallest **whole**-number ratio of atoms.

So let's continue the example of a substance which is 30.43% N and 60.56% O and determine the empirical formula:

$$30.43 \, g\,N \times \frac{1\,mol\,N}{14.01\,g\,N} = 2.172 \, mol\,N$$

$$69.56 \, g\,O \times \frac{1\,mol\,O}{16.00\,g\,O} = 4.348 \, mol\,O$$

| Examples 3.9, 3.10 |
| Exercises 12 - 14 |

Then divide by the smaller of the two, divide both by 2.172 mol:

For N:  2.172 ÷ 2.172 = 1
For O:  4.348 ÷ 2.172 = 2

The empirical formula is therefore $NO_2$.

***Molecular Formula.*** To determine the molecular formula from the empirical formula, one more piece of information must be known: the molar mass. The molecular formula is a whole- number multiple of the empirical formula. To determine the molecular formula, divide the molar mass of the compound by the molar mass of the empirical formula.  Going back to the nitrogen oxide compound, the molar mass was determined to be 92.00 by mass spectrometry.

$$92/(14.01+2\times16.00)=2$$

Therefore, the molecular formula is 2 times the empirical formula, or $N_2O_4$.

| Example 3.11 |
| Exercises 15 |

## Chemical Reactions and Chemical Equations (Section 3.7)

***Chemical Equations.***  A chemical reaction is a process in which one or more chemical substances is changed into one or more new substance. A chemical equation is a symbolic shorthand for representing a chemical reaction. The chemical formulas of the reactants (starting materials) are written on the left side of the equation, and the formulas of the products on the right side. Equations are written according to the following format:

1.  Reactants and products are separated by an arrow. The arrow is read as "produces" or "yields."
2.  Plus signs are placed between reactants and between products. A plus sign between reactants means that all the reactants are required for reaction. A plus between products implies that a mixture of two or more products is formed by the reaction. The plus sign is read as "plus" or "and."
3.  Abbreviations are sometimes included in parentheses to indicate the physical states of the reactants and products.
4.  Balanced equations reflect the law of conservation of mass. That is, the number of atoms of a certain element appearing in the reactants must be equal to the number of atoms of that element appearing in the products.
5.  Differences in the relative amounts of reactants consumed or products generated are reflected by stoichiometric coefficients. A coefficient is a number placed before a chemical formula in an equation that is a multiplier for the formula. For example, $3H_2O$ means three molecules of water, a total of 6 hydrogen atoms and three oxygen atoms. Subscripts, 2 in $H_2O$, represent the number of atoms in a compound. Absence of a coefficient or element subscript is understood to mean one.

***Balancing Equations.***  When the hydrocarbon pentane ($C_5H_{12}$) is burned or combusted (reacted with $O_2$), carbon dioxide and water are produced. The unbalanced chemical equation representing the reaction is:

$$C_5H_{12} + O_2 \rightarrow CO_2 + H_2O$$

Eventually, you will not need a procedure to balance chemical reactions. The basic idea is to change the coefficients of the reactants and products until the numbers of each type of atom are the same on each side of the reaction. It is important that you <u>NEVER</u> change the subscripts of the reactants or products; changing the subscripts changes the chemical compound. The following can be used to balance equations. For simplicity, the physical state subscripts are omitted during this process.

1.  Count and list the number of each type of atom on each side of the arrow.
2.  Look for elements that occur once on both sides of the equation; change the coefficients so that these elements are balanced, if necessary.
3.  Change the coefficient of any remaining components to accommodate the changes made in step 2.
4.  Check the balanced equation to be sure the numbers of each atom are equal on both sides of the equation.

Examples
3.12 – 3.15

Exercises
16 – 18

Consider the combustion (burning) of pentane, a petroleum product:

| Step # | Equation | # reactant atoms | | | # product atoms | | | Balanced? |
|---|---|---|---|---|---|---|---|---|
| 1 | $C_5H_{12} + O_2 \rightarrow CO_2 + H_2O$ | 5 C | 12 H | 2 O | 1 C | 2 H | 3 O | No |
| | Count the number of atoms on each side. No element has same number of atoms on each side. | | | | | | | |
| 2 | $C_5H_{12} + O_2 \rightarrow 5CO_2 + H_2O$ | 5 C | 12 H | 2 O | 5 C | 2 H | 11 O | No |
| | The coefficient 5 is placed in front of $CO_2$ and now carbon is balanced but other elements are not. | | | | | | | |
| 3 | $C_5H_{12} + O_2 \rightarrow 5CO_2 + 6H_2O$ | 5 C | 12 H | 2 O | 5 C | 12 H | 16 O | No |
| | The coefficient 6 is placed in front of $H_2O$. Hydrogen as well as carbon is balanced. | | | | | | | |
| 4 | $C_5H_{12} + 8O_2 \rightarrow 5CO_2 + 6H_2O$ | 5 C | 12 H | 16 O | 5 C | 12 H | 16 O | Yes |
| | The coefficient 8 is placed in front of $O_2$ to balance the oxygen atoms. Atoms of all elements are now balanced. | | | | | | | |

## Amounts of Reactants and Products (Section 3.8)

*Mass Relationships in Reactions.* Chemical equations contain information about the relative numbers of reactants and products involved in the reaction. The molar masses of the reactants and products link the numbers of reactants and products to their masses. Therefore, the balanced equation can be used to determine how many grams of one substance will be needed to react with a given mass of another, or how many grams of product can be produced by the reaction of a specific mass of a reactant. For instance, the balanced equation

$$
\begin{array}{ccccccc}
2\,Al & + & Fe_2O_3 & \rightarrow & Al_2O_3 & + & 2\,Fe \\
2\ mol & & 1\ mol & & 1\ mol & & 2\ mol \\
53.96\ g & & 159.70\ g & & 101.96\ g & & 111.70\ g
\end{array}
$$

states that 2 moles of Al, (53.96 g), will react completely with 1 mole of $Fe_2O_3$, (159.70 g), and will yield 1 mole, (101.96 g), of $Al_2O_3$ and 2 moles, (111.70 g), of Fe.

The quantitative relationship between elements and compounds in chemical reactions is a part of chemistry called stoichiometry. The word stoichiometry derives from two Greek words: *stoicheion*, meaning "element" and *metron*, meaning "measure". J. B. Richter (1762–1807) was the first to lay down the principles of stoichiometry. He said "stoichiometry is the science of measuring the quantitative proportions or mass ratios in which chemical elements stand to one another." In all stoichiometry problems, the balanced chemical equation provides the "bridge" that relates the amount of one reactant to the amount of another; and the amounts of products to reactants.

*Mole Ratios.* The balanced equation describing the reaction of aluminum and iron oxide defines the quantitative relationships between the reactants and products. We can see, for example, that 2 moles of Al are transformed to 1 mole of $Al_2O_3$. We can write these relationships in shorthand using mole ratios. Mole ratios are conversion factors that relate the number of reactants consumed to each other or to the number of product molecules formed.

| Conversion between | Tool |
|---|---|
| mol A $\longleftrightarrow$ mol B* | coefficients of balanced equation |

\* A and B represent different substances in the reaction equation.

| $2Al + Fe_2O_3 \rightarrow Al_2O_3 + 2Fe$ | | |
|---|---|---|
| 2 mol Al $\rightleftharpoons$ 1 mol $Al_2O_3$ | $\dfrac{2 \text{ mol Al}}{1 \text{ mol } Al_2O_3}$  or | $\dfrac{1 \text{ mol } Al_2O_3}{2 \text{ mol Al}}$ |
| 2 mol Al $\rightleftharpoons$ 2 mol Fe | $\dfrac{2 \text{ mol Al}}{2 \text{ mol Fe}}$  or | $\dfrac{2 \text{ mol Fe}}{2 \text{ mol Al}}$ |
| 2 mol Al $\rightleftharpoons$ 1 mol $Fe_2O_3$ | $\dfrac{2 \text{ mol Al}}{1 \text{ mol } Fe_2O_3}$  or | $\dfrac{1 \text{ mol } Fe_2O_3}{2 \text{ mol Al}}$ |
| 1 mol $Al_2O_3$ $\rightleftharpoons$ 2 mol Fe | $\dfrac{1 \text{ mol } Al_2O_3}{2 \text{ mol Fe}}$  or | $\dfrac{2 \text{ mol Fe}}{1 \text{ mol } Al_2O_3}$ |

\* the symbol $\rightleftharpoons$ means "stoichiometrically equivalent to"

Mole ratios are used to answer stoichiometry questions such as

> How many grams of iron are produced when 25.0 g $Fe_2O_3$ reacts with aluminum?
> How many grams of Al are required to react completely with 25.0 g $Fe_2O_3$?

Stoichiometry problems are solved in three steps:

1.  Convert the amounts of given substances into moles, if not already given as moles.
2.  Use the appropriate mole ratio constructed from the balanced equation to calculate the number of moles of the needed or unknown substance.
3.  Convert the number of moles of the needed substance into mass units if required.

---

How many grams of iron are produced when 25.0 g $Fe_2O_3$ reacts with aluminum?

1.  $n_{Fe_2O_3} = 25.0 \text{ g } Fe_2O_3 \times \dfrac{1 \text{ mol } Fe_2O_3}{159.70 \text{ g } Fe_2O_3} = 0.157 \text{ mol } Fe_2O_3$

2.  $n_{Fe} = 0.157 \text{ mol } Fe_2O_3 \times \dfrac{2 \text{ mol Fe}}{1 \text{ mol } Fe_2O_3} = 0.314 \text{ mol Fe}$

3.  $m_{Fe} = 0.314 \text{ mol Fe} \times \dfrac{55.85 \text{ g Fe}}{1 \text{ mol Fe}} = 17.5 \text{ g Fe}$

With practice, it will be easier to combine these three steps into a single calculation:

$$m_{Fe} = 25.0 \text{ g } Fe_2O_3 \times \dfrac{1 \text{ mol } Fe_2O_3}{159.70 \text{ g } Fe_2O_3} \times \dfrac{2 \text{ mol Fe}}{1 \text{ mol } Fe_2O_3} \times \dfrac{55.85 \text{ g Fe}}{1 \text{ mol Fe}} = 17.5 \text{ g Fe}$$

---

How many grams of Al are required to react completely with 20.0 g $Fe_2O_3$?

$$m_{Al} = 25.0 \text{ g } Fe_2O_3 \times \dfrac{1 \text{ mol } Fe_2O_3}{159.70 \text{ g } Fe_2O_3} \times \dfrac{2 \text{ mol Al}}{1 \text{ mol } Fe_2O_3} \times \dfrac{26.98 \text{ g Al}}{1 \text{ mol Al}} = 8.45 \text{ g Al}$$

Example
3.16

Exercises
19 – 22

## Limiting Reagents & Reaction Yield (Sections 3.9 – 3.10)

***Limiting Reagents.*** Usually the reactants are not present in the exact ratio prescribed by the balanced chemical equation. For example, a large excess of a less expensive reagent is used to ensure complete reaction of the more expensive reagent. In this situation, one reactant will be completely consumed before the other runs out. When this occurs, the reaction will stop and no more products will be made. The reactant that is consumed first is called the limiting reagent because it limits, or determines the amount of product formed. The reactant that is not completely consumed is called the excess reagent. The figure below illustrates the relationships between reagents and products in this case.

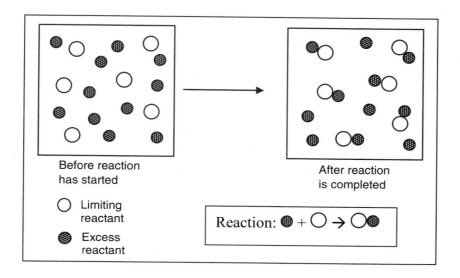

Reconsidering the aluminum/iron oxide reaction, suppose that 1.2 moles of Al are reacted with 1 mole of $Fe_2O_3$. In any stoichiometry problem, it is important to determine which reactant is the limiting one, to correctly predict the yield of products.

| Step # | # reactant amounts | | # product amounts | |
|---|---|---|---|---|
| | 2 Al | + $Fe_2O_3$ → | $Al_2O_3$ | + 2 Fe |
| initial | 1.2 mol | 1.0 mol | 0 mol | 0 mol |
| change | -1.2 mol | -0.6 mol | +0.6 mol | +1.2 mol |
| final | 0 mol | 0.4 mol | 0.6 mol | 1.2 mol |

TOOL

Explanation of the I.C.F table above:
- **I**nitial amounts are the initial mole amounts of each reactant. If given grams, convert to moles to put into the table.
- **C**hange amounts are how much reactant will be consumed and how much product will be formed. The balanced equation in this example shows that you have to have twice as much Al as $Fe_2O_3$. Since 1.2 mol Al isn't twice the 1.0 mol $Fe_2O_3$, then the Al is the limiting reactant and ½ as much $Fe_2O_3$ will form. Note that the limiting reactant (1.2 mol Al) is not the reactant in lesser molar amount (1.0 mol $Fe_2O_3$).
- **F**inal amount is what is left over after the reaction takes place.

If the problem then wants to know grams of product formed, or grams of excess reactant remaining, simply convert.

Example
3.17

Exercise
23

Note: Solving limiting reactant problems can be accomplished in a number of ways, this study guide gives an alternative procedure than in your textbook. Choose the method that works best for you.

***Percent Yield.*** The product amounts predicted from the balanced reaction are best case scenarios, the perfect world result. In practice, the actual yield is usually less than the calculated, or "theoretical" yield. The theoretical yield is calculated based on the assumption that all the limiting reagent reacts according to the balanced equation. The theoretical yield for the aluminum/iron oxide reaction above is 0.6 mol $Al_2O_3$ or 61.18 g of $Al_2O_3$. The percent yield is based on the actual amount of product produced in the reaction. The percent yield is defined as

$$\% \text{ percent yield} = \frac{\text{actual yield}}{\text{theoretical yield}} \times 100\%$$

So if the aluminum/iron oxide reaction above actually produced 52.5 g of $Al_2O_3$, the percent yield is

$$\% \text{ percent yield} = \frac{52.5\,g}{61.18\,g} \times 100\% = 85.8\%$$

Example
3.18

Exercise
24, 25

## GLOSSARY LIST

| | | |
|---|---|---|
| **atomic mass unit** | **chemical reaction** | **limiting reagent** |
| **atomic mass** | **chemical equation** | **excess reagent** |
| **molecular mass** | **reactant** | **theoretical yield** |
| **Avogadro's number** | **product** | **actual yield** |
| **mole** | | **percent yield** |
| **molar mass** | **stoichiometry** | |
| | **stoichiometric amount** | |
| **percent composition** | **mole method** | |

## EQUATIONS

| Algebraic Equation | English Translation |
|---|---|
| $\% \text{ composition} = \dfrac{n \times \text{element molar mass}}{\text{compound molar mass}} \times 100\%$ | The percent composition is the fraction of the mass of the compound that comes from element X. |
| $\% \text{ percent yield} = \dfrac{\text{actual yield}}{\text{theoretical yield}} \times 100\%$ | The percent yield is the fraction of product which actually forms compared to the calculated amount, |

## SUMMARY OF CONVERSION TOOLS

| Conversion between | Tool |
|---|---|
| g ←→ kg (as an example) | prefix meanings |
| g ←→ mL | density |
| mol ←→ actual number | Avogadro's number |
| g ←→ mol | molar mass |
| mol compound ←→ mol atom in compound | subscript |
| mol A ←→ mol B | coefficients of balanced equation |

## WORKED EXAMPLES

---

**EXAMPLE 3.1  Average Atomic Mass**

The element boron (B) consists of two stable isotopes with atomic masses of 10.012938 amu and 11.009304 amu. The average atomic mass of B is 10.81 amu. Which isotope is more abundant, boron-10 or boron-11?

**• Solution**

The atomic mass is a weighted average of the masses of the naturally occurring isotopes of an element. Each isotopic mass is multiplied by its relative abundance. If both isotopes in this example had a natural abundance of 50 percent, then the atomic mass would be the average of the two masses, 10.5 amu. Since the atomic mass of boron (10.81 amu) is greater than 10.5, the abundance of the boron-11 isotope must be larger than that of the boron-10. The abundances of boron-10 and boron-11 are 19.6% and 80.4%, respectively.

---

**Work EXERCISES & PROBLEMS: 1, 2**

---

**EXAMPLE 3.2  Molecular Mass**

Calculate the molecular mass of carbon tetrachloride ($CCl_4$).

**• Solution**

The molecular mass is the sum of the atomic masses of all the atoms in the molecule:

$$\text{molecular mass } CCl_4 = (\text{atomic mass of C}) + 4(\text{atomic mass of Cl})$$
$$= (12.01 \text{ amu}) + 4(35.45 \text{ amu})$$
$$\text{molecular mass } CCl_4 = 153.81 \text{ amu}$$

---

**Work EXERCISES & PROBLEMS: 3, 4**

---

**EXAMPLE 3.3  The Molar Mass**

Naturally occurring lithium consists of 7.5 percent Li-6 atoms, and 92.5 percent Li-7 atoms. Complete the following sentences:

a. The number of Li atoms in one mole of Li is _____.

b. One mole of Li atoms consists of _____ Li-6 atoms and _____ Li-7 atoms and will have a mass of _____ g.

**• Solution**

a. One mole of Li contains $6.022 \times 10^{23}$ Li atoms (a mixture of Li-6 atoms and Li-7 atoms).

b. One mole of Li atoms contains $(0.075)(6.022 \times 10^{23})$ Li-6 atoms, which is equal to $\underline{4.5 \times 10^{22}}$ Li-6 atoms; and $(0.925)(6.022 \times 10^{23})$ Li-7 atoms, which is $\underline{5.57 \times 10^{23}}$ Li-7 atoms. Note that the total number of Li atoms is $6.022 \times 10^{23}$, and thus will have a mass of $\underline{6.94}$ g (obtained from the periodic table.)

---

**EXAMPLE 3.4  Mass of a Given Number of Moles**

What is the mass of 2.25 moles of iron (Fe)?

**• Solution**

To convert moles of Fe to grams we use the molar mass as the conversion tool. The molar mass of Fe is 55.85 g/mol or

$$? \text{ g Fe} = 2.25 \text{ mol Fe} \times \frac{55.85 \text{ g}}{1 \text{ mol Fe}} = 126 \text{ g Fe}$$

**Work EXERCISES & PROBLEMS: 4, 5**

**EXAMPLE 3.5  Number of Atoms in a Given Mass**

How many zinc atoms are present in 20.0 g Zn?

**• Solution**

Using tools we know we can do the following: g Zn → mol Zn → atoms Zn. Step one uses molar mass of Zn (65.39 g/mol), step two uses Avogradro's number.

Step 1: Starting with the given quantity:  20.0 g Zn, convert to moles:

$$? \text{ Zn mol} = 20.0 \text{ g Zn} \times \frac{1 \text{ mol Zn}}{65.39 \text{ g Zn}} = 0.3059 \text{ mol Zn}$$

Step 2: Here we will carry one more digit than the correct number of significant figures. To convert moles of zinc to atoms of zinc, we use Avogadro's number:

$$? \text{ Zn atoms} = 0.3059 \text{ mol Zn} \times \frac{6.022 \times 10^{23} \text{ Zn atoms}}{1 \text{ mol Zn}} = 1.84 \times 10^{23} \text{ Zn atoms}$$

**•Comment**

Instead of calculating the number of moles separately, we could have strung the conversion factors together into one calculation:

$$? \text{ Zn atoms} = 20.0 \text{ g Zn} \times \frac{1 \text{ mol Zn}}{65.39 \text{ g Zn}} \times \frac{6.022 \times 10^{23} \text{ Zn atoms}}{1 \text{ mol Zn}}$$

$$= 1.84 \times 10^{23} \text{ Zn atoms}$$

---

### EXAMPLE 3.6  Moles of a Molecular Compound

a.   How many molecules of ethane ($C_2H_6$) are present in 50.3 g of ethane?
b.   How many atoms each of H and C are in this sample?

### • Solution

a.  Road map: g $C_2H_6$ → mol $C_2H_6$ → molecules $C_2H_6$. Step one uses the molar mass of
$C_2H_6$, step 2 uses Avogadro's number.

The molecular mass of $C_2H_6$ is:

$$2(12.01 \text{ amu}) + 6(1.008 \text{ amu}) = 30.07 \text{ amu}  \text{ or } 30.07 \text{ g/mol}$$

Step 1:

$$? \text{ mol } C_2H_6 = 50.3 \text{ g } C_2H_6 \times \frac{1 \text{ mol } C_2H_6}{30.07 \text{ g } C_2H_6} = 1.67 \text{ mol } C_2H_6$$

Step 2:

$$? \ C_2H_6 \text{ molecules} = 1.67 \text{ mol } C_2H_6 \times \frac{6.022 \times 10^{23} \text{ molecules}}{1 \text{ mol } C_2H_6}$$

$$= 1.01 \times 10^{24} \ C_2H_6 \text{ molecules}$$

b.  Road map: molecules $C_2H_6$ → atoms C; molecules $C_2H_6$ → atoms H. Both of these steps
require you to look at the subscripts in the formula, giving the following conversion factors.

$$\frac{6 \text{ H atoms}}{C_2H_6 \text{ molecule}} \quad \text{and} \quad \frac{2 \text{ C atoms}}{C_2H_6 \text{ molecule}}$$

Multiplying the number of $C_2H_6$ molecules by the number of atoms of each kind per
molecule gives the number of each kind of atom present.

$$1.01 \times 10^{24} \text{ molecules} \times \frac{6 \text{ H atoms}}{C_2H_6 \text{ molecule}} = 6.06 \times 10^{24} \text{ H atoms}$$

$$1.01 \times 10^{24} \text{ molecules} \times \frac{2 \text{ C atoms}}{C_2H_6 \text{ molecule}} = 2.02 \times 10^{24} \text{ C atoms}$$

**Work EXERCISES & PROBLEMS: 6 – 9**

**EXAMPLE 3.7  Percent Composition**

Calculate the percent composition of glucose ($C_6H_{12}O_6$).

**• Solution**

First, the molar mass of glucose must be found from the atomic masses of each element.

$$\text{molar mass of } C_6H_{12}O_6 = \left[ 6 \text{ mol C} \times \frac{12.01 \text{ g}}{1 \text{ mol C}} \right] + \left[ 12 \text{ mol H} \times \frac{1.008 \text{ g}}{1 \text{ mol H}} \right] +$$

$$\left[ 6 \text{ mol O} \times \frac{16.00 \text{ g}}{1 \text{ mol O}} \right]$$

$$= 72.06 \text{ g} + 12.10 \text{ g} + 96.00 \text{ g} = 180.16 \text{ g}$$

Then, the percent of each element present is found by dividing the mass of each element in 1 mole of glucose by the total mass of 1 mole of glucose.

$$\text{mass percent of C} = \frac{\text{mass of C in 1 mol } C_6H_{12}O_6}{\text{mass of 1 mole } C_6H_{12}O_6} \times 100\%$$

$$\%C = \frac{6 \text{ mol C} \times (12.01 \text{ g/mol C})}{180.16 \text{ g } C_6H_{12}O_6} \times 100\% = \frac{72.06 \text{ g}}{180.16 \text{ g}} \times 100\% = 40.00\% \text{ C}$$

After practicing the calculation several times, you can simplify the notation:

$$\%H = \frac{12 \times 1.008 \text{ g}}{180.16 \text{ g}} \times 100\% = \frac{12.10 \text{ g}}{180.16 \text{ g}} \times 100\% = 6.716\% \text{ H}$$

$$\%O = \frac{6 \times 16.00 \text{ g}}{180.16 \text{ g}} \times 100\% = \frac{96.00 \text{ g}}{180.16 \text{ g}} \times 100\% = 53.29\% \text{ O}$$

Thus, glucose ($C_6H_{12}O_6$) contains 40.0% C, 6.72% H, and 53.3% O by mass.

**• Comment**

A good way to check the results is to sum the percentages of the elements, because their total must add to 100%. Here we get 100.01%.

---

**EXAMPLE 3.8  Carbon–Hydrogen Analysis**

When a 0.761-g sample of a compound of carbon and hydrogen is burned in a C–H combustion apparatus (see Figure 3.6 in the text), 2.23 g $CO_2$ and 1.37 g $H_2O$ are produced. Determine the percent composition of the compound.

### • Solution

Here we want the grams of carbon and hydrogen in the original compound. All the carbon is now present in 2.23 g of $CO_2$. All the hydrogen is now present in 1.37 g of $H_2O$. How many grams of C are there in 2.23 g of $CO_2$, and how many grams of H are there in 1.37 g of $H_2O$? We know there is 1 mole of C atoms (12.01 g) in every mole of $CO_2$, or 44.01 g.

The solution road map is:

$$g\ CO_2 \rightarrow mol\ CO_2 \rightarrow mol\ C \rightarrow g\ C$$

$$?\ g\ C = 2.23\ g\ CO_2 \times \frac{1\ mol\ CO_2}{44.01\ g\ CO_2} \times \frac{1\ mol\ C}{1\ mol\ CO_2} \times \frac{12.01\ g\ C}{1\ mol\ C} = 0.609\ g\ C$$

$$g\ H_2O \rightarrow mol\ H_2O \rightarrow mol\ H \rightarrow g\ H$$

$$?\ g\ H = 1.37\ g\ H_2O \times \frac{1\ mol\ H_2O}{18.0\ g\ H_2O} \times \frac{2\ mol\ H}{1\ mol\ H_2O} \times \frac{1.008\ g\ H}{1\ mol\ H} = 0.153\ g\ H$$

The percentage composition is:

$$\%H = \frac{0.153\ g\ H}{0.761\ g\ compound} \times 100\% = 20.1\%\ H$$

$$\%C = \frac{0.609\ g\ C}{0.761\ g\ compound} \times 100\% = 80.0\%\ C$$

**Work EXERCISES & PROBLEMS: 10, 11**

---

### EXAMPLE 3.9  Using Percent Composition

Calculate the mass of carbon in 10.00 g of glucose ($C_6H_{12}O_6$).

### • Solution

In Example 3.7 we calculated that glucose contains 40.00% carbon. Therefore,

$$mass\ C = 40.00\%\ of\ 10.00\ g\ C_6H_{12}O_6 = 10.00\ g\ C_6H_{12}O_6 \times \frac{40.00\ g\ C}{100\ g\ C_6H_{12}O_6} = 4.000\ g\ C$$

This calculation would have to be done from scratch if we did not already know the percentage of carbon. The road map is:

mass of glucose → mol glucose → mol carbon → g carbon

$$?\ mass\ C = 10.00\ g\ C_6H_{12}O_6 \times \frac{1\ mol\ C_6H_{12}O_6}{180.16\ g\ C_6H_{12}O_6} \times \frac{6\ mol\ C}{1\ mol\ C_6H_{12}O_6} \times \frac{12.01\ g\ C}{1\ mol\ C} = 4.000\ g\ C$$

**EXAMPLE 3.10  Empirical Formula**

An elemental analysis of a new compound reveals its percent composition to be:  50.7% C, 4.23% H, and 45.1% O. Determine the empirical formula.

**• Solution**

The empirical formula gives the relative numbers of atoms of each element in a formula unit. To start, assume you have exactly 100 g of compound, and then determine the number of moles of each element present. Then, find the ratios of the number of moles.

$$50.7 \text{ g C} \times \frac{1 \text{ mol C}}{12.01 \text{ g C}} = 4.22 \text{ mol C}$$

$$4.23 \text{ g H} \times \frac{1 \text{ mol H}}{1.008 \text{ g H}} = 4.20 \text{ mol H}$$

$$45.1 \text{ g O} \times \frac{1 \text{ mol O}}{16.00 \text{ g O}} = 2.82 \text{ mol O}$$

This gives the mole ratio for C : H : O of 4.22 : 4.20 : 2.82.
Dividing by the smallest of the molar amounts gives:

$$\frac{4.22 \text{ mol C}}{2.82 \text{ mol O}} = 1.50 \text{ mol C to } 1.00 \text{ mol O} \quad \text{and} \quad \frac{4.20 \text{ mol H}}{2.82 \text{ mol O}} = 1.49 \cong 1.5 \text{ mol H to } 1.00 \text{ mol O}$$

Therefore, $C_{1.5}H_{1.5}O_{1.0}$ is a possible formula.

This is not an acceptable formula because it does not contain all whole numbers. To convert these fractions to whole numbers, multiply each subscript by the same number, in this case a 2. The empirical formula therefore is $C_3H_3O_2$.

**Work EXERCISES & PROBLEMS: 12-14**

**EXAMPLE 3.11  Molecular Formula**

Mass spectrometer experiments on the compound in Example 3.10 show its molecular mass to be about 140 amu. What is the molecular formula?

**• Solution**

The molar mass of the empirical formula $C_3H_3O_2$ is 71 g. Molar mass ÷ empirical formula mass gives 140/71 = 2. Therefore, the molecular formula must have twice as many atoms as the empirical formula.

molecular formula = $C_6H_6O_4$

**Work EXERCISES & PROBLEMS: 15**

---

**EXAMPLE 3.12 Balancing Chemical Equations**

Balance the following reaction. In this displacement reaction, zinc displaces $H_2$ from chemical combination with chlorine.

$$Zn(s) + HCl(aq) \rightarrow ZnCl_2(aq) + H_2(g)$$

• **Solution**

**Step 1:** Identify elements that occur in only two compounds in the equation. In this case, each of the three elements Zn, H, and Cl occurs in only two compounds.
**Step 2:** Of these, balance the element with the largest subscript. Both Cl and H have the subscript 2, so balance either one of them first. Multiplying HCl by 2 to balance H:

$$Zn + 2\,HCl \rightarrow ZnCl_2 + H_2$$

Taking stock:

| Reactants | Products |
|-----------|----------|
| H (2) | H (2) |
| Cl (2) | Cl (2) |
| Zn (1) | Zn (1) |

At this point, both H and Cl are balanced. Further inspection reveals that Zn is also balanced. Thus, the equation is balanced!

---

**EXAMPLE 3.13 Balancing Chemical Equations**

The reaction of a hydrocarbon with oxygen is an example of a combustion reaction. Balance the equation for the reaction of pentane with oxygen gas:

$$C_5H_{12} + O_2 \rightarrow CO_2 + H_2O$$

• **Solution**

Following step 1 given in the previous example, we note that both H and C appear in only two compounds. According to step 2 (balancing the element with the highest subscript) H has the largest subscript. Balance H first by placing the coefficient 6 in front of $H_2O$. Next balance the carbon atoms by placing a 5 in front of $CO_2$.

$$C_5H_{12} + O_2 \rightarrow 5\,CO_2 + 6\,H_2O$$

Taking stock:

| Reactants | Products |
|-----------|----------|
| C (5) | C (5) |
| H (12) | H (12) |
| O (2) | O (16) |

Finally, the oxygen atoms can be balanced. The products already have their coefficients determined, and so the reactants must be adjusted to supply 16 O atoms. Eight molecules of $O_2$ contain 16 O atoms, and so write the coefficient 8 in front of $O_2$.

$$C_5H_{12} + 8\,O_2 \rightarrow 5\,CO_2 + 6\,H_2O$$

| Reactants | Products |
|-----------|----------|
| C (5) | C (5) |
| H (12) | H (12) |
| O (16) | O (16) |

## EXAMPLE 3.14  Balancing Chemical Equations

Balance the following decomposition reaction:

$$NaClO_3(s) \rightarrow NaCl(s) + O_2(g)$$

• **Solution**

According to steps 1 and 2, O should be balanced first.  There are 3 O atoms on the reactant side and 2 on the product side. When an element that occurs in only two substances in the equation and the number of atoms of that element is even on one side and odd on the other side of the equation, use two coefficients that increase the number of atoms on each side of the equation to the least common multiple.

$$NaClO_3 \rightarrow NaCl + O_2$$

| Reactants | Products |
|-----------|----------|
| O (3) | O (2) |

The least common multiple for balancing O is $3 \times 2 = 6$. The two coefficients are 2 and 3.

$$2\,NaClO_3 \rightarrow NaCl + 3\,O_2$$

| Reactants | Products |
|-----------|----------|
| O (6) | O (6) |

Balancing the Na and Cl atoms gives:

$$2\,NaClO_3 \rightarrow 2\,NaCl + 3\,O_2$$

## EXAMPLE 3.15  Balancing Chemical Equations

Balance the following precipitation reaction in which sulfuric acid reacts with barium chloride in aqueous solution to yield hydrochloric acid and a precipitate of barium sulfate.

$$H_2SO_4(aq) + BaCl_2(aq) \rightarrow HCl(aq) + BaSO_4(s)$$

• **Solution**

Here, it helps to recognize certain groups of atoms that maintain their identity during the reaction. In this equation the sulfate ion, $SO_4^{2-}$, appears as a unit.

$$H_2SO_4 + BaCl_2 \rightarrow HCl + BaSO_4$$

| Reactants | Products |
|-----------|----------|
| H (2) | H (1) |
| $SO_4^{2-}$ (1) | $SO_4^{2-}$ (1) |
| Ba (1) | Ba (1) |
| Cl (2) | Cl (1) |

Either H or Cl could be balanced first. Balance chlorine by multiplying HCl by 2:

$$H_2SO_4 + BaCl_2 \rightarrow BaSO_4 + 2\ HCl$$

A check of the H atoms shows they are now balanced along with the other elements.

**Work EXERCISES & PROBLEMS: 16-18**

---

**EXAMPLE 3.16  The Mole Method**

Sulfur dioxide can be removed from refinery stack gases by reaction with quicklime, CaO.

$$SO_2(g) + CaO(s) \rightarrow CaSO_3(s)$$

If 975 g of $SO_2$ is to be removed from stack gases by the above reaction, what mass of CaO is required to completely react with it?

**• Solution**

First be certain that the equation is correctly balanced, and it is.  Then plan your road map.

$$\begin{array}{ccccccc} & \text{step 1} & & \text{step 2} & & \text{step 3} & \\ \text{Road map:} & \text{g } SO_2 & \rightarrow & \text{mol } SO_2 & \rightarrow & \text{mol CaO} & \rightarrow & \text{g CaO} \end{array}$$

1.  Convert the 975 g of $SO_2$ to moles using its molar mass.
2.  Use the chemical equation to find the number of moles of CaO that reacts per mole of $SO_2$.
    From the balanced equation we can write:  1 mol CaO $\rightleftharpoons$ 1 mol $SO_2$
3.  Convert moles of CaO to grams of CaO using the molar mass of CaO.

String the conversion factors from the three steps one after the other.

$$?\ \text{g CaO} = 9.75 \times 10^2\ \text{g } SO_2 \times \frac{1\ \text{mol } SO_2}{64.07\ \text{g } SO_2} \times \frac{1\ \text{mol CaO}}{1\ \text{mol } SO_2} \times \frac{56.08\ \text{g CaO}}{1\ \text{mol CaO}}$$

$$= 8.53 \times 10^2\ \text{g CaO}$$

**Work EXERCISES & PROBLEMS: 19-22**

## EXAMPLE 3.17  Limiting Reagent

Phosphine ($PH_3$) burns in oxygen to produce phosphorus pentoxide ($P_2O_5$) and water.

$$2\ PH_3(g) + 4\ O_2(g) \rightarrow P_2O_5(s) + 3\ H_2O(\ell)$$

How many grams of $P_2O_5$ will be produced when 17.0 g of phosphine is mixed with 16.0 g of $O_2$ and reaction occurs?

## • Solution

When the amounts of **both reactants** are given, it is possible that one will be completely used up before the other. This is your indication that you are dealing with a limiting reactant problem. It is important that you look for this so you can plan your strategy. For limiting reactant problems, one tool is to construct an I.C.F. table, placing mole amounts into the initial line.

$$17.0\ g\ PH_3 \times \frac{1\ mol\ PH_3}{33.99\ g\ PH_3} = 0.500\ mol\ PH_3$$

$$16.0\ g\ O_2 \times \frac{1\ mol\ O_2}{32.00\ g\ O_2} = 0.500\ mol\ O_2$$

| Step # | reactant amounts | | product amounts | |
|---|---|---|---|---|
| | $2\ PH_3(g) + 4\ O_2(g) \rightarrow$ | | $P_2O_5(s)\ +\ 3\ H_2O(\ell)$ | |
| initial | 0.500 mol | 0.500 mol | 0.000 mol | 0.000 mol |
| change | −0.250 mol | −0.500 mol | + 0.125 mol | +0.375 mol |
| finish | 0.250 mol | 0.000 mol | 0.125 mol | 0.375 mol |

The key to the I.C.F. table is the change line. Looking at the mole ratio of $PH_3$ to $O_2$ we see that twice as many moles of $O_2$ are needed to react with the $PH_3$. There are not twice as many moles of $O_2$ as there are $PH_3$. Therefore, the limiting reactant is $O_2$. All of it will react. All of the other boxes in the change line are determined by the coefficients in the balanced equation. There is ½ as much $PH_3$ in the balanced equation; therefore a 0.250 is subtracted under $PH_3$. There is ¼ as much $P_2O_5$ produced as $O_2$ used; therefore a 0.125 is added to the product side. And last, there is three times as much $H_2O$ formed as $P_2O_5$; therefore, 3(0.125) = 0.375 is placed in the change box under $H_2O$. Finish filling out the table.

The problem is finished by solving for grams of $P_2O_5$ following the road map:

mol $P_2O_5 \rightarrow$ g $P_2O_5$

$$0.125\ mol\ P_2O_5 \times \frac{141.74\ g\ P_2O_5}{1\ mol\ P_2O_5} = 17.7\ g\ P_2O_5$$

---

**EXAMPLE 3.17  Limiting Reagent (Same example, different way to solve)**

Phosphine ($PH_3$) burns in oxygen to produce phosphorus pentoxide ($P_2O_5$) and water.

$$2\ PH_3(g) + 4\ O_2(g) \rightarrow P_2O_5(s) + 3\ H_2O(\ell)$$

How many grams of $P_2O_5$ will be produced when 17.0 g of phosphine is mixed with 16.0 g of $O_2$ and reaction occurs?

- **Solution; method 2**

Using the method presented in your textbook, determine which of the reactants produce the least number of moles of product following the road maps:

g $PH_3$ → mol $PH_3$ → mol $P_2O_5$  and
g $O_2$ → mol $O_2$ → mol $P_2O_5$

$$17.0\ g\ PH_3 \times \frac{1\ mol\ PH_3}{33.99\ g\ PH_3} \times \frac{1\ mol\ P_2O_5}{2\ mol\ PH_3} = 0.250\ mol\ P_2O_5$$

$$16.0\ g\ O_2 \times \frac{1\ mol\ O_2}{32.00\ g\ O_2} \times \frac{1\ mol\ P_2O_5}{4\ mol\ O_2} = 0.125\ mol\ P_2O_5$$

Finish the problem based upon the least amount of $P_2O_5$.

$$0.125\ mol\ P_2O_5 \times \frac{141.9\ g\ P_2O_5}{1\ mol\ P_2O_5} = 17.7\ g\ P_2O_5$$

**Work EXERCISES & PROBLEMS: 22**

---

**EXAMPLE 3.18  Percent Yield**

In the reaction of 4.0 moles of $N_2$ with 6.0 moles of $H_2$, a chemist obtained 1.6 moles of $NH_3$. What is the percent yield of $NH_3$?

$$3\ H_2 + N_2 \rightarrow 2\ NH_3$$

• **Solution**

The actual yield is given (1.6 moles of $NH_3$); therefore we must calculate the theoretical yield of $NH_3$. That means we must calculate the number of moles $NH_3$ which should form. Notice that once again, this is a limiting reactant problem because data is provided for both reactants. This time let's see the textbook method (and not the ICF table method) for determining the theoretical yield.

Road map:

mol $H_2$ → mol $NH_3$
mol $N_2$ → mol $NH_3$
then choose the one that produces the least amount of $NH_3$. Note: Since the actual yield is given in moles, we can stop at moles as well.

$$6.0 \text{ mol H}_2 \times \frac{2 \text{ mol NH}_3}{3 \text{ mol H}_2} = 4.0 \text{ mol NH}_3$$

Smallest amount of product; therefore the theoretical yield

$$4.0 \text{ mol N}_2 \times \frac{2 \text{ mol NH}_3}{1 \text{ mol N}_2} = 8.0 \text{ mol NH}_3$$

The percent yield is found by dividing the actual yield by the theoretical yield and multiplying by 100 percent.

$$\% \text{ yield NH}_3 = \frac{\text{actual yield NH}_3}{\text{theoretical yield NH}_3} \times 100\%$$

$$= \frac{1.6 \text{ mol NH}_3}{4.0 \text{ mol NH}_3} \times 100\%$$

$$\% \text{ yield NH}_3 = 40\%$$

**Work EXERCISES & PROBLEMS: 24, 25**
**Work the rest of the EXERCISES & PROBLEMS**

## EXERCISES & PROBLEMS

1. An atom of oxygen-18 is 1.49992 times heavier than an atom of carbon-12. What is the mass of an oxygen-18 atom in amu?

2. The element silver consists of two isotopes: $^{107}_{47}\text{Ag}$ with an atomic mass of 106.905 amu and a natural abundance of 51.83%, and $^{109}_{47}\text{Ag}$ with an atomic mass of 108.905 amu and natural abundance of 48.17%. Calculate the average atomic mass of silver.

3. Calculate the molecular mass of the following:
   a. carbon tetrachloride, $CCl_4$
   b. formaldehyde, $H_2CO$
   c. xenon difluoride, $XeF_2$

4. Calculate the molar mass of the following:
   a. $Na_2SO_4$, sodium sulfate
   b. $FeCl_3$, iron(III) chloride
   c. $Ba(OH)_2$, barium hydroxide
   d. $SO_3$, sulfur trioxide

5. What is the mass of 1 mole of each of the following?
   a. Cu  b. CuO  c. $CuSO_4$  d. $CuSO_4 \cdot 5H_2O$

6. a. How many moles of silver are in 5.00 g of Ag?
   b. How many moles of sodium are in 5.00 g of Na?

7. a. How many silver atoms are in 5.0 g of Ag?
   b. How many $H_2$ molecules are there in 4.0 g of $H_2$?
   c. How many molecules are in 25.0 g of methane ($CH_4$)?

8. a. What is the mass of 2.0 moles of $H_2$?
   b. What is the mass of $2.79 \times 10^{22}$ atoms of Ag?
   c. What is the average mass in grams of one atom of aluminum?

9.  a. How many moles of $NaNO_3$ are in 8.72 g of $NaNO_3$?
    b. How many moles of O atoms are in 8.72 g of $NaNO_3$?

10. Calculate the percent composition by mass of the following compounds:
    a. $CO_2$           b. $H_3AsO_4$           c. $CHCl_3$
    d. $NaNO_2$         e. $H_2SO_4$

11. What is the empirical formula of each of the following compounds?
    a. $H_2O_2$   b. $CaF_2$   c. $C_2H_4O_2$   d. $B_2H_6$

12. Determine the empirical formula of each compound from its percent composition.
    a. 46.7% N, 53.3% O          c. 55.3% K, 14.6% P, 30.1% O
    b. 63.6% N, 36.4% O          d. 26.6% K, 35.4% Cr, 38.1% O

13. When 2.65 mg of the substance responsible for the green color on the yolk of a boiled egg is analyzed, it is found to contain 1.42 mg of Fe and 1.23 mg of S. What is the empirical formula of the compound?

14. Ascorbic acid is a compound consisting of three elements: C, H, and O. When a 0.214 g sample is burned in oxygen, 0.320 g of $CO_2$ and 0.0874 g of $H_2O$ are formed. What is the empirical formula of ascorbic acid?

15. Cyclohexane has the empirical formula $CH_2$. Its molecular mass is 84.16 amu. What is its molecular formula?

16. Rust ($Fe_2O_3$) forms readily when iron is exposed to air. Write a balanced chemical equation for the formation of rust.

17. Write a balanced chemical equation for the reaction between hydrogen gas and carbon monoxide to yield methanol ($CH_3OH$).

18. Balance the following equations:
    a. __ $CH_4$ + __ $H_2O$ → __ $CO_2$ + __ $H_2$
    b. __ $H_2SO_4$ + __ $NaOH$ → __ $Na_2SO_4$ + __ $H_2O$
    c. __ $NH_3$ + __ $O_2$ → __ $NO$ + __ $H_2O$

19. Silicon tetrachloride ($SiCl_4$) can be prepared by heating silicon in the presence of chlorine gas:

    $Si(s) + 2Cl_2(g) → SiCl_4(l)$

    a. How many moles of $SiCl_4$ are formed when 4.24 moles of $Cl_2$ gas react?
    b. How many grams of $SiCl_4$ are produced when 4.24 moles of $Cl_2$ gas react?

20. The following reaction can be used to prepare hydrogen gas:

    $CaH_2 + 2H_2O → Ca(OH)_2 + 2H_2$

    How many grams of $H_2$ will result from the reaction of 100 g of $CaH_2$ with excess $H_2O$?

21. Oxygen gas can be produced by the decomposition of mercury(II) oxide, HgO. How many grams of $O_2$ will be produced by the reaction of 24.2 g of the oxide?

    $2 HgO(s) → 2 Hg(l) + O_2$

22. Carbon dioxide in the air of a spacecraft can be removed by its reaction with lithium hydroxide.

$$CO_2(g) + 2\ LiOH(aq) \rightarrow Li_2CO_3(aq) + H_2O(l)$$

On average, a person will exhale about one kg of $CO_2$ per day. How many kg of LiOH are required to react with 1.0 kg of $CO_2$?

23. Given 5.00 mol of KOH and 2.00 mol of $H_3PO_4$, how many moles of $K_3PO_4$ can be prepared?

$$H_3PO_4 + 3\ KOH \rightarrow K_3PO_4 + 3\ H_2O$$

24. The reaction of iron ore with carbon follows the equation:

$$2\ Fe_2O_3 + 3\ C \rightarrow 4\ Fe + 3\ CO_2$$

a. How many grams of Fe can be produced from a mixture of 200 g of $Fe_2O_3$ and 300 g of C?
b. If the actual yield of Fe is 110 g, what is the percent yield of iron?

25. Hydrofluoric acid (HF) can be prepared according to the reaction:

$$CaF_2 + H_2SO_4 \rightarrow 2\ HF + CaSO_4$$

In one experiment 42.0 g of $CaF_2$ was treated with excess $H_2SO_4$ and a yield of 14.2 g of HF was obtained.
a. What is the theoretical yield of HF?
b. Calculate the percent yield of HF.

26. The density of silver is 10.5 $g/cm^3$. How many Ag atoms are present in a silver bar that measures 0.10 m × 0.05 m × 0.01 m?

27. The pesticide malathion has the chemical formula $C_{10}H_{19}O_6PS_2$.
a. What is the molar mass of malathion?
b. The dose that is lethal to 50 percent of a human population is about 1.25 g per kilogram of body mass. How many molecules are in a dose lethal to an adult male weighing 70 kg?  To an adult female weighing 58 kg?

28. An unknown element M combines with oxygen to form a compound with a formula $MO_2$. If 25.0 g of the element combines with 4.50 g of oxygen, what is the atomic mass of M?

29. Describe why the atomic mass unit scale is referred to as a relative scale.

30. Sulfur atoms, on average, have twice the mass of O oxygen atoms. Why is it important to stress the wording "on average"?

## PRACTICE TEST QUESTIONS
See notes on taking practice test in the Preface

1. If the standard of the atomic mass unit scale had been defined such that a fluorine atom was 10.0 amu, what would the atomic masses of oxygen and argon be?

2. How many times heavier are bromine atoms, on average, than neon atoms?

3. A sample of magnesium could one day be brought back from a planet circling a distant star. What is the atomic mass of this magnesium if it has two isotopes with the following masses and abundances?

   magnesium-24  23.985 amu;  18.05%
   magnesium-26  25.982 amu;  81.95%

4. a. What is the mass, in grams, of $1.5 \times 10^{22}$ molecules of $PCl_5$?

   b. What is the mass, in grams, of $10^{12}$ carbon atoms?

5. What is the mass of a single atom of fluorine in grams?

6. How many O atoms are in each of the following amounts?

   a. 1 mol ozone, $O_3$
   b. 1 mol CuO
   c. 20 g $CuSO_4 \cdot 5 \, H_2O$

7. What mass of $Ca_3(PO_4)_2$ would contain 0.500 mol of Ca?

8. a. How many moles of $CaSO_4$ are there in 600 g of $CaSO_4$?
   b. How many moles of oxygen are there in 600 g of $CaSO_4$?
   c. How many oxygen atoms are in 600 g of $CaSO_4$?

9. How many antimony atoms are there in 5.00 mol of $Sb_2O_5$?

10. Answer the following questions concerning sulfur trioxide.
    a. What is the mass of 17.5 mol of $SO_3$?
    b. What is the mass of $7.5 \times 10^{20}$ $SO_3$ molecules?

11. What is the empirical formula of each of the following compounds?
    a. $C_6H_8O_6$
    b. $P_4O_{10}$
    c. $MgCl_2$

12. Calculate the percent composition by mass of the following compounds:
    a. $Al_2O_3$
    b. $HNO_3$
    c. $CCl_2F_2$

13. Calculate the percentage of fluorine by mass in the refrigerant HFC-134a. Its formula is $C_2H_2F_4$.

14. The empirical formula of styrene is CH, and its molar mass is 104 g/mol. What is its molecular formula?

15. Chromic oxide is a compound of chromium and oxygen. Determine the empirical formula of chromic oxide given that 2.00 g of the compound contains 1.04 g of Cr.

16. The compound glycerol contains three elements: C, H, and O. When a 0.740 g sample is burned in oxygen, 1.06 g of $CO_2$ and 0.580 g of $H_2O$ are formed. What is the empirical formula of glycerol?

17. Balance the following equations:
    a. $P_4O_{10} + H_2O \rightarrow H_3PO_4$
    b. $Ga + H_2SO_4 \rightarrow Ga_2(SO_4)_3 + H_2$
    c. $C_4H_{10} + O_2 \rightarrow CO_2 + H_2O$

18. Complete lines a, b, and c of the following table for the reaction:
    $$2\ SO_2 + O_2 \rightarrow 2\ SO_3$$

|   | mol $SO_2$ | grams $O_2$ | mol $SO_3$ | grams $SO_3$ |
|---|---|---|---|---|
| a | 1.50 | | | |
| b | | 20.0 | | |
| c | | | 5.21 | |

19. Sodium thiosulfate can be made by the reaction:
    $$Na_2CO_3 + 2\ Na_2S + 4\ SO_2 \rightarrow$$
    $$3\ Na_2S_2O_3 + CO_2$$

    a. How many grams of $Na_2CO_3$ (sodium carbonate) are required to produce 321 g of sodium thiosulfate, $Na_2S_2O_3$?
    b. How many grams of $Na_2S$ are required to react with 25.0 g of $Na_2CO_3$ if $SO_2$ is present in excess?

20. Hydrofluoric acid (HF) can be prepared according to the following equation:
    $$CaF_2 + H_2SO_4 \rightarrow 2\ HF + CaSO_4$$
    a. How many grams of HF can be prepared from 75.0 g of $H_2SO_4$ and 63.0 g of $CaF_2$?
    b. How many grams of the excess reagent will remain after the reaction ceases?
    c. If the actual yield of HF is 26.2 g, what is the percent yield?

# Chapter 4.  Reactions in Aqueous Solutions

**General Properties of Aqueous Solutions (Section 4.1)**
**Precipitation Reactions (Section 4.2)**
**Acid-Base Reactions (Section 4.3)**
**Oxidation-Reduction Reactions (Section 4.4)**
**Solution Concentration (Section 4.5)**
**Gravimetric Analysis (Section 4.6)**
**Volumetric Analysis - Titrations (Sections 4.7 – 4.8)**


## SUMMARY

### General Properties of Aqueous Solutions (Section 4.1)

***Solutions.*** Many chemical reactions occur in aqueous solutions. Solutions are homogeneous mixtures of two or more substances. Solutions have two components, the solute, which is the substance present in the smaller amount, and the solvent, which is the substance present in the larger amount. A solution can be a solid, liquid, or gaseous substance. In aqueous solutions, water is the solvent.

***Electrolytes and Nonelectrolytes.*** Many aqueous solutions of ionic compounds conduct electricity well, whereas pure water conducts very poorly. These solutions are called electrolytic solutions. An electrolyte is a substance that, when dissolved in water, produces ions that are capable of carrying an electrical current through the solution, yielding an electrically conducting solution.

   For example, salts such as KCl(s), which already consist of ions, separate into individual ions when dissolved in water. *Dissociation* is the term used to describe salts when they separate into ions.

$$KCl(aq) \rightarrow K^+(aq) + Cl^-(aq)$$

Acids can separate into ions in water as well. *Ionization* is the term used for acids and bases as they separate. Hydrochloric acid, even though it is a molecule, will completely ionize in water.

$$HCl(aq) \rightarrow H^+(aq) + Cl^-(aq)$$

Thus a solution of HCl, for instance, contains $H^+$ ions and $Cl^-$ ions, but <u>no</u> HCl molecules! *Strong* electrolytes are 100% dissociated (or ionized) into ions in solution.

   Weak electrolytes are substances which ionize only *partially* to yield a few ions capable of conducting a weak electrical current. Compounds that are incompletely dissociated into ions are called weak electrolytes. Acetic acid ($CH_3COOH$) is an example of a weak electrolyte. You will see in Chapter 16 that in a solution of $CH_3COOH$ about 99% of the $CH_3COOH$ molecules are not ionized. Only 1% are ionized into $H^+$ and $CH_3COO^-$ ions as shown below.

| Example 4.1 |
| :---: |
| Exercises 1, 2 |

$$CH_3COOH(aq) \rightleftharpoons CH_3COO^-(aq) + H^+(aq)$$

   Recall that a double arrow means the reaction is reversible. This means that the reaction can occur in the reverse direction as well as the forward direction. The reverse reaction removes ions from the solution and leads to incomplete ionization. Table 4.1 of the text lists examples of substances that are strong electrolytes, weak electrolytes, and nonelectrolytes.

A nonelectrolyte is a substance that does not ionize in aqueous solution. Such a solution does not conduct an electrical current. Sugar is an example of a nonelectrolyte. In solution, sugar exists as $C_6H_{12}O_6$ molecules. These are neutral (no charge) and cannot conduct an electrical current. All nonelectrolytes exist as molecules in aqueous solution.

## Precipitation Reactions (Section 4.2)

*Solubility Rules.* The solubility of a solute is the amount that can be dissolved in a given quantity of solvent at a given temperature. For example, the solubility of $Pb(NO_3)_2$ is 56 g per 100 mL $H_2O$ at 20 °C. The solubilities of ionic solids in water vary over a wide range. For convenience, we divide compounds into three categories called *soluble*, *slightly soluble*, and *insoluble*. Compounds are classified as soluble if a good deal of solid visibly dissolves when added to water. On the other hand, if only a small amount of solid dissolves, the solid is slightly soluble. If no solid can be observed to dissolve, the substance is described as insoluble. Be aware that insoluble is a relative term and does not mean that no solute dissolves. An ionic compound can have a low solubility, but still be a strong electrolyte, i.e., the units which dissolve will dissociate 100%. The solubilities of various compounds in water at 25 °C are summarized below and in Table 4.2 in your text.

---

**Solubility Rules**

1. All alkali metal (Group 1A) salts and all ammonium ($NH_4^+$) salts are soluble.
2. All salts containing nitrate ($NO_3^-$), chlorate ($ClO_3^-$), and bicarbonate ($HCO_3^-$) are soluble.
3. Most hydroxides ($OH^-$) are insoluble, except those of the alkali metals and barium hydroxide $Ba(OH)_2$. Calcium hydroxide $Ca(OH)_2$, is slightly soluble.
4. Most compounds containing chlorides, bromides, and iodides are soluble, except those containing $Ag^+$, $Hg_2^{2+}$, and $Pb^{2+}$ ions.
5. All carbonates ($CO_3^{2-}$), phosphates ($PO_4^{3-}$), and sulfides ($S^{2-}$) are insoluble, except those covered by rule 1.
6. Most sulfates ($SO_4^{2-}$) are soluble. Both $Ag_2SO_4$ and $CaSO_4$ are slightly soluble. $Ba^{2+}$, $Sr^{2+}$, and $Pb^{2+}$ sulfates are insoluble.

Example 4.2

Exercises 3 – 6

---

*Precipitation Reactions.* When certain solutions of electrolytes are mixed, the cation of one solute may combine with the anion of the other solute to form an insoluble compound. This leads to the formation of a precipitate. A precipitate is an insoluble solid that settles from a solution. For instance, when a solution of $Pb(NO_3)_2$ is mixed with a solution of HCl, a precipitate of insoluble $PbCl_2(s)$ forms. This occurs because combining the solutions results in a new solution containing more lead chloride than is soluble. A condition referred to as supersaturation. In this case, solid $PbCl_2$ must settle out of the solution. Chemical equations for precipitation reactions can be written in several ways.

In the molecular equation the formulas of the compounds are written as their usual chemical formulas.

$$Pb(NO_3)_2(aq) + 2\ HCl(aq) \rightarrow PbCl_2(s) + 2\ HNO_3(aq)$$

The precipitation reaction is more accurately represented by the ionic equation. The ionic equation shows the compounds as being dissociated in solution.

$$Pb^{2+}(aq) + 2\ NO_3^-(aq) + 2\ H^+(aq) + 2\ Cl^-(aq) \rightarrow PbCl_2(s) + 2\ H^+(aq) + 2\ NO_3^-(aq)$$

If you examine the ionic equation, you will notice that the $H^+$ ions and $NO_3^-$ ions are not involved in the formation of the precipitate. We refer to the ions that are not involved in the reaction as spectator ions. Identical species on both sides of the equation can be omitted from the equation. The net ionic equation shows only the species that actually undergo a chemical change.

$$Pb^{2+}(aq) + 2\ Cl^-(aq) \rightarrow PbCl_2(s)$$

In order to predict whether a precipitate will form when two solutions are mixed, you need to know the solubility rules for ionic compounds.

| Example 4.3 |
| --- |
| Exercises 7, 8 |

## Acid-Base Reactions (Section 4.3)

***The Arrhenius Acid-Base Theory.*** Acids and bases are general classifications of compounds that are based on their properties. In general, solutions of acids have a sour taste, change blue litmus to red, dissolve certain metals, and react with carbonates to produce carbon dioxide gas. On the other hand, bases are compounds that form solutions that have a bitter taste, feel slippery, and change red litmus to blue. The properties of an acid are neutralized by the addition of a base, and vice versa.

An acid-base theory or definition is an attempt to explain what happens to molecules of acidic and basic substances in solution that gives rise to their characteristic properties. Svante Arrhenius proposed that an acid is a substance that produces hydrogen ions ($H^+$, also called protons) in aqueous solution, and a base is a substance that produces hydroxide ions ($OH^-$) in aqueous solution. Nitric acid is an example of an acid:

$$HNO_3(aq) \rightarrow H^+(aq) + NO_3^-(aq)$$

Calcium hydroxide is an example of a base:

$$Ca(OH)_2(aq) \rightarrow Ca^{2+}(aq) + 2\ OH^-(aq)$$

We can conclude that all acids and bases are electrolytes undergoing dissociation in aqueous solution.

Arrhenius explained neutralization as the combination of $H^+$ ions and $OH^-$ ions to form water.

$$H^+(aq) + OH^-(aq) \rightarrow H_2O(l)$$

Notice that $H^+(aq)$ is called the hydrogen ion or a proton. To understand the name proton, remember that a hydrogen atom consists of a proton as the nucleus and an electron which travels around the nucleus. The positive ion results from the loss of the electron. Thus, a hydrogen ion is just a proton.

***The Brønsted Theory.*** J. N. Brønsted proposed a more general theory of acid-base reactions than that of Arrhenius. According to Brønsted's theory, an acid is defined as a substance that is able to donate a proton, and a base is a substance that accepts a proton, in a chemical reaction. For example, when HCl gas dissolves in water, it can be viewed as donating a proton to molecules of the solvent. Hydrochloric acid is a Brønsted acid.

$$HCl(aq) + H_2O(l) \rightarrow H_3O^+(aq) + Cl^-(aq)$$
$$\text{acid} \qquad\quad \text{base}$$

The water molecule is the proton acceptor, and so water is a Brønsted base in this reaction. All acid-base reactions, according to Brønsted, involve the transfer of a proton. In the above reaction $H_3O^+$ is called the hydronium ion. It is a close representation of the proton in solution. All protons are hydrated, which means that they are surrounded with water molecules.

When ammonia dissolves in water, it accepts a proton and so is classified as a Brønsted base. In this reaction, water is the acid and ammonia is the base.

$$NH_3(aq) + H_2O(l) \rightleftharpoons NH_4^+(aq) + OH^-(aq)$$

Ammonia is a weak electrolyte because only a small fraction of dissolved $NH_3$ molecules are actually ionized at any one time.

***Monoprotic, Diprotic, and Triprotic Acids.*** Some acids are capable of donating more than one proton to a base. Acids are referred to as monoprotic, diprotic, and triprotic depending on whether they can donate one, two, or three protons. Sulfuric acid ($H_2SO_4$) and carbonic acid ($H_2CO_3$) are diprotic acids. Phosphoric acid ($H_3PO_4$) is a triprotic acid.

***Writing Equations for Acid-Base Neutralization Reactions.*** Acids and bases react with each other in a reaction called neutralization. A neutralization reaction is a reaction between an acid and a base that produces water and a salt.

| Example 4.4 |
| --- |
| Exercises 9 - 12 |

acid + base → water + salt

In a neutralization reaction the hydrogen ion from the acid and the hydroxide ion from the base combine to form water. The metal ion from the base and the nonmetal ion from the acid constitute a solution of a salt. The neutralization reaction of hydrochloric acid and barium hydroxide, for example, is represented by the following molecular equation.

$$2\ HCl(aq) + Ba(OH)_2(aq) \rightarrow 2\ H_2O(l) + BaCl_2(aq)$$
acid          base                    water          salt

Note that one mole of base contains 2 moles of $OH^-$ ions. This means that two moles of HCl are needed to neutralize one mole of base.

Since the acid and the base are both completely ionized in solution we can write the ionic equation:

$$2\ H^+(aq) + 2\ Cl^-(aq) + Ba^{2+}(aq) + 2\ OH^-(aq) \rightarrow 2\ H_2O(l) + Ba^{2+}(aq) + 2\ Cl^-(aq)$$

Notice that $Ba^{2+}$ and $Cl^-$ ions are spectator ions. The net ionic equation is:

$$2\ H^+(aq) + 2\ OH^-(aq) \rightarrow 2\ H_2O(l)$$

The coefficients can be cancelled. The neutralization equation for any strong acid and strong base is:

$$H^+(aq) + OH^-(aq) \rightarrow H_2O(l)$$

If we had carried out this reaction with two moles of HCl for each mole of base, there would be no left over acid or base, and the solution would be described as neutral.

## Oxidation-Reduction Reactions (Section 4.4)

***Oxidation and Reduction.*** In this section we review an important type of chemical process called the oxidation-reduction reaction, or just redox reaction for short. In redox reactions electrons are transferred from one atom to another. Oxidation is a chemical change in which an atom, ion, or molecule loses electrons. Reduction is a chemical change in which an atom, ion, or molecule gains electrons. In the reaction of metallic sodium with chlorine gas, an electron is transferred from the sodium atom to a chlorine atom.

$$2\ Na + Cl_2 \rightarrow 2\ Na^+ + 2\ Cl^-$$

Sodium is oxidized and chlorine is reduced. Oxidation and reduction always occur simultaneously.

> A memory devise to remember is LEO says GER. "Loss of Electrons – Oxidation" and "Gain of Electrons – Reduction."

In a redox reaction the substance oxidized is called the reducing agent because it supplies electrons which cause reduction in another substance. The substance being reduced is called the oxidizing agent because it acquires electrons from another substance. In the sodium-chlorine reaction, sodium is the reducing agent and chlorine is the oxidizing agent.

> A musician's agent doesn't play the music, he books the venues and sees that the musician gets paid, that is, it enables the musician to do his thing. In the same way, an oxidizing agent doesn't get oxidized, it enables the other substance to be oxidized.

Redox reactions are conveniently separated into two half-reactions: the oxidation half-reaction and the reduction half-reaction. For the sodium-chlorine reaction, the half-reactions are:

oxidation      $2\ Na \rightarrow 2\ Na^+ + 2\ e^-$
reduction      $2\ e^- + Cl_2 \rightarrow 2\ Cl^-$

Summation of the two half-reactions yields the overall redox reaction shown above. Half-reactions are used for convenience, but you must not forget that oxidation and reduction occur simultaneously.

The concepts of oxidation and reduction have been extended from cases of complete electron gain or loss to reactions involving varying degrees of electron transfer. When tin ore is reduced to metallic tin by heating tin oxide with charcoal, tin ions are reduced and carbon is oxidized.

$$SnO_2(s) + C(s) \rightarrow Sn(l) + CO_2(g)$$

The reduction of tin is obvious, but in what sense is carbon oxidized? In the context of the original definition of oxidation, carbon is oxidized because it combines with oxygen. Carbon loses its electrons in the sense that oxygen is more electronegative. In charcoal, each carbon atom shares its four valence electrons equally with other carbon atoms. But, in $CO_2$ the bonding electrons are shared unequally, lying closer to the oxygen atoms. Even though a $C^{4+}$ ion is not formed in $CO_2$, the C atoms do relinquish their electrons to some extent to the O atoms. Thus, carbon is oxidized in the reaction. In this reaction oxygen also is reduced.

***Oxidation Numbers.*** The oxidation number or oxidation state of the reactive substances is used to help keep account of electrons in chemical reactions. The oxidation number is the charge an atom if it is an ion or would have in a molecular compound if the electrons were completely transferred instead of shared. The oxidation number of an atom designates its positive or negative character. A positive oxidation number indicates that an atom has lost electrons compared to an uncombined atom. A negative oxidation number indicates that an atom has gained electrons. Oxidation numbers are useful in writing formulas, in recognizing redox reactions, and in balancing redox reactions. An element is oxidized in a reaction whenever its oxidation number *increases* as a result of the reaction. When the oxidation number of an element *decreases* in a reaction, it is reduced.

**Oxidation Number Rules**

1.  *The oxidation number of an uncombined element is zero.*  The atoms in free elements are neutral. Thus the atoms in Fe, Ne, $O_2$, $H_2$, and $P_4$, for example, have oxidation numbers of zero.
2.  *The oxidation number of an element in a simple monatomic ion is the same as the charge of the ion.*  The atoms in the monatomic ions $Na^+$, $Mg^{2+}$, and $Al^{3+}$ have oxidation numbers of +1, +2, and +3, respectively. The oxidation number of chlorine in $Cl^-$ is −1, and of sulfur in $S^{2-}$ is −2.
3.  *Certain elements have the same oxidation numbers in all (or almost all) of their compounds.*
    a.    The oxidation number of oxygen is −2 in almost all of its compounds, except when it is in a peroxide, where it is −1.
    b.    The oxidation number of hydrogen is +1, except when it is bonded to a metal in binary compounds. In metal hydrides, hydrogen has an oxidation number of −1. An example is $AlH_3$.
    c.    Fluorine has an oxidation number of −1 in all of its compounds.
    d.    The oxidation numbers of Group 1A and Group 2A metals in compounds are always +1 and +2, respectively. Aluminum is always +3.
4.  *In a neutral compound, the algebraic sum of the oxidation numbers of all the atoms is zero.* For example, in sodium oxide, $Na_2O$
    $$2(\text{ox. no. Na}) + (\text{ox no. O}) = 0$$
    $$2(+1) + (-2) = 0$$
5.  *The sum of the oxidation numbers of all atoms in a polyatomic ion is equal to the net charge of the ion.*  For example, in nitrate ion, $(NO_3^-)$ which has a net charge of −1:
    $$(\text{ox. no. N}) + 3(\text{ox. no. O}) = -1$$
    $$(+5) + 3(-2) = -1$$

***Recognizing Redox Reactions.***  It is easy to spot the equation of a redox reaction when you see one. First assign oxidation numbers to atoms of the elements in the reactants and products. Then look to see whether any of the elements have changed their oxidation states as they changed from reactants to products. In the reaction of tin oxide on the previous page, the oxidation number of tin in $SnO_2$ is +4, and in Sn it is 0. Therefore, Sn is reduced in the reaction. Carbon on the other hand must be oxidized. Inspection indicates that the oxidation number of C as the element is 0, and that of C in $CO_2$ is +4. Therefore, carbon is oxidized as we thought.

Examples
4.5, 4.6

Exercises
13 – 17

## Solution Concentration (Section 4.5)

***Solution Concentration.***  Solutions are characterized by their concentrations, that is, the amount of solute dissolved in a given amount of solvent. The most common unit of concentration used for aqueous solutions is molarity. The molarity of a solution is the number of moles of solute per liter of the solution.

$$\text{molarity} = \frac{\text{moles of solute}}{\text{liters of solution}}$$

To prepare a 0.50 molar (0.50 *M*) solution of KCl, for instance, we first measure out 0.50 mole of solid KCl, (37.3 g). The KCl is then added to a 1.0 L volumetric flask. This is a flask with a calibrated ring etched around the neck to mark the point at which the flask contains 1.0 liter of liquid. When water is added up to this mark, the flask will contain 1.0 L of the KCl solution, and will contain 0.50 mole of KCl.

The number of moles of KCl in given amounts of the previous solution is easy to find. For instance, 0.10 L of 0.50 *M* KCl will contain

$$0.10 \text{ L soln} \times \frac{0.50 \text{ mol KCl}}{1 \text{ L soln}} = 0.050 \text{ mol KCl}$$

> Example
> 4.7, 4.8
>
> Exercises
> 18 – 21

Here molarity is being used as a conversion factor (mol per L). In general, solution volume (V) × molarity (*M*) = moles solute.

| Conversion Between | Tool |
|---|---|
| mol ←→ L (of solution) | molarity |

***Dilution.***  Dilution is a procedure used to prepare a less concentrated solution from a more concentrated solution. Stock solutions of laboratory acids and ammonia, in particular, are purchased in pint bottles as highly concentrated aqueous solutions. The key to understanding dilution is to remember that adding more water to a given amount of solution does not change the number of moles of solute present in solution. See Figure 4.19 in the text.

moles of solute before dilution = moles of solute after dilution.

Since the moles of solute equal the solution volume (V) times the molarity (*M*) we have

moles$_i$ = moles$_f$

$$M_iV_i = M_fV_f$$

> Example
> 4.9
>
> Exercises
> 22, 23

where the subscripts i and f, stand for initial solution, and final solution, respectively. The concentration after dilution is

$$M_f = M_i \times \frac{V_i}{V_f}$$

Note that $V_f$ is always larger than $V_i$.

> A common mistake is to use this dilution formula for titrations (covered in Sections 4.7 and 4.8). It should NOT be used for anything except *dilutions*.

***Electrolytes and Nonelectrolytes.*** It is important to keep in mind that molarity refers only to the amount of solute originally dissolved in water, and does not reflect any subsequent processes such as the dissociation of a salt or the ionization of an acid. The symbols 1.0 M HCl and 1.0 M $Ca(OH)_2$ mean that 1.0 mole of the solutes HCl and $Ca(OH)_2$ were dissolved to make 1 liter of solution. It, however, does not say anything about the actual chemical species in the solution. Since HCl and $Ca(OH)_2$ are strong electrolytes, there are no molecules of HCl or $Ca(OH)_2$ in these solutions. The HCl solution has 1.0 M $H^+$ ions and 1.0 M $Cl^-$ ions. While the $Ca(OH)_2$ solution has 1.0 M $Ca^{2+}$ ions and 2.0 M $OH^-$ ions.

## Gravimetric Analysis (Section 4.6)

***Gravimetric Analysis.*** Quantitative analysis is the determination of the amount or concentration of a substance in a sample. Precipitation reactions form the basis of a type of chemical analysis called gravimetric analysis. A gravimetric analysis experiment involves the formation of a precipitate, and measurement of its mass. By knowing the mass and chemical formula of the precipitate, we can calculate the mass (or concentration) of the cation or anion component in the original sample.

***Stoichiometry of a Precipitation Reaction.*** When solutions of $AgNO_3$ and NaCl are mixed, a precipitate of AgCl forms because silver chloride is insoluble in water. The equation is:

$$AgNO_3(aq) + NaCl(aq) \rightarrow AgCl(s) + NaNO_3(aq)$$

If 0.275 L of 0.100 *M* $AgNO_3$ is mixed with 0.135 L of 1.00 *M* NaCl, how many grams of AgCl would be formed? This example is just like the limiting reagent problems you worked earlier except that now the reaction occurs in solution. Therefore, let's find the number of moles of each reactant first from $M \times V$.

$$\text{mol } AgNO_3 = 0.275 \text{ L} \times \frac{0.100 \text{ mol}}{L} = 0.0275 \text{ mol } AgNO_3$$

$$\text{mol NaCl} = 0.135 \text{ L} \times \frac{0.100 \text{ mol}}{L} = 0.135 \text{ mol NaCl}$$

Since 0.0275 mol $AgNO_3$ will react with only 0.0275 mol NaCl, NaCl must be present in excess. The reaction will stop when all the $AgNO_3$ (specifically the $Ag^+$) has reacted. Since 1 mole of AgCl forms for each mole of $AgNO_3$, then the yield of AgCl will be 0.0275 mol. The amount of AgCl(s) formed is:

$$? \text{ g AgCl} = 0.0275 \text{ mol AgCl} \times \frac{143.4 \text{ g AgCl}}{1 \text{ mol AgCl}} = 3.94 \text{ g AgCl}$$

In gravimetric analysis the goal is to turn this procedure around. That is to isolate and weigh the precipitate from a precipitation reaction. This makes the mass of precipitate a known quantity. Then we calculate the number of moles and concentration of the limiting reagent.

| Example 4.10, 4.11 |
| Exercises 24 – 26 |

Given the amount of limiting reagent, calculate the yield of a precipitate,
or
Given the yield of a precipitate, calculate the amount of limiting reagent.

## Volumetric Analysis – Titrations (Sections 4.7 – 4.8)

***Acid-Base Titrations.*** A titration is a procedure for determining the concentration of an unknown acid (or base) solution using a known (standardized) concentration of a base (or acid) solution. In the titration of a measured volume of acid solution of unknown concentration, the volume of standardized base solution required to exactly neutralize the acid is carefully measured. The point at which exactly enough base has been added to neutralize the acid is called the equivalence point. Because the concentration (*M*) and the required volume (V) of the base are known, the number of moles of base needed to neutralize the acid can be calculated. The number of moles of acid that were neutralized is readily calculated using the mole ratios from the balanced neutralization reaction. If the volume of acid solution was measured then the concentration of acid is:

$$\text{acid concentration} = \frac{\text{moles of acid neutralized}}{\text{liters of acid solution}}$$

| Example 4.12 – 4.13 |
| Exercises 27 - 30 |

***Redox Titrations.*** In redox titrations the oxidizing agent and reducing agent react by a known reaction with exact stoichiometry. If the amount of one substance reacting is known, the amount of the other substance that reacts can be calculated. When the concentration of the oxidizing agent solution is known, this solution is placed in a buret. A known volume of the solution of unknown concentration of reducing agent is placed in the titration flask. For example, consider the titration of arsenic(III) of unknown concentration with standard potassium permanganate ($KMnO_4$). In the equation below, the $K^+$ of $KMnO_4$ is not shown. It is a spectator ion.

$$5\ H_3AsO_3 + 2\ MnO_4^- + 6\ H^+ \rightarrow 2\ Mn^{2+} + 5\ H_3AsO_4 + 3\ H_2O$$

$KMnO_4$ solution containing the purple permanganate ion is placed in a buret. The solution is added drop by drop to a known volume of $H_3AsO_3$ solution in a flask. The resulting mixture remains colorless as long as $MnO_4^-$ is consumed by the reaction. When enough permanganate has been added to oxidize all the $H_3AsO_3$, the next slight excess of $KMnO_4$ gives a pale pink color to the solution and signals the endpoint. At that point the number of moles of $KMnO_4$ reacted is equal to its concentration times the volume added from the buret.

$$\text{moles of } MnO_4^- = \text{volume} \times \text{molarity}$$

In this case the number of moles of $H_3AsO_3$ originally present can be calculated because we know the mole ratio from the balanced equation.

$$\text{moles of } MnO_4^- \times \frac{5 \text{ mol } H_3AsO_3}{2 \text{ mol } MnO_4^-} = \text{moles of } H_3AsO_3$$

| Example 4.14 |
| Exercises 31 - 32 |

The concentration of $H_3AsO_3$ is:

$$\text{molarity} = \frac{\text{mol } H_3AsO_3}{\text{volume of solution titrated, in L}}$$

## GLOSSARY LIST

| | | |
|---|---|---|
| solutions | hydronium ion | combustion reaction |
| solute | hydroxide ion | decomposition reaction |
| solvent | neutralization reaction | displacement reaction |
| aqueous solution | salts | disproportionation reaction |
| electrolyte | monoprotic acid | |
| nonelectrolyte | diprotic acid | |
| | triprotic acid | concentration of a solution |
| solubility | | molarity/molar concentration |
| precipitate | oxidation-reduction reaction | dilution |
| molecular equation | redox reaction | quantitative analysis |
| ionic equation | oxidation reaction | gravimetric analysis |
| net ionic equation | reduction reaction | titration |
| spectator ions | | standard solution |
| | half-reaction | equivalence point |
| Brønsted acid | oxidation number | indicator |
| Brønsted base | oxidation state | |

## EQUATIONS

| Algebraic Equation | English Translation |
|---|---|
| $MV$ = moles solute | The molar amount of a substance in a solution is the product of the concentration and volume. |
| $M_iV_i = M_fV_f$ | dilution formula |

## SUMMARY OF CONVERSION TOOLS

| Conversion between | Tool |
|---|---|
| g $\leftrightarrow$ kg (as an example) | prefix meanings |
| g $\leftrightarrow$ mL | density |
| mol $\leftrightarrow$ actual number | Avogadro's number |
| g $\leftrightarrow$ mol | molar mass |
| mol compound $\leftrightarrow$ mol atom in compound | subscript |
| mol A $\leftrightarrow$ mol B | coefficients of balanced equation |
| mol $\leftrightarrow$ L (of solution) | molarity |

TOOL

## WORKED EXAMPLES

---

### EXAMPLE 4.1  Identifying Electrolytes and Nonelectrolytes

Identify the following substances as strong, weak, or nonelectrolytes:
a.  $CH_3OH$ (methyl alcohol)   b.  $NH_3$        c.  $Na_2CO_3$

• **Solution**

Refer to Table 4.1 in the text.
a.  Various sugars and alcohols dissolve as neutral molecules. *Answer:* nonelectrolyte.
b.  Ammonia is a weak base. See Section 4.3 of the text. *Answer:* weak electrolyte.
c.  Sodium carbonate is an ionic compound. All ionic compounds dissociate completely in aqueous solution. *Answer:* strong electrolyte.

**Work EXERCISES & PROBLEMS: 1 – 3**

---

### EXAMPLE 4.2  Solubility Rules

According to the solubility rules, which of the following compounds are soluble in water?
a. $MgCO_3$     b. $AgNO_3$     c. $MgCl_2$     d. $Na_3PO_4$     e. KOH

• **Solution**

a.  According to rule 5, $MgCO_3$ is insoluble.
b.  According to rule 2, $AgNO_3$ is soluble.
c.  According to rule 4, $MgCl_2$ is soluble.
d.  According to rules 1 and 5, $Na_3PO_4$ is soluble.
e.  According to rules 1 and 3, KOH is soluble.

**Work EXERCISES & PROBLEMS: 4 – 6**

---

### EXAMPLE 4.3  Writing Ionic and Net Ionic Equations

Write balanced molecular, ionic, and net ionic equations for the reaction that occurs when a $CaCl_2$ solution is mixed with a $Na_2CO_3$ solution.

• **Solution**

The combined solution will contain $Na^+$, $CO_3^{2-}$, $Ca^{2+}$, and $Cl^-$ ions.  The two new possible ionic compounds are NaCl and $CaCO_3$ (swap partners).  If one (or both) is insoluble, it (they) will precipitate.  According to solubility rules 1 and 4, NaCl is soluble, and from rule 5, $CaCO_3$ is insoluble.  The molecular equation should show $CaCO_3(s)$ as a product along with NaCl(aq).

The molecular equation is:  $CaCl_2(aq) + Na_2CO_3(aq) \rightarrow CaCO_3(s) + 2\,NaCl(aq)$

The ionic equation:
$$Ca^{2+}(aq) + 2\,Cl^-(aq) + 2\,Na^+(aq) + CO_3^{2-}(aq) \rightarrow CaCO_3(s) + 2\,Na^+(aq) + 2\,Cl^-(aq)$$

The net ionic equation omits spectator ions:  $Ca^{2+}(aq) + CO_3^{2-}(aq) \rightarrow CaCO_3(s)$

**Work EXERCISES & PROBLEMS: 7 – 8**

## EXAMPLE 4.4  Neutralization Reactions

Write balanced molecular, ionic, and net ionic equations for the neutralization reaction that occurs when nitric acid and barium hydroxide are mixed.

• **Solution**

First, write the formulas of the reactants

$$HNO_3(aq) + Ba(OH)_2(aq) \rightarrow$$

Neutralization reactions of acids and metal hydroxides yield water and a salt. The cation of the salt produced comes from the base and the anion of the salt comes from the acid. In order to write proper equations, you need to be able to write the formulas of salts. Since the barium ion has a +2 charge ($Ba^{2+}$) and the nitrate ion has a −1 change ($NO_3^-$), the salt is $Ba(NO_3)_2$. It will take two moles of the $HNO_3$ to neutralize one mole of $Ba(OH)_2$. The molecular equation is:

$$2\,HNO_3(aq) + Ba(OH)_2(aq) \rightarrow 2\,H_2O(\ell) + Ba(NO_3)_2(aq)$$

To write the ionic equation recall that the acid and the base are 100% ionized in solution.

$$HNO_3(aq) \rightarrow H^+(aq) + NO_3^-(aq)$$

$$Ba(OH)_2(aq) \rightarrow Ba^{2+}(aq) + 2\,OH^-(aq)$$

The ionic equation is:

$$2H^+(aq) + 2NO_3^-(aq) + Ba^{2+}(aq) + 2\,OH^-(aq) \rightarrow 2\,H_2O(\ell) + Ba^{2+}(aq) + 2NO_3^-(aq)$$

The net ionic equation is the equation left after any spectator ions are cancelled:

$$H^+(aq) + OH^-(aq) \rightarrow H_2O(\ell)$$

Note: This equation is always the net ionic equation of the reaction between a strong acid and a strong base.

**Work EXERCISES & PROBLEMS: 9 – 12**

## EXAMPLE 4.5  Determining Oxidation Numbers

Assign oxidation numbers to all the atoms in the following compounds and ions:
a. $Na_2SO_4$
b. $CuCl$
c. $SO_3^{2-}$

• **Solution**

a. In $Na_2SO_4$ we can assign oxidation numbers to Na and O because their oxidation numbers are almost always the same in all compounds (rule 3). Na is an alkali metal and is +1. Oxygen is −2. The oxidation number of sulfur varies from compound to compound. Apply rule 4, The sum of the oxidation numbers of all atoms in the compound is zero.

$$2(\text{ox. no. Na}) + (\text{ox. no. S}) + 4(\text{ox. no. O}) = 0$$
$$2(+1) \quad + (\text{ox. no. S}) + 4(-2) \qquad = 0$$
$$(\text{ox. no. S}) = 8 - 2 = +6$$

b.  In CuCl, the chlorine ion has an oxidation number of −1. Cu must be +1 by rule 4.

c.  In $SO_3^{2-}$, we assume the oxidation number of oxygen is −2, and solve for the oxidation number of sulfur using rule 5.

(ox. no. S) + 3(ox. no. O) = −2
(ox. no. S) + 3(−2)      = −2
      (ox. no. S) = −2 + 6 = +4

**Work EXERCISES & PROBLEMS: 13, 14**

---

**EXAMPLE 4.6  Identifying Oxidation and Reduction**

Determine whether iron in the reactant species is oxidized or reduced during the following reactions?

a.  $2\,Fe + \frac{3}{2}\,O_2 + 2\,H_2O \rightarrow Fe_2O_3 \cdot 2H_2O$ (rust)

b.  $FeO + CO \rightarrow Fe + CO_2$

c.  $2\,Hg + 2\,Fe^{3+} + 2\,Cl^- \rightarrow Hg_2Cl_2 + 2\,Fe^{2+}$

**• Solution**

a.  There are two ways to tell. First, assign oxidation numbers to iron in Fe and in $Fe_2O_3$ (ignore the waters of hydration). The oxidation number changes form zero in Fe to +3 in $Fe_2O_3$. Since Fe becomes more positive, it must have lost electrons. Thus, iron is oxidized. In the second way, when a metal combines with oxygen, its electrons are always transferred to the oxygen atom. All elements except fluorine are oxidized when they combine with $O_2$, and oxygen is reduced in these reactions.

b.  Using the oxidation number change as a guide, the oxidation number of iron changes from +2 in FeO to zero in Fe. Therefore, iron is gaining electrons in this reaction, and so it is reduced.

c.  The oxidation number of iron is more negative in the product, and so iron must gain an electron in this reaction. Fe is reduced.

**Work EXERCISES & PROBLEMS: 15 – 17**

---

**EXAMPLE 4.7  Calculation of Molarity**

What is the molarity of a solution made from 32.1 g $KNO_3$ dissolved in enough water to make 500 mL of solution?

**• Solution**

First, write the equation that defines molarity.

$$\text{molarity} = \frac{\text{moles } KNO_3}{\text{liters of solution}}$$

Then, substitute into the equation the number of moles of $KNO_3$ in 32.1 g, and the volume of solution.

$$\text{moles } KNO_3 = 32.1 \text{ g } KNO_3 \times \frac{1\,\text{mol } KNO_3}{101.2 \text{ g } KNO_3}$$

$$= 0.317 \text{ mol } KNO_3$$

Convert 500 mL to 0.500 L

$$\text{molarity} = \frac{0.317 \text{ mol KNO}_3}{0.500 \text{ L}} = 0.634 \ M$$

This is normally written 0.634 $M$ KNO$_3$.

**Work EXERCISES & PROBLEMS: 18 – 20**

---

**EXAMPLE 4.8  Solution Composition**

How many moles of solute are in 250 mL of 0.100 M KCl?

**• Solution**

Road map: mL → L → mol

To obtain the number of moles, we will use the concentration as a conversion factor.

$$250 \text{ mL} \times \frac{1 \text{L}}{1000 \text{ mL}} \times \frac{0.100 \text{ mol KOH}}{1 \text{L}} = 0.0250 \text{ mol KOH}$$

**Work EXERCISES & PROBLEMS: 18 – 21**

---

**EXAMPLE 4.9  Dilution**

What volume of a 6.0 $M$ hydrochloric acid stock solution is required to prepare 725 mL of 0.12 $M$ HCl?

**• Solution**

Using the dilution formula:

$$M_i V_i = M_f V_f$$

Rearranging to solve for the initial volume, we get:

$$V_i = V_f \frac{M_f}{M_i}$$

$V_i = ?$
$V_f = 725$ mL
$M_f = 0.12 \ M$
$M_i = 6.0 \ M$

$$V_i = 725 \text{ mL} \times \frac{0.120 \ M}{6.0 \ M} = 15 \text{ mL}$$

**• Comment**
Note: That 0.725 L of 0.12 $M$ HCl contains 0.087 mol HCl.  The exact amount supplied by 0.0145 L of 6.0 $M$ HCl.  Also we were able to use milliliter units in the calculation because the units of molarity cancelled.

**Work EXERCISES & PROBLEMS: 22, 23**

**EXAMPLE 4.10  Solution Stoichiometry**

Calculate the number of moles of NaCl that must be added to 0.752 L of 0.150 $M$ AgNO$_3$ in order to precipitate all the Ag$^+$ ions as AgCl.

**• Solution**

First, we need to determine the chemical equation for the precipitation reaction. Recall that NaCl is a strong electrolyte and so dissociates entirely into Na$^+$ and Cl$^-$ ions in solution. AgNO$_3$ is also a strong electrolyte yielding Ag$^+$ and NO$_3^-$ ions. At first the mixed solution contains Ag$^+$, NO$_3^-$, Na$^+$, and Cl$^-$ ions. Since AgCl is insoluble, the net ionic reaction is:

$$Ag^+(aq) + Cl^-(aq) \rightarrow AgCl(s)$$

The ionic equation is:

$$Ag^+(aq) + NO_3^-(aq) + Na^+(aq) + Cl^-(aq) \rightarrow AgCl(s) + Na^+(aq) + NO_3^-(aq)$$

And the molecular equation is:

$$AgNO_3(aq) + NaCl(aq) \rightarrow AgCl(s) + NaNO_3(aq)$$

Road Map: L AgNO$_3$ → mol AgNO$_3$ → mol NaCl

$$0.752\,L \times \frac{0.150\,mol\,AgNO_3}{1L} \times \frac{1\,mol\,NaCl}{1\,mol\,AgNO3} = 0.113\,mol\,NaCl$$

Obtained from coefficients of the balanced equation.

**• Comment**

In this case 0.113 mol of NaCl supplies enough Cl$^-$ ion to react with the given amount of Ag$^+$ ion.

**Work EXERCISES & PROBLEMS: 24**

**EXAMPLE 4.11  Gravimetric Analysis**

Calculate the molar concentration of Ba$^{2+}$ ions in a 500.0 mL sample of an unknown aqueous solution if 2.47 g BaSO$_4$ is formed upon the addition of excess Na$_2$SO$_4$?

**• Solution**

First, write the chemical equation for the precipitation reaction.

Barium sulfate being insoluble is formed in the following reaction.

$$Ba^{2+}(aq) + SO_4^{2-}(aq) \rightarrow BaSO_4(s)$$

Therefore, the number of moles of $Ba^{2+}$ originally equals the number of moles of $BaSO_4$ precipitate.

Road Map: g $BaSO_4$ → mol $BaSO_4$ → mol $Ba^{2+}$ → M $Ba^{2+}$

$$2.47\ g\,BaSO_4 \times \frac{1\,mol\,BaSO_4}{233.4\ g\,BaSO_4} \times \frac{1\,mol\,Ba^{2+}}{1\,mol\,BaSO_4} = 0.0106\ mol\,Ba^{2+}$$

Molar mass of $BaSO_4$

This takes care of steps 1 and 2 of road map.

To find the original concentration of $Ba^{2+}$ ions, substitute into the equation for molarity:

$$\text{molarity of } Ba^{2+} = \frac{mol\,Ba^{2+}}{L\ soln} = \frac{0.0106\ mol}{0.500\ L\ soln}$$

$$= 0.0212\ M\ Ba^{2+}\ \text{ions}$$

Or, alternately, you can do all step together like this:

$$2.47\ g\,BaSO_4 \times \frac{1\,mol\,BaSO_4}{233.4\ g\,BaSO_4} \times \frac{1\,mol\,Ba^{2+}}{1\,mol\,BaSO_4} \times \frac{1}{0.500\ L\ soln} = \frac{0.0212\ molBa^{2+}}{L\ soln}$$

$$= 0.0212\ M\ Ba^{2+}$$

**Work EXERCISES & PROBLEMS: 25, 26**

---

**EXAMPLE 4.12  Stoichiometry of an Acid-Base Reaction**

What volume of 0.900 *M* HCl is required to completely neutralize 25.0 g of $Ca(OH)_2$?

• **Solution**

Write and balance the neutralization equation

$$2\ HCl + Ca(OH)_2 \rightarrow CaCl_2 + 2\ H_2O$$

Since the balanced equation relates moles of one reactant to moles of another reactant, first determine the number of moles of $Ca(OH)_2$ in 25.0 g.

Road Map:    g $Ca(OH)_2$ → mol $Ca(OH)_2$ → mol HCl → volume HCl

$$25.0\ g\,Ca(OH)_2 \times \frac{1\,mol\,Ca(OH)_2}{74.10\ g\,Ca(OH)_2} \times \frac{2\,mol\,HCl}{1\,mol\,Ca(OH)_2} \times \frac{1L\ soln}{0.900\ mol\,HCl} = 0.750\ L\ HCl\ soln$$

Molar mass $Ca(OH)_2$

From balanced equation

Use molarity as a conversion factor.

**EXAMPLE 4.13 Acid-Base Titration**

A volume of 128 mL of 0.650 $M$ $Ba(OH)_2$ was required to completely neutralize 50.0 mL of nitric acid solution ($HNO_3$). What was the concentration of the acid solution?

**• Solution**

This is similar to the other problems. Our goal is to fill in the equation for molarity:

$$\text{molarity of nitric acid} = \frac{\text{mol } HNO_3}{\text{L acid soln}}$$

First we need the balanced chemical equation to give us the important mole-ratio. Then we construct a roadmap:

$$Ba(OH)_2(aq) + 2\ HNO_3(aq) \rightarrow Ba(NO_3)_2(aq) + 2\ H_2O(\ell)$$

Road map: mL $Ba(OH)_2$ → L $Ba(OH)_2$ → mol $Ba(OH)_2$ → mol $HNO_3$ → $M$ $HNO_3$

Steps 1 through 3 are accomplished by the following:

$$128\ \text{mL } Ba(OH)_2 \times \frac{1\ \text{L } Ba(OH)_2}{1000\ \text{mL } Ba(OH)_2} \times \frac{0.650\ \text{mol } Ba(OH)_2}{1\ \text{L } Ba(OH)_2} \times \frac{2\ \text{mol } HNO_3}{1\ \text{mol } Ba(OH)_2} = 0.166\ \text{mol } HNO_3$$

Molarity of barium hydroxide

From coefficients of balanced equation

Step 4 can then be accomplished separately:

$$\text{molarity} = \frac{\text{mol } HNO_3}{1\ \text{L soln}} = \frac{0.166\ \text{mol}}{0.050\ \text{L}} = 3.33\ M\ HNO_3$$

**Work EXERCISES & PROBLEMS: 27 – 30**

---

**EXAMPLE 4.14 A Redox Titration**

In aqueous solution 31.0 mL of 0.150 $M$ $KMnO_4$ was required to completely oxidize 50.0 mL of $H_3AsO_3$. What is the concentration of the $H_3AsO_3$ solution?

**• Solution**

The balanced equation was given previously.

$$5\ H_3AsO_3 + 2\ MnO_4^- + 6\ H^+ \rightarrow 2\ Mn^{2+} + 5\ H_3AsO_4 + 3\ H_2O$$

The concentration of $KMnO_4$ = the concentration of $MnO_4^-$ (which is in the balanced equation).

Road map: mL $MnO_4^-$ → L $MnO_4^-$ → mol $MnO_4^-$ → mol $H_3AsO_3$ → $M$ $H_3AsO_3$

Notice that this road map is virtually identical to the road map in the previous example. A titration problem doesn't change between redox titrations and acid-base titrations.

$$31.0\,mL\,MnO_4^- \times \frac{1L\,MnO_4^-}{1000\,mL\,MnO_4^-} \times \frac{0.150\,mol\,MnO_4^-}{1L\,MnO_4^-} \times \frac{5\,mol\,H_3AsO_3}{2\,mol\,MnO_4^-} = 1.16 \times 10^{-2}\,mol\,H_3AsO_3$$

The concentration of $H_3AsO_3$ is:

$$molarity = \frac{mol\,H_3AsO_3}{liters\,soln}$$

$$= \frac{1.16 \times 10^{-2}\,mol}{50.0\,mL \times \dfrac{1L}{1000\,mL}} = 0.232\,M$$

**Work EXERCISES & PROBLEMS: 31, 32**
**Work the rest of the EXERCISES & PROBLEMS**

## EXERCISES & PROBLEMS

1.  Which of the following are ionic compounds and therefore are strong electrolytes?
    a. $CaF_2$   b. $C_6H_{12}O_6$   c. KBr   d. $LiNO_3$   e. $CuSO_4$

2.  Which of the following compounds are strong electrolytes?
    a. KOH   b. Sucrose   c. $MgI_2$   d. Methyl alcohol   e. HBr   f. AgCl

3.  What ionic species are present in aqueous solutions of the following?
    a. $Mg(NO_3)_2$   b. KOH   c. $CaF_2$

4.  Which of the following compounds are soluble in water?
    a. $BaCl_2$   b. $PbSO_4$   c. $Ni(OH)_2$   d. $Ca_3(PO_4)_2$   e. $NH_4NO_3$

5.  According to the solubility rules, which of the following compounds are insoluble in water?
    a. $NH_4CO_3$   b. AgBr   c. $CaCO_3$   d. $FeCl_3$   e. ZnS

6.  Predict whether a precipitate will form when aqueous solutions of the following are mixed. Write the formulas for any precipitates.
    a. $MgBr_2$ and NaOH   b. NaI and $AgNO_3$

7.  Identify the spectator ions in parts (a) and (b) of number 6.

8.  Write net ionic equations for the reactions that occur when solutions of the following compounds are mixed.
    a. $MgBr_2$ and $Pb(NO_3)_2$   b. NaBr and $AgNO_3$

9.  Write the equations for the ionization of the following acids in aqueous solution.
    a. HBr   b. $HClO_4$   c. $H_2SO_4$

10. Write the equations for the dissociation of the following bases in aqueous solution.
    a. $Ca(OH)_2$   b. $HN(CH_3)_2$   c. $CsOH$

11. Write balanced molecular equations for the reaction of $NaOH(aq)$ with the following.
    a. $HClO_4(aq)$   b. $H_2SO_4(aq)$

12. Complete and balance the following molecular equations.
    a. $HNO_3(aq) + LiOH(aq) \rightarrow$
    b. $Ca(OH)_2(aq) + HCl(aq) \rightarrow$
    c. $KOH(aq) + H_3PO_4(aq) \rightarrow$

13. Assign oxidation numbers to the elements specified in the following molecules and ions.
    a. $\underline{C}O_2$   b. $K_2\underline{Cr}O_4$   c. $H_2\underline{S}$   d. $\underline{H}_2S$   e. $\underline{H}_2$
    f. $Ca\underline{H}_2$   g. $\underline{C}O_3^{2-}$   h. $\underline{Fe}Br_3$   i. $\underline{Cr}O_4^{2-}$   j. $K\underline{Cl}O_4$

14. What is the oxidation number of Mn before and after the reaction?

    $$5\ H_3AsO_3 + 2\ MnO_4^- + 6\ H^+ \rightarrow 2\ Mn^{2+} + 5\ H_3AsO_4 + 3\ H_2O$$

15. In the following reactions, identify the elements oxidized and the elements reduced.
    a. $3\ Ag + 4\ HNO_3 \rightarrow 3\ AgNO_3 + NO + 2\ H_2O$
    b. $6\ Fe^{2+} + Cr_2O_7^{2-} + 14\ H^+ \rightarrow 6\ Fe^{3+} + 2\ Cr^{3+} + 7\ H_2O$
    c. $Cu^{2+} + Zn \rightarrow Cu + Zn^{2+}$

16. In the following reactions, identify the oxidizing agents and reducing agents.
    a. $3\ Ag + 4\ HNO_3 \rightarrow 3\ AgNO_3 + NO + 2\ H_2O$
    b. $6\ Fe^{2+} + Cr_2O_7^{2-} + 14\ H^+ \rightarrow 6\ Fe^{3+} + 2\ Cr^{3+} + 7\ H_2O$
    c. $Cu^{2+} + Zn \rightarrow Cu + Zn^{2+}$

17. Identify which of the following reactions are oxidation-reduction reactions.
    a. $Ca(s) + HCl(aq) \rightarrow CaCl_2(aq) + H_2(g)$
    b. $HNO_3(aq) + NaOH(aq) \rightarrow NaNO_3(aq) + H_2O(l)$
    c. $H_2O_2 + 2\ HI \rightarrow I_2 + 2\ H_2O$
    d. $Pb^{2+}(aq) + S^{2-} \rightarrow PbS(s)$

18. What is the molarity of a solution consisting of 11.8 g of NaOH dissolved in enough water to make 300 mL of solution?

19. How many moles of $H_2SO_4$ are contained in 225 mL of 0.750 $M$ $H_2SO_4$?

20. How many grams of HCl are contained in 250 mL of 0.500 $M$ HCl?

21. Consider 125 mL of 0.110 $M$ $MgCl_2$ solution.
    a. How many moles of $MgCl_2$ are contained in 125 mL of 0.110 $M$ $MgCl_2$?
    b. What is the concentration of $Mg^{2+}$ ion?
    c. What is the concentration of $Cl^-$ ion?

22. When 147 mL of 4.25 $M$ $NH_3$ solution is mixed with 353 mL of $H_2O$, what is the concentration of the final solution?

23. Describe how you would prepare $2.50 \times 10^2$ mL of 0.666 $M$ KOH solution, starting with a 5.0 $M$ KOH solution.

24. Calculate the number of moles of NaBr that must be added to 450 mL of 0.250 $M$ $AgNO_3$ solution in order to precipitate all the $Ag^+$ ions as AgBr.

25. In order to precipitate AgCl, excess $AgNO_3$ was added to 10.0 mL of a solution containing $Cl^-$ ion. If 0.339 g of AgCl was formed, what was the concentration of $Cl^-$ in the original solution?

26. A 0.198 g sample of an ionic compound containing the $Br^-$ ion was dissolved in water and treated with excess $AgNO_3$. If the mass of silver bromide precipitate that forms was 0.0964 g, what is the percent of $Br^-$ by mass in the original compound?

27. How many milliliters of 1.00 $M$ $H_2SO_4$ solution are required to neutralize 2.10 g KOH?

28. What volume of 0.210 $M$ $H_2SO_4$ solution is needed to exactly neutralize 50.0 mL of 0.082 $M$ NaOH?

29. What volume of 0.0824 $M$ NaOH solution is needed to titrate 9.8 mL of 0.210 $M$ $H_2SO_4$ to the equivalence point?

30. What is the molarity of an oxalic acid ($H_2C_2O_4$) solution if 22.50 mL of this solution requires 35.72 mL of 0.198 $M$ NaOH for complete neutralization?

31. What volume of 0.200 $M$ $K_2Cr_2O_7$ will be required to oxidize 4.0 g of $H_3AsO_3$?

$$14\ H^+ + Cr_2O_7^{2-} + 3\ H_3AsO_3 \rightarrow 2\ Cr^{3+} + 3\ H_3AsO_4$$

32. In an experiment, 52.0 mL of $K_2Cr_2O_7$ solution was required to oxidize 4.0 g of $H_3AsO_3$. What was the molarity of the dichromate solution?

$$14\ H^+ + Cr_2O_7^{2-} + 3\ H_3AsO_3 \rightarrow 2\ Cr^{3+} + 3\ H_3AsO_4$$

33. Explain how an ionic compound can have a low solubility, but be a strong electrolyte.

34. What makes molarity a convenient concentration unit for use in chemistry?

35. Suggest a precipitation reaction and a redox reaction by which barium sulfate can be prepared.

## PRACTICE TEST QUESTIONS
See notes on taking practice test in the Preface

1. Identify each of the following substances as a strong electrolyte, a weak electrolyte or a nonelectrolyte.
   a. NaBr          b. HCl
   c. $CH_3COOH$    d. $NH_3$
   e. glucose

2. Write balanced *net ionic equations* for the reactions between:
   a. HCl(aq) and $Mg(OH)_2$(aq)
   b. $CH_3COOH$(aq) and NaOH(aq)
   c. $Ca(OH)_2$(aq) and $H_2SO_4$(aq)

3. Write balanced net ionic equations for the reactions that occur when solutions of the following solutes are mixed:
   a. $Ba(NO_3)_2$ and $Na_2CO_3$
   b. RbCl and $AgNO_3$
   c. $Pb(NO_3)_2$ and $K_2S$

4. Characterize the following compounds as soluble or insoluble in water.
   a. $Mg(OH)_2$        b. AgCl
   c. $BaSO_4$          d. $CaCO_3$
   e. $Pb(NO_3)_2$      f. $Na_2CO_3$

5. Write formulas for the acid and base whose reactions produce the following salts.
   a. $CuSO_4$(aq)
   b. KBr(aq)
   c. $Ca_3(PO_4)_2$(s)

6. Write ionic and net ionic equations for the reactions that occur when solutions of the following compounds are mixed.
   a. NaBr and $AgNO_3$
   b. $MgBr_2$ and $Pb(NO_3)_2$

7. Assign oxidation numbers to the underlined elements in the following molecules and ions:
   a. $\underline{N}O_2^-$         b. $\underline{N}H_3$
   c. $\underline{Cl}O_4^-$        d. $H_2\underline{S}O_3$
   e. $Na_2\underline{S}_2O_3$     f. $\underline{Hg}_2Cl_2$

8. Identify which of the following reactions are oxidation-reduction reactions.
   a. Mg(s) + HCl(aq) → $MgCl_2$(aq) + $H_2$(g)
   b. HCl(aq) + $NH_3$(aq) → $NH_4Cl$(aq)
   c. Mg(s) + $CO_2$(g) → MgO(s) + C(s)
   d. $Pb^{2+}$(aq) + $S^{2-}$ → PbS(s)

9. Identify the oxidizing agents and reducing agents in the following reactions:
   a. S + $O_2$ → $SO_2$
   b. $BrO_3^-$ + 6 $I^-$ + 6 $H^+$ → 3 $I_2$ + 3 $H_2O$ + $Br^-$
   c. As + $H^+$ + $NO_3^-$ + $H_2O$ → $H_3AsO_3$ + NO

10. What is the molarity of a solution consisting of 11.8 g of NaOH dissolved in enough water to make exactly 300 mL of solution?

11. How many grams of NaCl are present in 45.0 mL of 1.25 $M$ NaCl?

12. How many liters of 0.50 $M$ glucose, $C_6H_{12}O_6$, solution will contain exactly 100 g glucose?

13. If 30 mL of 0.80 $M$ KCl is mixed with water to make a total volume of 0.400 L, what is the final concentration of KCl?

14. When aqueous solutions of $Pb(NO_3)_2$ and $Na_2SO_4$ are mixed a precipitate of $PbSO_4$ is formed. Calculate the mass of $PbSO_4$ formed when 655 mL of 0.150 $M$ $Pb(NO_3)_2$ and 525 mL of 0.0751 $M$ $Na_2SO_4$ are mixed.

15. Calculate the molar concentration of $Pb^{2+}$ ions in 500.0 mL unknown aqueous solution if 1.07 g $PbSO_4$ is formed upon the addition of excess $Na_2SO_4$?

16. How many mL of 0.10 $M$ $H_2SO_4$ would be required to neutralize 2.5 mL of 1.0 $M$ NaOH?

17. In a titration experiment, a student finds that 23.6 mL of 0.755 $M$ $H_2SO_4$ solution are required to completely neutralize 30.0 mL of NaOH solution. Determine the concentration of the NaOH solution.

18. If 10 mL of 1.0 $M$ HCl are required to neutralize 50 mL of a NaOH solution, how many mL of 1.0 $M$ $H_2SO_4$ will neutralize another 50 mL of the same NaOH solution?

19. Calculate the volume of 0.0300 $M$ $Ce^{4+}$ required to reach the equivalence point when titrating 41.0 mL of 0.0200 $M$ $C_2O_4^{2-}$ (oxalate ion).

$$2Ce^{4+} + C_2O_4^{2-} \rightarrow 2Ce^{3+} + 2CO_2$$
(acidic solution)

20. A sample of iron ore weighing 1.824 g is analyzed by converting the Fe to $Fe^{2+}$ and then titrating with standard potassium dichromate.

$$6Fe^{2+} + Cr_2O_7^{2-} + 14H^+ \rightarrow 6\ Fe^{3+} + 2Cr^{3+} + 7H_2O$$

If 37.21 mL of 0.0213 $M$ $K_2Cr_2O_7$ was required to reach the equivalence point, what was the mass percent of iron in the ore?

21. $KMnO_4$ solution can be standardized against $As_2O_3$. A 0.2661 g sample of $As_2O_3$ was dissolved in acidic solution.

$$As_2O_3 + 2\,H_2O \rightarrow 2\,H_3AsO_3$$

If 38.22 mL of $KMnO_4$ solution was required to react with this amount of $As_2O_3$, what is the molarity of the $KMnO_4$ solution?

$$2\,MnO_4^- + 6\,H^+ + 5\,H_3AsO_3 \rightarrow 2\,Mn^{2+} + 5\,H_3AsO_4 + 3\,H_2O$$

4

# Chapter 5.  Gases

**Properties of Gases (Sections 5.1 – 5.2)**
**Gas Laws (Section 5.3)**
**Ideal Gas Equation (Section 5.4)**
**Gas Stoichiometry (Section 5.5)**
**Dalton's Law of Partial Pressure (Section 5.6)**
**Kinetic Molecular Theory of Gases (Section 5.7)**
**Nonideal Gases (Section 5.8)**


## SUMMARY

## Properties of Gases (Sections 5.1 – 5.2)

***Properties of Gases.***  All matter exists as either a solid, a liquid, or a gas, depending on the temperature and the pressure.  The elements that are gases at 25 °C and 1 atm, are shaded in the periodic table below.

| 1 1A | | | | | | | | | | | | | | | | | 18 8A |
|---|---|---|---|---|---|---|---|---|---|---|---|---|---|---|---|---|---|
| 1 **H** 1.008 | 2 2A | | | | | | | | | | | 13 3A | 14 4A | 15 5A | 16 6A | 17 7A | 2 **He** 4.003 |
| 3 **Li** 6.941 | 4 **Be** 9.012 | | | | | | | | | | | 5 **B** 10.81 | 6 **C** 12.01 | 7 **N** 14.01 | 8 **O** 16.00 | 9 **F** 19.00 | 10 **Ne** 20.18 |
| 11 **Na** 22.99 | 12 **Mg** 24.31 | 3 3B | 4 4B | 5 5B | 6 6B | 7 7B | 8 | 9 8B | 10 | 11 1B | 12 2B | 13 **Al** 26.98 | 14 **Si** 28.09 | 15 **P** 30.97 | 16 **S** 32.07 | 17 **Cl** 35.45 | 18 **Ar** 39.95 |
| 19 **K** 39.10 | 20 **Ca** 40.08 | 21 **Sc** 44.96 | 22 **Ti** 47.88 | 23 **V** 50.94 | 24 **Cr** 52.00 | 25 **Mn** 54.94 | 26 **Fe** 55.85 | 27 **Co** 58.93 | 28 **Ni** 58.69 | 29 **Cu** 63.55 | 30 **Zn** 65.39 | 31 **Ga** 69.72 | 32 **Ge** 75.59 | 33 **As** 74.92 | 34 **Se** 78.96 | 35 **Br** 79.90 | 36 **Kr** 83.80 |
| 37 **Rb** 85.47 | 38 **Sr** 87.62 | 39 **Y** 88.91 | 40 **Zr** 91.22 | 41 **Nb** 92.91 | 42 **Mo** 95.94 | 43 **Tc** (98) | 44 **Ru** 101.1 | 45 **Rh** 102.9 | 46 **Pd** 106.4 | 47 **Ag** 107.9 | 48 **Cd** 112.4 | 49 **In** 114.8 | 50 **Sn** 118.7 | 51 **Sb** 121.8 | 52 **Te** 127.6 | 53 **I** 126.9 | 54 **Xe** 131.3 |
| 55 **Cs** 132.9 | 56 **Ba** 137.3 | 57 **La** 138.9 | 72 **Hf** 178.5 | 73 **Ta** 180.9 | 74 **W** 183.9 | 75 **Re** 186.2 | 76 **Os** 190.2 | 77 **Ir** 192.2 | 78 **Pt** 195.1 | 79 **Au** 197.0 | 80 **Hg** 200.6 | 81 **Tl** 204.4 | 82 **Pb** 207.2 | 83 **Bi** 209.0 | 84 **Po** (210) | 85 **At** (210) | 86 **Rn** (222) |
| 87 **Fr** (223) | 88 **Ra** (226) | 89 **Ac** (227) | 104 **Rf** (257) | 105 **Db** (260) | 106 **Sg** (263) | 107 **Bh** (262) | 108 **Hs** (265) | 109 **Mt** (266) | 110 **Ds** (281) | 111 **Rg** (272) | 112 | 113 | 114 | 115 | 116 | 117 | 118 |

Gaseous molecules include CO, $CO_2$, HCl, $NH_3$, $CH_4$, $NO_2$, $N_2O_4$, $SO_2$ (See Table 5.1, pg 171 in the text).  What all these species have in common are small attractive forces between molecules. This is discussed in detail in Chapter 11. For these molecules, the attractions are very weak and it keeps them from condensing to liquids or solids under atmospheric conditions.  Consequently, very different gases can have very similar physical properties:

1. *Expandability.*  Gases expand indefinitely to fill the space available to them.
2. *Indefinite shape.*  Gases fill all parts of a container evenly and so have no definite shape of their own.
3. *Compressibility.*  Gases are the most compressible of the states of matter.
4. *Miscibility.*  Two or more gases will mix evenly and completely when confined to the same container.

5. *Low density.* Gases have much lower densities than liquids and solids. Typically, densities of gases are about 1/1000 those of liquids and solids. Consequently, the units of g/L are used instead of g/mL for the units of density for gases.

**Gas Pressure.** Gas particles are in constant motion. Their collisions with the walls of the container exert pressure on the container. Pressure is force per unit area. Therefore, the SI unit for pressure (force/area) is the pascal (Pa = $N/m^2$). The pascal is not the most convenient pressure unit; the pressure of a small puff of air is > 100,000 Pa. (The most useful units produce values near 1 for ordinary amounts.) The atmosphere is a more convenient unit:

$$1 \text{ atm} = 101,325 \text{ Pa}$$

The torr is based on using the barometer to measure atmospheric pressure. The simplest barometer consists of a tube of mercury inverted in a small pool of mercury. The height of the mercury column rises and falls in response to the pressure the atmosphere exerts on the mercury in the pool. Atmospheric pressure of 1 atm will support a mercury column that is 760 mm high.

| Examples 5.1 – 5.2 |
| Exercises 1, 2 |

$$1 \text{ atm} = 760 \text{ mmHg} = 760 \text{ torr}$$

## Gas Laws (Sections 5.3)

**The Pressure-Volume Relationship.** Of the three states, the gaseous state is the simplest to describe mathematically. In the seventeenth century, Robert Boyle made many observations of the relationship between gas pressure and gas volume. He discovered that the volume occupied by molecules in a container decreases as the pressure increases at fixed temperature. This conclusion is summarized in Boyle's Law:

$$P = \frac{k_{PV}}{V} \text{ or } PV = k_{PV} \quad \text{for constant amounts of gas at constant temperature.}$$

| Example 5.3 |
| Exercise 3 |

One of the most useful applications of Boyle's law is the prediction of pressure or volume changes of fixed amounts of gases at constant temperature. The form of Boyle's law for this application is

$$P_{initial}V_{initial} = P_{final}V_{final} \quad \text{often abbreviated } P_1V_1 = P_2V_2 .$$

**The Temperature-Volume Relationship.** In the nineteenth century Jacques Charles and Joseph Gay-Lussac observed that the volume occupied by molecules in a container increases as the temperature increases at fixed pressure. In other words, the volume occupied by a fixed amount of gas is proportional to the temperature at fixed pressure. Their observations are summarized by Charles' Law:

| Example 5.4 |
| Exercises 4 - 6 |

$$V = k_{VT}T \text{ or } \frac{V}{T} = k_{VT} \quad \text{for constant amounts of gas at constant pressure.}$$

Applications of Charles' Law (Gay-Lussac often gets omitted because his name is long) include the prediction of temperature changes with volume, such as the cooling that occurs when gases expand. The form of Charles' law for this application is

$$\frac{V_{initial}}{T_{initial}} = \frac{V_{final}}{T_{final}} \text{ abbreviated } \frac{V_1}{T_1} = \frac{V_2}{T_2} \text{ or } V_1 T_2 = V_2 T_1$$

***The Volume-Amount Relationship.*** Around the same time, Amadeo Avogadro observed that equal volumes of different gases contained the same number of molecules. In other words, that the volume occupied by molecules in a container increases as the number of gas molecules in the vessel increases at fixed pressure and temperature. These observations are summarized as Avogadro's Law:

$$V = k_{nv} n \text{ at constant temperature and pressure.}$$

One consequence of this relationship is the fact that, ideally, one mole quantities of different gases have the same volumes when the temperature and pressure are constant. The molar volume is the volume of one mole of gas at 0 °C and 1 atm (called standard temperature and pressure, STP): 22.41 L. This is a very important observation, because a reaction like

$$2\,H_2(g) + O_2(g) \rightarrow 2\,H_2O(g)$$

reduces the volume by a factor of 2/3. Avogadro's law was useful to scientists attempting to understand the stoichiometry of chemical reactions. This subject is discussed in more detail in Section 5.5.

## Ideal Gas Equation (Section 5.4)

***The Ideal Gas Equation.*** All the gas relationships can be combined into a single equation called the Ideal Gas Equation (IGE):

$$PV = nRT$$

where n is the number of moles and $R$ is the ideal gas constant, $R = 0.0821$ L·atm/K·mol. The dots in the units are to remind us that both L and atm are in the numerator, and K and mol are in the denominator. The units of $R$ may also be written, $R = 0.0821$ L·atm·$K^{-1}$·$mol^{-1}$. The molecules of an ideal gas

- have no intermolecular interactions (attraction or repulsion between molecules)
- take up no volume (compared to container volume)

The properties of many real gases can be described by the IGE at high temperatures (>0 °C) and low pressures (<10 atm).

What can you do with the IGE?

1. Given any three of the four properties related by the IGE, the value of the fourth can be calculated. For example, a common quantity to know in order to work gas problems is the number of moles. Pressure, volume and temperature are all easily measured quantities. P, V and T can be plugged into the equation and you can solve for n. Keep this in your bag of

Example
5.6

Exercises
9 – 13

conversion tools. It doesn't use dimensional analysis, but you can plug in the known and solve for the unknown (lovingly called "plug and chug.")

| Conversion between | Tool |
|---|---|
| measured gas properties ←→ mol | PV=nRT |

TOOL

2.  Solving *PV= nRT* for *R* gives:

$$\frac{PV}{nT} = R$$

Remember, *R* is a constant. Therefore in problems where gas properties are changing, the change can be predicted by this equation.

Example 5.5

Exercises 7, 8

$$\frac{V_{initial}V_{initial}}{n_{initial}T_{initial}} = \frac{P_2V_2}{n_2T_2} \quad \text{or} \quad \frac{P_1V_1}{n_1T_1} = \frac{P_2V_2}{n_2T_2}$$

If any of the properties are held constant, the appropriate gas law can be derived from this one equation (so you do not have to memorize the individual laws.) For example, if in a problem it states, "a sample of gas at constant temperature," then *n* and *T* are constant and they can be removed from the above equation giving:

$$P_1V_1 = P_2V_2$$

3.  Another IGE rearrangement relates gas properties to the gas density. Remembering that the number of moles is equal to the ratio of the sample mass to its molar mass,

$$n = \frac{m}{\mathcal{M}}, \quad \text{then}$$

Example 5.7

Exercises 21

$$d = \frac{m}{V} = \frac{P\mathcal{M}}{RT}$$

One of the applications of this relationship is that gas density, which can be measured experimentally, can be combined with pressure and temperature to determine the molar mass.

## 4. Gas Stoichiometry (Section 5.5)

The "4" above gives reference to the fact that stoichiometry of gases is the 4$^{th}$ type of problem which involves the ideal gas equation.

Chapter 3 presented the idea that balanced chemical reactions describe the relationships between moles (or molecules) of reactants and products. The IGE relates the number of moles to gas volumes or pressures, making it possible to establish relationships between the amounts of reactants and products without converting volumes to moles first. You may treat the coefficients of a balanced equation as volume units instead of moles. Consider the reaction equation:

$$2H_2(g) + O_2(g) \rightarrow 2H_2O(g)$$

We already know that the equation states that **two moles** of hydrogen gas react with **one mole** of oxygen gas to produce **two moles** of water vapor. Due to the proportionality of *V* and *n*, we can now say that **two liters** of hydrogen gas react with **one liter** of oxygen gas to produce **two liters** of water vapor. (Provided temperature and pressure are constant.)

| Conversion between | Tool |
|---|---|
| *V* of A(g) $\longleftrightarrow$ *V* of B(g) | Coefficients of balanced equation |

TOOL

We can show the use of this tool using dimensional analysis. For the above reaction, how many liters of oxygen would be needed to react with 4.30 L hydrogen at STP?

$$4.30\,L\,H_2 \times \frac{1\,L\,O_2}{2\,L\,H_2} = 2.15\,L\,O_2$$

Note, this tool can be used with any other volume unit besides liters. *But, the substances must be gases at constant temperature and pressure.* If there is a gas present and these conditions do not apply, you can use *PV=nRT* to convert to moles of the gas and then work the problem using the same tools of stoichiometry you learned in chapters 3 and 4.

Example 5.8

Exercises 18 - 20

## Dalton's Law of Partial Pressures (Section 5.6)

Another consequence of the IGE is that the pressure of a mixture of gases is the sum of the pressures of the individual components. The IGE reduces to $\frac{P}{n} = k_{Pn}$ because the volume and temperature are the same for all the components. For example, the pressure of air is the sum of the pressures of several gases, mostly nitrogen, air and carbon dioxide:

$$P_{air} = P_{N_2} + P_{O_2} + P_{CO_2} + P_{trace\ compounds}$$

Another way to state this relationship is based on the mole fraction, $X_i$, the ratio of the number of moles of a component to the total number of moles of gas, $\frac{n_i}{n_T}$.

$$P_T = P_1 + P_2 + ... + P_C \qquad P_i = \frac{n_i}{n_T}P_T$$

$$P_T = \frac{n_1}{n_T}P_T + \frac{n_2}{n_T}P_T + ... + \frac{n_C}{n_T}P_T$$

$$P_T = X_1P_T + X_2P_T + ... + X_CP_T$$

$$P_T = (X_1 + X_2 + ... + X_C)P_T$$

Example 5.9

Exercise 21

## Kinetic Molecular Theory of Gases (Section 5.7)

The gas laws let us predict gas behavior, but they don't explain gas behavior. The kinetic molecular theory of gases provides explanations of gas behavior at the molecular level. The principal hypotheses of the kinetic theory of gases are:

H-0. Gas particles obey Newton's laws of motion

H-1. Gas molecules are separated by distances much larger than their dimensions and can be considered as points (mass, but no volume). H-1 explains compressibility.

H-2. Gas molecules are in constant motion in random directions and often collide. Gas collisions are perfectly elastic, i.e., total amount of energy in the colliding molecules is the same after collision. H-2 explains pressure-volume relationship.

H-3. There are no intermolecular forces (attraction or repulsion) between gas molecules. H-3 explains Dalton's Law of Partial Pressures.

H-4. The energy of gas molecules is primarily kinetic (internal E is small) and depends on mass and speed of molecules: $\overline{KE} = \frac{1}{2}m\overline{u}^2$, where $m$ is the mass and $\overline{u}^2$ is the average of the squares of the speeds of all the molecules. The kinetic energy is also proportional to the absolute temperature (temperature in Kelvins):

$\overline{KE} = \frac{3}{2}k_BT$, where $k_B$ is the Boltzmann constant (the real gas constant for individual molecules rather than a mole of them). H-4 explains Charles' Law.

The kinetic molecular theory has several other applications. The molecules in a container of gas move randomly, but it is possible to determine the distribution of molecular speeds from kinetic molecular theory. As the temperature increases, the range of speed narrows. The average speed of the molecules in a gas can be calculated by combining the two features of H-4:

$$N_A\left(\frac{1}{2}m\overline{u}^2\right) = N_A\frac{3}{2}k_BT$$

$$\frac{1}{2}\mathcal{M}\overline{u}^2 = \frac{3}{2}RT$$

$$u_{rms} = \sqrt{\overline{u}^2} = \sqrt{\frac{3RT}{\mathcal{M}}}$$

Example 5.10

Exercises 22 – 24

$\mathcal{M}$ is the molar mass of the substance. When using this equation, you must use the value for $R = 8.314$ J/mol·K and molar mass must be in units of kg/mol so units cancel properly leaving units of m/s for speed.

If two gases $u_{rms}$ are compared under the same conditions of temperature, the following ratio can be derived:

$$\frac{u_{rms}(A)}{u_{rms}(B)} = \sqrt{\frac{\mathcal{M}(B)}{\mathcal{M}(A)}}$$

## Nonideal Gases (Section 5.8)

The kinetic theory and ideal gas law describe gas molecules as independent of one another, not exerting attractive or repulsive forces on one another. A plot of *PV/RT* versus *P* is a horizontal line for an ideal gas, but this relationship is valid for real gases only at low pressures (less than about 10 atm) as Figure 5.22 of the text shows. The condensation of real gases to liquids at low temperatures is a second indication of the non-ideality of real gases. At low temperatures, the gas molecules lack the energy they need to break away from intermolecular forces.

***van der Waals Equation.*** In the late nineteenth century J.D. van der Waals modified the ideal gas law to account for two molecular properties that produce deviations from ideal gas behavior: intermolecular forces and finite molecular volume. According to van der Waals, the pressure of an ideal gas would be larger than that of a real gas because attraction to neighboring molecules reduces the impact real molecules make with the container wall. Similarly, the volume occupied by an ideal gas will be reduced by the small, but finite, volume of the molecules. Incorporating these corrections into the IGE, the formula becomes

$$\left( P_{obs} + a\frac{n^2}{V_{obs}^2} \right)\left( V_{obs} - nb \right) = nRT$$

| |
|---|
| Example 5.11 |
| Exercises 25 – 26 |

where $P_{obs}$ and $V_{obs}$ are the observed pressure and volume, respectively, and the constants *a* and *b* are specific for a particular gas. The larger the values of *a* and *b*, the less ideal the behavior of the gas. Table 5.4 of the text gives value for *a* and *b*. The larger the values of a and b, the greater the deviation from ideality.

## GLOSSARY LIST

| | | |
|---|---|---|
| pressure | Avogadro's law | kinetic molecular theory |
| Pascal | | kinetic energy (KE) |
| atmospheric pressure | ideal gas | root-mean-square speed |
| barometer | ideal gas equation | joule (J) |
| manometer | gas constant (R) | diffusion |
| Newton | Standard temperature and | effusion |
| standard atmospheric pressure | pressure (STP) | Graham's law of diffusion |
| | Dalton's law of partial pressure | van der Waals equation |
| Boyle's law | partial pressure | |
| Charles's law | mole fraction | |
| Kelvin temperature scale | | |
| Absolute zero | | |
| Absolute temperature scale | | |

## EQUATIONS

| Algebraic Equation | English Translation |
|---|---|
| $P_{initial}V_{initial} = P_{final}V_{final}$ | The energy of a gas ($PV$) is constant at constant temperature |
| $\dfrac{V_{initial}}{T_{initial}} = \dfrac{V_{final}}{T_{final}}$ | Volume and temperature of a sample of gas are directly proportional at constant $P$. |
| $\dfrac{P_{initial}}{T_{initial}} = \dfrac{P_{final}}{T_{final}}$ | Pressure and temperature of a sample of gas are directly proportional at constant $V$. |
| $PV = nRT$ | ideal gas equation |
| $\dfrac{P_{initial}V_{initial}}{T_{initial}} = \dfrac{P_{final}V_{final}}{T_{final}}$ | combined gas law |
| $X_i = \dfrac{n_i}{n_T}$ | mole fraction |
| $P_i = X_i P_T$ | partial pressure as a function of mole fraction and total pressure |
| $d = \dfrac{m}{V} = \dfrac{PM}{RT}$ | density of a gas as a function of molar mass |
| $\bar{E}_K = \dfrac{1}{2}m\overline{u^2} = \dfrac{3}{2}k_B T$ | kinetic energy is proportional to speed-squared and also temperature |
| $u_{rms} = \sqrt{\overline{u^2}} = \sqrt{\dfrac{3RT}{M}}$ | root-mean-square speed as a function of temperature and molar mass. |
| $\left(P_{obs} + a\dfrac{n^2}{V_{obs}^2}\right)(V_{obs} - nb) = nRT$ | van der Waals equation |

## SUMMARY OF CONVERSION TOOLS

| Conversion between | Tool |
|---|---|
| g $\longleftrightarrow$ kg (as an example) | prefix meanings |
| g $\longleftrightarrow$ mL | density |
| mol $\longleftrightarrow$ actual number | Avogadro's number |
| g $\longleftrightarrow$ mol | molar mass |
| mol compound $\longleftrightarrow$ mol atom in compound | subscript |
| mol A $\longleftrightarrow$ mol B | coefficients of balanced equation |
| mol $\longleftrightarrow$ L (of solution) | molarity |
| gas properties $\longleftrightarrow$ mol | $PV=nRT$ |
| mol of gas (A) $\longleftrightarrow$ mol of gas (B) | coefficients of balanced equation |

## WORKED EXAMPLES

### EXAMPLE 5.1 Pressure Measurement

What is the pressure, in mmHg, of the gas trapped in the J-tube shown below if the atmospheric pressure is 0.803 atm?

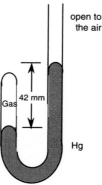

• **Solution**

The pressure of the gas is sufficient to support a column of mercury 42 mm high and hold back the pressure of the atmosphere as well. Therefore the pressure of the gas is:

$$P_{gas} = P_h + P_{atm}$$

The pressure from the mercury column is just the difference in heights of the two mercury surfaces, 42 mmHg. Atmospheric pressure must be converted to the same units:

$$0.803 \text{ atm} \times \frac{760 \text{ mmHg}}{1 \text{ atm}} = 610 \text{ mmHg}$$

$P_{gas} = $ 42 mmHg + 610 mmHg
$P_{gas} = 652$ mmHg

---

### EXAMPLE 5.2 Converting Pressure Units

The atmospheric pressure on Mars is about 0.22 inches of Hg as compared to about 30 inches of Hg on earth. Express the Martian pressure in mmHg, atmospheres, and pascals.

• **Solution**

Treat this like any other factor label problem. First, state the problem:

? mmHg = 0.22 in Hg

Recall that there are 2.54 cm/in, therefore there are 25.4 mm/in.
The unit factor needed is

$$\frac{25.4\ mm}{1\ in}$$

Converting inches to millimeters:

$$?\ mmHg = 0.22\ in\ Hg \times \frac{25.4\ mm}{1\ in} = 5.6\ mmHg$$

The number of atmospheres is:

$$?\ atm = 5.6\ mmHg \times \frac{1\ atm}{760\ mmHg} = 7.4 \times 10^{-3}\ atm$$

The pressure in pascals is:

$$?\ pascals = 7.4 \times 10^{-3}\ atm \times \frac{1.013 \times 10^5\ Pa}{1\ atm} = 750\ Pa$$

**Work EXERCISES & PROBLEMS: 1, 2**

---

**EXAMPLE 5.3 Boyle's Law**

At constant temperature, a sample of gas was compressed to 1/3 its original volume. How was the pressure changed?

**• Solution**

Boyle's law is expressed by the equation $P_1V_1 = P_2V_2$. Pressure and volume are inversely proportional so if the volume is decreased by a factor of 3, the pressure must increase by a factor of 3. Therefore, the pressure is tripled.

To verify this, let's choose some actual numbers and plug them into the equation. $P_1 = 1\ atm$; $V_1 = 3\ L$; $V_2 = 1\ L$ (volume is compressed to 1/3 its original volume)

$$\frac{P_1V_1}{V_2} = P_2$$

$$\frac{1\ atm\ (3\ L)}{1L} = 3\ atm$$

**Work EXERCISES & PROBLEMS: 3**

---

**EXAMPLE 5.4  Charles' Law Calculation**

When 2.0 L of chlorine gas ($Cl_2$) at STP is warmed at constant pressure to 100 °C, what is the new volume?

• **Solution**

The volume of a fixed amount of a gas at constant pressure is proportional to the absolute temperature according to Charles's law, where $T_1$ = 273 K and $T_2$ = (100 °C + 273 °C) K = 373 K.

$$\frac{V_1}{T_1} = \frac{V_2}{T_2}$$

Rearranging gives:

$$V_2 = V_1 \times \frac{T_2}{T_1}, \text{ where } V_1, \text{ the initial volume, is 2.0 L.}$$

Substituting, we get:

$$2.0\,L \times \frac{373\,K}{273\,K} = 2.7\,L$$

• **Comment**

Note that we expect $V_2 > V_1$ because of the temperature increase. This provides us with a simple check of our answer. If we calculated that $V_2 < V_1$, we would know our answer was wrong.

**Work EXERCISES & PROBLEMS: 4 – 6**

---

**EXAMPLE 5.5  Combined Gas Law Calculation**

Given 10.0 L of neon gas at 5 °C and 630 mmHg, calculate the new volume at 400 °C and 2.5 atm.

• **Solution**

First, note that both the pressure and the temperature of the gas are changed, but that the number of moles is constant. Write the ideal gas equation with all the constant terms on one side.

$$\frac{PV}{T} = nR = \text{a constant}$$

We see that PV/T is a constant. Therefore

$$\frac{P_1V_1}{T_1} = \frac{P_2V_2}{T_2}$$

where the subscripts 2 refer to the final state, and the subscripts 1 refer to the initial state.

Rearranging to solve for $V_2$ gives

$$V_2 = V_1 \times \frac{P_1}{P_2} \times \frac{T_2}{T_1}$$

We need pressure in the same unit, so convert 630 mmHg to atm. (Equally correct would be to convert 2.5 atm to mmHg.)

where $P_1 = 630$ mmHg $\times \dfrac{1 \text{ atm}}{760 \text{ mmHg}} = 0.829$ atm

$P_2 = 2.5$ atm
$T_1 = (273 \text{ °C} + 5 \text{ °C})$ K $= 278$ K
$T_2 = (273 \text{ °C} + 400 \text{ °C})$ K $= 673$ K

Substituting

$$V_2 = 10.0 \text{ L} \times \frac{0.829 \text{ atm}}{2.5 \text{ atm}} \times \frac{673 \text{ K}}{278 \text{ K}} = 8.0 \text{ L}$$

• **Comment**

Note that $P_1$ and $P_2$ must be expressed in the same units and both temperatures in Kelvin.

**Work EXERCISES & PROBLEMS: 7, 8**

---

**EXAMPLE 5.6  Ideal Gas Equation**

What volume will be occupied by 0.833 mole of fluorine ($F_2$) at 645 mmHg and 15.0 °C?

• **Solution**

The ideal gas law relates the gas volume to temperature, pressure and number of moles.

$$V = \frac{nRT}{P}$$

First, convert the pressure into atmospheres:

$$P = 645 \text{ mmHg} \times \frac{1 \text{ atm}}{760 \text{ mmHg}} = 0.849 \text{ atm}$$

$$= \frac{(0.833 \text{ mol})(0.0821 \text{ L} \cdot \text{atm} / \text{K} \cdot \text{mol})(288 \text{ K})}{0.849 \text{ atm}} = 23.2 \text{ L}$$

• **Comment**

Even though it takes longer to write out the units for each term in the equation, doing so is helpful because when the units cancel and leave you with the desired units (in this case liters) you are probably closer to the right answer.

**Work EXERCISES & PROBLEMS: 9 – 13**

**EXAMPLE 5.7  Molar Mass of a Gas**

A gaseous compound has a density of 1.69 g/L at 25 °C and 714 torr. What is its molar mass?

**• Solution**

The density of an ideal gas is directly proportional to its molecular mass. We use the equation

$$d = \frac{m}{V} = \frac{PM}{RT}$$

Convert pressure into units of atmospheres.

$$P = 714 \text{ torr} \times \frac{1 \text{ atm}}{760 \text{ torr}} = 0.939 \text{ atm}$$

Rearranging and substituting, we get:

$$M = \frac{dRT}{P} = \frac{(1.69 \text{ g/L})(0.0821 \text{ L} \cdot \text{atm/K} \cdot \text{mol})(298 \text{ K})}{0.939 \text{ atm}} = 44.0 \text{ g/mol}$$

**Work EXERCISES & PROBLEMS: 14 – 17**

---

**EXAMPLE 5.8  Gas Stoichiometry**

Calculate the volume of ammonia gas, measured at 645 torr and 21 °C, that is produced by the complete reaction of 25.0 g of quicklime, CaO, with excess ammonium chloride, $NH_4Cl$, solution.

**• Solution**

First, write the balanced chemical equation.

$$CaO(s) + 2NH_4Cl(aq) \rightarrow 2NH_3(g) + CaCl_2(aq) + H_2O(\ell)$$

Here we must determine the number of moles of $NH_3$ formed in the reaction and convert this into the volume of an ideal gas. Three steps are required:  (1) convert g CaO to moles CaO, (2) convert moles CaO to moles $NH_3$ produced, and (3) use the ideal gas equation to calculate the volume of $NH_3$.

Road map:     grams CaO $\xrightarrow{1}$ moles CaO $\xrightarrow{2}$ moles $NH_3$ $\xrightarrow{3}$ volume $NH_3$

We already know how to find conversion factors 1 and 2; therefore let's first find the number of moles of $NH_3$ formed.
Stating the problem:

Molar mass used here.

Coefficients of balanced equation used here.

? mol $NH_3$ = 25.0 g CaO

$$? \text{ mol } NH_3 = 25.0 \text{ g CaO} \times \frac{1 \text{ mol CaO}}{56.1 \text{ g CaO}} \times \frac{2 \text{ mol } NH_3}{1 \text{ mol CaO}} = 0.891 \text{ mol } NH_3$$

For step 3, the volume of 0.891 mol $NH_3$ can be calculated from the ideal gas equation:

$$V = \frac{nRT}{P}$$

Remember that that this conversion tool cannot be put in your chain of converters above. You must stop and use the IGE.

Convert pressure into units of atm before substituting into the ideal gas equation.

$$P = 645 \text{ torr} \times \frac{1 \text{ atm}}{760 \text{ torr}} = 0.849 \text{ atm}$$

$$V = \frac{0.891 \text{ mol}(0.0821 \text{ L} \cdot \text{atm/K} \cdot \text{mol})(294 \text{ K})}{0.849 \text{ atm}} = 25.3 \text{ L}$$

**Work EXERCISES & PROBLEMS: 18 – 20**

## EXAMPLE 5.9  Partial Pressures

When oxygen gas is collected over water at 30 °C and the total pressure is 645 mmHg:
a.  What is the partial pressure of oxygen? Given the vapor pressure of water at 30 °C is 31.8 mmHg.
b.  What are the mole fractions of oxygen and water vapor?

**• Solution**

a. Mixtures of gases obey Dalton's law of partial pressures which says that the total pressure is the sum of the partial pressures of oxygen and water vapor.

$$P_t = P_{O_2} + P_{H_2O}$$
$$P_{O_2} = P_t - P_{H_2O} = 645 \text{ mmHg} - 31.8 \text{ mmHg}$$
$$P_{O_2} = 613 \text{ mmHg}$$

b. Recall that the partial pressures of $O_2$ and $H_2O$ are related to their mole fractions,

$$P_{O_2} = X_{O_2}P_t \qquad P_{H_2O} = X_{H_2O}P_t$$

Therefore, on rearranging, we get:

$$X_{O_2} = \frac{P_{O_2}}{P_T} \qquad X_{H_2O} = \frac{P_{H_2O}}{P_T}$$

$$X_{O_2} = 613/645 = 0.950, \text{ and } X_{H_2O} = 31.8/645 = 0.0493$$

---

**• Comment**

Note also that the sum of the mole fractions is 1.0, within the number of significant figures, given:

$$X_{O_2} + X_{H_2O} = 0.950 + 0.0493 = 0.999$$

---

**Work EXERCISES & PROBLEMS: 21**

---

**EXAMPLE 5.10  Root-Mean-Square Speed**

Calculate the root-mean-square speed of gaseous argon atoms at STP.

**• Solution**

The root-mean-square speed is given by the equation

$$u_{rms} = \sqrt{\overline{u^2}} = \sqrt{\frac{3RT}{M}}$$

where R = 8.314 J/K·mol. T is the absolute temperature 273 K; and *M* is the molar mass of Ar in kilograms, 0.0399 kg/mol. Note that standard pressure is irrelevant here because the average kinetic energy depends only on the temperature. Before substituting into the equation, we recall that 1 J = 1 kg m$^2$/s$^2$.

$$u_{rms} = \sqrt{\frac{3(8.314 \text{ J/K} \cdot \text{mol})(273 \text{ K})}{(3.99 \times 10^{-2} \text{ kg/mol})} \times \frac{1 \text{ kg m}^2/\text{s}^2}{1 \text{ J}}}$$

$$u_{rms} = (1.71 \times 10^5 \text{ m}^2/\text{s}^2)^{1/2} = 413 \text{ m/s}$$

---

**Work EXERCISES & PROBLEMS: 22 – 24**

---

**EXAMPLE 5.11  van der Waals Equation**

The volume of 1.0 mol isopentane, $C_5H_{12}$, is 1.0 L at 503 K and 30.0 atm:
a. Determine the predicted pressure if the gas behaves ideally. Is the gas ideal?
b. Given that a = 17.0 L$^2$ atm/mol$^2$ and b = 0.136 L/mol, calculate the pressure of isopentane as predicted by the van der Waals equation.

**• Solution**

a. According to the ideal gas equation the pressure of 1 mole of a gas at 503 K that occupies 1.0 L is:

$$P = \frac{nRT}{V} = \frac{(1.0 \text{ mol})(0.082 \text{ L} \cdot \text{atm/K} \cdot \text{mol})(503 \text{ K})}{1.0 \text{ L}}$$

$$P = 41 \text{ atm}$$

This calculated result differs considerably from the observed pressure of 30 atm. In fact, the percent error which is the difference between the two values divided by the actual pressure is:

$$\% \text{ error} = \frac{41\,\text{atm} - 30.0\,\text{atm}}{30.0\,\text{atm}} \times 100\%$$

% error = 37%

We conclude that under these conditions $C_5H_{12}$ behaves in a nonideal manner.

b.  In this case, write the van der Waals equation

$$\left(P_{real} + \frac{an^2}{V^2}\right)(V - nb) = nRT$$

and substitute into it, but first calculate the correction terms.

$$\frac{an^2}{V^2} = \frac{(17.0\,L^2 \cdot atm/mol^2)(1.0\,mol)^2}{1.0\,L^2} = 17\,\text{atm}$$

nb = 1.0 mol (0.136 L/mol) = 0.14 L

nRT = (1.0 mol)(0.0821 L·atm/K·mol)(503 K) = 41 L atm

Now substitute using the van der Waals equation.

$$(P + 17.0\,\text{atm})(1.0\,L - 0.14\,L) = 41\,L\,\text{atm}$$

$$(P + 17.0\,\text{atm})(0.9\,L) = 41\,\text{atm}$$

$$P + 17.0\,\text{atm} = 50\,\text{atm}$$

$$P = 33\,\text{atm}$$

Thus, the pressure calculated by the van der Waals equation is much closer to the actual value of 30.0 atm.  The percent error is only 10 percent.

**Work EXERCISES & PROBLEMS: 25, 26**
**Work the rest of the EXERCISES & PROBLEMS**

## EXERCISES & PROBLEMS

1.  Convert a pressure of 645 mmHg into its value in
    a. atmospheres   b. kilopascals

2.  Do the following unit conversions.
    a. 125 mmHg to torr
    b. 725 mmHg to kilopascals

3.  A sample of gas has a volume of 200 cm$^3$ at 25 °C and 700 mmHg.  If the pressure is reduced to 280 mmHg, what volume would the gas occupy at the same temperature?

4.  A quantity of 2.00 L of oxygen gas at −15 °C are heated and the volume expands.  At what temperature will the volume reach 2.31 L?

5. A 20.0-mL sample of a gas is enclosed in a gas-tight syringe at 50.0 °C. What will the volume of the gas be at the same pressure after the syringe has been immersed in ice water?

6. If 2.00 L of oxygen at –15 °C are allowed to warm to 25 °C at constant pressure, what is the new volume of oxygen gas?

7. The gas pressure in an aerosol can is 1.5 atm at 25 °C. What pressure would develop in the can if it were heated to 450 °C?

8. A sample of gas occupies 155 mL at 21.5 °C, and at 305 mmHg. What is the pressure of the gas sample when it is placed in a $5.00 \times 10^2$-mL flask at a temperature of –10 °C?

9. What is the pressure in atmospheres of $1.20 \times 10^4$ moles of methane, $CH_4$, when stored at 22 °C in a $3.00 \times 10^3$ L tank?

10. A 1.75 g sample of $CO_2$ is contained in a $7.50 \times 10^2$-mL flask at 35 °C. What is the pressure of the gas?

11. How many moles of $CO_2$ gas are required to fill a 5.00-L balloon to a pressure of 1.05 atm at 5.0 °C?

12. How many grams of helium are required to fill a 10.0-L balloon to a pressure of 1250 torr at 25 °C?

13. How many $O_2$ molecules occupy a 1.00-L flask at 75 °C and 777 mmHg?

14. What is the density of $H_2(g)$ at 35 °C and 650 torr?

15. Which one of the following gases will have the greatest density when they are all compared at the same temperature and pressure?

   $O_2$   $CO_2$   $NO_2$   $CF_4$

16. When 2.96 g of mercuric chloride is vaporized in a 1.00 liter bulb at 680 K, the pressure is 450 mmHg. What is the molar mass and molecular formula of mercuric chloride?

17. Determine the molar mass of Freon-11 gas if a sample weighing 0.597 g occupies $1.00 \times 10^2$ mL at 95 °C, and 1000 mmHg.

18. The discovery of oxygen resulted from the decomposition of mercury(II) oxide.

   $2\ HgO \rightarrow 2\ Hg + O_2(g)$

   a. What volume of oxygen will be produced by the decomposition of 25.2 grams of the oxide, if the gas is measured at STP?
   b. How many grams of mercury(II) oxide must be decomposed to yield 10.8 L of $O_2$ gas at 1 atm and 298 K?

19. Sodium azide decomposes according to the equation.

    $$2\ NaN_3(s) \rightarrow 2\ Na(s) + 3\ N_2(g)$$

    What volume of $N_2$ at 1.1 atm and 50.0 °C will be produced by the decomposition of 5.0 g $NaN_3$?

20. Consider the reaction of 20.0 g calcium oxide with carbon dioxide.

    $$CaO(s) + CO_2(g) \rightarrow CaCO_3(s)$$

    If you have 5.5 L of $CO_2$ at 7.50 atm and 22 °C, will you have enough carbon dioxide to react with all the CaO?

21. Hydrogen and helium are mixed in a 20.0 L flask at room temperature (20 °C). The partial pressure of hydrogen is 250 mmHg and that of helium is 75 mmHg. How many grams of $H_2$ and He are present?

22. Calculate the root-mean-square speed of ozone molecules ($O_3$) in the stratosphere where the temperature is –83 °C.

23. If $NO_2$ molecules have a root-mean-square speed of 290 m/s, what is the temperature corresponding to this average speed?

24. If the root-mean-square speed of an $O_2$ molecule is $4.2 \times 10^2$ m/s at some temperature, what is the average speed of a $Cl_2$ molecule at the same temperature?

25. 0.50 mol $CCl_4$ gas is introduced into a 10.0-L flask. What fraction of the total volume of the flask is occupied by $CCl_4$ molecules? See Table 5.3 in the text for the value of the van der Waals constant b.

26. By looking at the van der Waals constants in Table 5.3 of the text, determine which gas should have the higher strength of attractive forces between its molecules: ammonia or carbon tetrachloride?

27. Why is the density of a gas much lower than that of a solid or liquid?

28. Which sample contains more molecules: a. 1.0 L of $O_2$ gas at 20 °C and 2.0 atm or b. 1.0 L of $SF_4$ gas at 20 °C and 2.0 atm? Which sample has more mass?

29. How does the kinetic molecular theory explain Charles's law?

30. A 0.356-g sample of $XH_2(s)$ reacts with water according to the following equation:

    $$XH_2(s) + 2H_2O(\ell) \rightarrow X(OH)_2(s) + 2H_2(g)$$

    The hydrogen evolved is collected over water at 23 °C and occupies a volume of 431 mL at 746 mmHg total pressure. Find the number of moles of $H_2$ produced and the atomic mass of X. Vapor pressure of $H_2O = 21$ mmHg.

31. A certain noble gas compound contains 68.8 percent Kr and 31.2 percent F by mass. Its density at STP is 5.44 g/L. What is the molecular formula of the compound?

## PRACTICE TEST QUESTIONS
See notes on taking practice test in the Preface

1. A barometer reads 695 mmHg. Calculate the pressure in units of:
   a. atm
   b. torr
   c. Pa

2. The pressure of $H_2$ gas in a 0.50-L cylinder is 1775 mmHg at 70 °F. What volume would the gas occupy at 1 atm and the same temperature?

3. If 30.0 L of oxygen is cooled at constant pressure from 200 °C to 0 °C, what is the new volume of oxygen?

4. A balloon filled with helium has a volume of 1500 L at 0.925 atm and 23 °C. At an altitude of 20 km the temperature is –50 °C and the pressure is 151.6 mmHg. What is the volume of this balloon at 20 km?

5. What is the pressure in mmHg of the gas trapped in the apparatus shown at the right when the atmospheric pressure is 0.950 atm?

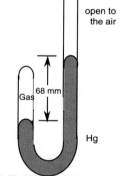

6. Given that the Martian atmosphere is mostly $CO_2$, at a pressure of 5.5 mmHg at a temperature of –31.4 °C. What is the density of the atmosphere?

7. How many grams of chlorine ($Cl_2$) occupy a 0.716-L cylinder when the pressure is 10.9 atm at 30 °C?

8. Fill in the blank spaces in the table.

|   | P | V | n | T |
|---|---|---|---|---|
| a | 7.25 atm | 40.0 L | 10.5 mol | |
| b | 451 torr | 150 mL | $2.50 \times 10^{-3}$ mol | |
| c | 14.2 atm | 12.0 L | | 325 K |
| d | 152 kPa | 120 mL | | 25 °C |
| e | 2.50 atm | | 12.0 mol | 501 K |
| f | 2280 torr | | $2.00 \times 10^2$ mol | 450 °C |
| g | | 22.4 L | 1.25 mol | 301 K |
| h | | 10.0 mL | 0.0625 mol | 25 °C |

9. Calculate the volume occupied by 15.2 g of $CO_2$ at 0.74 atm and 24 °C.

10. What is the density of uranium hexafluoride gas, $UF_6$, at STP?

11. When 1.48 g of mercuric chloride is vaporized in a 1.00-L bulb at 680 K, the pressure is 225 mmHg. What is the molar mass and molecular formula of mercuric chloride vapor?

12. Determine the molar mass of chloroform gas if a sample weighing 0.495 g is collected as a vapor (gas) in a flask of volume 127 $cm^3$ at 98 °C. The pressure of the chloroform vapor at this temperature in the flask was determined to be 754 mmHg.

13. A 150-mL sample of $O_2$ gas is collected over water at 20 °C and 758 torr. What volume will the same sample of oxygen occupy at STP when it is dry? The vapor pressure of water at 20 °C is 17.54 torr.

14. A sample of nitrogen gas was bubbled through liquid water at 25 °C and then collected in a volume of 750 $cm^3$. The total pressure of the gas, which was saturated with water vapor, was found to be 740 mmHg at 25 °C. The vapor pressure of water at this temperature is 24 mmHg. How many moles of nitrogen were in the sample?

15. The volume of carbon monoxide gas (CO) collected over water at 25 °C was 680 $cm^3$ with a total pressure of 752 mmHg. The vapor pressure of water at 25 °C is 23.8 mmHg. Determine the partial pressure and mole fraction of CO in the container.

16. The partial pressures of $N_2$, $O_2$, and Ar in dry air are 570, 153, and 6 torr, respectively. What are the mole fractions of these three gases?

17. A mixture of 40.0 g of $O_2$ and 40.0 g of He has a total pressure of 0.900 atm. What are the partial pressures of $O_2$ and He in the mixture?

18. a. What volume of $CO_2$ at 1 atm and 225 °C would be produced by the reaction of 12.0 g $NaHCO_3$?

$$2\ NaHCO_3 + H_2SO_4 \rightarrow Na_2SO_4 + 2\ CO_2 + 2\ H_2O$$

b. On cooling to 20 °C, what volume would the $CO_2$ occupy?

5

5

5

5

5

5

5

5

5

5

19. How many liters of ammonia at a constant pressure and pressure of 10 atm and 500 °C can be produced by the reaction of 6.0 g of hydrogen with excess $N_2$?

$$3 H_2(g) + N_2(g) \rightarrow 2 NH_3(g)$$

20. In the oxidation of ammonia

$$4 NH_3(g) + 5 O_2(g) \rightarrow 4 NO(g) + 6 H_2O(\ell)$$

how many liters of $O_2$, measured at 18 °C and 1.10 atm, must be used to produce 50 liters of NO at the same conditions?

21. Write the reaction for the combustion of hexane, $(C_6H_{14})$, a component of gasoline. How many liters of oxygen, measured at STP, are required for the complete combustion of 72.0 grams of hexane?

22. If the root-mean-square speed of an $N_2$ molecule is 475 m/s at some temperature, what is the root-mean-square speed of a He atom at 25 °C?

23. In each of the following pairs, which gas would you expect to deviate more from the value PV/nRT = 1 expected for an ideal gas?
    a. $N_2$ or $SF_6$
    b. He or $O_2$
    c. $CO_2$ or $SO_2$

24. Calculate the pressure of 200 moles of $NH_3$ in a 10.0-L container at 500 °C using (a) the ideal gas law, and (b) the van der Waals equation. The values of the van der Waals constants are a = 4.17 atm·$L^2$·$mol^{-2}$; b=0.0371 L·$mol^{-1}$.

# Chapter 6. Thermochemistry

The Nature & Types of Energy (Section 6.1)
Energy Changes in Chemical Reactions (Section 6.2)
Introduction to Thermodynamics (Section 6.3)
Enthalpy of Chemical Reactions (Section 6.4)
Calorimetry (Section 6.5)
Standard Enthalpy of Formation and Reaction (Section 6.6)
Heat of Solution and Dilution (Section 6.7)


## SUMMARY

This chapter describes energy changes associated with chemical reactions and the techniques used to measure them experimentally.

## The Nature & Types of Energy (Section 6.1)

*Energy.* Energy is one of those terms that is difficult to understand because it is an abstract concept that is used in many different ways. The standard scientific definition is a system's capacity to do work (move an object against an opposing force). There are three principal types of energy, some with a number of sub-classes. The types of energy that are important in our discussion of chemical reactions are

| | |
|---|---|
| kinetic (dynamic) energy: | energy an object possesses by virtue of its motion |
| potential (static) energy: | energy an object possesses by virtue of its position |
| radiant (electromagnetic) energy: | energy a wave possesses by virtue of its frequency |
| sub-classes: | chemical energy (kinetic and potential energy E associated with bond breakage & formation) thermal energy (kinetic and potential energy E associated with random particle motion) heat (thermal energy transferred between objects at different temperatures) |

The law of conservation of energy states that energy is neither destroyed nor created; it is transferred from one part of the universe to another or it is converted from one form to another.

## Energy Changes in Chemical Reactions (Section 6.2)

In order to study energy changes in chemical reactions we divide the universe into the system (where the reaction occurs) and the surroundings. In an open system, mass and energy in the system are exchanged with the surroundings. For example, a burning fire is an open system. The system consists of the reactants (wood and oxygen) and products ($CO_2$, water vapor, ashes and so on). The surroundings are everything else. In a closed system, only energy from the system is exchanged with the surroundings. In an isolated system, neither mass nor energy from the system is exchanged with the surroundings.

When a reaction in the system produces heat that is released to the surroundings, that is when energy is a product, the reaction is called exothermic. When a reaction in the system consumes heat from the surroundings (heat is a reactant) the reaction is called endothermic.

## Introduction to Thermodynamics (Section 6.3)

The theories that allow us to describe energy changes quantitatively are part of thermodynamics, the study of the interconversion of different types of energy. The first law of thermodynamics (conservation of E) limits the energy absorbed or released by a reaction to two parts: work done on or by the system, w, and heat absorbed or released by the system, q. By definition, work done on the system is positive (work done by the system is negative) and heat absorbed by the system is positive (heat released from the system is negative). For the hypothetical reaction: A →B, the conservation law is

$$\Delta E_{sys} = -\Delta E_{surr} \qquad \Delta E_{sys} = q + w$$

where $\Delta E_{sys}$ is the change in the internal energy of the system and $\Delta E_{surr}$ is the change in the energy of the surroundings. If we could relate q and w directly to the reactants and products (A and B), we'd have the quantitative description we're looking for. The problem is that q and w are path dependent properties, in other words their values change as the conditions and procedures used to carry out the reaction change. We need path-independent properties that only depend on A and B and stay the same regardless of how the reaction is carried out. It is a little more work, but we can relate q and w to state functions, which is the technical term for path-independent properties.

   Work done by gases is -PV (work = force × distance; pressure = force/area; volume = area × distance), so we can define an energy state function especially for reactions in which the pressure is constant. The state function is called the enthalpy, H. The formula for the enthalpy is

$$H = E + PV$$

The following equations show that enthalpy changes ($\Delta H$) equal the heat transferred from the system by the reaction at constant pressure, $q_p$.

$$\Delta H = H_{final} - H_{initial} = \left(E_{final} + PV_{final}\right) - \left(E_{initial} + PV_{initial}\right)$$
$$= \left(E_{final} - E_{initial}\right) + P\left(V_{final} - V_{initial}\right)$$
$$= \Delta E + P\Delta V = \Delta E - w = q_p$$

> Examples
> 6.1 – 6.3
>
> Exercises
> 1 – 3

## Enthalpy of Chemical Reactions (Section 6.4)

Most reactions are carried out at constant pressure, so the enthalpy is equal to the heat transferred by the reaction. When the enthalpy is included in a chemical equation it is called a thermochemical equation. For exothermic reactions, (heat is released by the system) the enthalpy change is negative:

$$CH_4(g) + 2\,O_2(g) \rightarrow CO_2(g) + 2\,H_2O(g) \qquad \Delta H = -802.4 \text{ kJ/mol}$$

For endothermic reactions, (heat is absorbed by the system) the enthalpy change is positive:

$$CaCO_3(s) \rightarrow CaO(s) + CO_2(g) \qquad \Delta H = 177.8 \text{ kJ/mol}$$

When a reaction is reversed, the sign of ΔH is reversed

$$H_2O(s) \rightarrow H_2O(\ell) \qquad \Delta H = 6.01 \text{ kJ/mol}$$
$$H_2O(\ell) \rightarrow H_2O(s) \qquad \Delta H = -6.01 \text{ kJ/mol}$$

The enthalpy is an extensive property, its value depends on the amount of material in the system. Therefore, scaling the reaction (multiplying by a constant) also scales ΔH.

$$H_2O(s) \rightarrow H_2O(\ell) \qquad \Delta H = 6.01 \text{ kJ/mol}$$
$$2 H_2O(s) \rightarrow 2 H_2O(\ell) \qquad \Delta H = 12.0 \text{ kJ/mol}$$

The physical states of all reactants and products are crucial to enthalpy values

$$CH_4(g) + 2 O_2(g) \rightarrow CO_2(g) + 2 H_2O(g) \qquad \Delta H = -802.4 \text{ kJ/mol}$$
$$CH_4(g) + 2 O_2(g) \rightarrow CO_2(g) + 2 H_2O(\ell) \qquad \Delta H = -890.4 \text{ kJ/mol}$$

The balanced thermochemical equation can be used to convert between amount of reactant or product and the amount of heat. For example, let's consider the reaction:

$$CH_4(g) + 2 O_2(g) \rightarrow CO_2(g) + 2 H_2O(g) \qquad \Delta H = -802.4 \text{ kJ/mol}$$

The -802.4 kJ/mol means 802.4 kJ of heat is released per mole of reaction as it is balanced. This provides a series of conversion factors. Study the equation and each of the following conversion factors.

$$\frac{-802.4 \text{ kJ}}{1 \text{ mol } CH_4} \qquad \frac{-802.4 \text{ kJ}}{2 \text{ mol } O_2} \qquad \frac{-802.4 \text{ kJ}}{1 \text{ mol } CO_2} \qquad \frac{-802.4 \text{ kJ}}{2 \text{ mol } H_2O}$$

| Conversion between | Tool |
|---|---|
| kJ ←→ mol | balanced thermochemical equation |

TOOL

To calculate the amount of heat produced when 80.0 g of CH₄ is burned, first convert grams to moles, then use the conversion appropriate mol → kJ conversion tool.

$$80.0 \text{ g } CH_4 \times \frac{1 \text{ mol } CH_4}{16.032 \text{ g } CH_4} \times \frac{-802.4 \text{ kJ}}{1 \text{ mol } CH_4} = 4.00 \times 10^3 \text{ kJ}$$

Example
6.4

Exercises
5 - 7

The next part of this chapter contains a discussion as to how the ΔH of a reaction can be determined. There are three ways discussed in the chapter.
1. Using calorimetry (section 6.5)
2. Direct method (using heats of formation - section 6.6)
3. Indirect method (using Hess's Law - section 6.6)

## Calorimetry (Section 6.5)

*Heat Capacity and Specific Heat.*  A calorimeter is a device for measuring the heat changes associated with chemical reactions.  All calorimetry experiments involve carrying out a reaction in a vessel immersed in a medium such as water and measuring the temperature change ($\Delta t$) of the medium. The heat capacity, C, of a substance is the amount of heat, q, required to raise the temperature of a given mass of substance by 1 °C.  The heat capacity of 100 g $H_2O$ is 418.4 J/°C.  This means that 418.4 J of heat will raise the temperature of 100 g water 1 degree Celsius.  The heat capacity is a proportionality constant that relates the amount of heat absorbed or released by a material to its change in temperature:

$$q = C\left(t_{final} - t_{initial}\right) = C\Delta t$$

The specific heat, s, is a related term which is the amount of heat required to raise the temperature of one gram of a substance by 1 °C.  In terms of the specific heat, the heat absorbed or released is:

$$q = ms\Delta t$$

*Constant-Volume Calorimetry.*  Bomb calorimeters, as shown in Figure 6.8 in the textbook, are used to measure the heat evolved in combustion reactions. A high-pressure steel vessel, called a bomb, is loaded with a small amount of a combustible substance and $O_2$ at 30 atm of pressure.  The loaded bomb is immersed in a known amount of water.  The heat evolved during combustion is absorbed by the water and the calorimeter:

$$q_{sys} = -q_{surr}$$
The system is the reaction, the surroundings is made of the calorimeter parts as well as the water.

$$q_{sys} = -(q_{cal} + q_{water})$$
The change in the temperature of the water and calorimeter are related to the heat evolved by the reaction, q, and to the enthalpy, $\Delta H$.

$$\Delta H \approx q_{rxn} = -(C_{cal}\Delta t + ms_{water}\Delta t)$$

Reactions in a bomb calorimeter occur under constant volume rather than constant pressure conditions, and so the heat released does not equal $\Delta H$ exactly.  For most reactions, the difference is small and can be neglected.  For instance, for the combustion of 1 mole of pentane, the difference is only 7 kJ out of 3500 kJ.

*Constant-Pressure Calorimeter.*  The heat evolved in non-combustion reactions is measured in a constant-pressure calorimeter.  Coffee cup calorimeters, shown in Figure 6.9 in the textbook, are very inexpensive examples of these devices.  Water-soluble reactants are added to water in the calorimeter.  The reaction occurs in the solution, absorbing or liberating heat that changes the temperature of the solution (mostly water), and the calorimeter:

$$q_{rxn} = -(q_{cal} + q_{water})$$

These calorimeters are constructed of low heat capacity materials, such as Styrofoam, so the heat absorbed by the calorimeter can be neglected:

$$q_{rxn} = -q_{water}$$

Reactions in a coffee cup calorimeter occur under constant pressure conditions, and so the heat released is equal to $\Delta H$.

$$\Delta H = q_{rxn} = -ms_{water}\Delta t$$

Examples<br>
6.5 – 6.7<br><br>
Exercises<br>
8 – 13

## Standard Enthalpy of Formation and Reaction (Section 6.6)

Recapping the information presented so far, we can calculate $\Delta H$ for any reaction for which $H_{reactants}$ and $H_{products}$ are known and measure $\Delta H$ for most reactions. The problem is that there is no way to measure $H_{reactants}$ and $H_{products}$. Absolute energy values, such as $H_{reactants}$ and $H_{products}$, can't be measured because there is no way to define absolute zero energy. Relative energy values can be measured because the reference energy is arbitrary. The reference point for enthalpy values is the standard of enthalpy of formation of the elements which is defined as zero at 1 atm for gases or 1 $M$ for solutions and 25 °C (in their standard state). The standard enthalpy of allotropes and single atoms of diatomic elements have nonzero values. For example,

$$\Delta H_f^\circ(O_2) = 0 \text{ kJ/mol} \qquad \Delta H_f^\circ(C_{graphite}) = 0 \text{ kJ/mol} \qquad \Delta H_f^\circ(H_2) = 0 \text{ kJ/mol}$$

$$\Delta H_f^\circ(O_3) = 142 \text{ kJ/mol} \qquad \Delta H_f^\circ(C_{diamond}) = 1.9 \text{ kJ/mol} \qquad \Delta H_f^\circ(H) = 218.2 \text{ kJ/mol}$$

Tables of standard heats of formation are given in Table 6.4 and Appendix 3 in the textbook.

***Direct Calculation of $\Delta H_{rxn}$.*** Now we can calculate $\Delta H$ for any reaction using $\Delta H_f^\circ$ of the reactants and products. The enthalpy change for any reaction is the difference of the *sum* of the standard enthalpies of formation of the products multiplied by their stoichiometric coefficients and the *sum* of the standard enthalpies of formation of the reactants multiplied by their stoichiometric coefficients. For example, the enthalpy of the decomposition of $CaCO_3$,

$$CaCO_3(s) \rightarrow CaO(s) + CO_2(g)$$

can be determined by

Example<br>
6.8<br><br>
Exercises<br>
14

$$\Delta H_{rxn}^\circ = \sum n\Delta H_f^\circ(\text{products}) - \sum n\Delta H_f^\circ(\text{reactants})$$

$$= \Delta H_f^\circ(CaO) + \Delta H_f^\circ(CO_2) - \Delta H_f^\circ(CaCO_3)$$

$$= (1)(-635.6 \text{kJ/mol}) + (1)(-393.5 \text{kJ/mol}) - (1)(-1206.9 \text{kJ/mol}) = 177.8 \text{kJ/mol}$$

***Indirect Calculation of $\Delta H_{rxn}$.*** Some reactions don't proceed as written. For example, the reaction may proceed too slowly or generate side-products. In other cases, the enthalpy of formation of one or more reactant or products may not be available. The alternative comes from Hess's law of heat summation. When a reaction is the sum of several reaction steps, then the enthalpy change for the reaction is equal to the sum of the enthalpy changes of the reaction steps.  For example, it's not convenient to measure the enthalpy of the conversion of graphite to diamond.  This reaction takes thousands of years at high pressure in the earth.  However, it is easy to measure the enthalpies of combustion of graphite and diamond:

$$C_{graphite} + O_2(g) \rightarrow CO_2(g) \qquad \Delta H° = -393.51 \, kJ/mol$$

$$C_{diamond} + O_2(g) \rightarrow CO_2(g) \qquad \Delta H° = -395.40 \, kJ/mol$$

Notice that when we reverse the diamond thermochemical reaction, the sum of the combustion reactions is the graphite to diamond conversion:

$$C_{graphite} + O_2(g) \rightarrow CO_2(g) \qquad \Delta H° = -393.51 \ kJ/mol$$

$$\underline{CO_2((g) \rightarrow C_{diamond} + O_2(g) \qquad \Delta H° = \ \ \ 395.40 \ kJ/mol}$$

$$C_{graphite} \rightarrow C_{diamond} \qquad \Delta H° = \ \ \ \ \ 1.89 \, kJ/mol$$

| Example 6.9 |
|:---:|
| Exercises 15 |

## Heat of Solution and Dilution (Section 6.7)

*Heat of Solution.*  Dissolving ionic solutes in a solvent often produces temperature changes as the extent of solute hydration changes.  We can think of solute dissolution as requiring 2 steps:  solute dissociation to gaseous ions followed by ion hydration to form the solution.  This is not what actually happens, but these are steps for which $\Delta H$ values are available.  The energy required to separate 1 mole of an ionic compound to gaseous ions (step 1) is called the lattice energy, U.  The enthalpy change of the hydration of 1 mole of gaseous ions (step 2) is the enthalpy of hydration, $\Delta H_{hydr}$.  The enthalpy of solution is the sum of these two terms. For the dissolution of LiCl:

$$LiCl(s) \qquad \rightarrow Li^+(g) + Cl^-(g) \qquad U \ \ = \ 853 \ \ kJ/mol$$

$$Li^+(g) \ + Cl^-(g) \ \rightarrow Li^+(aq) + Cl^-(aq) \qquad \Delta H_{hydr} = -890.1 \ kJ/mol$$

---

$$LiCl(s) \qquad \rightarrow Li^+(aq) + Cl^-(aq) \ \Delta H_{sol} \ = \ U + \Delta H_{hydr} = -37.1 \ kJ/mol$$

*Heat of Dilution.*  When U (always positive) is smaller in magnitude than $\Delta H_{hydr}$ (always negative), the solvation reaction is exothermic. If more water is added to such a solution, more heat will be released by the system. (This is why you never add water to acid, but always add acid to water with vigorous stirring.  If the water boils it is much less dangerous than if concentrated acid boils!) Using the same reasoning, if the solvation reaction is endothermic, dilution will cause more heat to be absorbed from the surroundings.

| Example 6.10 |
|:---:|
| Exercises 16, 17 |

## GLOSSARY LIST

| | | |
|---|---|---|
| energy | 1<sup>st</sup> law of thermodynamics | heat capacity |
| potential energy | work | specific heat |
| chemical energy | heat | calorimetry |
| thermal energy | state function | standard enthalpy |
| | enthalpy | of formation |
| radiant energy | | standard enthalpy of |
| | endothermic | of reaction |
| thermodynamics | exothermic | Hess's law |
| system | thermochemical equation | lattice energy |
| surroundings | thermochemistry | heat of hydration |
| open system | enthalpy of reaction | heat of solution |
| closed system | standard state | enthalpy of solution |
| isolated system | | heat of dilution |

## EQUATIONS

| Algebraic Equation | English Translation |
|---|---|
| $\Delta E = q + w$ | The energy of a system consists of heat and work. |
| $w = -P\Delta V$ | PV work |
| $H = E + PV$ | definition of enthalpy |
| $\Delta H = \Delta E + P\Delta V$ | |
| $\Delta E = \Delta H - RT\Delta n$ | |
| $C = ms$ | heat capacity |
| $q = C\Delta t$ | heat of a system calculated from heat capacity |
| $q = ms\Delta t$ | heat of a system calculated form specific heat |
| $\Delta H^\circ_{rxn} = \sum n\Delta H^\circ_f(\text{products}) - \sum m\Delta H^\circ_f(\text{reactants})$ | direct method for calculation of standard enthalpy of a reaction |

## SUMMARY OF CONVERSION TOOLS

| Conversion between | Tool |
|---|---|
| g ⟷ kg (as an example) | prefix meanings |
| g ⟷ mL | density |
| mol ⟷ actual number | Avogadro's number |
| g ⟷ mol | molar mass |
| mol compound ⟷ mol atom in compound | subscript |
| mol A ⟷ mol B | coefficients of balanced equation |
| mol ⟷ L (of solution) | molarity |
| gas properties ⟷ mol | PV=nRT |
| mol of gas (A) ⟷ mol of gas (B) | coefficients of balanced equation |
| kJ ⟷ mol | balanced thermochemical equation |

## WORKED EXAMPLES

### EXAMPLE 6.1  Work Done on the System

Calculate the work done on the system when 6.0 L of a gas is compressed to 1.0 L by a constant external pressure of 2.0 atm.

• **Solution**

The work done is:

$$w = -P\,\Delta V = -P(V_2 - V_1)$$
$$= -2.0 \text{ atm } (1.0 \text{ L} - 6.0 \text{ L}) = +10 \text{ L} \cdot \text{atm}$$

The answer can be converted to joules.

$$10\,\text{L}\cdot\text{atm} \times \frac{101.3\,\text{J}}{1\text{L}\cdot\text{atm}} = 1.0 \times 10^3\,\text{J}$$

• **Comment**

Obtaining a positive value for work means that work is done *on* the system by the surroundings in a compression. A positive work value means the system gains energy.

### EXAMPLE 6.2 Gas Expansion Work

A gas, initially at a pressure of 10.0 atm and having a volume of 5.0 L, is allowed to expand at constant temperature against a constant external pressure of 4.0 atm until the new volume is 12.5 L. Calculate the work done by the gas on the surroundings.

• **Solution**

In this problem the system does work on the surroundings as it expands, and by convention (Table 6.5 in the text) the sign is negative:

$$w = -P\Delta V = -P(V_2 - V_1)$$

P is the pressure opposing the expansion, $\Delta V$ is the change in volume of the system.

$$w = -4.0 \text{ atm } (12.5 \text{ L} - 5.0 \text{ L}) = -30 \text{ L·atm}$$

This quantity can be expressed in units of joules.

$$w = -30 \text{ L·atm} \times \frac{101.3 \text{ J}}{1 \text{ L · atm}} = -3.0 \times 10^3 \text{ J}$$

## EXAMPLE 6.3  The First Law of Thermodynamics

A gas is allowed to expand at constant temperature from a volume of 10.0 L to 20.0 L against an external pressure of 1.0 atm. If the gas also absorbs 250 J of heat from the surroundings, what are the values of q, w, and $\Delta E$?

• **Solution**

The work done by the system is:

$$w = -P \, \Delta V = -P(V_2 - V_1)$$
$$= -1.0 \text{ atm } (20.0 \text{ L} - 10.0 \text{ L}) = -10 \text{ L·atm}$$

$$w = -10 \text{ L·atm} \times \frac{101.3 \text{ J}}{1 \text{ L · atm}} = -1.0 \times 10^3 \text{ J}$$

The amount of heat absorbed was 250 J, and so q = 250 J. (Heat is a positive value, since it is endothermic.)

Substituting into the first law of thermodynamics gives the energy change, $\Delta E$.

$$\Delta E = q + w$$
$$= 250 \text{ J} - 1000 \text{ J}$$
$$\Delta E = -750 \text{ J}$$

• **Comment**

In this example, the system did more work than the energy absorbed as heat; therefore, the internal energy E decreased.

**Work EXERCISES & PROBLEMS: 1 - 3**

## EXAMPLE 6.4  Thermochemical Equations

The thermochemical equation for the combustion of propane is:

$$C_3H_8(g) + 5 \, O_2(g) \rightarrow 3 \, CO_2(g) + 4 \, H_2O(\ell) \qquad \Delta H^{\circ}_{rxn} = -2220 \text{ kJ/mol}$$

a. How many kJ of heat are released when 0.50 mole of propane reacts?
b. How much heat is released when 88.2 g of propane reacts?

• **Solution**

a. Let q = the heat absorbed or released by the reaction. The heat released by an exothermic reaction is an extensive property. This means that q depends on the amount of propane consumed. The equation indicates that 2220 kJ of heat is released per mole of propane burned.

$$q = 0.50 \text{ mol } C_3H_8 \times \frac{-2220 \text{ kJ}}{1 \text{ mol } C_3H_8}$$

$$q = -1.1 \times 10^3 \text{ kJ}$$

Obtained from the balanced equation.

b. Since we know the heat of reaction per mole, we convert the number of grams of $C_3H_8$ to moles. Road map: g → mol → kJ

$$q = 88.2 \text{ g } C_3H_8 \times \frac{1 \text{ mol } C_3H_8}{44.1 \text{ g } C_3H_8} \times \frac{-2220 \text{ kJ}}{1 \text{ mol } C_3H_8}$$

$$q = -4440 \text{ kJ}$$

**Work EXERCISES & PROBLEMS: 5 – 7**

---

**EXAMPLE 6.5  Specific Heat**

a.  What is the heat capacity of a block of lead if the temperature of a 425-g block increases 2.31 °C when it absorbs 492 J of heat?

b.  What is the specific heat of lead?

• **Solution**

a. The heat capacity of the block of lead is the heat absorbed divided by the temperature rise. Since q=CΔT, then:

$$C = \frac{q}{\Delta t} = \frac{492 \text{ J}}{(2.31 \text{ °C})} = 213 \text{ J/°C}$$

b. In terms of the specific heat (s), the amount of heat absorbed when an object of mass m is heated from $T_i$ to $T_f$ is:

$$q = ms \, \Delta t$$

Substituting into the equation the given quantities for q, m, and Δt gives:

$$492 \text{ J} = (425 \text{ g}) \text{ s } (2.31 \text{ °C})$$

Rearranging to solve for the specific heats:

• **Comment**

Since C = ms we could have solved part b from s = C/m using our answer from part a.

$$s = \frac{492 \text{ J}}{(425 \text{ g})(2.31 \text{ °C})} = 0.501 \text{ J/g·°C}$$

**Work EXERCISES & PROBLEMS: 8, 9**

**EXAMPLE 6.6 Determining The Calorimeter Constant**

The combustion of benzoic acid is often used as a standard source of heat for calibrating combustion bomb calorimeters. The heat of combustion of benzoic acid has been accurately determined to be 26.42 kJ/g. When 0.8000 g of benzoic acid was burned in a calorimeter containing 950 g of water, a temperature rise of 4.08 °C was observed. What is the heat capacity of the bomb calorimeter (the calorimeter constant)?

**• Solution**

The combustion of 0.8000 g of benzoic acid produces a known amount of heat.

$$q_{rxn} = 0.8000 \text{ g} \times \frac{26.42 \text{ kJ}}{1 \text{ g}} = -21.14 \text{ kJ} = -2.114 \times 10^4 \text{ J}$$

And since $\Delta t$ and the amount of water are known, $C_{cal}$, the heat capacity of the calorimeter, can be calculated. All the heat from the combustion reaction is absorbed by the bomb calorimeter and water.

$$q_{rxn} = -q_{surr}$$

$$q_{rxn} = -(q_{cal} + q_{water})$$

$$q_{rxn} = -(C_{cal} \Delta t + ms \Delta t)$$

We know all of the variables in the above equation, except $C_{cal}$.

$$q_{rxn} = -2.114 \times 10^4 \text{ J}$$
$$m = 950 \text{ g water}$$
$$s = 4.184 \text{ J/g·°C}$$
$$\Delta t = 4.08 \text{ °C}$$

Solve for $C_{cal}$ and plug in all of the other variables. (Plug and Chug)

$$q_{rxn} = -(C_{cal} \Delta t + ms \Delta t)$$

$$-q_{rxn} - ms\Delta T = C_{cal}\Delta T$$

$$\frac{-q_{rxn} - ms\Delta T}{\Delta T} = C_{cal}$$

$$C_{cal} = \frac{-(-2.114 \times 10^4 \text{ J}) - (950 \text{ g})(4.184 \text{ J/g·° C})(4.08°C)}{4.08°C}$$

$$C_{cal} = 1.21 \times 10^3 \text{ J/°C}$$

**EXAMPLE 6.7  Determining the Heat of Combustion**

The thermochemical equation for the combustion of pentane ($C_5H_{12}$) is:

$$C_5H_{12}(\ell) + 8\ O_2(g) \rightarrow 5\ CO_2(g) + 6\ H_2O(\ell) \qquad \Delta H° = -3509\ kJ/mol$$

From the following information calculate the heat of combustion, $\Delta H°$, and compare your result to the value given above. 0.5521 g of $C_5H_{12}$ was burned in the presence of excess $O_2$ in a bomb calorimeter. The heat capacity of the calorimeter was $1.800 \times 10^3$ J/°C, the temperature of the calorimeter and $1.000 \times 10^3$ g of water rose from 21.22 °C to 25.70 °C.

**• Solution**

Again, as with all calorimetry problems:

$$q_{rxn} = -q_{surr}$$

The heat evolved by the combustion reaction is absorbed by the water and the calorimeter assembly (the surrounding).

$$q_{rxn} = -(q_{cal} + q_{water})$$
where
$$q_{rxn} = -(C_{cal}\Delta t + ms\Delta t) \qquad\qquad \Delta t = 25.70\ °C - 21.22\ °C = 4.48\ °C$$

Substituting:

$$q_{rxn} = -[(1.800 \times 10^3\ J/°C)(4.48°C) + (1.000 \times 10^3\ g)(4.184\ J/g°C)(4.48°C)]$$

$$= -2.68 \times 10^4\ J \quad \text{or} \quad -26.8\ kJ$$

$\Delta H°$ refers to the reaction of 1 mole of pentane Therefore, if we can determine the amount of heat released for 1 mole of pentane, we will have the $\Delta H$ for the reaction as balanced.

- The 26,800 J was evolved by 0.5521 g $C_5H_{12}$.
- The molar mass of pentane is 72.15 g.

$$\Delta H° = q$$
$$= 1\ mol\ C_5H_{12} \times \frac{72.15\ g\ C_5H_{12}}{1\ mol\ C_5H_{12}} \times \frac{26.8\ kJ}{0.5521\ g} = -3500\ kJ$$

$\Delta H° = -3.50 \times 10^3$ kJ/mol (where mol represents the mol of reaction as balanced.)

**• Comment**

To three significant figures the result compares well with the value given in the thermochemical equation. Small differences will arise because the heat evolved at constant volume is not quite the same as the heat evolved at constant pressure (which is equal to $\Delta E$). Only the heat evolved at constant pressure exactly equals $\Delta H$, the enthalpy change.

**Work EXERCISES & PROBLEMS: 10 – 13**

**EXAMPLE 6.8  Using Enthalpies of Formation**

Using $\Delta H_f^\circ$ values in Appendix 3 of the textbook calculate the standard enthalpy change for the incomplete combustion of ethane ($C_2H_6$).

$$C_2H_6(g) + \frac{5}{2}O_2(g) \rightarrow 2\,CO(g) + 3\,H_2O(\ell)$$

• **Solution**

The enthalpy change for this chemical reaction in terms of enthalpies of formation is:

$$\Delta H_{rxn}^\circ = \sum n\Delta H_f^\circ \text{ (products)} - \sum m\,\Delta H_f^\circ \text{ (reactants)}$$

$$\Delta H_{rxn}^\circ = [2 \times \Delta H_f^\circ \text{ (CO)} + 3 \times \Delta H_f^\circ \text{ (H}_2\text{O)}] - [1 \times \Delta H_f^\circ \text{ (C}_2\text{H}_6) - \frac{5}{2} \times \Delta H_f^\circ \text{ (O}_2)]$$

To avoid cumbersome notation, the physical states of the reactants and products were omitted from this equation. Be careful to obtain from Appendix 3 the appropriate value of $\Delta H_f^\circ$. Note that the coefficients from the chemical equation are equal to the number of moles of each substance and that <u>all</u> the terms involving $\Delta H_f^\circ$ of the *reactants* are preceded by a negative sign. Substituting the values from Appendix 3:

$$\Delta H_{rxn}^\circ = (2)(-110.5 \text{ kJ/mol}) + (3)(-285.8 \text{ kJ/mol})] - [(1)(-84.68 \text{ kJ/mol}) + (5/2)(0 \text{ kJ/mol})]$$

$$= (-221.0 \text{ kJ/mol}) + (-857.4 \text{ kJ/mol}) - (-84.68 \text{ kJ/mol})$$

$$= -221.0 \text{ kJ/mol} - 857.4 \text{ kJ/mol} + 84.68 \text{ kJ/mol}$$

$$\Delta H_{rxn}^\circ = -993.7 \text{ kJ/mol}$$

**Work EXERCISES & PROBLEMS: 14**

**EXAMPLE 6.9  Hess's Law**

The standard enthalpy change for the combustion of 1 mole of ethanol is:

$$C_2H_5OH(\ell) + 3\,O_2(g) \rightarrow 2\,CO_2(g) + 3\,H_2O(\ell) \qquad \Delta H_{rxn}^\circ = -1367 \text{ kJ/mol}$$

What is $\Delta H_{rxn}^\circ$ for the following reaction in which $H_2O$ is formed as a gas, rather than as a liquid?

$$C_2H_5OH(\ell) + 3\,O_2(g) \rightarrow 2\,CO_2(g) + 3\,H_2O(g)$$

Given the heat of vaporization of water

$$H_2O(\ell) \rightarrow H_2O(g) \qquad \Delta H_{vap} = 44 \text{ kJ/mol}$$

**• Solution**

We can imagine a two-step path for this reaction. In the first step, ethanol undergoes combustion to form liquid $H_2O$, followed by the second step in which $H_2O(\ell)$ is vaporized. The sum of the two steps gives the desired overall reaction, and according to Hess's law the sum of the two $\Delta H$'s gives the overall $\Delta H$.

$$C_2H_5OH(\ell) + 3\ O_2(g) \rightarrow 2\ CO_2(g) + 3\ H_2O(\ell) \quad \Delta H^\circ_{rxn} = -1367\ \text{kJ/mol}$$

$$\underline{\phantom{C_2H_5}3\ H_2O(\ell) \rightarrow 3\ H_2O(g) \phantom{C_2H_5OHOO(g)} \quad \Delta H^\circ_{rxn} = \phantom{-}132\ \text{kJ/mol}}$$

$$C_2H_5OH(\ell) + 3\ O_2(g) \rightarrow 2\ CO_2(g) + 3\ H_2O(g) \quad \Delta H^\circ_{rxn} = -1235\ \text{kJ/mol}$$

**Work EXERCISES & PROBLEMS: 15**

---

**EXAMPLE 6.10  Heat of Hydration**

The heat of solution of NaBr is −1.0 kJ/mol. The lattice energy of NaBr is 735 kJ/mol. Determine the heat of hydration of NaBr and write the equation for the hydration reaction.

**• Solution**

Recall that the heat of solution is the sum of the lattice energy (U) and the heat of hydration ($\Delta H_{hydr}$):

$$\Delta H_{soln} = U + \Delta H_{hydr}$$

Rearrranging for the heat of hydration:

$$\Delta H_{hydr} = \Delta H_{soln} - U$$
$$= -1.0\ \text{kJ/mol} - 735\ \text{kJ/mol}$$
$$= -736\ \text{kJ/mol}$$

$\Delta H_{hydr}$ refers to the heat liberated or absorbed when the gas phase ions are dissolved in water: The equation is:

$$Na^+(g) + Br^-(g) \xrightarrow{\ H_2O\ } Na^+(aq) + Br^-(aq)$$

**• Comment**

For NaBr, U and $\Delta H_{hydr}$ have essentially the same values, but are opposite in sign. Therefore, when NaBr dissolves in water, the lattice energy required to break up the crystal lattice is "paid back" by the hydration of the $Na^+$ and $Br^-$ ions. Also, NaBr does not dissolve in nonpolar solvents. Molecules of nonpolar solvents can only interact weakly with ions in the crystal lattice, and so "solvation" is not sufficient to compensate for the lattice energy.

**Work EXERCISES & PROBLEMS: 16, 17**
**Work the rest of the EXERCISES & PROBLEMS**

## EXERCISES & PROBLEMS

1.   Calculate the work done on a gas when 22.4 L of the gas is compressed to 2.24 L under a constant external pressure of 10.0 atm.

2.   A system does 975 J of work on its surroundings while at the same time it absorbs 625 J of heat. What is the change in energy, $\Delta E$ for the system?

3.   A system does 975 J of work on the surroundings. How much heat does the system absorb at the same time, if its energy change is −350 J?

4.   Given the thermochemical equation:

$$SO_2(g) + \tfrac{1}{2}O_2(g) \rightarrow SO_3(g) \qquad \Delta H^{\circ}_{rxn} = -99.0 \text{ kJ}$$

how much heat is evolved when 75 g $SO_2$ undergoes combustion?

5.   How many grams of $SO_2$ must be burned to yield 251 kJ of heat?  Given:

$$2\,SO_2(g) + O_2(g) \rightarrow 2\,SO_3(g) \qquad \Delta H^{\circ}_{rxn} = -198.0 \text{ kJ}$$

6.   The reaction that occurs when a typical fat, glyceryl trioleate, is metabolized by the body is:

$$C_{57}H_{104}O_6(s) + 80\,O_2(g) \rightarrow 57\,CO_2(g) + 52\,H_2O(\ell) \qquad \Delta H^{\circ}_{rxn} = -3.35 \times 10^4 \text{ kJ/mol.}$$

How much heat is evolved when 1.00 gram of this fat is completely oxidized?

7.   Given $2\,Al(s) + \dfrac{3}{2}O_2(g) \rightarrow Al_2O_3(s) \qquad \Delta H^{\circ} = -1670 \text{ kJ/mol.}$

What is $\Delta H^{\circ}$ for the reaction?

$$2\,Al_2O_3(s) \rightarrow 4\,Al(s) + 3\,O_2(g) \qquad \Delta H^{\circ} = ?$$

8.   A piece of iron initially at a temperature of 25.2 °C absorbs 16.9 kJ of heat. If its mass is 82.0 g, calculate the final temperature. The specific heat of iron is 0.444 J/g·°C.

9.   How much heat is absorbed by 52.0 g of iron when its temperature is raised from 25 °C to 275 °C? The specific heat of iron is 0.444 J/g • °C.

10.   A 25.0-g piece of aluminum at 4.0 °C is dropped into a beaker of water. The temperature of water drops from 75.0 °C to 55.0 °C. What amount of heat did the aluminum absorb?  The specific heat of Al is 0.902 J/g • °C.

11.   Benzoic acid ($C_6H_5CO_2H$) was used to calibrate a bomb calorimeter. Its enthalpy of combustion is accurately known to be −3226.7 kJ/mol. When 1.0236 g of benzoic acid was burned in a bomb calorimeter, the temperature of the calorimeter and the $1.000 \times 10^3$ g of water surrounding it rose from 20.66 °C to 24.47 °C. What is the heat capacity of the calorimeter?

12. To determine the heat capacity of a bomb calorimeter, a student added 150 g of water at 50.0 °C to the calorimeter. The calorimeter initially was at 20.0 °C. The final temperature of the water and calorimeter was 32.0 °C. What is the heat capacity of the calorimeter in J/°C?

13. Octane, $C_8H_{18}$, a constituent of gasoline burns according to the equation:

$$C_8H_{18}(\ell) + \frac{25}{2}O_2(g) \rightarrow 8\ CO_2(g) + 9\ H_2O(\ell)$$

0.1111 g of $C_8H_{18}$ was burned in the presence of excess $O_2$ in a bomb calorimeter. The heat capacity of the calorimeter was $1.726 \times 10^3$ J/°C. The temperature of the calorimeter and $1.200 \times 10^3$ g of water rose from 21.22 °C to 23.05 °C. Calculate the heat of combustion per gram of octane.

14. A reaction used for rocket engines is:

$$N_2H_4(\ell) + 2\ H_2O_2(\ell) \rightarrow N_2(g) + 4\ H_2O(\ell)$$

What is the enthalpy of reaction in kJ/mol? Given the following enthalpies of formation:

$\Delta H_f^\circ\ (N_2H_4) = \quad 95.1$ kJ/mol

$\Delta H_f^\circ\ (H_2O_2) = -187.8$ kJ/mol

$\Delta H_f^\circ\ (H_2O)\ = -285.8$ kJ/mol

15. From the following thermochemical equations, calculate the enthalpy of formation of $CH_4$.

$CH_4(g) + 4\ F_2(g) \rightarrow CF_4(g) + 4\ HF(g) \quad \Delta H_{rxn}^\circ = -1942$ kJ/mol

$C(graphite) + F_2(g) \rightarrow CF_4(g) \quad \Delta H_{rxn}^\circ = -933$ kJ/mol

$H_2(g) + F_2(g) \rightarrow 2\ HF(g) \qquad\qquad \Delta H_{rxn}^\circ = -542$ kJ/mol

16. Given that the lattice energy of LiCl is 828 kJ/mol and the heat of hydration of $Li^+$ and $Cl^-$ ions is –865 kJ/mol, calculate the heat of solution of LiCl(s).

17. Given that the heat of solution of LiF is +32 kJ/mol and that the lattice energy for LiF is 1006 kJ/mol, calculate the heat of hydration of $Li^+$ and $F^-$ ions.

18. Hess's law of heat summation works because enthalpy is a state function. Explain.

19. Why is the lattice energy always a positive quantity? Why is the hydration energy always negative?

20. Define energy, work, and heat. Consider a 9 V "alkaline" battery. Which of these three does it contain? Name a device that uses a battery to do work. Name a device that uses a battery to generate heat.

## PRACTICE TEST QUESTIONS
See notes on taking practice test in the Preface

1. What is a state function? Name three state functions used in thermochemistry.

2. Given the thermochemical equation:

   $$SO_2(g) + \frac{1}{2}O_2(g) \rightarrow SO_3(g)$$
   $$\Delta H^\circ_{rxn} = -99 \text{ kJ/mol}$$

   How much heat is liberated when
   a. 0.50 mole of $SO_2$ reacts
   b. 2.0 moles of $SO_2$ reacts

3. Given the thermochemical equation:

   $$H_2 + I_2 \rightarrow 2HI \qquad \Delta H^\circ_{rxn} = 52 \text{ kJ/mol}$$

   What is $\Delta H$ for the reaction

   $$HI \rightarrow \frac{1}{2}H_2 + \frac{1}{2}I_2 \qquad \Delta H^\circ_{rxn} = ?$$

4. Given the following reactions and their associated enthalpy changes:

   $$2 H(g) \rightarrow H_2(g) \qquad \Delta H^\circ_{rxn} = -436 \text{ kJ/mol}$$
   $$Br_2(g) \rightarrow 2 Br(g) \qquad \Delta H^\circ_{rxn} = +224 \text{ kJ/mol}$$
   $$H_2(g) + Br_2(g) \rightarrow 2 HBr(g)$$
   $$\Delta H^\circ_{rxn} = -72 \text{ kJ/mol}$$

   Calculate $\Delta H^\circ_{rxn}$ in kJ for the reaction:
   $$HBr(g) \rightarrow H(g) + Br(g)$$

5. Nitrogen and oxygen react according to the following thermochemical equation:

   $$N_2(g) + O_2(g) \rightarrow 2 NO(g)$$
   $$\Delta H^\circ_{rxn} = 180 \text{ kJ/mol}$$

   How many kJ of heat are absorbed when 50.0 g of $N_2$ reacts with excess $O_2$ to produce 107 g of NO?

6. How much energy is required to raise the temperature of 180 g of graphite from 25 °C to 500 °C? The specific heat of graphite is 0.720 J/g • °C

7. 0.500 g of ethanol, $C_2H_5OH(\ell)$, was burned in a bomb calorimeter containing 2.000 × $10^3$ g of water. The heat capacity of the bomb calorimeter was 950 J/°C, and the temperature rise was found to be 1.60 °C.
   a. Write a balanced equation for the combustion of one mole ethanol.
   b. Calculate the amount of heat transferred to the calorimeter.
   c. Calculate $\Delta H$ for the reaction as written in (a) above. .

8. The enthalpy of combustion of sulfur is:

   $$S(rhombic) + O_2(g) \rightarrow SO_2(g)$$
   $$\Delta H^\circ_{rxn} = -296 \text{ kJ/mol}$$

   What is the enthalpy of formation of $SO_2$?

9. The combustion of methane occurs according to the equation

   $$CH_4(g) + 2 O_2(g) \rightarrow CO_2(g) + 2 H_2O(\ell)$$
   $$\Delta H^\circ_{rxn} = -890 \text{ kJ/mol}$$

   The standard enthalpies of formation for $CO_2$ is -393.5 kJ/mol and for $H_2O$ is -285.8 kJ/mol. Determine the enthalpy of formation of methane.

6

6

6

6

6

6

6

6

6

10. During expansion of its volume from 1.00 L to 10.00 L against a constant external pressure of 2.00 atm, a gas absorbs 200 J of energy as heat. Calculate the change in internal energy of the gas.

11. How much work must be done on or by the system in a process in which the internal energy remains constant and 322 J of heat is transferred from the system?

6                                    6

# Chapter 7.   Quantum Theory and the Electronic Structure of Atoms

**Atomic Structure Revisited (Sections 7.1 – 7.2)**
**Bohr's Theory of the Hydrogen Atom (Section 7.3)**
**The Dual Nature of the Electron (Section 7.4)**
**Quantum Mechanics and Quantum Numbers (Sections 7.5 – 7.7)**
**Electron Configurations and the Aufbau Principle (Sections 7.8 – 7.9)**

## SUMMARY

### Atomic Structure Revisited (Sections 7.1 – 7.2)

The experiments conducted to study the structure of atoms at the end of the nineteenth and beginning of the twentieth centuries were described in Chapter 2.  The results of these experiments depict an atom with an exceedingly small, positively charged nucleus surrounded by negative electrons.  When scientists attempted to interpret the results of these experiments using the accepted theories of the day, such as Newton's laws of motion and Maxwell's electromagnetic wave theory of radiation, they did not have much success.  For example, it was well known that the wavelength of radiation emitted by hot solids decreases with increasing temperature, but the correlation could only be predicted for either short or long wavelengths.  The results of other experiments couldn't be predicted at all.  The same theory that postulated that atomic stability required electrons to be in motion, predicted that electrons moving in the electric field of the positive nucleus would emit radiation, lose energy and quickly collapse onto the nucleus.  On the basis of the same theory, gases subjected to electric fields in discharge tubes were expected to emit a continuous range of energies, but they emit narrow bands of radiation (selected wavelengths) instead.  Equally perplexing results were observed in a very different experiment involving metals.  When metals were irradiated by some frequencies of light, electrons were ejected from the metal surface.  This was called the photoelectric effect.  Efforts to correlate the incidence and rate of electron ejection to the properties of the radiation using nineteenth century physics were futile.

  The solution to all of these problems started with the work of Max Planck in 1900.  Planck successfully described the entire wavelength range of radiation emitted by hot solids by constraining energy to be absorbed and emitted in discrete packets called quanta.  A single quantum of energy is defined as proportional to the frequency, $\nu$ (nu), of the radiation

$$E = h\nu.$$

The constant, h, is Plank's constant and has a value of $6.63 \times 10^{-34}$ J•s. The importance of the idea that energy is transferred in quanta was reinforced in 1905 by the work of Albert Einstein, which established that electromagnetic radiation exists as quanta.  This was a great surprise because a staggering amount of work confirmed the fact that electromagnetic radiation consists of mutually perpendicular, oscillating electric and magnetic fields, i.e., waves.  The distance between the maxima in the waves is called the wavelength, $\lambda$ (lambda).  Remember, the SI unit for distance is the meter (m), though it is more common for wavelengths to be reported in nanometers (nm).  The frequency of a wave is the number of crests that pass a fixed point per second.  The unit of frequency is cycles per second which is written as /s or $s^{-1}$. The SI unit for frequency is the hertz (Hz).  The product of the wavelength and the frequency of any wave is the velocity or speed of the wave, usually in meters per second.  All electromagnetic waves travel at the same speed through a vacuum. This speed, known as the *speed of light*, is $3.00 \times 10^8$ m/s and has the symbol c:

$$c = \lambda \nu.$$

Long wavelength radiation, such as radio waves, is low frequency; short wavelength radiation, such as x-rays, is high frequency.  Visible radiation covers a very small range of wavelengths (400 – 700 nm) in the middle of the infinite electromagnetic spectrum.

In wave theory, the energy carried by an electromagnetic wave (intensity) depends on the square of the electric field amplitude (height of the wave measured from its midpoint). This is why the photoelectric effect was impossible to understand before Einstein applied Planck's idea of quantized energy to the situation.  Electrons are only ejected from surfaces by radiation with a frequency equal to or greater than some threshold value (the threshold frequency is different for each metal).  According to wave theory, this should not be the case; it should be possible to increase the amplitude of any wave to a large enough value to induce electron ejection.  What scientists observed was that changing the intensity (amplitude) of the wave changed the rate of electron ejection only if the radiation frequency was above the threshold frequency.  A very intense radiation below the threshold frequency would melt the metal rather than induce electron ejection.  Each electron is held to the metal by a force called the binding energy (BE).  Einstein realized that if the energy of the radiation was quantized, the radiation was behaving like a particle that could interact with a single electron on the metal surface.  Today, we call radiation particles photons.  According to Einstein, below the threshold frequency, a photon does not have enough kinetic energy to overcome the binding energy and eject an electron from the metal surface, but above the threshold frequency it does.  Since the binding energy is constant, as the photon energy (frequency) increases, more of its energy is available to become kinetic energy (KE) in the electrons ejected from the metal.

$$KE = h\nu - BE$$

This theory described the experimental results completely, but it was difficult for scientists to embrace the idea that radiation could behave as waves in some circumstances and as particles in others.  Today, this wave-particle duality is widely accepted and known to apply to all energy, including matter.  More on this later.

| Examples |
| 7.1, 7.2 |
| Exercises |
| 1 – 6 |

Within the course of this chapter it will be important to know the energy of a photon, wavelength and/or frequency in order to work a problem. If given one of the three, the other two can be found using the equations below.

| Conversion Between | Tool |
|---|---|
| wavelength ←→frequency | $c = \lambda \nu$ |
| energy of photon ←→ frequency | $E = h\nu$ |
| energy of photon ←→wavelength | $E = hc/\lambda$ |

## Bohr's Theory of the Hydrogen Atom (Section 7.3)

One of the experimental observations that perplexed scientists at the end of the nineteenth century was the emission of narrow bands of radiation by gases excited in discharge tubes. Scientists had known for years that each element emits a unique spectrum (series of radiation frequencies), but the source of the emission was not understood until 1913 when Niels Bohr predicted the emission spectrum of the hydrogen atom. Bohr kept the popular solar system-inspired picture of an electron making a circular orbit around the nucleus, but restricted the electron to a finite set of orbits, each associated with a specific radius and energy. When energy is absorbed by the atom, the electron must jump to a higher energy orbit. The lowest energy configuration occurs when the electron is closest to the nucleus, when n = 1. This is called the ground state. Hydrogen atoms that have an electron in a higher energy orbit are called excited state atoms. Bohr was able to calculate the radii of the allowed orbits and their energies. The energies of the H atom are given by:

$$E_n = -R_H \left( \frac{1}{n^2} \right)$$

where $R_H$ is the Rydberg constant which has a value equal to $2.18 \times 10^{-18}$ J. The integer n is a label for each electron orbit and is called the principal quantum number. The minus sign in the equation does not signify negative energy; it means that the energy of the hydrogen atom is lower than that of a completely separated proton and electron for which the force of attraction is zero.

Bohr's stroke of genius was to equate an energy change of the atom to the energy of a photon of emitted light. Bohr attributed the lines of the emission spectrum to the radiation emitted by the atom as the electron drops from a higher energy orbit to a lower energy one closer to the nucleus. In order for energy to be conserved, the energy lost as radiation must equal the difference in the energies of the initial and final electron orbits. For the emission process, $E_i$ and $n_i$ (i stands for initial) represent the higher atomic energy and larger orbit radius; $n_f$ and $E_f$ (f stands for final) represent the lower energy and smaller radius. Substituting the expression for each energy level, the change in energy of the atom $\Delta E_{atom}$ is

$$\Delta E_{atom} = E_f - E_i = -R_H \left( \frac{1}{n_f^2} - \frac{1}{n_i^2} \right)$$

Keep in mind that this equation is only good for the *hydrogen* atom. The energy of the photon emitted as the electron transitions from a higher level to a lower level in the hydrogen atom. The energy of the photon is always positive. Once the energy of a photon is known, the wavelength and frequency can be determined from the equations you already know. See the Tool Table on the preceding page.

The Balmer series is the portion of the hydrogen atom spectrum that falls in the visible region. It corresponds to transitions to the second energy level of the hydrogen atom.

| Example |
| 7.3 |
| Exercises |
| 7 – 10 |

## The Dual Nature of the Electron (Section 7.4)

Bohr's theory works for H atoms but fails to predict spectra of any other elements or even for $H_2$. Moreover, no one, including Bohr, could justify the quantization of electron orbits or atomic energy levels. In 1924 Louis de Broglie hypothesized that if photons can have the properties of particles as well as waves, electrons could have the properties of waves as well as particles! He realized that the "wavelength" of the electron would have to be a multiple of the orbit circumference to avoid canceling itself out by destructive interference. De Broglie deduced that the wavelength $\lambda$ of a particle depends on its mass, m, and velocity, u:

$$\lambda = \frac{h}{mu}$$

Experimental evidence of the wave properties of electrons was provided in 1927 by diffraction experiments. Diffraction is a phenomenon that can be explained only by wave motion. Since electrons were observed to produce diffraction patterns, the electrons were proven to behave as waves. According to de Broglie's equation, the wavelength of a particle increases as its mass decreases, so wave properties are significant for atomic and subatomic particles. As the mass of a particle reaches macroscopic size (say, $> 10^{-12}$ g) its wavelength becomes extremely short and wave properties cannot be observed.

> Example
> 7.4
>
> Exercise
> 11

## Quantum Mechanics and Quantum Numbers (Sections 7.5 – 7.7)

Werner Heisenberg realized that if a subatomic particle behaves like a wave, it is impossible to know its position and momentum precisely and simultaneously. The position of a moving particle is most accurately determined over an infinitely short time interval. However, it is impossible to describe an infinitely short time interval using a single frequency. Similarly, the energy of a wave is most accurately determined when the extent of the wave is infinitely long. The idea of position makes no sense for an infinitely long wave. The Heisenberg uncertainty principle distills this relationship to a simple statement:

$$\Delta x \Delta p = \Delta t \Delta E \geq \frac{h}{4\pi}$$

Erwin Schrödinger developed a theory for computing electron energies based on the idea that we don't have to specify the position or path of the electron in order to compute its energy and predict spectra. The Schrödinger equation (SE) is beyond our ability to solve, but we will use the SE *solutions*, energies ($E$) and wave functions ($\psi$) to describe the state of electrons around the nucleus. The square of a wave function, $\psi^2$, is proportional to the probability that an electron can be found at specific spatial coordinates around the nucleus. (The dependence on $\psi^2$ makes sense when we remember that wave intensity depends on the square of the amplitude.) The probability that an electron can be found in a particular region around an atom also is called the electron density.

The solutions of the Schrödinger equation specify the energy levels and spatial distributions of the electron in the hydrogen atom. In contrast to the circular orbits of the Bohr solar model, the wave function specifies an atomic orbital, a representation of the volume an electron can occupy. The Schrödinger equation has an exact solution for the hydrogen atom, but can only be solved using approximation methods for atoms with more than one electron. The approximate methods use the Schrödinger equation solutions for the hydrogen atom as approximations for descriptions of electrons in larger atoms.

Each solution of the Schrödinger equation is specified by the values of three parameters called quantum numbers. We use the quantum numbers as symbols of the wave functions, the atomic orbitals they represent or the electrons that reside in them.

1.  Principal quantum number, n.  The principal quantum number can take any positive integer value: $1 \geq n > \infty$. The principal quantum number reflects the average distance of an electron from the nucleus.  All the known ground state elements have n between 1 and 7.
2.  Angular momentum quantum number, $\ell$. The angular momentum quantum number can take integer values between zero and (n-1).  The angular momentum quantum

number reflects the shape of the electron distribution. (The boundaries of an electron cloud need not be well defined for an orbital to have an overall shape.) The angular momentum quantum number, $\ell$, defines a group of orbitals composing a specific sublevel or subshell. The following table correlates the value of $\ell$ to the letter designation and shape.

| $\ell$ | letter | shape |
|---|---|---|
| 0 | s | spherical |
| 1 | p | dumbell |
| 2 | d | crossed dumbells |
| 3 | f | multi-lobed |

3.  Magnetic quantum number, $m_\ell$. The magnetic quantum number can take integer values between $-\ell$ and $\ell$. The number of $m_\ell$ values is important because it designates the number of orbitals within a subshell. For example, when $\ell = 1$, there are three $m_\ell$ values. Consequently, p orbitals always occur in groups of three orbitals. These three p orbitals all have the same energy, but have different orientations in space (each aligned with one of the Cartesian axes, for example).

The fourth quantum arises during the extension of the Schrödinger equation to many electron atoms. During his efforts to predict emission spectra, Wolfgang Pauli proposed a spin quantum number for electrons and later postulated an exclusion principle: no two electrons in an atom can have all four quantum numbers the same. This principle limits the number of electrons each atomic orbital can hold to no more than two electrons. The electron spin quantum number, $m_s$, designates one of two possible spin directions for electrons. (Remember that spinning charges induce magnetic fields.) The two possible values of $m_s$ are $+1/2$ and $-1/2$.

Examples
7.5 – 7.8

Exercises
12 – 17

The diagram to the right shows the shells and the subshells in each shell. It also shows the quantum numbers associated with the shells and subshells. For example, in the 2nd principle energy level (n=2) there are s and p subshells ($\ell$=0, 1). One orbital is in the s subshell and three orbitals are in the p subshell.

As we will see in the next section, each orbital can hold a maximum of 2 electrons. One with a +1/2 spin and one with a -1/2 spin.

Study the diagram and get a feel for the connections between quantum numbers and the information the quantum numbers give.

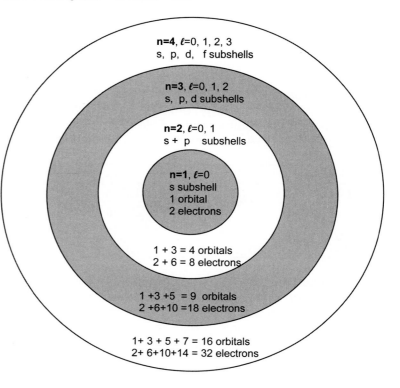

## Electron Configurations and the Aufbau Principle (Sections 7.8 – 7.9)

The energies of all the orbitals in a subshell are the same for a hydrogen atom (with 1 electron moving around). The electron density distributions are different for p and s orbitals, but they have the same energy. In other words, the energies of the orbitals increase with n:

$$1s < 2s = 2p < 3s = 3p = 3d < 4s = 4p = 4d = 4f < 5s \cdots$$

The energies of orbitals of atoms containing 2 or more electrons are shifted by electron repulsion and electrostatic shielding of the nucleus by the lower energy electrons. Consequently, the orbital energies depend on n and $\ell$. For example, the energy of an atom is lower when the 4s orbital is filled before the 3d. The order of the energy levels in many electron atoms becomes:

$$1s < 2s < 2p < 3s < 3p < 4s < 3d < 4p < 5s < 4d < 5p < 6s < 4f < 5d \ldots$$

With this knowledge, we can write an atomic formula that reflects the energy and distribution for every electron in every element on the periodic chart. We will see clearly what has been implied since Chapter 2: the similarity in chemical and physical properties associated with the elements in a group arise from similarities in the arrangement of the electrons around the nucleus of those atoms. We call this arrangement the electron configuration. The ground state electron configuration of H is $1s^1$, as illustrated.

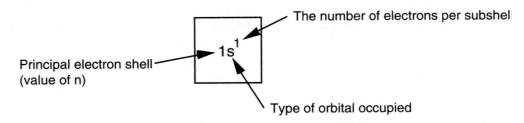

The electron configurations of many-electron atoms are constructed by adding an electron to the next (lowest energy) empty orbital. This is called the Aufbau or building up principle. So, the electron configuration of He is $1s^2$. The electron configurations of the first five elements in the 1A (alkali metals) and 8A (noble gas) groups are:

| H | $1s^1$ | He | $1s^2$ |
|---|---|---|---|
| Li | $1s^2 2s^1$ | Ne | $1s^2 2s^2 2p^6$ |
| Na | $1s^2 2s^2 2p^6 3s^1$ | Ar | $1s^2 2s^2 2p^6 3s^2 3p^6$ |
| K | $1s^2 2s^2 2p^6 3s^2 3p^6 4s^1$ | Kr | $1s^2 2s^2 2p^6 3s^2 3p^6 4s^2 3d^{10} 4p^6$ |
| Rb | $1s^2 2s^2 2p^6 3s^2 3p^6 4s^2 3d^{10} 4p^6 5s^1$ | Xe | $1s^2 2s^2 2p^6 3s^2 3p^6 4s^2 3d^{10} 4p^6 5s^2 4d^{10} 5p^6$ |

As you can see, the electron configurations for the large elements are long and unwieldy. The electron configurations of the noble gas elements can be used as abbreviations when writing electron configurations. Neon has the electron configuration $1s^2 2s^2 2p^6$. All elements beyond neon have this configuration for the first 10 electrons. We can use the symbol [Ne] to represent the configuration of the first 10 electrons and call it a neon core. Similarly, 18 electrons arranged in the ground state of an argon atom ($1s^2 2s^2 2p^6 3s^2 3p^6$) is written [Ar] and called an argon core. Krypton, xenon and radon cores can also be used. In practice, we select the noble gas that most nearly precedes the element being considered. Therefore, the electron configurations of the first five elements in the 1A (alkali metals) and 8A (noble gas) groups are:

| H | $1s^1$ | He | $1s^2$ |
|---|---|---|---|
| Li | $[He]2s^1$ | Ne | $[He]2s^2 2p^6$ |
| Na | $[Ne]3s^1$ | Ar | $[Ne]3s^2 3p^6$ |
| K | $[Ar]4s^1$ | Kr | $[Ar]4s^2 3d^{10} 4p^6$ |
| Rb | $[Kr]5s^1$ | Xe | $[Kr]5s^2 4d^{10} 5p^6$ |

As predicted, the electron configurations illustrate the electronic basis of the similarities of chemical and physical properties observed in the groups of the periodic table, especially when the nobel gas core is abbreviated. On the other hand, electron configurations tend to hide information about the number of electrons in any one outer orbital. The 2p subshell consists of three 2p orbitals: $2p_x$, $2p_y$, and $2p_z$. Therefore, the capacity of the subshell is 6 electrons. In the case of elements with a partially filled orbital, carbon for example, a choice arises about where to place the electrons. There are 2 electrons in the p subshell of a carbon atom. Are both in the $2p_x$ orbital, or are they distributed so that the $2p_x$ and the $2p_y$ orbitals each have one electron? An orbital diagram can simplify the choice. An orbital diagram groups boxes that represent individual orbitals, to designate subshells.

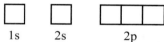

An arrow pointing up ↑ stands for an electron spinning in one direction, and an arrow pointing down ↓ stands for an electron spinning in the opposite direction. Electrons are placed into orbitals according to Hund's rule, which states that electrons entering a subshell containing more than one orbital will have the most stable arrangement when the electrons occupy the orbitals singly, rather than in pairs. Thus, carbon has one electron in the $2p_x$ and one in the $2p_y$, rather than two electrons in the $2p_x$.

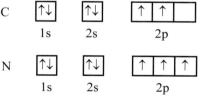

Similary, Hund's rule predicts that nitrogen atoms have the electrons in the 2p subshell distributed with one each in the $2p_x$, $2p_y$, and $2p_z$. The orbital diagram indicates the number of unpaired electrons in an atom. The presence, or absence, of unpaired electrons is indicated experimentally by the behavior of an element when placed in a magnetic field. Paramagnetic elements are attracted to a magnetic field because the atoms of paramagnetic substances contain unpaired electrons. Diamagnetic elements are repulsed by a magnetic field because the atoms of diamagnetic substances contain only paired electrons.

Recapping the rules for writing electron configurations –

1. Each shell (principal level) contains n subshells and a maximum total of $2n^2$ electrons.
2. Each subshell contains $2\ell+1$ orbitals.
   - Electrons fill degenerate (equal energy) orbitals 1 at a time
3. Each orbital can hold up to 2 electrons.
4. Pairs of electrons in same orbital have opposite spins.
5. The symbol of the preceding noble gas can be used to abbreviate the electron configuration of the core electrons.

In most cases, you should be able to write the electron configuration of any element on the periodic table. Transition metals, lanthanides and actinides often exhibit non-sequential ordering of sublevels; however, the rules above will give correct electron configurations in most cases.

Examples
7.9 – 7.11

Exercises
18 – 21

## GLOSSARY LIST

| | | |
|---|---|---|
| emission spectrum | line spectra | excited state |
| quantum | photon | Pauli exclusion principle |
| electromagnetic radiation | Heisenberg uncertainty principle | Hund's rule |
| frequency | quantum number | Aufbau principle |
| wavelength | boundary surface diagram | noble gas core |
| amplitude | atomic orbital | paramagnetic |
| wave | electron density | diamagnetic |
| electromagnetic radiation | many electron system | |
| electromagnetic wave | electron configuration | noble gas |
| | ground state | transition metal |
| photoelectric effect | | rare earth series |
| emission spectra | | actinide series |
| | | lanthanide series |

## EQUATIONS

| Algebraic Equation | English Translation |
|---|---|
| $c = \lambda v$ | The wavelength is inversely proportional to its frequency. |
| $E = h v$ | Energy of a photon is proportional to the frequency of the radiation. |
| $E_n = -R_H\left(\dfrac{1}{n^2}\right)$ | energy of an electron in a hydrogen atom |
| $\Delta E = R_H\left(\dfrac{1}{n_i^2} - \dfrac{1}{n_f^2}\right)$ | Difference in energy as an electron in a hydrogen atom transitions from one energy level to another. |
| $\lambda = \dfrac{h}{mu}$ | wavelength of a particle |
| $\Delta x \Delta p \geq \dfrac{h}{4\pi}$ | |

## WORKED EXAMPLES

### EXAMPLE 7.1 Wavelength and Frequency

Domestic microwave ovens generate microwaves with a frequency of 2.45 GHz. What is the wavelength of this microwave radiation?

• **Solution**

The equation relating the wavelength to the frequency is

$$c = \lambda \nu$$

where c is the speed of light.
Rearranging and substituting, we get:

The unit Hz can also be described as the unit 1/s. Therefore 1 GHz is $10^9$ Hz.

$$\lambda = \frac{c}{\nu} = \frac{3.00 \times 10^8 \, \text{m/s}}{2.45 \, \text{GHz}} \times \frac{1 \, \text{GHz}}{10^9 \, / \text{s}}$$

$$= 0.122 \, \text{m}$$

**Work EXERCISES & PROBLEMS: 1, 2**

### EXAMPLE 7.2 Planck's Equation

The orange light given off by a sodium vapor lamp has a wavelength of 589 nm. What is the energy of a single photon of this radiation?

• **Solution**

The energy of a quantum is proportional to its frequency. Since wavelength is given here, substitute $c/\lambda$ into Planck's equation for $\nu$.

$$E_{photon} = h\nu = \frac{hc}{\lambda}$$

Substitute values for Planck's constant and the speed of light into the equation. The units will not cancel unless the wavelength is converted from nanometers into meters.

$$E_{photon} = \frac{(6.63 \times 10^{-34} \, \text{J})(3.00 \times 10^8 \, \text{m/s})}{(589 \times 10^{-9} \, \text{m})} = 3.38 \times 10^{-19} \, \text{J}$$

• **Comment**

We can compare this energy to that of a mole of substance. The energy of a mole of photons is the product of $E_{photon} \times N_A$, where $N_A$ is Avogadro's number.

$$E = (3.38 \times 10^{-19} \, \text{J/photon})(6.02 \times 10^{23} \, \text{photon/mol})$$

$$= 203,000 \, \text{J/mol} = 203 \, \text{kJ/mol}$$

**Work EXERCISES & PROBLEMS: 3 – 6**

### EXAMPLE 7.3  The Hydrogen Atom

a.  What amount of energy, in joules, is lost by a hydrogen atom when an electron transition from n = 3 to n = 2 occurs in the atom?
b.  What is the wavelength of the light emitted by the transition described in part a?

### • Solution

a.  The energy of the atom is quantized and depends on the orbit of the electron. Each orbit is assigned a principal quantum number n:

$$E_n = -R_H \left( \frac{1}{n^2} \right)$$

For this transition $n_i$ = 3 and $n_f$ = 2, the change in energy of the electrons can be determined using:

$$\Delta E = R_H \left( \frac{1}{n_i^2} - \frac{1}{n_f^2} \right)$$

Where $n_i$ = 3 and $n_f$ = 2. Substitution gives:

$$\Delta E = 2.18 \times 10^{-18} \, J \left( \frac{1}{3^2} - \frac{1}{2^2} \right) = -3.03 \times 10^{-19} \, J$$

b.  The energy lost by the atom appears as a photon of radiation with its respective frequency and wavelength. Since the energy of the photon is positive, we will drop the negative sign from the calculation of the wavelength.

$$\Delta E_{atom} = E_{photon} = \frac{hc}{\lambda}$$

$$\lambda = \frac{hc}{\Delta E_{atom}} = \frac{hc}{3.03 \times 10^{-19} \, J}$$

$$\lambda = \frac{(6.63 \times 10^{-34} \, J \cdot s)(3.00 \times 10^8 \, m/s)}{3.03 \times 10^{-19} \, J}$$

$$\lambda = 6.56 \times 10^{-7} \, m = 656 \, nm$$

**Work EXERCISES & PROBLEMS: 7 - 10**

### EXAMPLE 7.4  De Broglie Wavelength

When an atom of Th-232 undergoes radioactive decay, an alpha particle which has a mass of 4.0 amu is ejected from the Th nucleus with a velocity of $1.4 \times 10^7$ m/s. What is the de Broglie wavelength of the alpha particle?

• **Solution**

The de Broglie wavelength depends on the mass and velocity of the particle:

$$\lambda = \frac{h}{mu}$$

Because Planck's constant has units of J·s, and 1 J = 1 kg·m²/s², the mass of the alpha particle should be expressed in kg. If 1 alpha particle has a mass of 4.0 amu, then 1 mole of alpha particles has a mass of 4.0 g. You need the mass of one alpha particle in kg and this is calculated as follows:

$$\frac{4.0\,g}{1\,mol} \times \frac{1\,kg}{1000\,g} \times \frac{1\,mol}{6.022 \times 10^{23}\,particles} = 6.64 \times 10^{-27}\,kg/particle$$

The wavelength of this alpha particle is:

$$\lambda = \frac{h}{mu} = \frac{(6.63 \times 10^{-34}\,J \cdot s) \times \left(\dfrac{1\,kg \cdot m^2 / s^2}{1\,J}\right)}{(6.64 \times 10^{-27}\,kg)(1.4 \times 10^7\,m/s)}$$

Notice that all the units cancel except meters. The wavelength is:

$$\lambda = 7.1 \times 10^{-15}\,m$$

• **Comment**

The wavelength is smaller than the diameter of the thorium nucleus, about $2 \times 10^{-14}$ m.

**Work EXERCISES & PROBLEMS: 11**

---

**EXAMPLE 7.5  Quantum Numbers**

If the principal quantum number of an electron is n = 2, what are allowed values for
a.  Its $\ell$ quantum number?
b.  Its $m_\ell$ quantum number?
c.  Its $m_s$ quantum number?

• **Solution**

a.  Recall that the angular momentum quantum number $\ell$ has values that depend on the value of the principal quantum number n. $\ell$ = 0, 1, 2, on up to the highest value (n − 1). When n = 2, then $\ell$ = 0, 1.
b.  The magnetic quantum number $m_\ell$ depends on the value of $\ell$ where $m_\ell = -\ell, -\ell +1, \dots, 0, \dots, \ell + 1, \ell$. There are two $\ell$ values and so two sets of $m_\ell$ values. When $\ell$ = 1, then $m_\ell$ = −1, 0, 1; and when $\ell$ = 0, then $m_\ell$ = 0.
c.  The $m_s$ quantum number for a single electron can be either +1/2 or −1/2.

**EXAMPLE 7.6  Quantum Numbers and Orbitals**

List all the possible types of orbitals associated with the principal energy level n = 4.

• **Solution**

The type of orbital is given by the angular momentum quantum number.  When n = 4; $\ell$ = 0, 1, 2, and 3.   These correspond to subshells consisting of s, p, d, f orbitals.

**EXAMPLE 7.7  Quantum Numbers**

For the following sets of quantum numbers for electrons, indicate which quantum numbers — n, $\ell$, $m_\ell$ — could not occur and state why.
a. 3, 2, 2         b. 2, 2, 2         c. 2, 0, –1

• **Solution**

a.   When n = 3, then $\ell$ = 0, 1, 2. When $\ell$ = 2; then $m_\ell$ = –2, –1, 0, 1, 2. Set (a) can occur.
b.   When n = 2, then $\ell$ = 0, 1.  The set in (b) cannot occur because $\ell \neq 2$ when n = 2. This means that the n = 2 principal level cannot have d orbitals.
c.   When n = 2, $\ell$ = 0, 1. When $\ell$ = 0, $m_\ell$ = 0.  The set in (c) cannot occur because $m_\ell \neq -1$ when $\ell$ = 0.

**Work EXERCISES & PROBLEMS: 12, 13**

**EXAMPLE 7.8  Quantum Numbers and Orbitals**

What are the possible values of $m_\ell$ for
a.   a 3d orbital?
b.   a 2s orbital?

• **Solution**

The values of $m_\ell$ depend only on $\ell$.  First, convert each orbital designation to the corresponding value of $\ell$, then list allowed values for $m_\ell$.
a.   For a d orbital $\ell$ = 2; therefore, possible $m_\ell$ values are –2, –1, 0, 1, and 2.
b.   For an s orbital $\ell$ = 0; therefore, $m_\ell$ = 0.

**Work EXERCISES & PROBLEMS: 14 – 17**

**EXAMPLE 7.9  Electron Configurations**

Which of the following electron configurations would correspond to ground states and which to excited states?
a. $1s^2 2s^2 2p^1$        b. $1s^2 2p^1$        c. $1s^2 2s^2 2p^1 3s^1$

• **Solution**

a.  In this configuration, electrons occupy the lowest possible energy states. It corresponds to a ground state.
b.  In this configuration electrons do not occupy the lowest possible energy states. The 2s orbital which lies lower than the 2p is vacant, and the last electron is in the 2p orbital. This is an excited state.
c.  In this case the last electron is in the 3s orbital while the lower energy 2p subshell is not completely filled. This is an excited state.

## EXAMPLE 7.10  Electron Configurations

Write the electron configuration for a potassium atom.

**• Solution**

Potassium has 19 electrons.  The first 10 of these will occupy the same orbitals as neon which is given above: $1s^2 2s^2 2p^6$.

This leaves 9 electrons to account for. Keeping in mind the order of increasing orbital energies, the next 2 electrons fill the 3s orbital. Then 6 electrons can be placed into the 3p subshell.  This leaves 1 electron. According to the order of orbital energies, the 3d does not fill next.  Rather, the 4s orbital is lower in energy than the 3d.  The last electron in potassium enters the 4s orbital.  The electron configuration of potassium is:

   K        $1s^2 2s^2 2p^6 3s^2 3p^6 4s^1$

or in terms of the argon core abbreviation:

   K        $[Ar]4s^1$

**Work EXERCISES & PROBLEMS: 18 - 21**

## EXAMPLE 7.11  Orbital Diagrams

a.  Write the electron configuration for arsenic.
b.  Draw its orbital diagram.
c.  Are As atoms diamagnetic or paramagnetic?

**• Solution**

a.  From the atomic number of As we see there are 33 electrons per arsenic atom. From Figure 7.23 in the text, the order of filling orbitals is 1s, 2s, 2p, 3s, 3p, 4s, etc. Placing electrons in the lowest energy orbitals until they are filled, the first 18 electrons are arranged as $1s^2 2s^2 2p^6 3s^2 3p^6$, corresponding to an Ar core. The next two enter the 4s, and the next ten enter the 3d. This leaves 3 electrons for the 4p subshell. The electron configuration of As is:

   As  $[Ar]4s^2 3d^{10} 4p^3$

b.  All of the orbitals are filled except for the 4p orbitals. The electrons must be placed into the 4p orbitals according to Hund's rule. The diagram for the outer orbitals is:

As  ⬛ [↑↓]     [↑↓][↑↓][↑↓][↑↓][↑↓]     [↑][↑][↑]

        4s                    3d                      4p

c.  Arsenic atoms are paramagnetic because they contain 3 unpaired electrons.

**Work the rest of the EXERCISES & PROBLEMS**

## EXERCISES & PROBLEMS

1.  What is the wavelength of electromagnetic radiation having a frequency of $5.00 \times 10^{14}$/s?

2.  What is the frequency of light that has a wavelength of 750 nm?

3.  What is the energy of a photon of radiation having a frequency of $6.2 \times 10^{14}$/s?

4.  The red line in the spectrum of lithium occurs at 670.8 nm. What is the energy of a photon of this light? What is the energy of 1 mole of these photons?

5.  A photon has an energy of $2.05 \times 10^{-19}$ J. Calculate the wavelength of the radiation.

6.  The binding energy for an electron on a surface of metallic sodium is $3.9 \times 10^{-19}$ J. Find the kinetic energy of the photoelectrons when sodium is illuminated with UV light of 305 nm.

7.  Consider the following energy levels for the hydrogen atom.

    _____ n = 4
    _____ n = 3
    _____ n = 2
    _____ n = 1

    Considering only these levels:
    a. How many emission lines are possible?
    b. Which transition produces photons of the greatest energy?
    c. Which transition for the H atom produces the emission line with the longest wavelength?

8.  What wavelength of radiation will be emitted when an electron in a hydrogen atom jumps from the n = 5 to the n = 1 principal energy level? Name the region of the electromagnetic spectrum corresponding to this wavelength.

9.  A hydrogen emission line in the ultraviolet region of the spectrum at 95.2 nm corresponds to a transition from a higher energy level n to the n = 1 level. What is the value of n for the higher energy level?

10. The second line in the Balmer series occurs at 486.1 nm. What is the energy difference between the initial and final energy levels involved in the electron transition? What is the $n_i$ value? Note that the Balmer series has $n_f = 2$.

11. Calculate the wavelength, in nanometers, of a proton (mass = $1.6725 \times 10^{-27}$ kg) that is moving at 10% of the speed of light.

12. Which of the following sets of quantum numbers is/are not allowed for describing an electron in an orbital?

|    | n | $\ell$ | $m_\ell$ | $m_s$ |
|----|---|--------|----------|-------|
| a. | 3 | 2 | −3 | 1/2 |
| b. | 2 | 3 | 0 | −1/2 |
| c. | 2 | 1 | 0 | −1/2 |

13. Which choice is a possible set of quantum numbers for the last electron added to make up an atom of gallium (Ga) in its ground state?

|    | n | $\ell$ | $m_\ell$ | $m_s$ |
|----|---|--------|----------|-------|
| a. | 4 | 2 | 0 | −1/2 |
| b. | 4 | 1 | 0 | 1/2 |
| c. | 4 | 2 | −2 | −1/2 |
| d. | 3 | 1 | +1 | 1/2 |
| e. | 3 | 0 | 0 | −1/2 |

14. Which of the following is/are incorrect designations for an atomic orbital?
    a. 3f   b. 4s   c. 2d   d. 4f

15. How many orbitals in an atom can have the following designations?
    a. 2s   b. 3d   c. 4p   d. n = 3

16. For each of the following, give the subshell designation, the $m_\ell$ values, and the number of possible orbitals.
    a. n = 3   $\ell$ = 2
    b. n = 4   $\ell$ = 3
    c. n = 5   $\ell$ = 1

17. For each of the following subshells, give the n and $\ell$ values, and the number of possible orbitals.
    a. 2s   b. 3p   c. 4f

18. What element has atoms with the electron configuration $[Xe]6s^2 4f^{14} 5d^{10} 6p^2$?

19. What third period element has atoms in the ground state with three unpaired electrons?

20. Write the electron configuration for the following atoms: Sb, V, Pb.

21. How many unpaired electrons do oxygen atoms have?

22. Briefly describe Bohr's theory of the hydrogen atom and discuss how the theory explains the production of an emission line spectrum.

23. Which of the following are observable?
    a. the position of an electron in an H atom
    b. the wavelength of light emitted by H atoms
    c. the nucleus of an atom
    d. an s orbital

24. Is the electron density distributed evenly inside an s orbital?

25. In many-electron atoms the 2s orbital has a lower energy than the 2p. Explain.

26. The energy to remove an electron from a ground state hydrogen atom (ionization energy) is determined by $E = R_h \left( \dfrac{1}{n^2} \right)$. If the energy required to ionize 1 mole of H atoms were used to raise the temperature of water, what mass of water could have its temperature increased by 50 °C?

27. If 5 percent of the energy supplied to an incandescent light bulb is radiated as visible light, how many "visible" photons per second are emitted by a 100-watt bulb? Assume the wavelength of all visible light to be 560 nm. Given: 1 watt = 1 J/s.

28. Silver bromide (AgBr) is the light-sensitive compound in most photographic films. Assume, when film is exposed, that the light energy absorbed dissociates the molecule into atoms. (The actual process is more complex.) If the energy of dissociation of AgBr is 100 kJ/mol, find the wavelength of light that is just able to dissociate AgBr.

## PRACTICE TEST QUESTIONS
See notes on taking practice test in the Preface

1. A certain AM radio station broadcasts at a frequency of 600 kHz. What is the wavelength of these radio waves, in meters?

2. How long, in seconds, would it take a radio wave of frequency $5.5 \times 10^5$ Hz to travel from the planet Venus to earth (28 million miles)?

3. How much energy is needed to remove an electron from a ground state H atom?

4. Calculate the energy (in kJ/mol) in 1.00 moles of photons of radiation with a wavelength of 670 nm.

5. A hydrogen emission spectrum has a green line at 486 nm. This line corresponds to an electron dropping from some level ($n_i$) to the n=2 level. Calculate the initial n.

6. The average kinetic energy of a neutron at 25 °C is $6.2 \times 10^{-21}$ J. What is its de Broglie wavelength (in nm)? The mass of a neutron is 1.008 amu. Hint: Kinetic energy = $(1/2)mu^2$ = $(mu)^2/2m$.

7. List the possible sets of four quantum numbers for an electron when n = 2.

8. What is the maximum number of electrons that can occupy an n = 3 energy level? The n = 4 energy level?

9. Which of the following subshells has a capacity of 10 electrons?
   a. 5s          b. 2p
   c. 4p          d. 3d
   e. 6s

10. How many orbitals are occupied by only one electron in a germanium atom?

11. Complete the sentence: An electron
    with $\ell = 2$ must
    a. have $m_\ell = -2$.
    b. be in an n = 3 energy level.
    c. be in a p orbital.
    d. be in a d orbital.

12. Write quantum numbers for:
    a. an electron in a 2s orbital.
    b. the outermost electrons in Ge.

13. Write the electron configuration for the following atoms: Ar, Se, Ag.

14. How many electrons are represented by the abbreviation [Kr]?

15. Which of the following is the correct orbital diagram for chromium?

(a)  3d: [↑↓][↑↓][ ][ ][ ]   4s: [↑↓]

(b)  3d: [↑↓][↑↓][↑↓][↑↓][ ]   4s: [ ]

(c)  3d: [↑][↑][↑][↑][ ]   4s: [↑↓]

(d)  3d: [↑][↑][↑][↑][↑]   4s: [↑]

16. Which of the following atoms has the greatest number of unpaired electrons?
    a. Ti        b. Ag
    c. O         d. P
    e. K

# Chapter 8.   Periodic Relationships Among the Elements

**Periodic Classification of the Elements (Sections 8.1 – 8.2)**
**Periodic Variation in Physical Properties (Section 8.3)**
**Ionization Energy and Electron Affinity (Sections 8.4 – 8.5)**
**Types of Elements (Section 8.6)**


## SUMMARY

### Periodic Classification of the Elements (Sections 8.1 – 8.2)

***The Periodic Table.*** When $19^{th}$ century chemists arranged the elements by increasing atomic mass, they notice that a pattern emerged in the chemical and physical properties of the elements. These trends were first correctly summarized in the periodic table by Dmitri Mendeleev in ~1870. The elements are ordered by chemical and physical properties as well as atomic mass. The table allowed him to predict the existence of elements that had not been discovered, for example, Ga and Ge. Although the table was very useful, no one understood *why* the order of the elements did not match the atomic weights, e.g.,

|   | Ar | K | Co | Ni | Te | I |
|---|----|----|-----|-----|------|------|
| M | 40.0 | 39.1 | 58.9 | 58.7 | 127.6 | 126.9 |

These discrepancies made sense after Moseley ordered the elements by the atomic number, Z (number of protons), rather than atomic mass. He found that wavelength of x-rays emitted by atoms depend on Z.

|   | Ar | K | Co | Ni | Te | I |
|---|----|----|-----|-----|------|------|
| Z | 18 | 19 | 27 | 28 | 52 | 53 |

> Examples
> 8.1 – 8.2
>
> Exercises
> 1 – 4

In fact, the electron configuration underlies the structure of the entire periodic table. All the members of a group will have the same valence (outer) electron configuration. Since these electrons are the ones involved in making chemical bonds, the chemical properties of the group members are similar. We use our understanding of the properties and trends of groups and periods to predict the behavior of elements that are not familiar.

| H | $1s^1$ | HCl | | | |
| Li | $[He]2s^1$ | LiCl | F | $[He]2s^22p^5$ | NaF |
| Na | $[Ne]2s^1$ | NaCl | Cl | $[Ne]3s^23p^5$ | NaCl |
| K | $[Ar]4s^1$ | KCl | Br | $[Ar]4s^23d^{10}4p^5$ | NaBr |
| Rb | $[Kr]5s^1$ | RbCl | I | $[Kr]5s^24d^{10}5p^5$ | NaI |

These kinds of trends are less clear for groups 3A-6A, consider 3A

| Al | $[Ne]3s^23p^1$ | metal | $AlCl_3$ |
| Ga | $[Ar]4s^23d^{10}4p^1$ | metalloid (semiconductor) | $Ga_2Cl_4$, $Ga_2Cl_6$ |
| In | $[Kr]5s^24d^{10}5p^1$ | metal | InCl, $InCl_2$, $InCl_3$ |
| Tl | $[Xc]6s^24f^{14}5d^{10}6p^1$ | metal | TlCl, $TlCl_2$, $TlCl_3$ |

Electron configurations also rationalize the charge states of the elements and stoichiometry of ionic compounds. Elements lose or gain electrons to achieve noble gas configuration:

| | | | | | | | | |
|---|---|---|---|---|---|---|---|---|
| $Be^{2+}$ | [He] | | $BeCl_2$ | $O^{2-}$ | $[He]2s^22p^6$ | = [Ne] | CaO | |
| $Mg^{2+}$ | [Ne] | | $MgCl_2$ | $S^{2-}$ | $[Ne]3s^23p^6$ | = [Ar] | CaS | |
| $Ca^{2+}$ | [Ar] | | $CaCl_2$ | $Se^{2-}$ | $[Ar]4s^23d^{10}4p^6$ | = [Kr] | | |
| $Sr^{2+}$ | [Kr] | | $SrCl_2$ | $Te^{2-}$ | $[Kr]5s^24d^{10}5p^6$ | = [Xe] | | |
| $Ba^{2+}$ | [Xe] | | $BaCl_2$ | $Po^{2-}$ | $[Xe]6s^25d^{10}6p^6$ | = [Rn] | | |

Isoelectronic species have the same number of electrons and have the same electron configuration.

| | | | | | | | | |
|---|---|---|---|---|---|---|---|---|
| Na | $[Na]3s^1$ | | $Na^+$ | [Na] | | F | $[He]2s^22p^5$ | $F^-$ $[He]2s^22p^5$ = [Ne] |

Notice that $Na^+$ and $F^-$ are isoelectronic. They both have the electron configuration of neon.

Cations of transition metals take a little extra care. Be sure to write the electron configuration of the metal and then remove the electrons from the highest principle quantum number (n). This means, remove the s before the d.

| | | | | |
|---|---|---|---|---|
| Mn | $[Ar]4s^23d^5$ | | $Mn^{2+}$ | $[Ar]3d^5$ |
| Cr | $[Ar]4s^13d^5$ | | $Cr^{3+}$ | $[Ar]3d^3$ |
| Cu | $[Ar]4s^13d^{10}$ | | $Cu^+$ | $[Ar]3d^{10}$ |

**Representative Elements.** The modern periodic table is arranged according to the type of subshells being filled with electrons (Figure 8.2 textbook). Groups 1A through 7A of the periodic table include elements that have incompletely filled sets of s or p orbitals. These are called representative elements, or main group elements. The groups are given numerals from one to seven, followed by the letter A, which stands for a group of representative elements. The noble gases have completely filled outer ns and np subshells, $ns^2np^6$, except for helium.

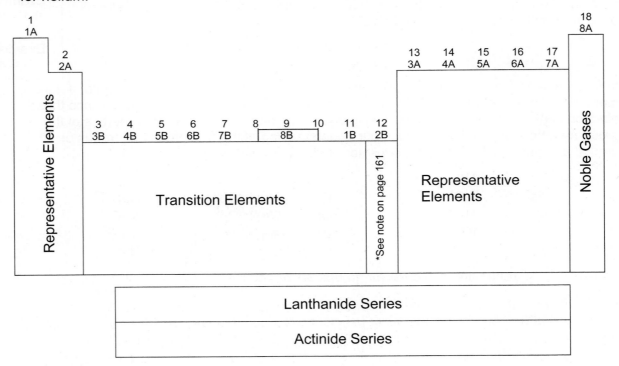

**Figure 8.1** General groups of elements.

***Transition Elements.*** The elements in the center of the periodic table, whose groups are labeled B, are the transition metals. Transition metals have incompletely filled d subshells or readily form cations having incompletely filled d subshells. In the fourth period, either the atoms or the ions of the elements Sc through Cu have incompletely filled 3d subshells. In the fifth period, the atoms or the ions of the elements Y to Cd have incompletely filled 4d subshells. *Because their d subshells are completed, the elements Zn, Cd, and Hg (Group 2B) are often not considered true transition elements. See Figure 8.1.

***Inner Transition Elements.*** Below the main part of the periodic table, there are two sets of 14 elements each depicted as horizontal rows. These groups are generally known as inner transition elements, but are also referred to as lanthanides (atomic numbers 58–71) and actinides (atomic numbers 90–103). For these elements, an inner f subshell is incompletely filled. Figure 8.1 shows the positions in the periodic table of the groups of elements discussed above.

| Examples 8.3 – 8.4 |
| Exercises 5 – 15 |

## Periodic Variation in Physical Properties (Section 8.3)

***Atomic Radius.*** Figure 8.5 of the textbook shows that the atomic radius decreases, in general, as the atomic number increases across any period of the periodic table. Thus, the alkali metals have the largest atoms and the noble gases the smallest. Furthermore, within a group, the atomic radius increases as the atomic number increases.

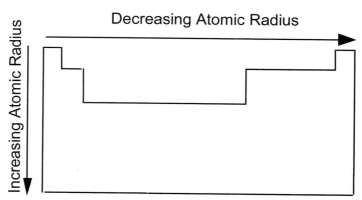

**Figure 8.2** Periodic trends in the atomic radius.

Another trend in chemical behavior is the diagonal relationship. This refers to similarities in some properties between a second row element and the third row element that is one space to the right in the table. Two such elements are on a diagonal (Figure 8.13 in the text). For example, Li and Mg are on a diagonal, and they have similar atomic radii. If you start at Li and go first to Be, the atomic radius decreases. Next, on moving down to Mg, the radius will increase again. If the decrease and increase are about the same, then Li and Mg will have approximately the same radius (155 pm for Li versus 160 pm for Mg). This relationship does not hold for all properties of two diagonally related elements.

***Ionic Radii.*** Because atoms and ions of the same element have different numbers of electrons, we should expect atomic radii and ionic radii to be different. The radii of cations are smaller than those of the corresponding neutral atoms. This should be expected because positive ions are formed by removing one or more electrons from the outermost shell. Since these electrons are furthest from the nucleus, their absence will make the cation significantly smaller. In addition, loss of an electron causes a decrease in the amount of electron-electron repulsion which also causes the cation to be smaller than the neutral atom.

In contrast, the radii of anions are larger than those of the corresponding neutral atom. When an electron is added to an atom to form an anion, there is an increase in the electron-electron repulsion. This causes electrons to spread out as much as possible, and so anions have a larger radius than the corresponding atoms. The radii of several metal and nonmetal ions are compared with their atomic radii in Table 8.1.

**Table 8.1** Atomic and Ionic Radii of Elements in Groups 1A and 7A

| Group 1A Radius (pm) | | | Group 7A Radius (pm) | | |
|---|---|---|---|---|---|
| Element | Atomic | Ionic, $X^+$ | Element | Atomic | Ionic, $X^-$ |
| Li | 123 | 60 | F | 64 | 136 |
| Na | 154 | 95 | Cl | 99 | 181 |
| K | 203 | 133 | Br | 111 | 195 |
| Rb | 216 | 148 | I | 128 | 216 |
| Ca | 174 | 99 | O | 73 | 140 |

We can also compare radii within an isoelectronic series. In the series in Table 8.2 the radii decrease steadily as atomic number increases.

**Table 8.2** Atomic and Ionic Radii in an Isoelectronic Series

| Species | $S^{2-}$ | $Cl^-$ | Ar | $K^+$ | $Ca^{2+}$ |
|---|---|---|---|---|---|
| Radius (pm) | 219 | 181 | 154 | 133 | 99 |

Each species has 18 electrons arranged in an argon configuration, and so the electron-electron repulsion is about the same in each ion. The reason for the decrease in radius is that the nuclear charge increases steadily within this series. This causes the electrons to be attracted more strongly toward the nucleus, and the ionic radius to contract as the nuclear charge increases.

**Effective Nuclear Charge.** The explanation of the trends in atomic size and ionization energy is based on the concept of the effective nuclear charge ($Z_{eff}$). The outermost electrons in an atom do not feel the full positive charge of the nucleus. Electrons in inner levels, between the nucleus and the outermost electrons, tend to shield the outermost electrons from the nuclear charge. In a sodium atom, for example, the inner 10 electrons ($1s^2 2s^2 2p^6$) shield the outer 3s electron from the positive charge of 11 protons. The $Z_{eff}$ experienced by the 3s electron is about +1. The $Z_{eff}$ is equal to the nuclear charge Z minus the shielding constant $\sigma$. The shielding constant is essentially equal to the number of inner shell electrons. Therefore, $Z_{eff} = Z - \sigma$ and for Na, $Z_{eff} = 11 - 10 = 1$.

**Atomic Radii.** The $Z_{eff}$ affects the atomic radius in the following way. In proceeding across a period, one proton at a time is added to the nucleus, and one electron is added to the outermost orbital. Electrons within the same energy level do not effectively shield each other from the nucleus. Thus, for magnesium the inner 10 electrons shield the electrons in the 3s orbitals from the +12 charge of the nucleus. $Z_{eff} = 12 - 10 = +2$. One of the 3s electrons does not effectively shield the other 3s electron because they are in the same orbital. Continuing from left to right across the third period, all the outermost electrons are in the n = 3 energy level, and the $Z_{eff}$ (listed below in parentheses) increases by about one unit per element.

Increasing $Z_{eff}$

→

Na(+1)  Mg(+2)  Al(+3)  Si(+4)  P(+5)  S(+6)  Cl(+7)  Ar(+8)

This means a greater force of attraction is experienced by the outermost electrons of atoms of elements on the right side of the periodic table than for those on the left side. Consequently, the atoms gradually decrease in size as we proceed from left to right across a period.

In proceeding from one element to another down a group, each successive element has its outer electrons in a principal energy level with a larger n value. The effective nuclear charge experienced by the outermost electrons is essentially the same for all elements within a group.  The result of this is that the size of the outer orbital is affected most by the value of n.  As you will recall, the size of an orbital increases as n increases. Therefore, atomic radius increases from top to bottom in a group.

***Transition Elements.***  Within a row of transition elements, in contrast to the representative elements, there is only a slight decrease in atomic radius when reading from left to right, as shown in Table 8.3.

This is because electrons are being added to an inner d subshell. In period four, for instance, the outermost electrons occupy a 4s orbital, but each successive electron is added to an inner 3d subshell. As we read across the period, the increasing nuclear charge is effectively shielded by the increase in the number of 3d electrons. Therefore, the outer 4s electrons within a series of the transition elements experience almost a constant $Z_{eff}$ . The effective nuclear charge, $Z_{eff}$ for Sc and Ti, for example are determined as follows:

Scandium: $Z_{eff}$ = 21 − 19 = +2 ($\sigma$ = 18 for [Ar] plus 1 for the 3d electron.)
Titanium:   $Z_{eff}$ = 22 − 20 = +2 ($\sigma$ = 18 for [Ar] plus 2 for the 3d electrons.)

This causes the atomic radius to change only a little across the period.

**Table 8.3**  Atomic Radius (pm)

| Sc | Ti | V | Cr | Mn | Fe | Co | Ni | Cu | Zn |
|----|----|----|----|----|----|----|----|----|----|
| 144 | 132 | 122 | 119 | 118 | 117 | 116 | 115 | 118 | 121 |

Examples
8.5 – 8.6

Exercises
16 – 19

## Ionization Energy and Electron Affinity (Sections 8.4 – 8.5)

***Ionization Energy.***  The minimum energy required to remove an electron from the ground state of an atom in the gas phase is its *ionization energy.* The magnitude of the ionization energy is a measure of how strongly the outermost electron is held by an atom. The greater the ionization energy, the more strongly/tightly the electron is held.

For a many-electron atom, the energies required to remove a second and a third electron are called the second ionization energy and the third ionization energy, respectively. The third ionization energy is always greater than the second, which in turn is always greater than the first ionization energy. The explanation of this trend is related to the electron repulsion among the remaining electrons. The first electron removed comes from a neutral atom. With an electron missing the repulsion among the remaining electrons decreases, and they move closer to the nucleus. This means that to remove the second electron requires a higher ionization energy. The third ionization energy is higher yet due to a further reduction in electron repulsion.

Figure 8.11 in the textbook shows that the ionization energy increases when moving from left to right across a period. Thus, the alkali metals have the lowest ionization energy and the noble gases the highest. Within a group the ionization energy decreases as atomic number increases (Figure 8.3).

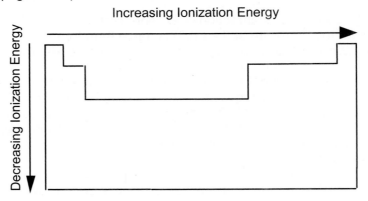

**Figure 8.3** Periodic trends in ionization energy.

***Ionization Energy Trends.*** The importance of the ionization energy is that it correlates closely with electron configuration and chemical properties. When considering ionization energy, it helps to recall that the first electron removed is the one farthest from the nucleus. Within a group of elements the ionization energy decreases with increasing atomic number. This trend is explained in the following way. Down a group of elements the effective nuclear charge is essentially constant. However, the outermost electron resides in increasingly higher energy levels. Therefore, as the atomic number increases within a group, the outermost electron is held more weakly and the ionization energy decreases.

The trend within a period should be related to the effective nuclear charge. In moving across the periodic table from left to right, the size of the atoms decreases due to the increase in effective nuclear charge. The outer electrons become more tightly held as we move from left to right, and the ionization energy must increase.

One characteristic of metals is the relative ease with which electrons can be removed from their atoms. Thus, the lower the ionization energy, the more metallic the element. The metals are located on the left hand side of the periodic table. Also, elements in all groups exhibit increased metallic behavior with increased atomic number.

> Examples
> 8.7 – 8.9
>
> Exercises
> 20 – 23

***Electron Affinity.*** Another property of atoms is one that relates to the formation of negative ions. The electron affinity of an element is related to the energy given off when a gaseous atom gains an additional electron to form a negatively charged ion. A fluorine atom, for example, gives off energy when it gains an electron to form a fluoride ion.

$$F(g) + e^- \rightarrow F^-(g) \quad \Delta H = -328 \text{ kJ/mol}$$

The **electron affinity** is defined as the opposite sign of the energy change. Therefore, for fluorine, the electron affinity is +328 kJ/mol because that is the amount of energy given off.

$$F(g) + e^- \rightarrow F^-(g) + 328 \text{ kJ/mol}$$

The greater the attraction of an atom for an electron, the greater the electron affinity value. The more energy released, the more stable the ion is compared to the atom.

Table 8.3 of the text shows the electron affinity values of elements arranged according to their positions in the periodic table. Figure 8.12 (textbook) is a plot of electron affinity values for the first 20 elements. A clear trend in electron affinity values of elements within a period is not evident. Some trends in electron affinity can be summarized as follows.

1. The electron affinities of the Group 6A and 7A elements are much greater than those of other elements.
2. The electron affinities of the noble gases are only slightly negative. Therefore the anions of these gases would be unstable.
3. The Group 1A metals have small positive electron affinities. The Group 2A metals have very small positive values and some negative electron affinities.
4. The electron affinities of metals are generally lower than those of nonmetals.

***Electron Affinity Trend.***  The trend in electron affinities is also related to effective nuclear charge, and to the energy of the orbital that the electron will enter. The Group 6A and 7A elements have the highest electron affinities. These elements have high $Z_{eff}$ and the added electron(s) enters the valence shell. The noble gases with even higher $Z_{eff}$ have no affinity for an additional electron. With filled s and p subshells, noble gas atoms have no tendency to add an electron because the next available orbital is in the next higher energy level beyond the valence shell.

## Types of Elements (Section 8.6)

All elements are classified in three broad groups according to their chemical and physical properties: 1) metals, 2) nonmetals, and 3) metalloids. Within this classification, certain groups of elements are known by common names, such as the alkali metals, alkaline earth metals, coinage metals, halogens, and noble gases.

***Metals.***  Most of the elements are metals. In general, they appear in the left and center of the periodic table. All metallic elements except mercury are solids at 25 °C. They have a lustrous appearance, are good conductors of heat and electricity, can be hammered or rolled into sheets (a property referred to as malleability), and can be drawn into wires (a property referred to as ductility). Metals generally have high densities and high melting points.

Metals have low ionization energies, low electron affinities, and relatively large atomic radii. All metallic elements combine with nonmetals such as oxygen and chlorine to form salts. The most reactive metals are at the left of the periodic table. The transition metals are less reactive than the Group 1A and 2A metals. Elements within a chemical group are more metallic as atomic mass increases.

***Nonmetals.***  Eighteen of the elements are nonmetals. These elements are on the right side of the periodic table. At 25 °C, eleven are gases, one (bromine) is a liquid, and the rest are brittle solids. Typically, their densities and melting points are low.

The atoms of nonmetal elements have high ionization energies and high electron affinities. Nonmetals combine with metals to form ionic compounds, and with other nonmetal elements to form molecular compounds. The nonmetals, except hydrogen, are located on the upper right hand side of the periodic table. Recall that hydrogen is a gas and exists as diatomic molecules, as do a number of nonmetals.

***Metalloids.***  Several elements have some properties that are characteristic of metals and some that are like those of nonmetals. These elements are called metalloids. Many periodic tables show a zig-zag line separating the metals from nonmetals. Refer to the periodic table on page 26 of this study guide. The elements that border this line on both sides are metalloids (except for Al, which is a metal). The metalloids have ionization energies and electron affinity values intermediate between metals and nonmetals.

***Alkali Metals.*** Certain groups of elements are known by common names (Figure 8.4). Some of the characteristics of the alkali metals, alkaline earth metals, coinage metals, halogens, and noble gases are described below.

The elements in Group 1A, with the exception of hydrogen, are called alkali metals. The $ns^1$ electron in the highest principal energy level is well shielded from the nucleus and is easily lost. Their low ionization energies make the alkali metals the most active family of metals. These elements are found in nature as +1 ions in chemical combination with nonmetal ions and polyatomic ions. The densities of these elements are low, in part, because of their large radii. Li, Na, and K are even less dense than water.

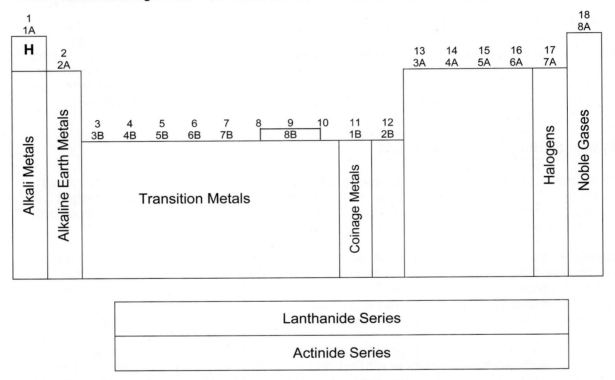

**Figure 8.4** Common names of several groups of elements.

***Alkaline Earths.*** Group 2A elements are called alkaline earth metals. The outermost electron configuration is $ns^2$. Although these two are not as easily lost as the $ns^1$ electron in alkali metal atoms, they are readily given up. Alkaline earth metals exist in nature as +2 ions in chemical combination with negative ions.

***Halogens.*** Group 7A elements are called the halogens. None of these elements is ever found free in nature. All have high electron affinities and so tend to acquire one electron to form −1 ions. The name halogen means "salt-formel." These nonmetals exist in the elemental form as diatomic molecules. At 25 °C, $F_2$ and $Cl_2$ are gases, $Br_2$ is a liquid, and $I_2$ is a solid.

***Noble Gases.*** Group 8A elements make up the noble gas elements. The term noble here means nonreactive. He, Ne, and Ar do not form any chemical compounds. Since 1960, a number of Kr and Xe compounds have been synthesized. The chemical inactivity of the noble gases is the result of the $ns^2np^6$ electron configuration of the outermost electrons in their atoms. The noble gas elements have negative electron affinities and high ionization energies.

***Coinage Metals.***  The elements of Group 1B, copper, silver, and gold, are generally nonreactive and are called the noble metals or coinage metals. These metals are excellent conductors of heat and electricity.

Atoms of the coinage metals have one electron in the outer s subshell and 10 electrons in the underlying d subshell. The electron configurations of their outermost electrons are $ns^1 (n - 1)d^{10}$. Note that this is an exception to the configuration expected from the Aufbau principle, which predicts $ns^2(n - 1)d^9$. It is as if the $(n - 1)d$ orbitals "borrowed" an electron from the higher energy ns orbital. For the coinage metal atoms, the "borrowed" electron is used to complete an inner subshell. The completed d subshell has an enhanced stability that corresponds to a lowering of the energy of the atom. Thus, $ns^1(n - 1)d^{10}$ is a lower energy configuration than $ns^2(n - 1)d^9$.

In copper, silver, and gold this one outer electron is held much more tightly than the outermost electron in the alkali metal atoms. This can be most easily seen by comparing first ionization energies as shown in Table 8.4.

| Examples 8.10, 8.11 |
| --- |
| Exercises 27 – 29 |

**Table 8.4**  Comparison of the First Ionization Energies of Coinage Metals and Alkali Metals

| Period | Element | $I_1$ (kJ/mol) |
| --- | --- | --- |
| 4 | K | 419 |
|   | Cu | 745 |
| 5 | Rb | 403 |
|   | Ag | 731 |
| 6 | Cs | 375 |
|   | Au | 890 |

## GLOSSARY LIST

| | | |
| --- | --- | --- |
| representative elements | ionic radius | electron affinity |
| valence electrons | isoelectronic | diagonal relationship |
| atomic radius | ionization energy | amphoteric oxide |

## WORKED EXAMPLES

---

**EXAMPLE 8.1  Element Groups**

Given the following formulas of compounds:

       Sodium sulfate $Na_2SO_4$
       Magnesium oxide $MgO$
       Aluminum chloride $AlCl_3$

Write the formulas of the following compounds using the given formulas and the periodic table:
a.  Calcium oxide
b.  Potassium sulfate
c.  Gallium bromide

• **Solution**

a.  Notice that Ca is in the same element group as Mg.  Among the given formulas, we find MgO. Since one Mg ion combines with one oxide ion, then one Ca ion will combine with one oxide ion. The formula is CaO.
b.  If sodium sulfate is $Na_2SO_4$, then potassium, being in the same chemical group as sodium, should form a sulfate with the same ratio of potassium atoms to sulfate ions. The formula is $K_2SO_4$.
c.  Gallium is in Group 3A along with Al, and bromine is a halogen, as is Cl. Therefore, since aluminum chloride has the formula $AlCl_3$, gallium will combine with any three halogen atoms. The formula of gallium bromide is $GaBr_3$.

---

**Work EXERCISES & PROBLEMS: 1 – 3**

---

**EXAMPLE 8.2  Predicting Properties of Elements**

Predict the missing value for the following:

a.

| Element | Density $(g/cm^3)$ |
|---------|---------------------|
| Ca      | 1.55                |
| Sr      | ?                   |
| Ba      | 3.5                 |

b.

| Element | Density $(g/cm^3)$ |
|---------|---------------------|
| Cl      | 0.099               |
| Br      | 0.114               |
| I       | ?                   |

• **Solution**

Within a group of the periodic table, the physical properties vary in a regular fashion as you read down, in the group.

---

a. From the positions of these elements in Group 2A of the periodic table, the density of Sr could be estimated to be half-way between the values for Ca and Ba. The average of 1.55 and 3.5 $g/cm^3$ is 2.5 $g/cm^3$. The observed value is 2.6 $g/cm^3$.

b. In this case, we must extrapolate rather than interpolate as above because I is farther down in Group 7A than Cl and Br. We could assume the change in the radius from Br to I will be the same as the change from Cl to Br. The difference between Cl and Br is 0.015 nm. Adding 0.015 to 0.114 gives the estimated radius for I as 0.129 nm. The observed radius of I is 0.133 nm.

**Work EXERCISES & PROBLEMS: 4**

---

**EXAMPLE 8.3 Electron Configurations and the Periodic Table**

Referring only to the periodic table:
a. Write the electron configuration for the outermost principal energy level (the valence shell) for any Group 5A element.
b. Identify the group whose outermost electrons have the configuration $ns^2np^1$.

**• Solution**

a. The elements in Group 5A are representative elements and so have all inner subshells filled, and the outermost electrons occupy s and p orbitals. The numeral 5 corresponds to five outermost electrons. The first two occupy the ns subshell, and the next three the np subshell. The valence electron configuration of any Group 5A element is $ns^2np^3$.
b. Among the representative elements, those in the first group to the right of the transition metals, Group 3A, have the electron configuration $ns^2np^1$.

---

**EXAMPLE 8.4 Electron Configurations of Ions**

Write the electron configurations of the following ions:
a. $Mg^{2+}$   b. $S^{2-}$   c. $Fe^{2+}$   d. $Fe^{3+}$

**• Solution**

a. A Mg atom has Z = 12, so the atom has 12 electrons, but the $Mg^{2+}$ ion contains only 10 electrons. Write the configuration for Mg.

Mg   $1s^22s^22p^63s^2$  or  $[Ne]3s^2$

Then, remove the two electrons in the outermost energy level to form the 2+ ion.

$Mg^{2+}$   $1s^22s^22p^6$ or $[Ne]$

$Mg^{2+}$ ions are *isoelectronic* with Ne atoms. That is, they have the same number of electrons.

b.  An S atom has 16 electrons and has the electron configuration $1s^2 2s^2 2p^6 3s^2 3p^4$. The sulfur atom becomes the negative $S^{2-}$ ion by gaining two electrons. Adding two electrons to the outermost subshell gives

$S^{2-}$ $1s^2 2s^2 2p^6 3s^2 3p^6$ or [Ar]

c.  As before, write the configuration of the atom.

Fe  $[Ar]4s^2 3d^6$

When forming a cation from an atom of a transition metal, electrons are always removed first from the ns orbital, and then from the (n − 1)d orbital. The first two electrons lost are removed from the outermost principal energy level (highest n value), which is the 4s orbital. This gives:

$Fe^{2+}$  $[Ar]3d^6$

d.  Next, remove one more electron from $Fe^{2+}$. This gives:

$Fe^{3+}$  $[Ar]3d^5$

• **Comment**

The stable ions of all but a few representative elements are isoelectronic to a noble gas. Keep in mind that most transition metals can form more than one cation, and that for the most part, these ions are not isoelectronic with the preceding noble gases.

**Work EXERCISES & PROBLEMS: 5 – 15**

---

**EXAMPLE 8.5  Trends in Atomic Radius**

Which one of the following has the smallest atomic radius?
a. Li     b. Na     c. Be     d. Mg

• **Solution**

Atomic radii increase from top to bottom in a group, therefore, Li atoms are smaller than Na atoms, and Be atoms are smaller than atoms of Mg. Next, compare Be and Li. The atomic radius decreases from left to right within a period of the periodic table, thus, Be atoms are smaller than Li atoms, as well as the other choices given.

---

**EXAMPLE 8.6  Trends in Ionic Radii**

In each of the following pairs, choose the ion with the largest ionic radius.
a.  $K^+$ or $Na^+$
b.  $K^+$ or $Ca^{2+}$
c.  $K^+$ or $Cl^-$

• **Solution**

a.  $K^+$ and $Na^+$ are in the same group in the periodic table. The outer electrons in $K^+$ occupy the third principal energy level, and those in $Na^+$ occupy the second. $K^+$ has the greater ionic radius.
b.  $K^+$ and $Ca^{2+}$ belong to an isoelectronic series; both have 18 electrons. Calcium ions with their greater nuclear charge attract their electrons more strongly than $K^+$ ions and so are smaller. $K^+$ is larger.
c.  Again $K^+$ and $Cl^-$ are isoelectronic species. The ion with the greater nuclear charge ($K^+$) will be smaller. $Cl^-$ has the greater ionic radius.

**Work EXERCISES & PROBLEMS: 16 – 19**

---

**EXAMPLE 8.7  Trends in Ionization Energy**

Which one of the following has the highest ionization energy?
a. K     b. Br     c. Cl     d. S

• **Solution**

Ionization energy increases from left to right within a period. Thus, the value for Cl is greater than for S, and the value for Br is greater than for K. In comparing Cl and Br, Cl has the higher ionization energy value because ionization energy decreases from top to bottom within a group.

---

**EXAMPLE 8.8  Trends in Ionization Energy**

Why does sodium have a lower first ionization energy than lithium?

• **Solution**

When a lithium atom is ionized the electron most easily removed comes from a 2s orbital, whereas in sodium the electron most easily removed comes from a 3s orbital. Because of shielding effects, the effective nuclear charge $Z_{eff}$ experienced by these electrons is about +1 in both atoms. The main reason for a difference in ionization energy in these atoms is the larger distance of separation between the electron and the nucleus in the case of Na atoms. As the principal quantum number n increases, so does the average distance of the electron from the nucleus. Consequently, the electron becomes easier to remove as you read down a group. Na will have a lower ionization energy than lithium.

---

**EXAMPLE 8.9  Ionization Energies**

The first, second, and third ionization energies for calcium are:

$$I_1 = 590 \text{ kJ/mol} \qquad I_2 = 1145 \text{ kJ/mol} \qquad I_3 = 4900 \text{ kJ/mol}$$

Explain why so much more energy is required to remove the third electron from Ca, as compared to removal of the first and second electrons.

**• Solution**

The successive ionization energy values follow the usual trend $I_1 < I_2 < I_3$. However, we must explain the very large difference between $I_2$ and $I_3$. The first two electrons are valence electrons and are removed from the 3s orbital. But the third electron must be removed from the inner 2p subshell. The $n = 2$ principal energy level lies much closer to the nucleus than the $n = 3$ energy level; therefore, its electrons are held much more strongly. Consequently, $I_3 \gg I_2$.

**Work EXERCISES & PROBLEMS: 20 – 23**

**EXAMPLE 8.10  Properties of Group 4A Elements**

Write the electron configurations for the atoms of all elements in Group 4A. Specify whether the element is a metal, a nonmetal, or a metalloid.

**• Solution**

First, see Figure 2.10 (text) for the type of element. The outermost principal energy level of atoms in Group 4A contains four electrons. The general electron configuration for elements in this group is $ns^2np^2$. Their electron configurations are

| | |
|---|---|
| C $1s^2 2s^2 2p^2$ | nonmetal |
| Si $[Ne]3s^2 3p^2$ | metalloid |
| Ge $[Ar] 4s^2 3d^{10} 4p^2$ | metalloid |
| Sn $[Kr] 5s^2 4d^{10} 5p^2$ | metal |
| Pb $[Xe] 6s^2 4f^{14} 5d^{10} 6p^2$ | metal |

**EXAMPLE 8.11  Properties of Potassium and Chlorine**

Compare the magnitudes of electron affinity and ionization energy for the alkali metal potassium with the halogen chlorine. Comment on their relative abilities to form ions.

**• Solution**

From Table 8.3 (text) the electron affinities of K and Cl are 48 and 349 kJ/mol, respectively. Table 8.2 (text) gives the ionization energies of K and Cl as 419 and 1251 kJ/mol, respectively. These values suggest that the metal atom K can lose an electron and form a $K^+$ ion much more easily than can the nonmetal Cl. And the nonmetal atom Cl can attract an electron to form a $Cl^-$ much more readily than can the metallic K atom.

**Work the rest of the EXERCISES & PROBLEMS**

## EXERCISES & PROBLEMS

1.   What is a representative element? Give three examples of representative elements.

2.   What is a transition element? Give three examples of transition elements.

3. What is an inner transition element? Give two examples of inner transition elements.

4. Estimate the melting point of $Br_2(s)$ given that the melting points of $Cl_2(s)$ and $I_2(s)$ are −101 °C and 114 °C, respectively.

5. Write the outer electron configurations for all the elements with the same general configuration as copper.

6. Write the outer electron configurations for:
   a. Group 3A elements.  b. Group 6A elements.

7. What group of elements has the outer electron configuration $ns^2np^5$?

8. What group of elements has the outer electron configuration $ns^2(n-1)d^2$?

9. Write the electron configuration of scandium (Sc). Classify the element as a representative, transition, or inner transition element. Classify the element as a metal or nonmetal.

10. Name the groups of the periodic table in which each of the following elements are found.
    a. $[Ar]4s^2$   b. $[Ar]4s^24p^3$   c. $[Kr]5s^24d^7$

11. Which of the following species are isoelectronic with each other?
    a. Kr   b. $Cl^-$   c. $Rb^+$   d. $K^+$   e. $Cu^{2+}$   f. $Se^{2-}$   g. $Zn^{2+}$

12. Write the ground state electron configurations of the following ions:
    a. $K^+$   b. $O^{2-}$   c. $Sr^{2+}$   d. $I^-$   e. $Li^+$   f. $Mg^{2+}$

13. Write the ground state electron configurations of the following transition metal ions.
    a. $Fe^{2+}$   b. $Fe^{3+}$   c. $Cu^{2+}$   d. $Cu^+$   e. $Ti^{2+}$   f. $Ti^{4+}$

14. Write the ground state electron configurations of the following ions.
    a. $Sc^{3+}$   b. $V^{5+}$   c. $Pb^{2+}$   d. $Pb^{4+}$

15. Identify any isoelectronic pairs among the following.
    $Na^+$, Ar, $Cl^-$, Ne, $Se^{2-}$

16. Which of the following has the largest radius?
    a. Na, Cl, Ne, S, Li
    b. C, Si, Sn, Pb

17. Which of the following has the largest radius?
    $S^{2-}$, Ar, $Se^{2-}$, $O^{2-}$, $Al^{3+}$

18. Which of the following has the largest radius?
    $Na^+$, $Mg^{2+}$, $Al^{3+}$, Ne

19. Which species is larger:  a. $Co^{2+}$ or $Co^{3+}$   b. S or $S^{2-}$

20. What group of elements in the periodic table have:
    a. the highest ionization energies?   b. the lowest ionization energies?

21. Based on periodic trends, which one of the following elements has the lowest ionization energy?

    K, S, Se, Li, Br

22. Based on periodic trends, which one of the following elements has the largest ionization energy?

    Cl, K, S, Se, Br

23. For silicon atoms, which ionization energy value (I) will show an exceptionally large increase over the preceding ionization energy value?

    $I_2$   $I_3$   $I_4$   $I_5$   $I_6$

24. Define electron affinity.

25. Which one of the following elements has both a low ionization energy and a positive electron affinity?

    K, Ne, Br, Fe, N

26. Which of the following atoms has both a high ionization energy and a large positive electron affinity?

    K, Ne, Br, Fe, N

27. Of the following, which is the most metallic element?

    V, Ge, Se, As, Zn

28. Which of the following is the least metallic element?

    V, Ge, Al, As, Ca

29. The first ionization energies of boron and silicon are 801 and 786 kJ/mol, respectively. The similarities of these values is an example of what relationship?

30. What charges do you expect for the ions of the alkali metals and the alkaline earth metals?

31. What charges do you expect for the ions of the Group 6A and the halogen elements?

32. Explain why the radii of atoms increase when proceeding down a group of the periodic table.

33. Explain why the ionization energies of atoms increase when proceeding from left to right across a period of the periodic table.

34. Why is an emphasis placed on the valence electrons in an atom when discussing its atomic properties?

## PRACTICE TEST QUESTIONS
See notes on taking practice test in the Preface

1. The element francium is extremely rare, and little is known about its chemical and physical properties. Use the following data to estimate its density and first ionization energy.

|  | K | Rb | Cs |
|---|---|---|---|
| Density (g/cm$^3$) | 0.86 | 1.53 | 1.87 |
| 1$^{st}$ IE (kJ/mol) | 419 | 403 | 375 |

2. The element technetium (Z = 43) does not occur naturally on earth. The densities of Mo and Ru are 10.2 and 12.4 g/cm$^3$, respectively. Estimate the density of Tc.

3. The periodic table has been extended to include the transuranium elements (Z > 92). What elements would element number 104 be similar to in chemical properties?

4. Which is the electron configuration for the outermost electrons of elements in Group 4A?

   a. $ns^1$        b. $ns^2$

   c. $ns^2np^4$     d. $ns^2np^2$

   e. $ns^2np^6nd^7$

5. In what group of the periodic table is each of the following elements found?

   a. $1s^22s^22p^6$       b. [Ar]$4s^1$

   c. [Xe]$6s^24f^{14}5d^5$     d. [Ne]$3s^23p^5$

6. Use the periodic table to write the electron configurations of Cr, Sb, and Pb.

7. How many valence electrons does an arsenic atom have?

8. Successive ionization energies—1st, 2nd, 3rd, etc.—always show an increasing trend. $I_1 < I_2 < I_3 < I_n$. For aluminum atoms, which ionization energy value will show an exceptionally large increase over the preceding ionization energy value?

   a. 2nd    b. 3d    c. 4th    d. 5th    e. 6th

9. Write the electron configurations for the following ions.

   a. $Ca^{2+}$        b. $Se^{2-}$

   c. $Cl^-$         d. $Mn^{2+}$

   e. $Co^{3+}$       f. $Sc^{3+}$

10. An Ar atom is isoelectronic with which one of the following?

   a. Ne        b. K

   c. $Sc^{3+}$      d. $Cl^-$

   e. $Na^+$

11. Which atom should have the largest radius?
    a. Br          b. Cl
    c. Se          d. Ge
    e. C

12. Which is the larger ion or atom in each pair?
    a.  I⁻ or $Cs^+$
    b.  Ne or $K^+$
    c.  Mg or $Mg^{2+}$

13. Which atom should have the greatest 1st ionization energy?
    a. Se          b. Te
    c. Na          d. Si
    e. S

14. Why does atomic radius decrease in going from left to right across a row in the periodic table?

15. Which two of the following would be most likely to have similar ionization energies?
    B, C, Si, Al, Ar

# Chapter 9.   Chemical Bonding I:  Basic Concepts

**Lewis Dot Symbols (Sections 9.1 – 9.2)**
**Ionic Bonding and the Lattice Energy (Section 9.3)**
**Covalent Bonding and Lewis Structures (Sections 9.4 & 9.6)**
**Electronegativity (Section 9.5)**
**Formal Charge (Section 9.7)**
**The Concept of Resonance (Section 9.8)**
**Exceptions to the Octet Rule (Section 9.9)**
**Bond Enthalpies (Section 9.10)**


## SUMMARY

### Lewis Dot Symbols (Sections 9.1 – 9.2)

***Lewis Symbols.***  Ionic and covalent bonds are important types of chemical bonds.  In this chapter, you will learn to use electron configurations and the periodic table to predict the type and the number of bonds an atom of a particular element can form.  Electron configurations can be used to write the Lewis dot symbols of the representative elements.  A Lewis dot symbol of an element consists of the chemical symbol with one or more dots placed around it.  Each dot represents a valence electron.  The orbital diagram and the Lewis symbol for the fluorine atom are shown in Figure 9.1.  Fluorine has 7 electrons in its outermost principal energy level (n = 2), and therefore, has 7 dots in its Lewis symbol.

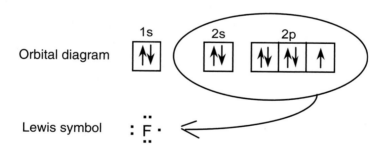

**Figure 9.1** Orbital diagram and Lewis symbol for the fluorine atom.

The Lewis dot symbols and the electron configurations of the outermost electrons of elements in the third period are as follows:

$$Na\cdot \quad \cdot Mg\cdot \quad \cdot \overset{\cdot}{Al}\cdot \quad \cdot \overset{\cdot}{\underset{\cdot}{Si}}\cdot \quad \cdot \overset{\cdot\cdot}{\underset{\cdot}{P}}\cdot \quad \cdot \overset{\cdot\cdot}{\underset{\cdot\cdot}{S}}\cdot \quad :\overset{\cdot\cdot}{\underset{\cdot\cdot}{Cl}}\cdot \quad :\overset{\cdot\cdot}{\underset{\cdot\cdot}{Ar}}:$$

$$3s^1 \quad\quad 3s^2 \quad\quad 3s^23p^1 \quad 3s^23p^2 \quad 3s^23p^3 \quad 3s^23p^4 \quad 3s^23p^5 \quad 3s^23p^6$$

For representative elements, the number of valence electrons is the same as the group number.

The representative metals form ions by losing all of their valence electrons. Lewis symbols can be used to represent the formation of positive magnesium and aluminum ions as follows:

179

$$\cdot Mg \cdot \ \rightarrow \ Mg^{2+} + 2e^-$$

$$\cdot \overset{\cdot}{Al} \cdot \ \rightarrow \ Al^{3+} + 3e^-$$

The nonmetal elements form negative ions by acquiring electrons until they are isoelectronic with a noble gas atom. Ions of nonmetal have 8 electrons in their valence shells.

$$: \overset{\cdot}{\underset{\cdot}{S}} : + 2e^- \ \rightarrow \ : \overset{\cdot\cdot}{\underset{\cdot\cdot}{S}} : ^{2-}$$

$$: \overset{\cdot\cdot}{\underset{\cdot\cdot}{Cl}} \cdot + e^- \ \rightarrow \ : \overset{\cdot\cdot}{\underset{\cdot\cdot}{Cl}} : ^-$$

| Examples 9.1, 9.2 |
| Exercises 1 – 2 |

Sulfide and chloride ions, for example, have net charges of –2 and –1 because the atoms needed $2e^-$ and $1e^-$, respectively, to become isoelectronic with the noble gas argon.

## Ionic Bonding and the Lattice Energy (Section 9.3)

***Coulomb's Law.*** The force that gives rise to the ionic bond is the electrical attraction existing between a positive ion and a negative ion. Chemical bonding lowers the energy of two interacting atoms (or ions). Coulomb's law states that the potential energy (E) of interaction of two ions is directly proportional to the product of their charges and inversely proportional to the distance between them:

$$E = k\frac{Q_{cation}Q_{anion}}{r}$$

where $Q_{cation}$ and $Q_{anion}$ are the charges of the two ions, r is their distance of separation, and k is a proportionality constant (its value will not be needed). When one ion is positive and the other negative, E will be negative. Bringing two oppositely charged particles closer together lowers their energy. The lower the value of the potential energy, the more stable is the pair of ions. Energy would need to be added to separate the two ions.

The factors that govern the stability of ion pairs are the magnitude of their charges, and the distance between the ion centers. The distance between ionic centers (r) is the sum of the ionic radii of the individual ions.

$$r = r_{anion} + r_{anion}$$

---

**The energy of attraction between two oppositely charged ions depends:**

- directly on the magnitude of the ion charges. The greater the ion charges, the stronger the attraction. The interaction of $Mg^{2+}$ and $O^{2-}$ ions is much greater than that of $Na^+$ and $Cl^-$ ions.

- inversely on the distance between ion centers. The distance between ions depends on the sizes of the ions involved. As the sum of the ionic radii increases, the interaction energy decreases.

***Lattice Energy.*** In an ionic crystal, the cations and anions are arranged in an orderly three-dimensional array, as illustrated in Figure 2.13 of the textbook. In a sodium chloride or rock salt type of crystal, the anion has six nearest neighbor cations. The ions are packed in such a way as to maximize attraction and minimize repulsion.

The lattice energy provides a measure of the attraction between ions and the strength of the ionic bond. The lattice energy is the energy required to separate the ions in 1 mole of a solid ionic compound into gaseous ions. Gaseous ions are far enough apart from one another that they do not interact. For example, the lattice energy of NaF(s) is equal to 908 kJ/mol. This means 908 kJ are required to vaporize one mole of NaF(s) and form one mole of $Na^+$ ions and one mole of $F^-$ ions in the gas phase.

$$NaF(s) \rightarrow Na^+(g) + F^-(g) \qquad \text{lattice energy} = 908 \text{ kJ}$$

The lattice energy is related to the stability of the ionic solid. The larger the lattice energy, the more stable the solid.

The value of the lattice energy depends on the charges of the ions and the ionic radii, in accordance with Coulomb's law. This can be seen in Table 9.1 by comparing the sodium halides. As the sum of the two ionic radii increases in going from NaF to NaI, the separation of ionic centers ($r = r_{cation} + r_{anion}$) increases and the lattice energy decreases. The effect of ionic charge on lattice energy can be seen by comparing the lattice energies of NaF and CaO in Table 9.1. The larger value for CaO is due to the stronger attraction of doubly charged cations ($Ca^{2+}$) for doubly charged anions ($O^{2-}$).

The lattice energies are reflected in the melting points of ionic crystals. During melting the ions gain enough kinetic energy to overcome the potential energy of attraction, and they move away from each other. The higher the melting point, the more energy the ions need to separate from one another. Therefore, as the lattice energy increases, so does the melting point of the compound.

**Table 9.1** Lattice Energies of Several Ionic Compounds

|       | $r_+ + r_-$* (pm) | Lattice Energy (kJ/mol) | Melting Point (°C) |
|-------|-------------------|--------------------------|---------------------|
| NaF   | 231               | 908                      | 1012                |
| NaCl  | 276               | 788                      | 801                 |
| NaBr  | 290               | 736                      | 747                 |
| NaI   | 314               | 686                      | 660                 |
| CaO   | 239               | 3540                     | 2580                |

\* See Figure 8.9 (textbook) for ionic radii values.

***Factors Favoring Formation of Ionic Bonds.*** It is important to identify any properties of atoms that affect their ability to form ionic compounds. In the formation of ions from atoms, the metal atom loses an electron and the nonmetal atom gains an electron. We will examine the formation of LiF as an example.

$$Li(g) \rightarrow Li^+(g) + e^-$$
$$e^- + F(g) \rightarrow F^-(g)$$

The overall change is: $\qquad Li(g) + F(g) \rightarrow Li^+(g) + F^-(g)$

One factor favoring the formation of an ionic compound is the ease with which the metal atom loses an electron. A second factor is the tendency of the nonmetal atom to gain an electron. Thus, ionic compounds tend to form between elements of low ionization energy and those of high electron affinity. The alkali metals and alkaline earth metals have low ionization energies. They tend to form ionic compounds with the halogens and Group 6A elements, both of which have high electron affinities.

A third factor is the lattice energy. Electrostatic attraction of the ions results in large amounts of energy being released when two kinds of gaseous ions are brought together to form a crystal lattice.

$$Li^+(g) + F^-(g) \rightarrow LiF(s) + energy \qquad \text{(Energy released = -lattice energy)}$$

In general, the smaller the ionic radius and the greater the ionic charge, the greater the lattice energy.

### Calculation of the Lattice Energy Using the Born-Haber Cycle.

Lattice energies cannot be measured directly and must be calculated using Hess's law. In Chapter 6, we saw that if a reaction can be broken down into a series of steps, the overall enthalpy of reaction is equal to the sum of the enthalpy changes for the individual steps. The series of steps used to calculate the lattice energies of ionic solids is called the Born-Haber cycle. Each step is one you've seen previously in relation to the properties of atoms. Here, we will illustrate the calculation of the lattice energy for potassium chloride.

$$KCl(s) \rightarrow K^+(g) + Cl^-(g) \qquad \text{lattice energy} = ?$$

The Born-Haber cycle starts by taking the overall equation to be the reaction in which the ionic compound is formed from the elements in their standard states. In this example, the enthalpy change is the same as the enthalpy of formation of potassium chloride from potassium and chlorine.

$$K(s) + \tfrac{1}{2} Cl_2(g) \rightarrow KCl(s) \qquad \Delta H^{\circ}_{overall} = \Delta H^{\circ}_f \text{ (KCl)}$$

This reaction is then envisioned to occur by a number of steps.

1. First, potassium sublimes:

$$K(s) \rightarrow K(g) \qquad\qquad \Delta H^{\circ}_1 = \Delta H_{subl}$$

2. Next, diatomic chlorine is dissociated into Cl atoms. The required energy is the bond enthalpy for 1/2 mole of Cl—Cl bonds.

$$\tfrac{1}{2} Cl_2(g) \rightarrow Cl(g) \qquad\qquad \Delta H^{\circ}_2 = \tfrac{1}{2} BE(Cl—Cl)$$

To this point, the focus has been on making the atoms of the elements potassium and chlorine. Next, these will be made into the appropriate ions.

3. Ionize 1 mole of potassium atoms. This is the 1$^{st}$ ionization energy of potassium.

$$K(g) \rightarrow K^+(g) + e^- \qquad\qquad \Delta H^{\circ}_3 = I_1$$

4. To form a Cl$^-$ ion, an electron is added to the chlorine atom. This will release an energy equal to the electron affinity of chlorine, EA. Because energy is given off, $\Delta H$ has the opposite sign of the electron affinity.

$$Cl(g) + e^- \rightarrow Cl^-(g) \qquad\qquad \Delta H^{\circ}_4 = -EA$$

5. Finally, 1 mole of gaseous $K^+$ ions, and 1 mole of gaseous $Cl^-$ ions are combined to make 1 mole of KCl(s). An amount of energy equal to the lattice energy will be released. The lattice energy must have the same magnitude as $\Delta H_5^\circ$, but an opposite sign.

$$K^+(g) + Cl^-(g) \rightarrow KCl(s) \qquad \Delta H_5^\circ = - \text{ lattice energy}$$

The summation of the steps gives the overall reaction above.

$$K(s) + \tfrac{1}{2} Cl_2(g) \rightarrow KCl(s) \quad \Delta H_{overall}^\circ = \Delta H_f^\circ (KCl)$$

Therefore, in general:

$$\Delta H_{overall}^\circ = \Delta H_1^\circ + \Delta H_2^\circ + \Delta H_3^\circ + \Delta H_4^\circ + \Delta H_5^\circ$$

or

$$\Delta H_f^\circ (KCl) = \Delta H_{subl} \text{(of K)} + \tfrac{1}{2} BE \text{ (of } Cl_2) + I_1 \text{ (of K)} - EA \text{ (of Cl)} - \text{lattice energy(of KCl)}$$

$$-\text{lattice energy} = \Delta H_5^\circ = \Delta H_{overall}^\circ - [\Delta H_1^\circ + \Delta H_2^\circ + \Delta H_3^\circ + \Delta H_4^\circ]$$

> Examples 9.3, 9.4
>
> Exercises 3 – 6

## Covalent Bonding and Lewis Structures (Sections 9.4 & 9.6)

***Octet Rule.*** In our study of the periodic table we saw that the valence electron configuration was related to the chemical and physical properties of an element. The noble gas elements are the least reactive, and therefore, the most stable group of elements. The Lewis dot symbols of the noble gas elements show eight valence electrons corresponding to filled s and p subshells. G. N. Lewis reasoned that when atoms enter into chemical combination they become more stable. He proposed that atoms gain or lose electrons until they have the same number of valence electrons as noble gas atoms, that is, eight. The octet rule states that when forming bonds, atoms of the representative elements tend to gain, lose, or share electrons until they have eight electrons in the valence shell.

Molecules are held together by bonds resulting from the sharing of electrons between two atoms in a manner that is consistent with the octet rule. A simple covalent bond is formed when two atoms in a molecule share a pair of electrons.

The formation of a covalent bond in hydrogen chloride can be represented with Lewis structures:

$$H\cdot + \cdot \overset{\cdot\cdot}{\underset{\cdot\cdot}{Cl}}\colon \rightarrow H\colon\overset{\cdot\cdot}{\underset{\cdot\cdot}{Cl}}\colon \quad or \quad H\!-\!\overset{\cdot\cdot}{\underset{\cdot\cdot}{Cl}}\colon$$

where the dash represents a covalent bond, or a pair of electrons shared by both the H atom and the Cl atom. By sharing the electron pair, the Cl atom has eight valence shell electrons. The stability of this bond results from both atoms acquiring noble gas configurations. Notice that hydrogen is an exception to the octet rule. Rather than achieving an octet, it needs only two electrons to achieve a filled outer energy level. The H atom becomes isoelectronic with helium. The electron pairs on the Cl atom that are not involved in bonding are called lone pairs, unshared pairs, or nonbonding electrons. The sharing of valence electrons in methane and carbon tetrachloride is shown in Figure 9.2. The circles represent the valence shells of the atoms. They help to point out that each atom achieves an octet of valence electrons by sharing one or more pairs of electrons.

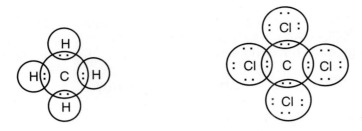

**Figure 9.2** Sharing of electron pairs in $CH_4$ and $CCl_4$.

In some cases two or three pairs of electrons are shared by two atoms in order to reach an octet. In these molecules, *multiple bonds* exist. A *double bond* is a covalent bond in which two pairs of electrons are shared between two atoms, as between C and O in formaldehyde.

$H_2CO$ (formaldehyde)          H : C :: O     or     H—C = O

In general, atoms joined by a double bond lie closer together than atoms joined by a single bond. The C = O bond length is shorter than the C—O bond length.

Nitrogen molecules ($N_2$) contain a triple bond.

:N :: N:     or     :N ≡ N:

***Lewis Structures.*** Lewis structures represent the covalent bonding and location of unshared electron pairs within molecules and polyatomic ions. The steps for writing Lewis structures are as follows:

1. Arrange the atoms in a reasonable skeletal form, placing the unique atom in the center. Determine what atoms are bonded to each other.
2. Count the valence electrons. For polyatomic anions remember to add one electron for each unit of negative charge.
3. Connect the central atom to the surrounding atoms with single bonds. Add unshared pairs to the atoms bonded to the central atom to complete their octets (except for hydrogen, of course).
4. Count the number of electrons used in step 3. If the octet rule is satisfied for each atom, and the total number of electrons is correct, stop here since the structure can be considered correct. If too many electrons were used in step 3, go on to step 5. If too few electrons are used this will be covered in the discussion of expanded octets on page 189.
5. In some cases, to satisfy the octet rule with single bonds, too many electrons are used. To correct this structure, we must write double or triple bonds between the central atom and the surrounding atoms. To make a double bond, move one of the unshared pairs from the surrounding atom to make the additional bond. For every 2 electrons too many used in step 4 (over then number actually contained in the molecule or polyatomic ion) determined in step 2, add one additional bond.

Example
9.5

Exercises
7 – 8

## Electronegativity (Section 9.5)

***Electronegativity.*** Chemical bonds are rarely pure covalent or completely ionic. Rather, most bonds exhibit some characteristics of both. In the previous chapter, we saw that atoms of the elements exhibit varying tendencies in their ability to attract and hold free electrons in the gas phase. In other words, electron affinity values show periodic variations. The term, electronegativity, is used to describe the ability of an atom within a molecule to attract a shared electron pair toward itself.

Linus Pauling developed a method for determining the relative electronegativities of the elements. These values are given in Figure 9.5 of the textbook. Pauling assigned the value 1.0 to Li and 4.0 to F. The values for second and third row elements are given in Table 9.2. Electronegativity values exhibit periodic behavior. In general, electronegativities increase from left to right across a period, and decrease within a group from top to bottom as shown in Figure 9.3.

**Table 9.2** Electronegativities of Second and Third Row Representative Elements

| Second Row | Li | Be | B | C | N | O | F |
|---|---|---|---|---|---|---|---|
| | 1.0 | 1.5 | 2.0 | 2.5 | 3.0 | 3.5 | 4.0 |
| Third Row | Na | Mg | Al | Si | P | S | Cl |
| | 0.9 | 1.2 | 1.5 | 1.8 | 2.1 | 2.5 | 3.0 |

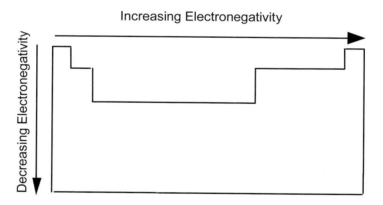

**Figure 9.3** Periodic trends in electronegativity.

As a consequence of the differing abilities of atoms in a bond to attract the shared electron pair, most electron pairs are not shared equally. This imbalance causes an electron pair to shift slightly toward the more electronegative atom, giving rise to a polar covalent bond. In the HCl molecule, for instance, the electronegativity $\chi$ of Cl is 3.0, and for H it is 2.1. The electronegativity difference $\Delta\chi$ is:

$$\Delta\chi = \chi_{Cl} - \chi_H$$
$$= 3.0 - 2.1 = 0.9$$

The chlorine atom with its higher electronegativity attracts the electron pair more strongly. This makes the Cl atom slightly negative and H slightly positive.

$$\overset{\delta+}{\text{H}}\!\!-\!\!\overset{\delta-}{\text{Cl}}$$

Here, $\delta$ denotes a partial charge, that is, a charge less than 1.0, as it would be in an ion.

Pure covalent bonding, which is the equal sharing of electron pairs, occurs in homonuclear diatomic molecules. Examples are $H_2$, $N_2$, and $Cl_2$. In a diatomic molecule with both atoms the same, $\Delta\chi$ must be zero, and the bonding electron pair is shared equally. Bonds of this type are described as nonpolar covalent bonds, or pure covalent bonds.

In bonds involving different atoms, the electronegativity difference will depend on the relative electronegativities of the two atoms. When $\Delta\chi \geq 2.0$, a bond is mostly ionic, for in this case one atom exerts so much more influence on electrons that they can be considered to be completely transferred to the more electronegative atom. Bonds between atoms such that $\Delta\chi < 2.0$ are classified as polar covalent bonds, or simply polar bonds. The "2.0 rule" is an approximation and does not apply to all ionic compounds.

Nonpolar covalent bonds and completely ionic bonds represent extreme situations in bonding. To refer to a bond as being "ionic" or "covalent" is an oversimplification. Sometimes, the term *"percent ionic character"* is used to describe the polar nature of a bond. A purely covalent bond has zero percent ionic character. Bonds that have $\Delta\chi \geq 2.0$ have at least 50% ionic character and are called ionic. There are no 100% ionic bonds. Compounds of Group 1A or 2A metals with a nonmetal usually have ionic bonds, whereas compounds of two nonmetals usually have polar covalent bonds. See Table 9.3 for examples.

**Table 9.3**  Bond Character of Some Common Bonds

| Ionic Compounds | $\Delta\chi$ |
|---|---|
| NaCl | 2.1 |
| NaBr | 1.9 |
| NaI | 1.6 |
| KCl | 2.2 |
| KBr | 2.0 |
| $MgCl_2$ | 1.8 |

| Polar Covalent Bonds | $\Delta\chi$ |
|---|---|
| C—Cl | 0.5 |
| N—H | 0.9 |
| O—H | 1.4 |
| S—H | 0.4 |
| S—O | 1.0 |
| N—O | 0.5 |

Examples
9.6 – 9.8

Exercises
9 - 13

## Formal Charge (Section 9.7)

*Formal Charges on Lewis Structures.*  The concept of formal charge provides a rational basis for choosing the more plausible Lewis structure from among several possibilities. The formal charge is the charge that an atom seems to have in a Lewis structure. When determining the formal charge, all nonbonding electrons count as belonging entirely to the atom in which they are found. All bonding electrons are divided equally between the bonded atoms. Thus, the formal charge of an atom in a Lewis structure is the number of valence electrons in an isolated atom minus the number of electrons assigned to that atom in a molecule. The formula for the formal charge of an atom is:

$$\text{formal charge} = \left(\begin{array}{c}\text{number of}\\\text{valence electrons}\end{array}\right) - \left(\begin{array}{c}\text{number of}\\\text{nonbonding electrons}\end{array}\right) - \frac{1}{2}\left(\begin{array}{c}\text{number of}\\\text{bonding electrons}\end{array}\right)$$

It will be good to keep in mind that the formal charge is really more a property of a structural formula than that of the species the formula represents. Formal charges do not indicate actual charges on atoms in the real molecule.

Two possible Lewis structures for $BF_3$ are:

$$: \ddot{F}-B-\ddot{F}: \qquad \text{and} \qquad : \ddot{F}-B=\ddot{F}$$
$$\quad\ \ | \qquad\qquad\qquad\qquad\qquad |$$
$$\quad\ \ :\ddot{F}: \qquad\qquad\qquad\qquad :\ddot{F}:$$

$$\qquad (1) \qquad\qquad\qquad\qquad\qquad (2)$$

The formal charges in (1) are:

> The boron atom:   formal charge = 3 − 0 − 1/2 (6) = 0
> The fluorine atom:   formal charge = 7 − 6 − 1/2 (2) = 0

The formal charges in (2) are:

> The boron atom:                               formal charge = 3 − 0 − 1/2 (8) = −1
> The fluorine atom (double bonded):   formal charge = 7 − 4 − 1/2 (4) = +1
> The other fluorine atoms:                  formal charge = 7 − 6 − 1/2 (2) = 0

The rule that is used to establish the more plausible structure is: A Lewis structure in which there are no formal charges is preferred over one where formal charges are present. Therefore, structure (1) is preferred over structure (2). Experiments indicate that each B—F bond is a single bond as shown in the first dot structure. This is true even though structure (1) does not obey the octet rule.

When formal charges are assigned in a Lewis structure, the sum of the formal charges must be zero in a neutral molecule. For a polyatomic ion the formal charges must add up to the charge of the ion. The Lewis structure for the chlorite ion ($ClO_2^-$), is

$$[: \ddot{O}-\ddot{C}l-\ddot{O} :]^-$$

Cover the figure below and see if you can determine the formal charge on each atom in the chlorite ion. Then uncover and check your work.

$$\quad\ \ -\quad\ +\quad\ -$$
$$[: \ddot{O}-\ddot{C}l-\ddot{O} :]^-$$

The sum of the formal charges is −1, matching the overall charge of the chlorite ion.

| Example 9.9 |
| Exercises 14, 15 |

The example brings out another feature of the formal charge concept. The most plausible Lewis structures will be those with negative formal charges on the more electronegative atoms. Since oxygen is more electronegative than chlorine, the formal negative charges on O atoms in this structure are more reasonable than in some other structure that would place a positive charge on the O atoms, and a negative charge on the Cl atoms.

This is also observed in the $BF_3$. $BF_3$ with a double bond to fluorine gives a +1 formal charge to fluorine – the most electronegative element, which is much less plausible.

## The Concept of Resonance (Section 9.8)

***Resonance Structures.*** It sometimes happens that one satisfactory electron dot structure for a molecule or polyatomic ion that reflects what we know about the substance cannot be drawn. When such a situation arises, a special procedure is invoked to arrive at a Lewis structure. For the nitrite ion $NO_2^-$, for instance, the following structure shows the correct number of valence electrons and satisfies the octet rule. The brackets are used to indicate that the –1 charge belongs to the entire nitrite ion, and not to just one atom in the structure.

$$[: \ddot{O}—\ddot{N}=\ddot{O}]^-$$

However, the structure does not accurately represent what is known about the bond lengths of the N—O bonds in $NO_2^-$. Both bond lengths are known to be the same, whereas, according to the structure, we expect the double bond to be shorter than the single bond. An N—O single bond length should be about 136 pm, and an N=O double bond length about 122 pm. However, the two bond lengths in $NO_2^-$ are actually equal and are intermediate between these two values.

It turns out to be impossible to draw a single satisfactory Lewis structure for $NO_2^-$. Situations like this can be handled by using the concept of resonance. First, draw two structures for $NO_2^-$ that reflect different choices of electron arrangements.

$$[: \ddot{O}—\ddot{N}=\ddot{O}]^- \longleftrightarrow [\ddot{O}=\ddot{N}—\ddot{O} :]^-$$

The nitrite ion is not adequately represented by either structure; but it may be described by a composite of these structures. This composite structure cannot be drawn using the rules for writing Lewis structures. Each of the structures that contribute to the composite structure is called a contributing or resonance structure. The symbol $\longleftrightarrow$ indicates that the structures shown are resonance structures.

In applying the concept of resonance, we assume that $NO_2^-$ is a composite or an average of the two structures. Thus, the N—O bonds are intermediate between single and double bonds. The term resonance was perhaps a poor choice because the word implies to some that the real molecule flips from one structure to the other. This is not what is meant here. The point is that the properties of $NO_2^-$ cannot be accounted for by a single Lewis structure. The term resonance means that there is a need for two or more Lewis structures to represent a particular molecule.

These structures have the same placement of atoms but different positions of electron pairs. The electron pairs do not actually move from one position to another within the molecule. Therefore, resonance structures are generated by moving electron pairs within the same skeletal structure.

| Example 9.10 |
| Exercise 16 |

## Exceptions to the Octet Rule (Section 9.9)

***Types of Exceptions.*** Lewis structures can be drawn for many compounds with the aid of the octet rule; however, structures of some compounds do not follow the rule. The text points out three types of molecules that are exceptions to the octet rule:

• Molecules in which the central atom has an incomplete octet.
• Molecules with odd numbers of electrons.
• Molecules in which the central atom has an expanded octet.

***Incomplete Octet.*** The boron halides $BX_3$ are well-known examples of molecules in which the central atom has an incomplete octet. They are all planar molecules in which the boron atom has only six valence electrons. The boron atom has an incomplete octet. This situation is typical in boron chemistry. Revisit the discussion on formal charge for an explanation of this fact.

***Odd Numbers of Electrons.*** Two common oxides of nitrogen, NO and $NO_2$, have odd numbers of electrons. Since an even number of electrons is required for complete pairing, the octet rule cannot be satisfied. Two additional odd-electron molecules that are known to exist in our atmosphere for very short periods of time are OH (hydroxyl radical) and $HO_2$ (hydroperoxyl radical).

$$\cdot \ddot{\underset{..}{O}}\text{—H} \qquad\qquad \text{H—}\ddot{\underset{..}{O}}\text{—}\ddot{\underset{..}{O}}\cdot$$

***Expanded Octets.*** Molecules in which the central atom exhibits an expanded octet (has more than eight valence electrons) require the presence of nonmetal atoms from the third period or beyond in the periodic table. Second period elements never exceed the octet rule. Third period elements are just as likely to exceed the octet rule as they are to follow it. Where for example, $PCl_3$ obeys the octet rule, gaseous $PCl_5$ has a phosphorus atom that is joined by single bonds to five chlorine atoms. The phosphorus atom has 10 electrons in its valence shell. This is called an expanded octet.

The central atoms in $SF_4$ and $SF_6$, and in the interhalogen compounds, $ClF_5$, $BrF_5$, and $IF_7$ exhibit expanded octets.

In Chapter 10, you will need to draw a number of dot structures of molecules exhibiting expanded octets. When the central atom is from the third period or beyond, complete the octets of the surrounding atoms first, and then complete the central atom. If extra electron pairs remain, place them on the central atom. See Example 9.11.

These exceptions tell us that an atom with a completed octet is not necessary for covalent bonding to occur. Within the Lewis framework, it is really the sharing of electron pairs that leads to covalent bonds. A shared pair of electrons acts to attract both atoms to each other.

| Example 9.11 |
| Exercises 9 – 17 |

# Bond Enthalpies (Section 9.10)

***Average Bond Enthalpies.*** Occasionally, the enthalpy change ΔH is needed for a reaction for which enthalpy of formation data does not exist. One approach that allows an estimate of ΔH° uses the concept of bond enthalpy. Consider the gas phase atomization process:

$$CH_4(g) \rightarrow C(g) + 4H(g) \qquad \Delta H^{\circ}_{rxn} = 1664 \text{ kJ/mol}$$

In this reaction, four C—H bonds are broken; therefore, we can define the average C—H bond enthalpy as one-fourth of $\Delta H^{\circ}_{rxn}$ for the reaction. Hence, the average C—H bond enthalpy in $CH_4$ is 416 kJ/mol.

The actual bond enthalpies of the individual C—H bonds in $CH_4$ are not the same as the average value. Even so, the use of average bond enthalpies makes it possible to estimate the enthalpy changes of certain reactions. Some of the bond enthalpies in Table 9.4 of the textbook are listed here in Table 9.4.

**Table 9.4** Bond Enthalpies of Some Common Bonds

| Bond | Bond Enthalpy (kJ/mol) |
|------|------------------------|
| H—H | 436 |
| C—H | 414 |
| N—H | 393 |
| O=O | 499 |
| C=O | 745 (799 in $CO_2$) |
| O—H | 460 |
| I—Cl | 210 |
| N≡N | 941 |
| N=O | 630 |

We can estimate $\Delta H$ for the combustion of methane as follows:

$$CH_4(g) + 2\ O_2(g) \rightarrow CO_2(g) + 2\ H_2O(g)$$

$$CH_4 + 2\ O_2 \xrightarrow{\ \sum BE\ (reactants)\ } C + 4\ H + 4\ O \xrightarrow{\ -\sum BE\ (products)\ } CO_2 + 2\ H_2O$$

First, we break 4 C—H bonds and 2 O=O bonds, where $\sum BE$ (reactants) is the sum of the bond enthalpies of bonds broken. The energy is *required* to break bonds and is positive in sign (endothermic).

Next, let the atoms recombine to form the 2 C=O and 4 O—H bonds of the products. $\sum BE$ (products) is the sum of the bond enthalpies of all bonds formed from the atoms. The formation of bonds releases energy and is negative in sign (exothermic.) Therefore,

$$\Delta H° = \sum BE\ (reactants) - \sum BE\ (products)$$

$$\Delta H° = [4\ BE\ (C—H) + 2\ BE\ (O=O)] - [2\ BE\ (C=O) + 4\ BE\ (O—H)]$$

Inserting values from Table 9.4 gives:

$$\Delta H° = [4(414\ kJ) + 2(499\ kJ)] - [2(802\ kJ) + 4(460\ kJ)]$$
$$= 2654\ kJ - 3444\ kJ/mol$$
$$\Delta H° = -790\ kJ/mol$$

If we compare this value with the standard enthalpy of combustion of methane, we get

$$CH_4(g) + 2\ O_2(g) \rightarrow CO_2(g) + 2\ H_2O(\ell) \qquad \Delta H°_{rxn} = -890\ kJ/mol$$

At first it seems there is a large discrepancy of 100 kJ. However, in our calculation, $H_2O$ is present as a gas! But $\Delta H°$ refers to the standard state of $H_2O$ as a liquid. Therefore, the 100 kJ difference is largely due to the $\Delta H$ of vaporization of two moles of $H_2O$, which is 81.4 kJ.

Two important conclusions can be drawn here:

Example
9.12

Exercises
18 – 19

(1) Bond enthalpies are used *to estimate* enthalpies of reaction of gas phase reactions.
(2) Enthalpies of reaction calculated from average bond enthalpies are only approximate
values.

## GLOSSARY LIST

| | | |
|---|---|---|
| Lewis dot symbol | Lewis structure | polar covalent bond |
| ionic bond | octet rule | electronegativity |
| lattice energy | single bond | formal charge |
| Coulomb's law | multiple bond | resonance |
| Born-Haber cycle | double bond | resonance structure |
| | triple bond | coordinate covalent bond |
| covalent bond | bond length | bond enthalpy |
| covalent compound | | |
| lone pair | | |

## EQUATIONS

| Algebraic Equation | English Translation |
|---|---|
| $$E = k\,\frac{Q_+ Q_-}{r}$$ | Coulombs Law: Energy is proportional to the charge and inversely proportional to the radius. |
| $$\text{formal charge} = \left(\begin{array}{c}\text{number of}\\ \text{valence electrons}\end{array}\right) - \left(\begin{array}{c}\text{number of}\\ \text{nonbonding electrons}\end{array}\right) - \frac{1}{2}\left(\begin{array}{c}\text{number of}\\ \text{bonding electrons}\end{array}\right)$$ | Calculation of formal charge from a Lewis structure using the number of valence electrons minus the electrons assigned to the atom in the Lewis structure. |
| $$\Delta H^\circ = \sum BE\ (\text{reactants}) - \sum BE\ (\text{products})$$ | Estimation of enthalpy of a reaction using bond energies. |

## WORKED EXAMPLES

---

**EXAMPLE 9.1  Lewis Dot Symbols**

Write Lewis symbols for the following elements:  a. Ca    b. O, S and Se

• **Solution**

The Lewis symbol of an element consists of the element symbol surrounded by between 1 and 8 dots, representing valence electrons.
a.  Ca is in Group 2A.  It has two electrons in the outermost energy level.  These are its valence electrons.  The Lewis symbol is  ·Ca·
b.  O, S, and Se are all in the same group, 6A.  Each has 6 valence electrons.

$$\cdot \overset{\cdot\cdot}{\underset{\cdot\cdot}{O}} \cdot \qquad \cdot \overset{\cdot\cdot}{\underset{\cdot\cdot}{S}} \cdot \qquad \cdot \overset{\cdot\cdot}{\underset{\cdot\cdot}{Se}} \cdot$$

---

**EXAMPLE 9.2  Lewis Dot Symbols for Ions**

Write Lewis symbols for the following ions:  a. $Ca^{2+}$    b. $Se^{2-}$

• **Solution**

a.  Removing two electrons (dots) from ·Ca· gives simply $Ca^{2+}$ as the Lewis symbol of a calcium ion. No electrons are shown around the $Ca^{2+}$ symbol because this ion has lost its valence shell electrons.

b.  Adding two electrons to · Se · gives $:\overset{\cdot\cdot}{\underset{\cdot\cdot}{Se}}:^{2-}$ as the Lewis symbol for a selenide ion.

---

**Work EXERCISES & PROBLEMS: 1, 2**

---

**EXAMPLE 9.3  Predicting Melting Points**

Which member of the pair will have the higher melting point?
a.  NaCl or CaO
b.  NaCl or NaI

• **Solution**

a.  According to Coulomb's law, the doubly charged ions in CaO will attract each other more strongly than do the singly charged ions in NaCl.  CaO will have a higher melting point.
b.  According to Coulomb's law, the closer the centers of two ions can approach each other, the stronger will be the attraction between them. The sum of the ionic radii in NaCl is smaller than in NaI.  NaCl will have the higher melting point.

## EXAMPLE 9.4  The Born-Haber Cycle

Given the following data calculate the lattice energy of potassium chloride:

Enthalpy of sublimation of potassium = 90.0 kJ/mol
Bond enthalpy (BE) of Cl—Cl = 242.7 kJ/mol
Ionization energy (I) of K = 419 kJ/mol
Electron affinity (EA) of Cl = +349 kJ/mol
$\Delta H_f^{\circ}$ (KCl) = −435.9 kJ/mol

• **Solution**

From the discussion in section 9.3 (page 182), we can calculate the lattice energy (or $-\Delta H_5^{\circ}$ ) from the equation:

$$\Delta H_5^{\circ} = \Delta H_{overall}^{\circ} - [\Delta H_1^{\circ} + \Delta H_2^{\circ} + \Delta H_3^{\circ} + \Delta H_4^{\circ}]$$

Next, we identify the appropriate energy changes:

$$K(s) \rightarrow K(g) \qquad\qquad \Delta H_1^{\circ} = \Delta H_{subl} = 90.0 \text{ kJ/mol}$$

$$\tfrac{1}{2} Cl_2(g) \rightarrow Cl(g) \qquad\qquad \Delta H_2^{\circ} = \tfrac{1}{2} BE = 121.4 \text{ kJ/mol}$$

$$K(g) \rightarrow K^+(g) + e^- \qquad\qquad \Delta H_3^{\circ} = I = 419 \text{ kJ/mol}$$

$$Cl(g) + e^- \rightarrow Cl^-(g) \qquad\qquad \Delta H_4^{\circ} = -EA = -349 \text{ kJ/mol}$$

$$K(s) + \tfrac{1}{2} Cl_2(g) \rightarrow KCl(s) \qquad\qquad \Delta H_{overall}^{\circ} = -435.9 \text{ kJ/mol}$$

Next, substitute into the equation:

$$\Delta H_5^{\circ} = \Delta H_{overall}^{\circ} - [\Delta H_1^{\circ} + \Delta H_2^{\circ} + \Delta H_3^{\circ} + \Delta H_4^{\circ}]$$

$$\Delta H_5^{\circ} = -435.9 \text{ kJ/mol} - [90.0 \text{ kJ} + 121.4 \text{ kJ} + 419 \text{ kJ} + (-349 \text{ kJ})] / mol$$

$$\Delta H_5^{\circ} = -435.9 \text{ kJ/mol} - 281.4 \text{ kJ/mol} = -717 \text{ kJ/mol}$$

lattice energy = 717 kJ/mol

**Work EXERCISES & PROBLEMS: 3 – 6**

## EXAMPLE 9.5  Drawing a Lewis Structure

Draw the Lewis structure for hydrazine, $N_2H_4$. How many unshared electron pairs (lone pairs) are there on each N atom?

• **Solution**

1.  Arrange the atoms in a reasonable skeletal form. H atoms form only one bond and so must be located on the outside of the atom.

        H N N H
          H H

2.  Count the valence electrons. Each N atom has 5 valence electrons and each H atom has 1. There are $2(5) + 4(1) = 14$ valence electrons.

3.  Connect the atoms with single bonds:

        H—N—N—H
          |   |
          H   H

    Normally we would add unshared pairs to complete all octets of surrounding atoms, but in this case the H atoms only need the two electrons shared in the bond to the N atom. Counting the number of electrons used; 5 pairs = 10 valence electrons. Now add unshared pairs to complete the octets of the N atoms.

           ..  ..
        H—N—N—H
          |   |
          H   H

4.  Count the electrons: 7 pairs = 14 valence electrons. This is the same number as given in step 2.

**Work EXERCISES & PROBLEMS: 7, 8**

---

**EXAMPLE 9.6  Bond Polarity**

Arranging the following bonds in order of increasing ionic character: C—O, C—H, and O—H.

• **Solution**

As the electronegativity difference increases, the bond becomes more polar and its ionic character increases. Using Figure 9.5 (textbook), we can determine the differences.

$$\text{for C—O} \quad \Delta\chi = 3.5 - 2.5 = 1.0$$
$$\text{for C—H} \quad \Delta\chi = 2.5 - 2.1 = 0.4$$
$$\text{for O—H} \quad \Delta\chi = 3.5 - 2.1 = 1.4$$

The ionic character increases in the order
$$\text{C—H} < \text{C—O} < \text{O—H}$$

• **Comment**

The electronegativity of hydrogen is unlike that of the other elements of Group 1A. In terms of electronegativity, hydrogen is similar to the nonmetal elements boron and carbon. The bonds of H to nonmetal atoms are polar covalent, rather than ionic (such as the bonds in LiCl and NaCl).

---

**EXAMPLE 9.7 Electronegativity Trends**

Using the trends within the periodic table, determine which of the following is the most electronegative element: As, Se, or S.

**• Solution**

Se and As are in the same period, and so the one further to the right has the higher electronegativity. That one is Se. Now, compare S and Se. They are in the same group. The one nearer the top of the group has the greater electronegativity, which is sulfur.

---

**EXAMPLE 9.8 Types of Bonds**

For the following pairs of elements, label the bonds between them as ionic, polar covalent, or pure covalent.
a.  Rb and Br
b.  S and S
c.  C and N

**• Solution**

a.  Bond type depends on electronegativity differences. The value of $\Delta\chi$ for a Rb—Br bond is 2.0. RbBr is ionic.
b.  For S—S, $\Delta\chi = 0.0$, and so the bond is a pure covalent bond.
c.  In the periodic table, carbon and nitrogen are adjacent to each other in period 2. The one on the right is N; it is more electronegative so we expect a polar covalent bond. Electronegativity values give $\Delta\chi = 0.5$.

**Work EXERCISES & PROBLEMS: 9 – 13**

---

**EXAMPLE 9.9 Assigning Formal Charges**

Assign formal charges to the atoms in the following Lewis structures:

a.  $: C \equiv O :$

b.  $\overset{..}{O} = \overset{..}{S} - \overset{..}{\underset{..}{O}} :$

**• Solution**

The formula used to calculate the formal charge of an atom is:

$$\begin{matrix} \text{formal} \\ \text{charge} \end{matrix} = \left( \begin{matrix} \text{number of} \\ \text{valence electrons} \end{matrix} \right) - \left( \begin{matrix} \text{number of} \\ \text{nonbonding electrons} \end{matrix} \right) - \frac{1}{2} \left( \begin{matrix} \text{number of} \\ \text{bonding electrons} \end{matrix} \right)$$

a.  The carbon atom: formal charge = $4 - 2 - 1/2 (6) = -1$
    The oxygen atom: formal charge = $6 - 2 - 1/2 (6) = +1$

b.   The sulfur atom:                   formal charge = 6 − 2 − 1/2 (6) = +1
     The oxygen atom on the right:   formal charge = 6 − 6 − 1/2 (2) = −1
     The oxygen atom on the left:    formal charge = 6 − 4 − 1/2 (4) = 0

## • Comment

Some chemists don't approve of the CO structure given in part (a) because it places a positive formal charge on the more electronegative oxygen atom.

**Work EXERCISES & PROBLEMS: 14, 15**

---

### EXAMPLE 9.10  Drawing a Lewis Structure with Resonance

Draw three resonance structures for $N_2O$. The skeletal structure is N—N—O.

## • Solution

1.  Arrange the atoms

    **N—N—O**

2.  Count the valence electrons.

    2(5) + 6 = 16 electrons

3.  Add unshared pairs to the atoms to give each atom an octet:

    :N̈—N̈—Ö:

4.  Count the electron pairs used. 10 electron pairs = 20 electrons. There are 4 more electrons than are counted in step 2.

5.  For every 2, too many used in step 3 than you have in step 2, and a multiple bond. In this case, two more bonds must be added and then the lone pair changed to give an octet to each atom.

    :N̈=N=Ö:

    This is a satisfactory structure because it completes the octets of all three atoms and uses the correct number of electrons. Additional resonance structures can be generated by moving electron pairs in such a way that the octet rule for each atom is always satisfied. The positions of the atoms in $N_2O$ cannot be altered.

Two possibilities are

:N≡N—Ö:   and   :N̈—N≡O:

## • Comment

The three structures for $N_2O$ are called resonance structures.  The actual bonding in $N_2O$ is a composite of these three structures.

**Work EXERCISES & PROBLEMS: 16**

**EXAMPLE 9.11 Exceptions to the Octet Rule**

Draw Lewis structures for:
a. $GaI_3$
b. $NO_2$ (all bonds are equivalent)
c. $ClF_3$

• **Solution**

a. Gallium is in Group 3A of the periodic table, a group well known for its electron deficient elements. $GaI_3$ has 3 + 3(7) = 24 valence electrons.

$$: \overset{..}{I} :$$
$$|$$
$$: \overset{..}{\underset{..}{I}} — Ga — \overset{..}{\underset{..}{I}} :$$

This structure shows 24 electrons, the correct number. Ga, with three electron pairs, is electron deficient. And while it is possible to follow the rules for writing Lewis structures and draw one that obeys the octet rule (moving one of the lone pairs on I and giving a Ga=I double bond), it is not likely to exist because it puts a negative formal charge on the central atom (a metal) and a positive formal charge on iodine. This is not probable. Draw the structure and calculate the formal charges on the atoms to verify this for yourself.

b. $NO_2$ has 17 valence electrons. With an odd number of electrons, it cannot obey the octet rule. The best we can do is start with 18 valence electrons as in $NO_2^-$ and then remove one from the nitrogen atom (because it is the unique atom). Two contributing structures are necessary.

$$\overset{..}{O} = \overset{.}{N} — \overset{..}{\underset{..}{O}} : \longleftrightarrow : \overset{..}{\underset{..}{O}} — \overset{.}{N} = \overset{..}{O}$$

c. $ClF_3$ has 28 valence electrons. 26 electrons are required to complete the octets of the four atoms. The remaining two electrons are placed on the central Cl atom because chlorine is in the third period, and has vacant 3d orbitals that can hold electrons in addition to an octet. Chlorine is said to have an expanded octet.

$$: \overset{..}{F} — \overset{.\,.}{Cl} — \overset{..}{F} :$$
$$|$$
$$: \overset{}{\underset{..}{F}} :$$

**Work EXERCISES & PROBLEMS: 17**

**EXAMPLE 9.12 Use of Bond Enthalpies to Estimate $\Delta H_{rxn}$**

Estimate $\Delta H_{rxn}$ for the reaction:

$$Cl_2(g) + I_2(g) \rightarrow 2ICl(g)$$

Use bond enthalpies given in Table 9.4 (text), and given that BE (I—Cl) = 210 kJ/mol.

---

**• Solution**

Recall that     $\Delta H° = \sum BE$ (reactants) $- \sum BE$ (products)
            $\sum BE$ (reactants) $= BE$ (Cl—Cl) $+ BE$ (I—I) $= 243 + 151 = 394$ kJ/mol
    and     $\sum BE$ (products) $= 2BE$ (I—Cl) $= 2(210) = 420$ kJ/mol

Subtraction yields:

      $\Delta H = 394$ kJ/mol $- 420$ kJ/mol
        $= -26$ kJ/mol

---

**Work EXERCISES & PROBLEMS: 18, 19**
**Work the rest of the EXERCISES & PROBLEMS**

## EXERCISES & PROBLEMS

1. Write Lewis dot symbols for atoms of the following elements:
   a. Mg   b. Se   c. Al   d. Br   e. Xe

2. Write Lewis dot symbols for the following ions:
   a. $K^+$   b. $S^{2-}$   c. $N^{3-}$   d. $I^-$   e. $Sr^{2+}$

3. Write a chemical equation for the process that corresponds to the lattice energy of MgO.

4. Which member of the pair will have the higher melting point?
   a. KCl or $CaCl_2$         b. RbI or NaI

5. Which member of the pair will have the higher lattice energy?
   a. NaCl or CaO         b. KI or KCl

6. Calculate the lattice energy of sodium bromide from the following information.

   $\Delta H_{subl}$ (Na) $= 109$ kJ/mol
   Ionization Energy (Na) $= 496$ kJ/mol
   Bond Enthalpy (Br—Br) $= 192$ kJ/mol
   Electron Affinity (Br) $= 324$ kJ/mol
   $\Delta H_f°$ (NaBr) $= -359$ kJ/mol

7. How many lone pairs are on the underlined atoms in the following compounds?
   a. $\underline{P}H_3$     b. $\underline{S}Cl_2$     c. $H_2C\underline{O}$

8. Write the Lewis structures for the following species.
   a. $NH_4^+$     b. $NCl_3$     c. $CF_2Cl_2$

9. Which atom is the most electronegative?
   Li    Cs    P    As    Ge

10. List the following bonds in order of increasing ionic character:
    N—O,   Na—O,   O—O,   S—O

11. Classify the O—H bond in $CH_3OH$ as ionic, polar covalent, or nonpolar covalent.

12. Which of the following is a nonpolar covalent bond (pure covalent)?
    H—Cl    Li—Br    Se—Br    Br—Br

13. Classify the following bonds as ionic, polar covalent, or nonpolar covalent.
    Se—Cl    Al—Cl    K—F    Cl—Cl

14. Assign formal charges to the atoms in the following Lewis structures:

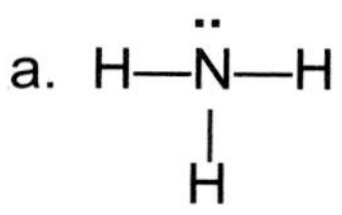

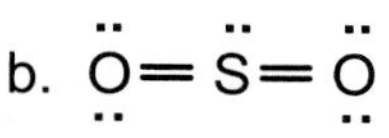

    a. H—N̈—H
           |
           H

    b. Ö=S̈=Ö

15. Assign formal charges to the atoms in the following possible Lewis structures.

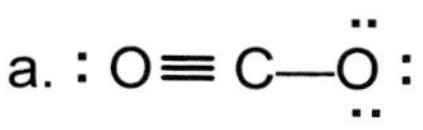

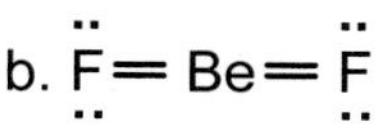

    a. :O≡C—Ö:    b. F̈=Be=F̈

16. Draw resonance structures for nitric acid:  H—O—N—O
                                                        |
                                                        O

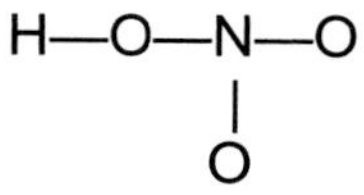

17. Write Lewis structures for the following molecules:
    a. $AlBr_3$    b. $SF_4$    c. NO

18. Given the N≡N and H—H bond enthalpies in Table 9.4 and standard enthalpy of reaction:

    $$\frac{1}{2}N_2(g) + \frac{3}{2}H_2(g) \rightarrow NH_3(g) \qquad \Delta H^\circ_{rxn} = -46.3 \text{ kJ/mol}$$

    calculate the average N—H bond enthalpy in ammonia.

19. Given the bond enthalpies in Table 9.4, calculate the enthalpy of formation for ammonia, $NH_3$.

    $$\frac{1}{2}N_2(g) + \frac{3}{2}H_2(g) \rightarrow NH_3(g) \qquad \Delta H^\circ_f (NH_3) = ?$$

20. Explain the difference between electron affinity and electronegativity.

21. Do formal charges represent actual charges? What do they represent?

22. Do resonance structures actually exist? Describe why the concept of resonance is needed.

## PRACTICE TEST QUESTIONS
See notes on taking practice test in the Preface

1. Arrange the following ionic compounds in order of increasing lattice energy: RbI, MgO, CaBr$_2$.

2. Which compound in each pair has the higher melting point?
   a. RbI or KI
   b. MgCl$_2$ or MgBr$_2$

3. Draw Lewis dot structures for:
   a. IBr     b. CS$_2$      c. PCl$_5$
   d. SO$_2$    e. CO       f. P$_2$Cl$_4$
   g. SO$_4^{2-}$    h. PO$_4^{3-}$

4. Draw resonance structures for SO$_3$.

5. Which of the following elements is the most electronegative?
   a. Ca       b. P
   c. As       d. K
   e. Si

6. Which one of the following is the most polar bond?
   a. B—C       b. B—N
   c. B—S       d. C—C

7. Which bond has the greatest percent ionic character?
   a. CO       b. SO
   c. NaI       d. NaBr

8. Which molecule has a Lewis structure that does not obey the octet rule?
   a. NO       b. PF$_3$
   c. CS$_2$       d. HCN
   e. BF$_4^-$

9. Which molecule has covalent bonding between atoms?
   a. NaF       b. K$_2$O
   c. ICl       d. SrI$_2$
   e. none of these

10. Indicate the molecule in each pair that does not follow the octet rule.
    a. PCl$_5$ or AsCl$_3$
    b. BCl$_3$ or CCl$_4$
    c. NH$_3$ or CH$_3$
    d. SbCl$_5$ or NCl$_3$

11. Consider the following Lewis structures for sulfate ion. Which is the more reasonable structure in terms of formal charges?

a. 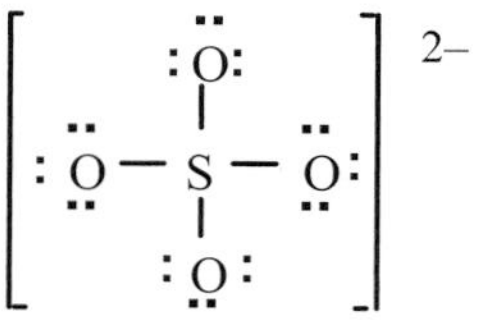
$$\left[ \begin{array}{c} :\ddot{O}: \\ | \\ :\ddot{O}-S-\ddot{O}: \\ | \\ :\ddot{O}: \end{array} \right]^{2-}$$

b. 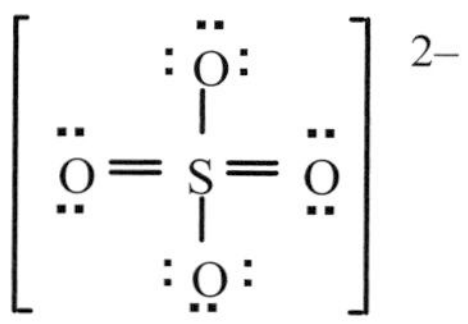
$$\left[ \begin{array}{c} :\ddot{O}: \\ | \\ \ddot{O}=S=\ddot{O} \\ | \\ :\ddot{O}: \end{array} \right]^{2-}$$

12. Assign formal charges to the atoms in the Lewis structures shown.

a.  $:N=N=\ddot{O}:$

b.  $[:\ddot{O}-H]^{-}$

c.  $[:\ddot{O}-\ddot{Cl}:]^{-}$

13. Write a mathematical expression that would allow you to calculate the lattice energy of $CaCl_2$ using Hess's law.

14. Calculate the $N\equiv N$ bond enthalpy, given that the standard enthalpy of formation of atomic nitrogen is 473 kJ/mol.

15. Estimate the enthalpy change for the following reaction using bond enthalpies:

$$N_2(g) + O_2(g) \rightarrow 2\ NO(g)$$

Hint: NO has a double bond.

# Chapter 10. Chemical Bonding II: Molecular Geometry and Hybridization of Atomic Orbitals

Molecular Geometry and the VSEPR Model (Section 10.1)
Polar Molecules (Section 10.2)
Valence Bond Theory and Hybrid Orbitals (Sections 10.3 – 10.5)
Molecular Orbital Theory (Sections 10.6 – 10.7)
Delocalized Molecular Orbitals (Section 10.8)

## SUMMARY

### Molecular Geometry and the VSEPR Model (Section 10.1)

***Electron-Pair Repulsion.*** The major features of the geometry of molecules and polyatomic ions can be predicted by applying a simple principle: "The valence shell electron pairs surrounding an atom are arranged such that they are as far apart as possible."

According to the valence-shell electron-pair repulsion (VSEPR) theory, a particular molecular geometry results from the orientation of electron pairs around a central atom. For example, consider the $BeH_2$ molecule in which Be is the central atom. Drawing its Lewis structure, we count two electron pairs in the valence shell of the Be atom. There are no lone pairs.

H : Be : H

In order to minimize repulsion, the two electron pairs are located as far apart as possible. The angle between two bonds to the same atom is called the bond angle. The H—Be—H bond angle is 180°. $BeH_2$ is an example of a molecule in which the atoms lie in a straight line. $BeH_2$ has a *linear geometry*.

When the central atom of a molecule has no lone pairs in its valence shell, the possible orientations in space around the central atom are determined by the number of atoms bonded to the central atom, called bonding electron pairs. Note: A double or triple bond to an atom counts the same as a single bond. Table 10.1 gives the shapes of molecules depending on the number of valence electron pairs about the central atom. This table gives the molecular geometry for molecules which have *no lone pairs* on the central atom.

***Three Electron Pairs.*** When there are three valence electron pairs around a central atom, they avoid each other best if they are positioned at the corners of a triangle. The central atom sits in the middle of the triangle, and all four atoms lie in the same plane. All bond angles are 120°. Such a molecule is called *trigonal planar*.

**Table 10.1** Bond Angles and Molecule Shapes

| Number of Electron Pairs about Central Atom | Bond Angles | Shape of Molecule (with no lone pairs) |
|---|---|---|
| 2 | 180° | Linear |
| 3 | 120° | Trigonal planar |
| 4 | 109.5° | Tetrahedral |
| 5 | 120°, 90° | Trigonal bipyramid |
| 6 | 90° | Octahedral |

***Four Electron Pairs.*** In chapter 9 we discussed the use of the octet rule to determine the correct formulas of covalent compounds. Lewis structures describe only the bonding within a molecule in a two-dimensional manner. They are not geometrical structures. For instance, the Lewis structure for methane $CH_4$ does not tell us whether the molecule is flat (planar) or tetrahedral. When the central atom has four bonding electron pairs (as C does in $CH_4$), these pairs try to keep as far as possible from each other, while maintaining their distance from the nucleus. The tetrahedral arrangement has less repulsion between the four electron pairs than the flat, two-dimensional geometry. This geometry is illustrated for methane in Figure 10.1. Four electron pairs around a central atom will be arranged at the corners of a tetrahedron. In methane the H—C—H bond angle is the tetrahedral angle, 109.5°.

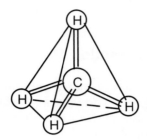

**Figure 10.1** The tetrahedral geometry of the methane molecule. The C atom is at the center of a four-sided three-dimensional figure called a tetrahedron. The four H atoms are found at the four corners. The H–C–H bond angle is the "tetrahedral" angle, 109.5°.

***Five Electron Pairs.*** Central atoms with expanded octets have 5 or 6 electron pairs occupying their valence shells. The geometrical figure defined by the preferred orientation of five electron pairs around the central atom is called a *trigonal bipyramid* (Figure 10.2).

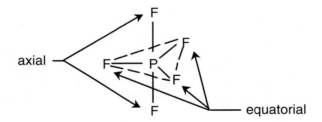

**Figure 10.2** The electron pairs that lie around the "equator" are called equatorial electrons. The electrons at the "north and south poles" are called axial electrons. The equatorial electron pairs lie in the same plane. In $PF_5$ the equatorial F—P—F bond angles are 120°. The axial atoms are at right angles to the equatorial plane, and so any bond angle formed by an axial atom, the central atom, and an equatorial atom is 90°.

***Six Electron Pairs.*** The most stable arrangement of 6 electron pairs around the central atom is called an octahedron (Figure 10.3). The example of $SF_6$ is shown on the next page. And although it is tempting to make a comparison with the trigonal bipyramid and assign axial and equatorial positions, the octahedron does not have these designations. All sites on the octahedron are identical.

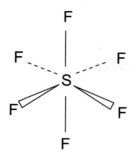

**Figure 10.3** The octahedron geometry. All F – S – F angels are 90°. There are no "equatorial" or "axial" positions.

***Effect of Lone Electron Pairs.*** When the central atom contains lone pairs (nonbonding pairs) in addition to bonding pairs, the situation becomes slightly more complicated. The positions of the lone electron pairs help to establish the geometry of a molecule. For instance, in ammonia ($NH_3$) the nitrogen atom has one lone electron pair.

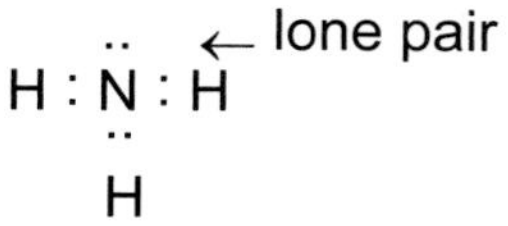

$$\overset{..}{\underset{\overset{..}{H}}{H : N : H}} \quad \leftarrow \text{lone pair}$$

There are 4 electron pairs in the valence shell of the central N atom; 3 atoms connected to the central atom plus one lone pair. To determine the molecular geometry, draw the 4 electron pairs as far apart from each other as possible (tetrahedral, as in Figure 10.1). But when describing the shape of a molecule, we consider *only the positions of the atoms*. Its geometry is described as a trigonal pyramid because the pyramid has three sides. See Figure 10.4 below The N atom sits at the top of the pyramid.

trigonal pyramid

**Figure 10.4** The N atoms sits at the top of a pyramid with a triangular base of hydrogen atoms. Therefore the name of this geometry is trigonal pyramidal.

In summary: when determining the molecular geometry of any molecule, follow the following procedure:

1. Draw the Lewis structure.
2. Count the number of bonding pairs around the central atom. Single, double and triple bonds count as only one bonding pair. So the bottom line is, you count the number of atoms connected to the central atom and the number of lone pairs.
3. Draw the bonding pairs as far apart from each other as possible, using the information in Table 10.1.
4. Place the lone pairs in the locations which have the most room. In cases where there are more than one lone pair, place lone pair electrons as far away from each other as possible.

5.  After all atoms and lone pairs are located around the central atom, by looking only at the atoms, determine the name. Names are assigned as in Table 10.2 below. Become familiar with the names and what they look like. Once you have drawn on paper where the atoms and lone pair are located, the name will be evident.
6.  Repeat for each central atom if there is more than one central atom in a molecule.

Table 10.2 lists the molecular shapes that result when the central atom has the symbol A, a pair of bonding electrons is B (i.e., the number of atoms bonded to the central atom), and a pair of nonbonding electrons is E. An example of each shape is drawn in Table 10.2 of the textbook.

**Table 10.2**  Molecule Shapes As Determined by VSEPR Theory

| Molecular Type | Shape of Molecule |
| --- | --- |
| $AB_2$ | Linear |
| $AB_3$ | Trigonal planar |
| $AB_2E$ | Bent (nonlinear) |
| $AB_4$ | Tetrahedral |
| $AB_3E$ | Trigonal pyramidal |
| $AB_2E_2$ | Bent (nonlinear) |
| $AB_5$ | Trigonal bipyramidal |
| $AB_4E$ | Distorted tetrahedron (see saw) |
| $AB_3E_2$ | T-shaped |
| $AB_2E_3$ | Linear |
| $AB_6$ | Octahedral |
| $AB_5E$ | Square pyramidal |
| $AB_4E_2$ | Square planar |

**Bond Angles.**  VSEPR theory also explains why many observed bond angles are not quite equal to their predicted angles. For instance, a true tetrahedral angle, as in $CH_4$, is 109.5°, but the measured H—N—H bond angle in $NH_3$ is 106.7°, and in water the H—O—H bond angle is 104.5°.

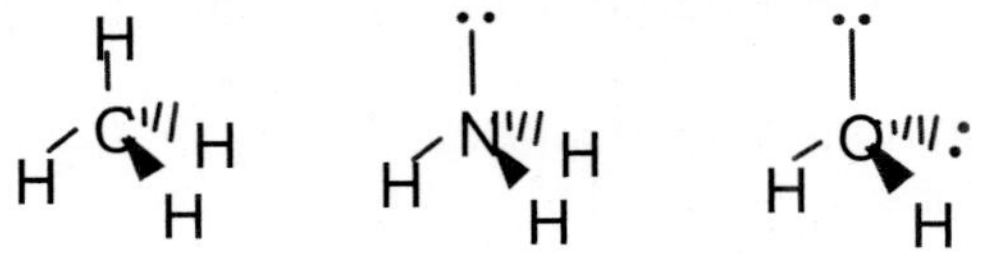

109.5°          106.7°          104.5°

The deviation from 109.5° arises because there are three types of electron-pair repulsions represented: (1) repulsion between bonding pairs, (2) repulsion between lone pairs, and (3) repulsion between a bonding pair and a lone pair. The force of repulsion is not the same among these three, but decreases as follows:

lone-pair vs. lone-pair  >  lone-pair vs. bonding-pair  >  bonding-pair vs. bonding-pair
      repulsion                     repulsion                          repulsion

In other words, lone pairs require more space than bonding pairs and tend to push the bonding pairs closer together. Bonding electron pairs are attracted by two nuclei at the same time. These pairs require less space than lone electron pairs.

In ammonia, the lone pair on the nitrogen atom repels the bonding pairs more strongly than the bonding pairs repel each other. The result is that the H—N—H bond angle is several degrees less than a true tetrahedral angle.

In water, the oxygen atom has two bonding pairs and two lone pairs. The greater lone-pair versus bonding-pair repulsion causes the two O—H bonds to be pushed in toward each other. The H—O—H angle in water should be less than the H—N—H angle in ammonia.

In general, when nonbonding electrons are present on the central atom, the shape will be close to, but not exactly, as predicted in Table 10.2.

| Example 10.1 |
| Exercises 1 – 4 |

## Polar Molecules (Section 10.2)

***Diatomic Molecules.*** In the preceding chapter you learned that polar bonds are those in which the centers of positive charge and negative charge do not coincide. This charge separation is called a dipole, and it results from electronegativity differences between two bonded atoms. A molecule with an electric dipole is said to be polar and to possess a dipole moment.

The dipole moment, $\mu$, is represented by an arrow, or vector, with the tail at the positive center and the head at the negative center. The length of the arrow represents the magnitude of the dipole moment.

$$\overset{\delta+ \quad \delta-}{\text{H—Cl}} \qquad \overset{\longmapsto}{\text{H—Cl}}$$

Quantitatively, the dipole moment is calculated from the equation

$$\mu = Qr$$

where Q is the magnitude of the charge in coulombs from either end of the dipole and r is the charge separation (the distance between the centers of positive and negative charge), in meters. In polar molecules, this charge is never as great as one unit of charge. Rather, the charge is a partial charge, $\delta+$ or $\delta-$, which signifies a charge of less than one unit. The SI unit for a dipole moment is the coulomb meter. Traditionally, dipole moments have been measured in Debye units, where 1 debye (D) = $3.33 \times 10^{-30}$ C m.

Molecules that possess a dipole moment are called polar molecules. Molecules without a dipole moment are called nonpolar molecules. All homonuclear diatomic molecules are nonpolar. In general, heteronuclear diatomic molecules are polar.

***Polyatomic Molecules.*** The dipole moment of a molecule containing more than one bond depends both on bond polarity and molecular geometry. In a polyatomic molecule the arrows representing polar bonds may add together to yield a polar molecule with a resultant dipole moment $\mu$. Conversely, the arrows may cancel each other when added, producing a zero resultant dipole moment. Table 10.3 lists dipole moments of some small molecules, in Debye units.

**Table 10.3**  Dipole Moments
of Some Molecules

| Molecule | $\mu$ (D) |
|----------|-----------|
| $H_2O$   | 1.87 D    |
| $H_2S$   | 1.10 D    |
| $NH_3$   | 1.46 D    |
| $CO_2$   | 0         |
| $SO_2$   | 1.60 D    |
| $BCl_3$  | 0         |

An important property of the water molecule is its dipole moment of 1.85 D. In the water molecule each bond is polar. The resultant of these bond dipoles yields a dipole moment for the molecule. The center of negative charge lies closer to the O atom than does the center of positive charge.

In $CO_2$, the two bond dipoles cancel and $\mu = 0$.

resultant  $\mu = 0$

When the bond dipoles are equal and point in opposite directions, the centers of + and − charge are both on the central atom. Thus, the absence of a molecular dipole in $CO_2$ suggests that the molecule is linear. Conversely, water has a dipole moment, and so it cannot be linear.

Example
10.2

Exercises
5 – 7

## Valence Bond Theory and Hybrid Orbitals (Sections 10.3 – 10.5)

***Covalent Bonds.***  The valence bond theory is one of two quantum mechanical descriptions of chemical bonding currently in use. The other is the molecular orbital theory which will be discussed in the next section. In the valence bond theory, the atomic orbitals of each bonded atom are essentially the same as in separate isolated atoms. A covalent bond forms when the orbitals from two atoms overlap and a pair of electrons occupies the region between the atoms. This overlap produces a region between the two nuclei where the probability of finding an electron is greatly enhanced. The presence of greater electron density between the two atoms tends to attract both atoms and bonding occurs. In HCl, for example, bonding results from the overlap of the hydrogen 1s orbital with the half-filled $2p_x$ orbital of the chlorine atom. A chlorine atom has only the $3p_z$ orbital available for bonding because its other orbitals are already filled. The valence electron configuration of Cl is $3p_x^2 3p_y^2 3p_z^1$.

Example
10.3

Exercise
8

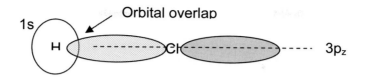

**Hybridization.** The concept of hybridization is used to account for bonding and geometry in molecules with trigonal planar, tetrahedral, and more complex shapes. The process of orbital mixing is called hybridization, and the new atomic orbitals are called hybrid orbitals. We can illustrate the main features of hybridization by examining the linear $BeCl_2$ molecule. To start with, an isolated Be atom has the ground state electron configuration $1s^2 2s^2$. The

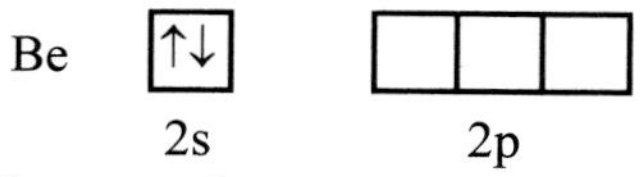

Be   $\boxed{\uparrow\downarrow}$   $\boxed{\phantom{x}}\boxed{\phantom{x}}\boxed{\phantom{x}}$

     2s       2p

orbital diagram for the valence electrons is shown above.

In a Be atom, the 2p orbitals lie quite close in energy to the 2s orbital. Orbitals on the same atom that lie close together in energy, have an ability to combine with one another to form *hybrid orbitals*. The presence of two Be—Cl bonds can be accounted for if, prior to bond formation, a 2s electron in a Be atom is promoted into an empty 2p orbital.

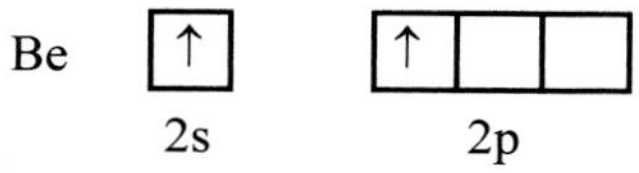

Be   $\boxed{\uparrow}$   $\boxed{\uparrow}\boxed{\phantom{x}}\boxed{\phantom{x}}$

     2s       2p

Now there are two unpaired electrons that could participate in two covalent bonds to two chlorine atoms. Since one electron is in an s orbital and one is in a p orbital, we would expect two types of Be—Cl bonds. However, both Be—Cl bonds are observed to be the same.

This equivalence is explained by hybridization or "mixing" of the 2s and the 2p orbitals to create two new equivalent orbitals called sp hybrid orbitals.

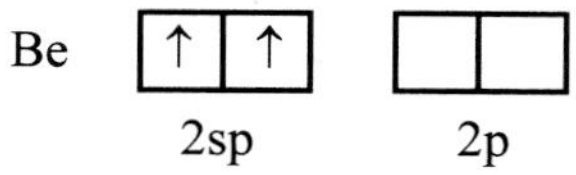

Be   $\boxed{\uparrow}\boxed{\uparrow}$   $\boxed{\phantom{x}}\boxed{\phantom{x}}$

     2sp      2p

These two hybrid orbitals point in opposite directions from the Be nucleus. The Be—Cl covalent bonds result from overlap of beryllium sp hybrid orbitals that contain one electron each with the half-filled $3p_z$ orbitals of the Cl atoms.

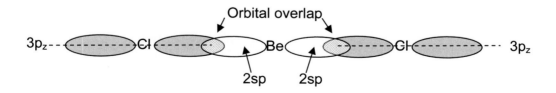

**We can summarize the rules of hybridization as follows:**

1. Hybridization is a process of mixing orbitals on a *single atom* in a molecule.
2. Only orbitals of similar energies can be mixed to form hybrid orbitals.
3. The number of hybrid orbitals obtained always equals the number of orbitals mixed together.
4. Hybrid orbitals are identical in shape to each other and they project outward from the nucleus with characteristic orientations in space.
5. To determine the hybridization, determine the arrangement of the electron pair (all atoms and lone pair included) about the central atom and use Table 10.4 below.

Table 10.4 shows several of the more important hybrid orbitals discussed in the textbook and their characteristic geometrical shapes.

**Table 10.4** Hybrid Orbitals and Their Geometries

| Designation | Arrangement | Example |
|---|---|---|
| $sp$ | linear (180° angle) | $BeCl_2$ |
| $sp^2$ | planar triangle (120° angle) | $BF_3$, $SO_2$ |
| $sp^3$ | tetrahedral (109.5° angle) | $CH_4$, $NH_4^+$, $H_2O$ |
| $sp^3d$ | trigonal bipyramidal (90° & 120°) | $PCl_5$, $SF_4$ |
| $sp^3d^2$ | octahedral (90° angle) | $SF_6$, $SbCl_6^-$ |

Examples
10.4 – 10.5

Exercises
9 – 12

***Multiple Bonds.*** Double and triple bonds can also be understood in terms of overlap of atomic orbitals. The C=C double bond in the ethylene molecule (see Figure 10.16, 10.17 and Section 10.5 of the textbook), consist of one sigma ($\sigma$) bond and one pi ($\pi$) bond.

A sigma bond is a covalent bond in which the electron density is concentrated in a cylindrical pattern around the line of centers between the two atoms. In the C=C bond, the $\sigma$ bond results from overlap of an $sp^2$ hybrid orbital from one of the carbon atoms with a $sp^2$ hybrid orbital from the other carbon atom (Figure 10.16 (a) text).

In a pi bond the electron density is concentrated above and below the C—C internuclear axis, but not cylindrically as in a sigma bond (Figure 10.5). The pi bond has two regions of electron density, but it should be kept in mind that it is only *one* bond. The $\pi$ bond in ethylene results from sideways overlap of the unhybridized $2p_z$ atomic orbitals of the two carbon atoms.

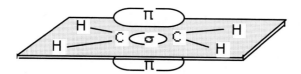

Example
10.6

Exercise
13

**Figure 10.5** The pi bond has electron density above and below the internuclear axis. The electron density is not symmetrical around the internuclear axis as in a σ bond. The two parts represent one π bond.

## Molecular Orbital Theory (Sections 10.6- 10.7)

***Molecular Orbitals.*** Until now we have considered covalent bonds in terms of the *valence bond theory*. In that theory, the electrons in a *molecule* occupy atomic orbitals of the individual *atoms*. Covalent bonds are visualized in terms of the overlap of two half-filled atomic orbitals. This overlap increases the electron density between both atomic centers.

The *molecular orbital model* has a different approach. In this model, the available atomic orbitals of all the atoms in the molecule are combined to form orbitals that belong to the *entire molecule*. Like atomic orbitals, these molecular orbitals have different shapes, sizes, and energies. Once the molecular orbitals have been derived, electrons are allotted to them in a manner analogous to allotting electrons to the orbitals of atoms (that is, Aufbau Principle, Pauli exclusion principle and Hund's rule).

Let's begin with the simplest case: a $H_2$ molecule. According to the molecular orbital (MO) theory, the result of the interaction of two 1s orbitals is the formation of two molecular orbitals. One of these is a bonding molecular orbital, which has a lower energy than the original 1s orbitals. The other molecular orbital is an antibonding orbital, which is of higher energy than the original 1s orbitals. It is the greater stability of the electrons in the bonding MO that results in covalent bond formation. In the bonding MO, electron density is concentrated between the two atoms. While the antibonding MO actually has a node (that is, a zero value) in the electron density midway between the two nuclei, there is no contribution to bonding when electrons enter antibonding MOs. An energy level diagram for the combination of two atomic orbitals of hydrogen to form the MOs in the $H_2$ molecule is shown in Figure 10.6.

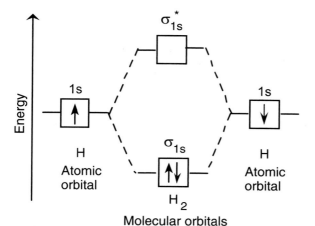

**Figure 10.6** Energy levels of bonding and antibonding molecular orbitals in the $H_2$ molecule. The two electrons of the molecule enter the bonding molecular orbital ($\sigma_{1s}$). Sigma antibonding and bonding molecular orbitals are formed from the combination of

1s atomic orbitals. The bonding MO is lower in energy than the atomic orbitals, and the antibonding MO is higher in energy than the atomic orbitals.

The antibonding and bonding MOs that result from combining 1s orbitals are examples of sigma molecular orbitals. The electron density of sigma MOs is distributed symmetrically about the line of centers between the two atoms. The sigma bonding MO and sigma antibonding MO resulting from combination of 1s atomic orbitals are designated $\sigma_{1s}$ and $\sigma_{1s}^*$, respectively. Sigma MOs also result from the combination of certain other atomic orbitals. For instance, the $2p_x$ orbitals on different atoms combine to form $\sigma_{2p}$ bonding and $\sigma_{2p}^*$ antibonding orbitals as shown in Figure 10.24(a) of the text.

Two $2p_y$ orbitals and two $2p_z$ orbitals on different atoms must approach each other sideways as shown in Figure 10.24(b) of the text. In the resulting molecular orbital, the electron density is concentrated above and below the line joining the two bonded atoms, rather than around the line of centers. Such an MO is called a pi molecular orbital. The symbol $\pi_{2p_z}$ stands for a bonding pi orbital formed by combination of two $2p_z$ atomic orbitals. The pi antibonding orbital is designated, $\pi_{2p_z}^*$.

Also, the combination of two $2p_y$ atomic orbitals from different atoms forms $\pi_{2p_y}$ bonding and $\pi_{2p_y}^*$ antibonding orbitals as well.

**Molecular Orbital Configurations.** For homonuclear diatomic molecules, any two similar atomic orbitals can merge to form two MOs, one bonding and one antibonding orbital. For $H_2$ we can illustrate the formation of the $\sigma_{1s}$ and $\sigma_{1s}^*$ MOs as the separate atoms are brought closer together in Figure 10.6. The $H_2$ molecule has two electrons that are positioned in the lowest energy orbital, the $\sigma_{1s}$. With both electrons in the bonding MO the molecule is stable. We write the ground state MO configuration

$$H_2 \ (\sigma_{1s})^2$$

where the superscript "2" means two electrons occupy the sigma-1s orbital.

The molecular orbital theory predicts $He_2$ to be unstable. A diatomic helium molecule would have four electrons. Using the same diagram as for $H_2$ we can see that the third and fourth electrons would enter the $\sigma_{1s}^*$ orbital, resulting in the configuration:

$$He_2 \ (\sigma_{1s})^2(\sigma_{1s}^*)^2$$

In this case, the stability gained by having two electrons in a bonding orbital is cancelled by the presence of two electrons in the antibonding orbitals. A stable $He_2$ molecule should not exist.

When comparing the stabilities of molecules, we use the concept of bond order.

$$\text{bond order} = \frac{1}{2}\left(\begin{array}{c}\text{number of electrons} \\ \text{in bonding MOs}\end{array} - \begin{array}{c}\text{number of electrons} \\ \text{in antibonding MOs}\end{array}\right)$$

The value of the bond order is a measure of the stability of the molecule. For $H_2$:

$$\text{bond order} = \frac{1}{2}(2 - 0) = 1$$

and for $He_2$:

$$\text{bond order} = \frac{1}{2}(2 - 2) = 0$$

A bond order of 1 represents a single covalent bond, and a bond order of 0 means the molecule is not stable. Bond orders of 2 and 3 result for molecules with double bonds and triple bonds, respectively.

For most homonuclear diatomic molecules containing atoms of second period elements, an approximate order of molecular energy levels is:

$$\sigma_{1s} < \sigma_{1s}^* < \sigma_{2s} < \sigma_{2s}^* < \pi_{2p_y} = \pi_{2p_z} < \sigma_{2p_x} < \pi_{2p_y}^* = \pi_{2p_z}^* < \sigma_{2p_x}^*$$

The electron configurations of diatomic molecules of the second period elements and some of their known properties are summarized in Table 10.5 in the textbook. The table shows the bond order of these molecules as predicted by MO theory and the corresponding bond energies and bond lengths. The number of unpaired electrons also correlates with the magnetic properties of these molecules. The MO theory satisfactorily predicts the magnetic properties of molecules, something that the valence bond theory does not do.

**A number of rules that apply to molecular orbitals and the stability of molecules are summarized below.**

1. The number of molecular orbitals formed is always equal to the number of atomic orbitals combined.
2. The more stable the bonding molecular orbital, the less stable is the corresponding antibonding molecular orbital.
3. In a stable molecule, the number of electrons in bonding molecular orbitals is always greater than that in antibonding molecular orbitals.
4. As with atomic orbitals, each molecular orbital can accommodate up to two electrons with opposite spins, in accordance with the Pauli exclusion principle.
5. When electrons are added to molecular orbitals having the same energy, the most stable arrangement is that predicted by Hund's rule. That is, as far as possible, electrons occupy these orbitals singly, and with parallel spins, rather than in pairs.
6. The number of electrons in the molecular orbitals is equal to the sum of all the electrons on the atoms.

Examples
10.7, 10.8

Exercises
14 – 17

## Delocalized Molecular Orbitals (Section 10.8)

Recall in chapter 9 the discussion of resonance. To understand the bonding in molecules that exhibit resonance, we must discuss delocalization of electrons. If a pair of electrons is not confined between two adjacent atoms, but is shared between three or more atoms, the bond formed is called a delocalized molecular orbital. To demonstrate the concept, we will examine $NO_2^-$. The resonance structures are as follows:

$$[\ddot{\text{O}}\!-\!\ddot{\text{N}}\!=\!\ddot{\text{O}}]^- \longleftrightarrow [\ddot{\text{O}}\!=\!\ddot{\text{N}}\!-\!\ddot{\text{O}}]^-$$

To describe the bonding, look at the central atom (of either resonance structure). The nitrogen has two bonded atoms and one lone pair for three electron pair groups. This is a trigonal planar geometry and thus, the nitrogen has sp² hybridization. Two of the sp² hybrid orbitals form the sigma bonds with the oxygen atoms and one houses the lone pair. The pi bond, is shown in the resonance structure to alternate between the two oxygen atoms. In

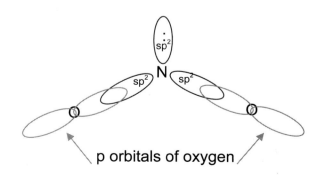

p orbitals of oxygen

delocalized molecular orbital theory, the two electrons which make up the pi bond are delocalized about the three atoms, by existing in the shared space formed by the overlap of the p orbitals of each atom, which lie perpendicular to the plane formed by the sigma bonds. This is illustrated in Figure 10.7

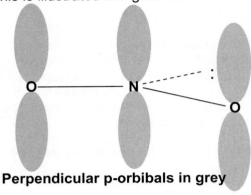

**Perpendicular p-orbibals in grey**

**Figure 10.7** The p orbitals perpendicular to the plane of the sigma bonds overlap. This gives electron density above and below the plane of the O–N–O bond. The two electrons which make up the pi bond are able to move about in the grey area depicted in the bottom diagram. In other words, the two electrons making the double bond (shown in the resonance structure) are now delocalized about the whole molecule.

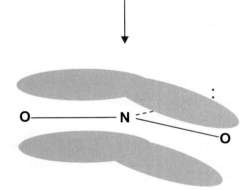

Anytime multiple resonance structures can be drawn for a molecule or ion, delocalization molecular orbitals can be used to describe the pi bonds in the molecule.

**Example 10.9**

**Exercise 18**

## GLOSSARY LIST

| | | |
|---|---|---|
| valence shell | hybrid orbital | bond order |
| **valence-shell** | **hybridization** | **homonuclear** |
| **diatomic** | | |
| **electron-pair** | | molecule |
| repulsion theory | sigma bond | delocalized molecular |
| | pi bond | orbital |
| dipole moment | molecular orbital | sigma molecular orbital |
| polar molecule | bonding molecular | pi molecular obital |
| nonpolar molecule | orbital | |
| | antibonding molecular | |

## EQUATIONS

| Algebraic Equation | English Translation |
|---|---|
| $\mu = Q \times r$ | Dipole moment as a function of charge and distance between charges. |
| bond order $= \frac{1}{2}\left[\left(\#\ \text{bonding e}^-\text{'s}\right) - \left(\#\ \text{antibonding e}^-\text{'s}\right)\right]$ | Determination of bond order using molecular orbital diagrams. |

## WORKED EXAMPLES

---

**EXAMPLE 10.1 VSEPR Theory**

Predict the geometrical shapes and bond angles for the following compounds using the VSEPR theory.
a.  $SnCl_4$
b.  $O_3$
c.  $IF_5$
d.  $XeF_2$

**• Solution**

a.  First, determine the number of valence shell electrons for the central Sn atom. A tin atom has four valence electrons, and each Cl atom contributes one electron which it shares with the Sn atom, so 4 + 4(1) = 8. The eight valence shell electrons are arranged as four electron pairs around the Sn atom. The Lewis structure is:

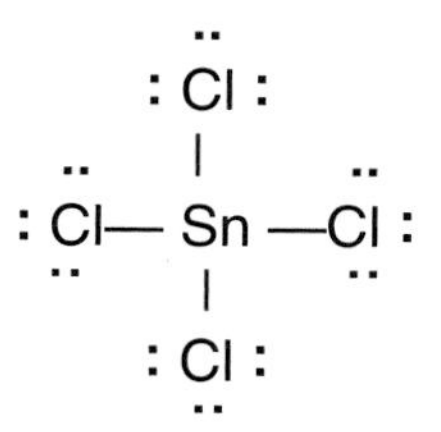

This structure shows that there are no lone pairs about the Sn atom. Thus, the four bonding electron pairs and the Cl atoms, as well, are oriented at the corners of a tetrahedron. $SnCl_4$ is an $AB_4$ type molecule, and has a tetrahedral structure.
The Cl—Sn—Cl bond angles are 109.5°.

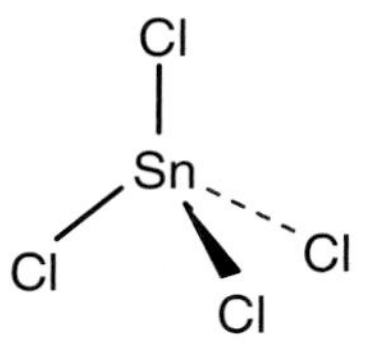

b.  The Lewis structure of ozone ($O_3$) is given below. Note, there are multiple resonance structures that can be drawn for ozone. You only need to draw one Lewis structure in order to determine the geometry.

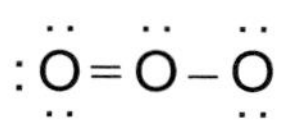

    Lewis structure 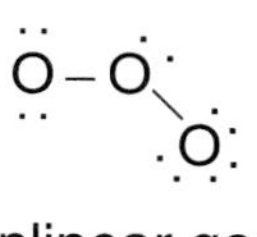    nonlinear geometry

We look at the central atom (the middle O). The number of bonded atoms is two, the number of lone pair is one, for a total of 3 electron groups. These three regions of charge are oriented at the corners of a triangle. However, one of these pairs is a lone pair. The molecule is of the type $AB_2E$. The positions of the three O atoms are described as nonlinear. The O—O—O bond angle should be close to 120°. However, the presence of lone-pair vs. bonding-pair repulsion compresses the bond angle. The observed angle in ozone is 117°.

c.   Now that we have had some practice we will abbreviate the explanation. To determine the shape of $IF_5$, count the valence shell electrons around the central I atom.

    from the I atom                  = 7
    from the F atoms (5 × 1)     = 5
    total valence shell electrons  = 12

    The Lewis structure is:

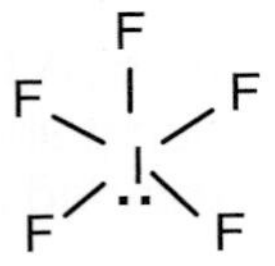

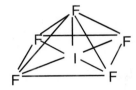

    valence shell electron pairs   = 6
    number of lone pairs           = 1
    number of bonded atoms         = 5
    molecular type $AB_5E$
    molecular shape:  square pyramid

    The F—I—F bond angles are 90°.

d.   To predict the shape of $XeF_2$. First, count the valence shell electrons around Xe.

    from the Xe atom               = 8
    from the F atoms 2 × 1         = 2
    total valence shell electrons  = 10

    The Lewis structure is:

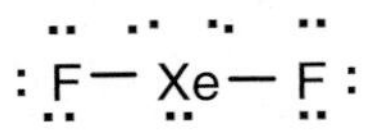

    :$\overset{..}{\underset{..}{F}}$— Xe —$\overset{..}{\underset{..}{F}}$:

    valence shell electron pairs   = 5
    number of lone pairs           = 3
    number of bonded atoms         = 2              F—Xe—F
    molecular type $AB_2E_3$
    molecular shape:  linear

The F—Xe—F bond angles are 180°.

**Work EXAMPLES & PROBLEMS: 1 – 4**

**EXAMPLE 10.2  Polar and Nonpolar Molecules**

Predict whether the following molecules are polar or nonpolar:
a.  CO          b.  $H_2CO$          c.  $CCl_4$

• **Solution**

Polarities of diatomic molecules depend on the electronegativity differences of the atoms.

a.  Oxygen is more electronegative than carbon. As a result, the $C\equiv O$ bond is polar with partial positive and negative charges on C and O, respectively.

$$\overset{\longmapsto}{C\equiv O}$$

The dipole moment of carbon monoxide is 0.1 D.

b.  The dipolar nature of a polyatomic molecule depends on both the bond polarity and molecular geometry. Formaldehyde is a planar molecule with a polar $C=O$ bond ($\Delta = 1.0$) and two C—H bonds of very low polarity ($\Delta = 0.4$). Neglecting the low polarity C—H bond dipoles, the charge distribution is:

$$\begin{array}{l} H \\ \phantom{H}\backslash \longmapsto \\ \phantom{H}C=O \\ \phantom{H}/ \\ H \end{array}$$

The net effect of all bond dipoles is that the center of positive charge is located near the carbon atom and the center of negative charge lies near the oxygen atom.

$$\begin{array}{l} H \\ \phantom{H}\backslash \; \delta+ \; \delta- \\ \phantom{H}C=O \\ \phantom{H}/ \\ H \end{array}$$

Formaldehyde has a measured dipole moment of about 2.5 D.

c.  Chlorine is more electronegative than carbon. Therefore, $CCl_4$ has four bond dipoles. However, these polar bonds are arranged in a symmetric tetrahedral fashion about the central carbon atom.

In this situation the centers of positive and negative charge are both on the carbon atom. The bond dipoles have canceled each other and the molecule as a whole does not possess a dipole moment.

**Work EXERCISES & PROBLEMS: 5 – 7**

**EXAMPLE 10.3  Valence Bond Theory**

Using *unhybridized* atomic orbitals, describe what atomic orbitals are used in the following molecules to form sigma bonds.

a.  $F_2$         b.  $H_2S$

**• Solution**

a.  From the orbital diagram of fluorine we can see that a F atom has one half-filled 2p orbital that can be used for bonding. If this orbital overlaps with the half-filled 2p orbital of another F atom, a pair of electrons (one from each atom) spend most of the time between the two atoms and bonding occurs.

    F   [↑↓]    [↑↓|↑↓|↑]
        2s        2p

b.  Hydrogen atoms have a $1s^1$ electron configuration. Therefore, they can use the half-filled 1s orbital for bonding. The orbital diagram of the valence electrons in an S atom is

    S   [↑↓]    [↑↓|↑|↑]
        3s        3p

    Overlap of the two half-filled 3p orbitals of a sulfur atom with the half-filled 1s orbitals of the two H atoms results in the formation of two covalent bonds.

**• Comment**

The directional properties of two p orbitals (say the $p_y$ and $p_z$) are such that the orbitals project out from the S atom at right angles to each other. The H—S—H bond angle should be 90°. The measured angle is 92.2°.

**Work EXERCISES & PROBLEMS: 8**

---

**EXAMPLE 10.4  Hybrid Orbitals**

Using orbital diagrams, outline the steps used to describe the formation of a set of $sp^2$ hybridized orbitals on a carbon atom.

**• Solution**

The ground state orbital diagram for the *valence electrons* of a carbon atom is:

    C   [↑↓]    [↑|↑| ]
        2s        2p

First, promote one 2s electron to the empty 2p orbital. Mixing or hybridization of the 2s orbital with two of the 2p orbitals creates three equivalent $sp^2$ hybrid orbitals. These hybrid orbitals point out from the carbon atom to the corners of a triangle. One electron remains in an unhybridized $2p^2$ atomic orbital.

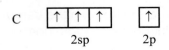

    C   [↑|↑|↑]    [↑]
        2sp         2p

**EXAMPLE 10.5  Hybrid Orbitals**

Describe the *hybridization* of the central atom in:
a.  $PH_3$                     b.  $PCl_5$

**• Solution**

First, use VSEPR theory to predict the geometry. Then, match the geometry to the corresponding hybridization.

a.  According to VSEPR theory, the $PH_3$ molecule has trigonal pyramidal geometry and a lone pair. There are four electron pairs around the P atom that occupy the corners of a tetrahedron.

Therefore, four equivalent orbitals project out from the central P atom at 109° from each other. The bonding orbitals of the P atom must be $sp^3$ hybrid orbitals.

b.  According to VSEPR theory, $PCl_5$ has the geometrical shape of a trigonal bipyramid.

$$Cl \overset{\displaystyle Cl}{\underset{\displaystyle Cl}{\overset{|}{\underset{|}{Cl - P \cdots Cl}}}}$$

The type of hybridization that gives five equivalent orbitals that point to the corners of a trigonal bipyramid is $sp^3d$.

**Work EXERCISES & PROBLEMS: 9 – 12**

**EXAMPLE 10.6  Sigma and Pi Bonds**

Account for the bonding in $H_2CO$, which is a triangular molecule.
a.  What is the hybridization of the carbon atom?
b.  What are the approximate bond angles about the carbon atom?
c.  How many sigma and pi bonds are there in the molecule?

**• Solution**

a.  The central carbon atom of $H_2CO$ must be using $sp^2$ hybrid orbitals, since the three hybrid orbitals of this type point to the corners of a triangle. There is also one electron in an unhybridized $2p$ atomic orbital.

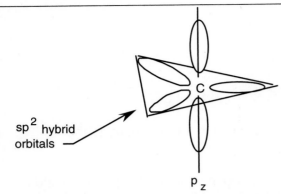

sp$^2$ hybrid
orbitals

b.  Overlap of the half-filled 1s orbital of a H atom with the half-filled sp² hybrid orbitals leads to formation of a C—H sigma bond. Two C—H bonds are formed in this way. The H—C—H bond angle should be 120°.

c.  Oxygen atoms have the valence electron configuration

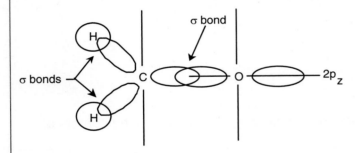

Overlap of one half-filled 2p orbital with the remaining sp² hybrid yields a C—O sigma bond. The two C—H sigma bonds and the C—O sigma bond are shown.

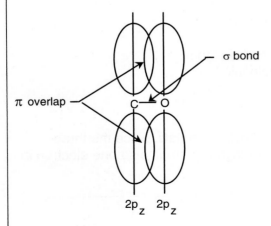

The sideways overlap of the carbon 2p$_z$ orbital with the 2p$_z$ orbital of the oxygen atom produces a C—O pi bond.

The C—O sigma and pi bonds constitute the C═O double bond.

The complete molecule consists of three sigma bonds and one pi bond.

**Work EXERCISES & PROBLEMS: 13**

---

**EXAMPLE 10.7  Molecular Orbital Diagram**

Represent bonding in the $O_2^-$ ion by means of a molecular orbital diagram. What is the bond order?

**• Solution**

The $O_2$ molecule has 16 electrons.  The $O_2^-$ ion will have 17 electrons. The energy level diagram is shown.  The 4 electrons in the $\sigma_{1s}$ and $\sigma_{1s}*$ orbitals are not shown because they are not involved in bonding.

The electron configuration of $O_2^-$ ion is

$$O_2^- \quad (\sigma_{1s})^2(\sigma_{1s}*)^2(\sigma_{2s})^2(\sigma_{2s}*)^2(\,_{2p_y})^2(\,_{2p_z})^2(\sigma_{2p_x})^2(\,_{2p_y}*)^2(\,_{2p_z}*)^1$$

$$\text{bond order} = \frac{1}{2}\left(\begin{array}{l}\text{number of electrons} \\ \text{in bonding MOs}\end{array} - \begin{array}{l}\text{number of electrons} \\ \text{in antibonding MOs}\end{array}\right)$$

$$= \frac{1}{2}(8-5) = 1.5$$

**Work EXERCISES & PROBLEMS: 14 – 16**

---

**EXAMPLE 10.8  Paramagnetism and Diamagnetism**

Which of the following species are paramagnetic and which are diamagnetic?
a.  $N_2$
b.  $N_2^+$
c.  $B_2$

• **Solution**

a.  The electron configuration of $N_2$ is:

  $N_2$ $(\sigma_{1s})^2(\sigma_{1s}*)^2(\sigma_{2s})^2(\sigma_{2s}*)^2(\ _{2p_y})^2(\ _{2p_z})^2(\sigma_{2p_x})^2$

  All orbitals contain pairs of electrons.  Because of the lack of unpaired electrons, $N_2$ is diamagnetic.

b.  $N_2^+$ will have one less electron than $N_2$.  Therefore, it has one unpaired electron and is paramagnetic.

c.  The electron configuration of $B_2$ should have 10 electrons:

  $B_2$ $(\sigma_{1s})^2(\sigma_{1s}*)^2(\sigma_{2s})^2(\sigma_{2s}*)^2(\ _{2p_y})^1(\ _{2p_z})^1$

  Because of the unpaired electrons, $B_2$ is paramagnetic.

**Work EXERCISES & PROBLEMS: 17**

---

**EXAMPLE 10.9  Delocalized Molecular Orbital Theory**

Which of the following species will contain delocalized molecular orbitals?
a.  $SO_2$                    b.  CO

•   **Solution**

First draw the Lewis structure, including any resonance structures

a.   For $SO_2$, the resonance structures are

$$:O - S = O: \quad \longleftrightarrow \quad :O = S - O:$$

The fact that resonance structures can be drawn determines that delocalized molecular orbitals exist.  The two electrons which make up the pi bond are delocalized in orbitals formed by the sideways overlap of the p orbitals on all three atoms.

> b. For CO, the Lewis structure is:
>
> $$:C\equiv O:$$
>
> Because there is only one way to draw this Lewis structure, there are no resonance structures; the molecule does not contain delocalized molecular orbitals.

**Work EXERCISES & PROBLEMS: 18**
**Work the rest of the EXERCISES & PROBLEMS**

## EXERCISES & PROBLEMS

1. Predict the shape of the following molecules.
   a. $BeCl_2$   b. $NF_3$   c. $SF_4$   d. $ClF_3$   e. $KrF_4$

2. Predict the shape of the following ions.
   a. $ClO_2^-$   b. $CO_3^{2-}$   c. $ZnCl_4^{2-}$

3. Which of the following molecules and ions have tetrahedral geometry?
   a. $CCl_4$   b. $SCl_4$   c. $AlCl_4^-$

4. Predict the following bond angles.
   a. O—C—O in $CO_2$   b. F—C—F in $CF_4$   c. H—O—O in $H_2O_2$

5. What two requirements must be met in order for a molecule to be polar?

6. Which of the following molecules are polar?
   a. $BCl_3$   b. $SO_3$   c. $PF_3$   d. $SF_6$

7. Which of the following molecules are polar?
   a. $BeH_2$   b. $CO_2$   c. $SO_2$   d. $NO_2$

8. Using only atomic orbitals, describe the bond formation in $F_2$ and HBr.

9. What is the angle between the following two hybrid orbitals on the same atom?
   a. two sp orbitals   b. two $sp^2$ orbitals   c. two $sp^3$ orbitals

10. What hybrid orbital set is used by the underlined atom in the following molecules?
    a. $\underline{C}S_2$   b. $\underline{N}O_2^-$   c. $\underline{B}F_3$   d. $\underline{Cl}F_3$   e. $H_2\underline{C}O$

11. What hybrid orbital set is used by the underlined atom in the following molecules?
    a. $\underline{S}F_2$   b. $\underline{S}F_4$   c. $\underline{Xe}F_4$

12. The sulfur atom in selenium hexachloride is $sp^3d^2$ hybridized. What is the geometry of $SeCl_6$?

13. How many sigma bonds and pi bonds are there in a molecule of $CO_2$?

14. a. Which has the higher energy, a bonding molecular orbital or its corresponding antibonding molecular orbital?
    b. Which has the higher energy, a bonding molecular orbital or the atomic orbitals from which it is created?

15. Compare the bond order in $Be_2$ to that in $Be_2^+$.

16. Give the electron configurations for $Li_2$ and $Li_2^+$. Which has the stronger Li—Li bond?

17. Which of the following diatomic species are paramagnetic?
    a. $O_2$    b. $O_2^-$    c. $Li_2$

18. What orbitals overlap to form a delocalized $\pi$ bond in ozone ($O_3$)?

19. In molecules having the trigonal bipyramidal geometry why does a lone pair occupy an equatorial position rather than an axial position?

20. Experimental evidence shows that carbon dioxide has no dipole moment. What does this suggest about its molecular shape?

21. Both $N_2$ and $O_2$ can be ionized to form $N_2^+$ and $O_2^+$, respectively. Explain why the bond order of $N_2$ is greater than that of $N_2^+$, but the bond order of $O_2$ is less than that of $O_2^+$.

## PRACTICE TEST QUESTIONS
See notes on taking practice test in the Preface

1. Use VSEPR theory to predict the shapes of the following molecules:
   a. $BrF_5$           b. HCN
   c. $BF_3$           d. $SO_2$
   e. $SCl_2$

2. What type of hybrid orbital is used by the central atom of each of the following molecules?
   a. $BrF_5$           b. HCN
   c. $BF_3$           d. $SO_2$
   e. $SCl_2$

3. Use VSEPR theory to predict the shapes of the following polyatomic ions:
   a. $BeF_3^-$           b. $AsF_4^-$
   c. $ClF_4^-$           d. $NO_3^-$
   e. $SO_4^{2-}$

4. What type of hybridization is used by the central atom of each of the following polyatomic ions?
   a. $BeF_3^-$           b. $AsF_4^-$
   c. $ClF_4^-$           d. $NO_3^-$
   e. $SO_4^{2-}$

5. Which of the following molecules would be tetrahedral?
   a. $SO_2$           b. $SiH_4$
   c. $SF_4$           d. $BCl_3$
   e. $XeF_4$

6. Predict the approximate bond angles in each of the following
   a. $GeCl_2$           b. $IF_4^-$
   c. $TeCl_4$

7. $CCl_4$ is a perfect tetrahedron, but $AsCl_4^-$ is a distorted tetrahedron. Explain.

8. $BeCl_2$ and $TeCl_2$ are both covalent molecules, yet $BeCl_2$ is linear while $TeCl_2$ is nonlinear (bent). Explain.

9. What types of hybrid orbitals can be formed by elements of the third period that cannot be formed by elements of the second period?

10. When we describe the formation of hybrid orbitals on a central atom, must all the available atomic orbitals enter into hybridization?

10

10

10

10

10

10

10

10

10

10

11. Which molecule should have the larger dipole moment, HBr or HI?

12. Which one of the following molecules has a dipole moment?
    a. $CCl_4$        b. $H_2S$
    c. $CO_2$        d. $BCl_3$
    e. $Cl_2$

13. Which of the following molecules have no dipole moment?
    a. ClF            b. $BCl_3$
    c. $BeCl_2$       d. $Cl_2O$
    e. $H_2CO$

14. Choose the best answer. For the water molecule:
    a. The bonds are polar, and the molecule is nonpolar.
    b. The bonds are nonpolar, and the molecule is polar.
    c. The bonds are polar, and the molecule is polar.
    d. The bonds are nonpolar, and the molecule is nonpolar.

15. Hydrogen peroxide ($H_2O_2$) has a dipole moment of 2.1 D. Which of the bonds in $H_2O_2$ are polar? Is the molecule linear?

16. Draw a molecular orbital energy level diagram and determine the bond order for each of the following:
    a. $H_2^+$        b. $He_2$
    c. $He_2^+$

17. The compound calcium carbide, $CaC_2$, contains the acetylide ion $C_2^{2-}$.
    a. Write molecular orbital electron configurations for $C_2$ and $C_2^{2-}$.
    b. Compare the bond orders in $C_2$ and $C_2^{2-}$.

18. The bond length in $N_2$ is 109 pm and in $N_2^+$ it is 112 pm. Explain why the bond lengths differ in this way.

10

10

10

10

10

10

10

10

# Chapter 11. Intermolecular Forces, Liquids & Solids

**The Kinetic Molecular Theory of Liquids and Solids (Section 11.1)**
**Intermolecular Forces (Section 11.2)**
**Properties of Liquids (Section 11.3)**
**The Solid State (Sections 11.4 – 11.7)**
**Phase Changes and Phase Diagrams (Sections 11.8 – 11.9)**

## SUMMARY

### The Kinetic Molecular Theory of Liquids and Solids (Section 11.1)

In this chapter, we examine the structure of liquids and solids, and discuss some of the fundamental properties of these two states of matter. Matter can exist in three phases: gas, liquid, and solid. Each phase can be described from a molecular viewpoint in terms of four characteristics: distance between molecules, attractive forces between molecules, the motion of molecules, and the orderliness of the arrangement of molecules. Figure 1.6 in the text gives a molecular representation of a solid, a liquid, and a gas.

In Chapter 5, we discussed the fact that a gas can be pictured as molecules that are far apart, in constant random motion, and exert almost no attractive forces on each other. The kinetic molecular theory was used to explain the behavior of gases.

Liquids and solids are quite different from gases. They are referred to as the condensed states because their molecules are very close together. Gases have low density, high compressibility, and completely fill a container. The condensed states have relatively high density, are almost incompressible, and have definite volumes. These properties indicate that the "molecules" of solids and liquids are close together and are held by strong intermolecular forces.

One difference between solids and liquids is that liquids, like gases, are *fluids*, but solids are not. In liquids, molecules can move past each other even though they cannot get very far from each other. In solids, molecules are fixed in position, and can only vibrate about that position. Many solids are characterized by the long-range order of their constituent molecules. This order gives rise to crystal structures. Table 11.1 summarizes some properties of the three states of matter.

**Table 11.1** Characteristic Properties of the States of Matter

| Gas | Liquid | Solid |
|---|---|---|
| Assumes the volume and shape of its container | Has a definite volume but assumes the shape of its container | Has a definite volume and shape |
| Is a fluid (flows readily) | Is a fluid | Is not a fluid |
| Very compressible | Only slightly compressible | Virtually incompressible |
| Low density | High density | High density |
| Molecules far apart | Molecules close together | Molecules close together |

## Intermolecular Forces (Section 11.2)

***van der Waals Forces.*** The forces that hold atoms together, such as covalent bonds, exist *within* molecules. In this chapter you will learn about the types of forces that act *between* individual molecules that hold them close together as in a solid or a liquid. Forces of attraction between molecules are called intermolecular forces. Intermolecular forces can be grouped for convenience into van der Waals forces, ion-dipole forces, and hydrogen bonds. The van der Waals forces cause a gas to deviate from ideal gas behavior and also are responsible for gases condensing to form liquids. They are attractive forces resulting from dipole-dipole forces, dipole-induced dipole forces, and dispersion forces. We shall discuss each of these interactions briefly.

***Dipole-Dipole Forces.*** Some molecules have a dipole moment $\mu$, and are referred to as polar molecules. In a polar substance, molecules tend to become aligned with the positive end of one molecule, directed toward the negative ends of neighboring dipoles. See Figure 11.1 in the text. The attraction is electrostatic. The larger the dipole moment, the stronger the force of attraction. Dipole-dipole forces are attractive forces between polar molecules.

***Dipole-Induced Dipole Forces.*** The presence of a polar molecule in the vicinity of another molecule (usually nonpolar) has the effect of polarizing the second molecule. This means that a dipole moment has been induced in the previously nonpolar molecule. The induced dipole can then interact with the dipole moment of the first molecule, and the two molecules are attracted toward each other. The induced dipole is a temporary dipole. Its interaction with a permanent dipole such as that in a polar molecule is called a dipole-induced dipole force.

***Dispersion Forces.*** Nonpolar molecules such as $O_2$, $SF_6$, and even the noble gases such as He, Ne, Ar, Kr, and Xe all show deviations from the ideal gas equation, and all condense at low temperatures to form liquids. We can ask, "What kind of attractive forces exist between nonpolar molecules?" In 1930, Fritz London proposed that although these molecules have no permanent dipole moments, their electron clouds are fluctuating. In a helium atom, for instance, the two electrons occupy a 1s orbital, which has a spherical shape. The electrons are in constant motion, and if for an instant they should both move to the same side of the nucleus, a short-lived dipole will exist. In an instant the electron will continue its motion and the dipole will be gone, but a new one will be formed. These short-lived dipoles are called instantaneous dipoles or temporary dipoles. An instantaneous dipole can polarize a neighboring molecule, thereby producing an induced dipole. The two dipoles will tend to stick together. The attractive interactions caused by instantaneous dipoles are known as dispersion forces.

The strength of dispersion forces depends on the polarizability of the molecule and can be as large as, or larger than, dipole-dipole forces. Polarizability is the tendency of an electron cloud to be distorted by the presence of an electrical charge such as that of an ion or the partial charge of a dipole. In general, the polarizability increases as the total number of electrons in a molecule increases. Since molecular mass and number of electrons are related, the polarizability of molecules and the strength of dispersion forces increases with increasing molecular mass.

***Ion-Dipole Forces.*** The electrostatic attraction between an ion and a polar molecule is called an ion-dipole force. Hydration of ions, which was discussed in Section 6.7 (in the discussion of heat of solution), is a good example of ion-dipole interactions. When an ionic compound such as NaBr dissolves, its cations and anions are attracted to water molecules.

Water molecules are polar and, therefore, have a negative end and a positive end. The $Na^+$ ions attract the negative end of the water molecule, and the $Br^-$ ions attract the positive end of the water molecule. As the charge of an ion increases, it attracts polar molecules more strongly. Thus, $Mg^{2+}$ ions attract water molecules more strongly than $Na^+$ ions, and $S^{2-}$ ions attract water molecules more strongly than $Br^-$ ions.

***Hydrogen Bonding.*** An additional type of intermolecular force is necessary to explain certain properties of ammonia, water, and hydrogen fluoride. A *hydrogen bond* is a special type of dipole-dipole interaction between the hydrogen atom in a polar bond and an O, N, or F atom. In water, for instance, the attractions are stronger than just the attractions of one dipole for another. Each hydrogen atom, with its partial positive charge, is attracted to one of the lone electron pairs of an oxygen atom of a neighboring molecule. This interaction is called a hydrogen bond, and is represented by the three dots in the following:

$$
\begin{array}{c}
\phantom{H-O:\cdots}H\phantom{-}\\
\phantom{..\;\;\;}\overset{..}{\phantom{H}}\quad\;\delta+\;\;\;\overset{|}{\phantom{H}}\delta-\\
H\!-\!O:\cdots H\!-\!O:\\
\phantom{H-}|\phantom{:\cdots H-O}\overset{..}{\phantom{:}}\\
\phantom{H-}H
\end{array}
$$

Hydrogen bonding is limited to compounds containing hydrogen bonded to nitrogen, oxygen, and fluorine. Hydrogen bonds are the strongest of the intermolecular forces, with energies of the order 10–40 kJ/mol. The energies of van der Waals interactions are between 2 and 20 kJ/mol.

---

**The requirements for hydrogen bond formation are:**

1. The element that is covalently linked to hydrogen must be sufficiently electronegative to attract bonding electrons and leave the hydrogen atom with a significant δ+ charge. Only N, O, and F are sufficiently electronegative.
2. The electronegative atom, bound to hydrogen by the hydrogen bond, must have a lone pair of electrons.
3. The small size of the hydrogen atom allows it to approach N, O, and F atoms in neighboring molecules very closely. It is significant that hydrogen bonding is limited to these three elements of the second period. Both sulfur and chlorine are highly electronegative, but do not form hydrogen bonds. These third period atoms are apparently too large and do not present a highly localized nonbonding electron pair for the H atom to be attracted to.

Examples
11.1 – 11.3

Exercises
1 – 5

---

## Properties of Liquids (Section 11.3)

***Surface Tension.*** The surface tension of a liquid is a property that has no direct counterpart in solids or gases. Surface tension is what makes water bead up on a freshly waxed surface and what makes soap bubbles round. Surface tension is a force that tends to minimize the surface area of a drop of liquid. Energy is required to expand the surface of a liquid, and the surface tension is the amount of energy required to increase the surface area of a liquid by a unit area. Liquids in which strong intermolecular forces exist, exhibit high surface tensions.

Water beads up on a freshly waxed surface because its cohesive forces are stronger than its adhesive forces. Cohesion is the intermolecular attraction between like molecules in the drop. Adhesion is the intermolecular attractions between unlike molecules, such as between water and wax. Since there is very little attraction between water and wax (adhesion), the strong cohesive forces tend to maintain the drop. The drop adopts a spherical shape because a sphere has the least surface area.

When water comes in contact with glass, adhesion is stronger than cohesion. Water is pulled against the glass surface, and we say that water 'wets" the surface. Wetting results in the spreading of a thin film of water on the glass surface. The strength of the cohesive forces overcomes the ability of the surface tension to make a spherical drop.

***Viscosity.***  One characteristic of liquids that we have all observed is related to how freely they flow. For example, water pours much more freely than motor oil, and motor oil more readily than molasses. The unique pouring characteristics of each liquid are the result of its viscosity. Viscosity is a measure of a fluid's resistance to flow. Liquids whose molecules have strong intermolecular forces have greater viscosities than liquids that have weaker intermolecular forces. Table 11.3 (in the text) lists viscosities of some common liquids.

The viscosity of water and glycerol is the result of strong hydrogen bonds. The viscosity of motor oil results from strong London dispersion forces.

***Properties of Water.***  Given its abundance and familiarity, we often fail to realize just how unusual water actually is. Some of the unusual properties of water are:

1.  Water has a considerably greater surface tension than most other liquids.
2.  At 4.184 J/g • °C, the specific heat of water is one of the highest of all substances.
3.  The heats of fusion and vaporization (discussed later in this chapter) are 6.0 and 40.8 kJ/mol, respectively. $H_2O$ is unusual because 40.8 kJ is a huge vaporization energy. Compare values in Table 11.6 (text).
4.  The boiling point of water is about 200 °C higher than might be reasonably expected. Boiling point tends to be related to the molar mass. Compare the boiling point of water with that of other low molar mass liquids in Table 11.2.

**Table 11.2**  Boiling Points of Some Small Molecules

| Liquid | Molar Mass (g/mol) | Boiling Pt ( °C) |
|---|---|---|
| $H_2O$ | 18 | 100° |
| $CH_4$ | 16 | −159° |
| $NH_3$ | 17 | −33° |

5.  Density versus temperature. As liquids cool they become more and more dense. As water is cooled from 100 °C down to 4 °C it does indeed get more dense. But from 4 °C (actually 3.98 °C) down to 0 °C, just the opposite happens:  it gets less dense (see Table 11.3).

**Table 11.3** Density of $H_2O$ Near 0 °C

| Temperature  °C | Density (g/cm³) |
|---|---|
| 10.0 | 0.9997 |
| 5.0 | 0.99999 |
| 3.98 | 1.00000 |
| 2.0 | 0.99997 |
| 0 | 0.9998 |

6.  Density of liquid versus density of solid. There are over eight million known chemical compounds. All, except a dozen or so, have a solid state that is more dense than the liquid. Water, of course, is one of the exceptions. The density of solid water at 0 °C is 0.917 g/cm³, which compares with 0.9998 g/cm³ for the liquid. Because of its lower density, ice floats on water.

> Example
> 11.4
>
> Exercises
> 6 – 7

## The Solid State (Sections 11.4 – 11.7)

***Crystal Structure.*** The atoms, molecules, or ions of a crystalline solid occupy specific positions in the solid, and possess long-range order. The unit cell is the smallest unit that when repeated over and over again, generates the entire crystal. Three types of unit cell are the simple cubic cell (scc), the body-centered cubic cell (bcc), and the face-centered cubic cell (fcc). These types are illustrated in Figure 11.17 of the text.

In a simple cubic cell, particles are located only at the corners of each unit cell. In a body-centered cubic cell, particles are located at the center of the cell as well as at the corners. In a face-centered cubic cell, particles are found at the center of each of the six faces of the cell as well as at the corners. In many calculations involving properties of a crystal, it is important to know how many atoms or ions are contained in each unit cell. Atoms at the corners of unit cells are shared by neighboring unit cells. For a cubic cell, each corner atom is shared by eight unit cells and is counted as 1/8 particle for each unit cell. A face-centered atom is shared by two unit cells and is counted as 1/2 of an atom for each unit cell. An atom located at the center belongs wholly to that unit cell.

The packing efficiency is the percentage of the cell space occupied by spheres. The empty space is the interstices or "holes" between spheres. There are many substances whose atomic arrangement can be pictured as a result of packing together identical spheres so as to achieve maximum density. Many metallic elements and many molecular crystals display these "closest packed" structures. The text describes these two "closest packed" structures in Section 11.4. Alternating layers of spheres in ABABAB... fashion yields a hexagonal close-packed (hcp) structure, while alternating layers of spheres in the ABCABC... fashion generates the cubic close-packed (ccp) structure. In both structures each sphere has a coordination number of 12 which is the maximum coordination number. For both structures the packing efficiency is 74%.

> Examples
> 11.5 – 11.7
>
> Exercises
> 8 – 11

***X-Ray Diffraction of Crystals.*** Crystal structures are analyzed experimentally by the technique of X-ray diffraction. Diffraction is a wave property. When an X-ray beam encounters a single crystal, the beam is scattered from the crystal at only a few angles, rather than randomly. The scattering angle, $\theta$, depends on the spacing, d, between layers or planes of atoms in the crystal, the wavelength of the X-rays, and the reflection order (n = 1, 2, 3, …). These quantities are related by the Bragg equation:

$$n\lambda = 2d \sin \theta$$

The spacing (d) between atomic planes in crystals can be determined by using X-rays of known wavelength ($\lambda$) and by experimental measurement of the angle $\theta$. See Figure 11.24 in text.

> Example
> 11.8
>
> Exercises
> 12 - 13

***Types of Crystals.*** In a solid, atoms, molecules, or ions occupy specific positions called lattice points. Crystalline solids are characterized by a regular three-dimensional arrangement of lattice points. In ionic solids, the lattice points are occupied by positive and negative ions. In all ionic compounds there are continuous three-dimensional networks of alternating positive and negative ions held together by strong electrostatic forces (ionic bonds). Most ionic crystals possess high lattice energies, high melting points, and high boiling points. In ionic solids there are no discrete molecules as are found in molecular or covalent substances.

In covalent crystals, the lattice points are occupied by atoms that are held by a network of covalent bonds. Diamond, graphite, and quartz are well known examples. Materials of this type have high melting points and are extremely hard because of the large number of covalent bonds that have to be broken to melt or break up the crystal. The entire crystal can be thought of as one giant molecule.

Covalent compounds form crystals in which the lattice positions are occupied by molecules. Such solids are called molecular crystals. Covalent crystals are soft and have low melting points. These properties are the result of the relatively weak intermolecular forces (van der Waals forces and hydrogen bonding) that hold the molecules in the crystal.

Crystals of polar compounds are held together by dipole-dipole forces and dispersion forces. Only dispersion forces occur in the lattices of nonpolar compounds. As a rule, polar compounds melt at higher temperatures than nonpolar compounds of comparable molecular mass.

Metallic crystals are quite strong. Most transition metals have high melting points and densities. In metals, the array of lattice points is occupied by positive ions. The outer electrons of the metal atoms are loosely held and move freely from ion to ion throughout the metallic crystal. This mobility of electrons in metals accounts for one of the characteristic properties of metals, namely the ability to conduct electricity.

> Example
> 11.9

Amorphous solids lack a well-defined arrangement and long range order. Glass is considered an amorphous solid because it lacks a regular three-dimensional arrangement of atoms.

## Phase Changes and Phase Diagrams (Sections 11.8 – 11.9)

***Liquid-Vapor Equilibrium.*** The different physical states of a substance that are present in a system are referred to as phases. A phase is a homogeneous part of the system that is in contact with other parts of the system, but is separated by a noticeable boundary. Ice, liquid water, and water vapor can exist together in a suitable container. Each of these is a phase. When a liquid substance is placed in a closed container, it will not be long before molecules of this substance can be found in the gas phase above the liquid. Vaporization is the process in which liquids become gases. When a portion of a liquid evaporates, the gaseous molecules exert a pressure called the vapor pressure. The term vapor is often applied to the gaseous state of a substance that is normally a liquid or solid at the temperature of interest.

In a closed container the vapor pressure does not just continually increase, rather a state of equilibrium is reached in which the vapor pressure becomes constant and the amount of liquid remains constant. This equilibrium vapor pressure results from two opposing processes. The opposing process to vaporization is condensation. As the concentration of the molecules in the vapor phase increases, some will strike the liquid surface and condense.

$$\text{liquid} \underset{\text{condensation}}{\overset{\text{vaporization}}{\rightleftharpoons}} \text{vapor}$$

When the rates of the opposing processes become equal, the vapor pressure remains constant and is called the equilibrium vapor pressure, or just vapor pressure. Note that in the state of equilibrium, even though the amounts of vapor and liquid do not change, there is considerable activity on the molecular level. Vaporization and condensation are constantly occurring, but at the same rates. Such an equilibrium state is referred to as a dynamic equilibrium.

Vapor pressure is a function of the temperature. As temperature increases the rate of vaporization increases, but the rate of condensation remains the same. Therefore, more molecules exist in the vapor phase at higher temperatures than at lower temperatures.

**Table 11.4** Vapor Pressure of Water

| Temp ( °C) | Pressure (mmHg) |
|---|---|
| 20 | 17.54 |
| 30 | 31.82 |
| 40 | 55.32 |
| 50 | 92.51 |
| 80 | 355.1 |
| 90 | 460.0 |
| 100 | 760.0 |

Boiling of a liquid is marked by bubble formation. In a container open to the atmosphere, boiling of a liquid will occur when the temperature is raised high enough. In order for a bubble to form, the vapor inside the bubble must be able to push back the atmosphere. This will not occur until a temperature is reached at which the vapor pressure is greater than the atmospheric pressure. Therefore, the boiling point of a liquid is the temperature at which the vapor pressure is equal to the atmospheric pressure. Since the boiling point of a liquid depends on the atmospheric pressure, and the atmospheric pressure varies daily, the boiling point is not a constant. A normal boiling point is defined for purposes of comparing liquid substances. The normal boiling point is the boiling temperature when the external pressure is 1 atm. Figure 11.1 shows the vapor pressures and normal boiling points of two liquids.

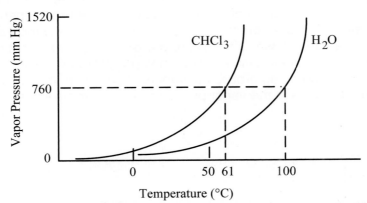

**Figure 11.1** The effect of temperature on the vapor pressure of chloroform and water. Notice that the normal boiling points for the liquids are the temperatures at which their vapor pressures are equal to 1 atm.

The energy required to vaporize one mole of a liquid is called the molar heat of vaporization ($\Delta H_{vap}$). The value of $\Delta H_{vap}$ is directly proportional to the strength of intermolecular forces. Liquids with relatively high heats of vaporization have low vapor pressures and high normal boiling points.

The quantity $\Delta H_{vap}$ can be determined experimentally. The vapor pressure of a liquid is proportional to the temperature. The quantitative relationship between the vapor pressure (P) and the absolute temperature (T) is given by the Clausius-Clapeyron equation:

$$\ln P = -\frac{\Delta H_{vap}}{RT} + C$$

where ln P is the natural logarithm of P, R is the ideal gas constant (8.314 J/K•mol), and C is a constant.

A useful form of this equation is

$$\ln\left(\frac{P_1}{P_2}\right) = \frac{\Delta H_{vap}}{R}\left(\frac{T_1 - T_2}{T_1 T_2}\right)$$

> Example
> 11.10 - 11.11
>
> Exercises
> 14 – 20

The $\Delta H_{vap}$ can be calculated if the vapor pressures $P_1$ and $P_2$ are measured at two temperatures $T_1$ and $T_2$. Alternatively, if $\Delta H_{vap}$ is already known and one vapor pressure $P_1$ is known at temperature $T_1$, you can calculate the vapor pressure $P_2$ at some new temperature $T_2$.

***Boiling Point as a Measure of Intermolecular Force.*** The boiling point and the heat of vaporization are convenient measures of the strength of intermolecular forces. In the liquid phase, molecules are very close together and are strongly influenced by intermolecular forces. In the gas phase, molecules are widely spaced, move rapidly, and have enough energy to overcome intermolecular forces. During vaporization, molecules are completely separated from one another. The heat of vaporization is the energy necessary to overcome intermolecular forces. The boiling point reflects the amount of kinetic energy that liquid molecules must achieve in order to overcome intermolecular forces and escape into the gas phase. When

> Example
> 11.12
>
> Exercises
> 21

comparing molecules, as the strength of intermolecular forces increase, the $\Delta H_{vap}$ and the normal boiling points increase.

***Critical Temperature and Pressure.*** When a liquid is heated in a closed container the vapor pressure increases, but boiling does not occur. The vapor cannot escape, and the vapor pressure continually rises. Eventually a temperature is reached at which the meniscus between liquid and vapor disappears. This temperature is called the critical temperature. The critical temperature is the highest temperature at which the substance can exist as a liquid. The critical pressure is the lowest pressure that will liquefy a gas at the critical temperature. Critical temperatures and pressures are listed in Table 11.7 (in the text).

As we saw previously, intermolecular forces are dominant in liquids and effectively determine many of their properties. The critical temperature and pressure are no exceptions. Substances with high critical temperatures have strong intermolecular forces of attraction.

***Liquid-Solid Equilibrium.*** The temperature at which the solid and liquid are in dynamic equilibrium at 1 atm pressure is called the normal melting point or normal freezing point. Melting is also called fusion. During melting the average distance between molecules is increased slightly as evidenced by the approximately 10% decrease in density for most substances.

$$\text{solid} \underset{\text{freezing}}{\overset{\text{fusion}}{\rightleftarrows}} \text{liquid}$$

The energy required to melt one mole of a solid is called the molar heat of fusion, $\Delta H_{fus}$. Upon freezing, the substance will release the same amount of energy. If this heat is removed from the system very quickly, the substance can be supercooled, that is cooled without solidifying. The $\Delta H_{fus}$ is always much less than $\Delta H_{vap}$ because during vaporization molecules are completely separated from each other, while melting only separates molecules to a small extent.

***Heating Curves.*** A heating curve is a convenient way to summarize the solid-liquid-gas transitions for a compound. The heating curve of water is shown in Figure 11.2. It is the result of an experiment in which a given amount of ice, 1 mole for instance, at some initial temperature below 0 °C is slowly heated at a constant rate.

The curve shows that the temperature of ice increases on heating (line 1) until the melting point is reached. No rise in temperature occurs while ice is melting (line 2); as long as some ice remains, the temperature stays at 0 °C. The length of line 2 is a measure of the heat necessary to melt 1 mole of ice, which is $\Delta H_{fus}$. Along line 3 the temperature of liquid water increases from 0 °C to 100 °C. The slope of the line $\Delta t$ per Joule of heat depends inversely on the specific heat of liquid water. The greater slope for line 1 than for line 3 means that less heat is required to raise the temperature of ice than of water. This is evidenced by the difference in specific heats of ice and liquid water. The specific heat of ice is 2.09 J/g• °C versus 4.18 J/g• °C for liquid $H_2O$.

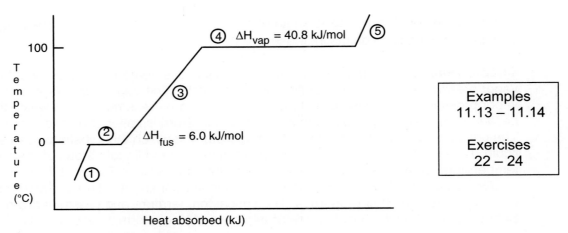

**Figure 11.2** The heating curve for water.

At 100 °C the liquid begins to boil, and the heat added is used to bring about vaporization. The temperature does not rise until all the liquid has been transformed to gas. The length of line 4 is the heat required to vaporize 1 mole of liquid. Line 4 will always be longer than line 2 because $\Delta H_{vap} > \Delta H_{fus}$. Line 5 corresponds to the heating of steam. Again the slope of line 5 depends on the specific heat of steam, which is about 1.98 J/g• °C.

***Solid-Vapor Equilibrium.*** Sublimation is the transition of a solid directly into the vapor phase. Dry ice and iodine are substances that sublime readily. Ice also sublimes to some extent. As with liquids, the vapor pressure of a solid increases as the temperature increases. The direct conversion of solid to vapor is equivalent to melting the solid first and then vaporizing the liquid. From Hess's law we obtain:

$$\Delta H_{sub} = \Delta H_{fus} + \Delta H_{vap}$$

***Phase Diagrams.*** From discussions in preceding sections we can see that the phase in which a substance exists depends on its temperature and pressure. In addition, two phases may exist in equilibrium at certain temperatures and pressures. Information about the stable phases for a specific compound is summarized by a phase diagram such as that shown in Figure 11.3.

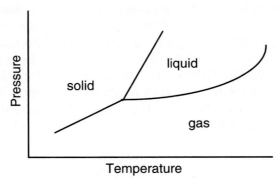

**Figure 11.3** A typical phase diagram.

A phase diagram is a graph of pressure versus temperature. The lines represent P and T at which two phases can coexist. The regions between lines represent P and T at which only one phase can exist. The line between the solid and liquid regions is made up of points of P and T at which the solid and liquid phases are in equilibrium. The line between the solid and gas regions is made up of points of P and T at which solid and gas are in equilibrium. And the line between the liquid and gas regions gives temperatures and pressures at which liquid and gas are in equilibrium. All three lines intersect at the triple point. This one point describes the conditions under which gas, liquid, and solid are all in equilibrium.

The phase diagram shows, at a glance, several properties of a substance: melting point, boiling point, and triple point. If a point (P, T) describing a system falls in the solid region, the substance exists as a solid. If the point falls on a line such as that between liquid and gas regions, the substance exists as liquid and vapor in equilibrium.

## GLOSSARY LIST

| | | |
|---|---|---|
| phase | crystalline solid | critical pressure |
| intermolecular forces | unit cell | critical temperature |
| van der Waals forces | closest-packed | boiling point |
| dipole-dipole forces | x-ray diffraction | melting point |
| induced dipole | coordination number | heat of vaporization |
| dispersion forces | | heat of fusion |
| ion-dipole forces | amorphous solid | freezing point |
| hydrogen bonds | | sublimation |
| intramolecular forces | | deposition |
| | phase change | heat of sublimation |
| surface tension | dynamic equilibrium | supercooling |
| adhesion | equilibrium vapor pressure | triple point |
| cohesion | condensation | phase diagram |
| viscosity | evaporation | |
| | vaporization | |

## EQUATIONS

| Algebraic Equation | English Translation |
|---|---|
| $n\lambda = 2d \sin \theta$ | Bragg equation |
| $\ln P = -\dfrac{\Delta H_{vap}}{RT} + C$ | Clausius-Clapeyron equation |
| $\ln \left( \dfrac{P_1}{P_2} \right) = \dfrac{\Delta H_{vap}}{R} \left( \dfrac{T_1 - T_2}{T_1 T_2} \right)$ | another form of the Clausius-Clapeyron equation |
| $\Delta H_{sub} = \Delta H_{fus} + \Delta H_{vap}$ | |

## WORKED EXAMPLES

---

**EXAMPLE 11.1  Types of Intermolecular Forces**

Which of the following substances should have the strongest intermolecular attractive forces:  $N_2$, Ar, $F_2$, or $Cl_2$?

**• Solution**

Note that none of these molecules (or atoms, in the case of Ar) is polar and that there is no chance for H bonding to occur. The only intermolecular forces existing in these molecules then are dispersion forces. Dispersion forces increase as the polarizability of the molecule increases, while polarizability increases with molecular mass. The most polarizable molecule will be $Cl_2$ (70.9 amu).

---

**EXAMPLE 11.2  Types of Intermolecular Forces**

Indicate all the different types of intermolecular forces that exist in each of the following substances.

a.  $CCl_4(\ell)$
b.  $HBr(\ell)$
c.  $CH_3OH(\ell)$

**• Solution**

The types of intermolecular forces depend primarily on the polarity of the molecule and the electronegativity of atoms.
a.  $CCl_4$ is nonpolar. The only type of intermolecular force between nonpolar molecules is dispersion forces.
b.  HBr is a polar molecule. The types of intermolecular forces are dipole-dipole and dispersion forces. There is no hydrogen bonding in HBr. The Br atom is not electronegative enough.
c.  $CH_3OH$ is polar and has a hydrogen atom bound to an oxygen atom. The intermolecular forces are dispersion forces and hydrogen bonding.

---

**EXAMPLE 11.3  Intermolecular Forces**

The dipole moment $\mu$ in HCl is 1.03 D, and in HCN it is 2.99 D. Which one should have the higher boiling point?

**• Solution**

The larger the dipole moment, the stronger the intermolecular force. The stronger the intermolecular force, the higher the temperature needed to provide molecules of the liquid with enough kinetic energy to overcome these attractive forces. The boiling point of HCN should be greater than that of HCl. Observed values are b.p. (HCl) = –85 °C, and b.p. (HCN) = 26 °C.

**Work EXERCISES & PROBLEMS: 1 – 5**

---

**EXAMPLE 11.4 Intermolecular Forces in Liquids**

The viscosity of ether is 0.000233 and of water is 0.00101 N s/m². Discuss their relative values in terms of the following molecular structures.

```
                    O                           CH3  O   CH3
   water           / \        diethyl ether       \  / \  /
                  H   H                           CH2   CH2
```

**• Solution**

Molecules with strong intermolecular forces have greater viscosities than those that have weak intermolecular forces. Here, both molecules are polar: water because of the O—H bonds, and ether because of the polar C—O bond. However, water has a higher viscosity than ether because of its ability to form hydrogen bonds between molecules.

**Work EXERCISES & PROBLEMS: 6, 7**

---

**EXAMPLE 11.5 Number of Atoms Per Unit Cell**

If atoms of a solid occupy a face-centered cubic lattice, how many atoms are there per unit cell?

**• Solution**

In a face-centered cubic cell there are atoms at each of the eight corners, and one in each of the six faces.

8 corners (1/8 atom per corner) + 6 faces (1/2 atom per face)
= 4 atoms per unit cell

---

**EXAMPLE 11.6 Dimensions of a Unit Cell**

Potassium crystallizes in a body-centered cubic lattice, and has a density of 0.856 g/cm³ at 25 °C.
a. How many atoms are there per unit cell?
b. What is the length of the side of the cell?

**• Solution**

a. A body-centered cubic structure has one K atom in the center, and eight other K atoms, one at each corner of the cube. The corner atoms, however, are shared by 8 adjoining cells. Each corner atom contributes 1/8 of an atom to the unit cell. The total number of K atoms per unit cell is

$$1 + \frac{1}{8}(8) = 2 \text{ atoms}$$

b.  Ideally, the crystal of potassium is made up of a large number of unit cells repeated over and over again. Thus, the density of the unit cell will be the same as the density of metallic K. Recall that density is an intensive property.

$$\text{density} = \frac{\text{mass}}{\text{volume}}$$

$$\text{density (unit cell)} = \frac{\text{mass of 2 K atoms}}{a^3}$$

where $\underline{a}$ is the length of the side of the unit cell.

The mass of a K atom is:

$$\frac{39.1\,g}{1\,mol} \times \frac{1\,mol}{6.02 \times 10^{23}\,atoms} = 6.50 \times 10^{-23}\,g/atom$$

Substituting into the density equation yields:

$$0.856\,g/cm^3 = \frac{2(6.50 \times 10^{-23}\,g/atom)}{a^3}$$

Rearranging yields:   $a^3 = 1.52 \times 10^{-22}\,cm^3$

$$a = 5.34 \times 10^{-8}\,cm = 534\,pm$$

---

### EXAMPLE 11.7  Atomic Radius of a Potassium Atom

Determine the radius of a K atom for the metallic potassium crystal in Example 11.6.

• **Solution**

From Figure 11.22 in the text, it is clear that we need the diagonal distance of the body-centered cubic cell because only along the diagonal do the atoms actually touch. The diagonal c will equal 4r where r is the atomic radius. From the figure we see that

$c = 4r$   and   $c = \sqrt{3}\,a$

Therefore,

$4r = \sqrt{3}\,a$

where a is the length of an edge. Solving for r, we get

$$r = \frac{a\sqrt{3}}{4} = \frac{534\,pm \times \sqrt{3}}{4}$$

$r = 231\,pm$

**Work EXERCISES & PROBLEMS: 8 – 11**

## EXAMPLE 11.8  Using the Bragg Equation

When X-rays with a wavelength of 0.154 nm are diffracted by a crystal of metallic copper, the angle corresponding to the first order diffraction (n = 1) is found to be 37.06°. What is the spacing between planes of Cu atoms that gives rise to this diffraction angle?

• **Solution**

The spacing between Cu atoms corresponds to d in the Bragg equation, which we rearrange to solve for d.

$$d = \frac{n\lambda}{2\sin\theta}$$

For a first-order diffraction n = 1. The sine of 37.06° can be determined with your calculator.

$$d = \frac{0.154 \text{ nm}}{2 \sin 37.06°} = \frac{0.154 \text{ nm}}{2(0.603)}$$

$$= 0.128 \text{ nm (128 pm)}$$

**Work EXERCISES & PROBLEMS 12 – 13**

## EXAMPLE 11.9  Types of Crystals

What type of force must be overcome in order to melt crystals of the following substances?
a.  Mg      b. $Cl_2$      c. $MgCl_2$      d. $SO_2$      e. Si

• **Solution**

a.  Magnesium is a metal and so it forms metallic crystals. Since the mobile electrons are shared between positive ions, the forces between Mg atoms could be described as covalent bonds, in this case called metallic bonds. The melting point is 1105 °C.
b.  $Cl_2$ forms molecular crystals that are held together by intermolecular forces, more specifically dispersion forces. The melting point is –101 °C.
c.  $MgCl_2$ is an ionic compound. When it melts, $Mg^{2+}$ ions and $Cl^-$ ions break away from their lattice positions. To melt magnesium chloride, ionic bonds must be broken. The melting point is 1412 °C.
d.  $SO_2$ is a polar molecule. Thus, dipole-dipole attractions and dispersion forces must be overcome. The melting point is –73 °C.
e.  Si crystallizes in a diamond structure with each silicon atom bound to four others by covalent bonds. It is a covalent crystal, and covalent bonds must be broken in order for it to melt. The melting point is 1410 °C.

**EXAMPLE 11.10  Heat of Vaporization**

How much heat is given off when 1.0 g of steam condenses at 100 °C?

**• Solution**

The heat of vaporization of water at 100 °C is 40.8 kJ/mol. Condensation is the reverse of vaporization so it releases the same amount of heat.

$$H_2O(g) \rightarrow H_2O(\ell) \quad \Delta H_{cond} = -40.8 \text{ kJ/mol}$$

Since $\Delta H_{cond}$ is in molar units, first, convert 1.0 g $H_2O$ into moles. The heat evolved when 1.0 g of water vapor condenses is:

$$q = \text{mol } H_2O \times \Delta H_{cond}$$

$$= 1.0 \text{ g} \times \frac{1 \text{ mol}}{18.0 \text{ g}} \times \frac{-40.8 \text{ kJ}}{1 \text{ mol}}$$

$$= -2.27 \text{ kJ}$$

**• Comment**

This is a lot of heat and shows us why burns from steam can be very serious.

---

**EXAMPLE 11.11  Clausius-Clapeyron Equation**

The vapor pressure of water is 55.32 mmHg at 40.0 °C, and 92.51 mmHg at 50.0 °C. (See Table 5.3 in the text.) Calculate the molar heat of vaporization of water.

**• Solution**

The Clausius-Clapeyron equation relates the vapor pressure of a liquid to its molar heat of vaporization.

$$\ln\left(\frac{P_1}{P_2}\right) = \frac{\Delta H_{vap}}{R}\left(\frac{T_1 - T_2}{T_1 T_2}\right)$$

Substituting the values given above

$$\ln\left(\frac{55.32 \text{ mmHg}}{92.51 \text{ mmHg}}\right) = \frac{\Delta H_{vap}}{8.314 \text{ J/K} \cdot \text{mol}} \times \left(\frac{313 \text{ K} - 323 \text{ K}}{(313 \text{ K})(323 \text{ K})}\right)$$

$$\ln 0.5980 = \frac{\Delta H_{vap}}{8.314 \text{ J/K} \cdot \text{mol}}\left(\frac{-10 \text{ K}}{1.01 \times 10^5 \text{ K}^2}\right)$$

Taking the logarithm and rearranging yields:

$$\Delta H_{vap} = \frac{-0.5142(8.314 \text{ J/K} \cdot \text{mol})(1.01 \times 10^5 \text{ K})}{-10 \text{ K}}$$

$\Delta H_{vap} = 43{,}000$ J/mol

### • Comment

Compare this result to the $\Delta H_{vap}$ listed in Table 11.6 of the text, 40,790 J/mol. The observed difference is real. The $\Delta H_{vap}$ is less at the boiling point than it is at a lower temperature. Over the entire temperature range of a liquid, $\Delta H_{vap}$, is close to, but not really, a constant.

**Work EXERCISES & PROBLEMS: 14 – 20**

### EXAMPLE 11.12  Vapor Pressure and Boiling Point

Using Figure 11.35 in the text estimate the boiling point of water in a pressure cooker at 2.0 atm pressure.

### • Solution

The boiling point of water at 2.0 atm pressure is the temperature at which the vapor pressure of water is 2.0 atm. According to the figure this should be approximately 120 °C.

**Work EXERCISES & PROBLEMS: 21**

### EXAMPLE 11.13  Heat of Fusion

The heat of fusion of aluminum is 10.7 kJ/mol. How much energy is required to melt one ton of Al at the melting point, 660 °C?

### • Solution

The heat of fusion is the energy required to melt 1 mole of a substance. Therefore, we need to convert 2000 lb of Al to moles.

$q = \text{mol Al} \times \Delta H_{fus}$

$$= 2000 \text{ lb} \times \frac{454 \text{ g}}{1 \text{ lb}} \times \frac{1 \text{ mol Al}}{27.0 \text{ g Al}} \times \frac{10.7 \text{ kJ}}{1 \text{ mol Al}} = 3.60 \times 10^5 \text{ kJ (360 MJ)}$$

### EXAMPLE 11.14  The Critical Temperature

Discuss the possibilities of liquefying oxygen and carbon dioxide at 25 °C by increasing the pressure. The critical temperature of $O_2$ is -119 °C and for $CO_2$ is 31 °C.

### • Solution

Above its critical temperature a substance cannot be liquefied by increasing the pressure. Substances having critical temperatures above 25 °C can be liquefied at 25 °C by application of sufficient pressure. With a critical temperature of 31 °C, $CO_2$ can be liquefied at 25 °C with application of enough pressure. The pressure required can be read off the phase diagram for $CO_2$ (Figure 11.41 in the text). Reading straight up from 25 °C you will cross the liquid-vapor equilibrium point at a pressure of 67 atm. This is the pressure required to liquefy $CO_2$ at 25 °C. Oxygen has a critical temperature of –119 °C; therefore, at any temperature above –119 °C oxygen cannot be liquefied, no matter how much pressure is applied. To liquefy $O_2$, its temperature must first be lowered below –119 °C and pressure applied.

## EXERCISES & PROBLEMS

1. Indicate the types of intermolecular forces that exist between molecules (or basic units) in each of the following.
   a. $PCl_3$    b. $CO_2$    c. $Cl_2$    d. $ICl$    e. $KCl$

2. Which of the following should have the strongest intermolecular forces?
   $CH_4$    $Cl_2$    $CO$    $CS_2$

3. Which of the following species are capable of hydrogen bonding with another like molecule?
   a. $CH_3F$    b. $HF$    c. $CH_3CH_2OH$    d. $CH_3NH_2$    e. $CH_4$

4. Which member of each pair should have the higher boiling point?
   a. $O_2$ or $CO$    b. $Br_2$ or $ICl$    c. $H_2O$ or $H_2S$    d. $PH_3$ or $AsH_3$

5. What type of intermolecular forces must be overcome in order to:
   a. vaporize water?    b. dissociate $H_2$ into H atoms?    c. boil liquid $O_2$?

6. What property of a liquid makes it possible to fill a glass of water to a level slightly above the rim?

7. Which of the two liquids has the higher viscosity at room temperature?
   $C_2H_5OH$ or $CH_3OCH_3$

8. Nickel crystallizes in a face-centered cubic lattice with an edge length of 352 pm. Calculate the density of Ni.

9. Nickel crystallizes in a face-centered cubic lattice with an edge length of 352 pm. Given that the density of Ni is 8.94 g/cm³, calculate Avogadro's number.

10. Copper crystallizes in a cubic system and the edge of the unit cell is 361 pm. The density of copper is 8.96 g/cm³. How many atoms are contained in one unit cell? What type of cubic cell does Cu form?

11. Copper crystallizes in a face-centered cubic lattice. The density of copper is 8.96 g/cm³. What is the length of the edge of a unit cell in nanometers?

12. In a diffraction experiment, X-rays of wavelength 0.154 nm were reflected from a gold crystal. The first-order reflection was at 22.2°. What is the distance between the planes of gold atoms, in pm?

13. The distance between the planes of gold atoms in a gold crystal is 204 pm. In a diffraction experiment the first-order reflection was observed at 22.2°. What was the wavelength of the X-rays in nanometers?

14. The molar heat of vaporization of bromine ($Br_2$) is 30.04 kJ/mol. What is the heat of vaporization per gram of bromine?

15. The heat of vaporization of iodine is 41.7 kJ/mol at the boiling point of iodine, 456 K. How much heat is required to vaporize 20.0 g of $I_2$?

16. The heat of vaporization of ammonia is 23.2 kJ/mol at its boiling point. How many grams of ammonia can be vaporized with 12.5 kJ of heat?

17. The vapor pressure of liquid potassium is 1.00 mmHg at 341 °C and 10.0 mmHg at 443 °C. Calculate the molar heat of vaporization of potassium.

18. The vapor pressure of ethanol is 100 mmHg at 34.9 °C and 400 mmHg at 63.5 °C. What is the molar heat of vaporization of ethanol?

19. The vapor pressure of liquid potassium is 10.0 mmHg at 443 °C. The heat of vaporization is 82.5 kJ/mol. Calculate the temperature in, °C, where the vapor pressure is 760 mmHg. This is the boiling point of liquid potassium.

20. The vapor pressure of ethanol is 400 mmHg at 63.5 °C. Its molar heat of vaporization is 41.7 kJ/mol. Calculate the vapor pressure of ethanol at 34.9 °C.

21. Which substance in each pair has the higher vapor pressure at a given temperature?
    a. $C_2H_5OH$ or $CH_3OH$
    b. $Cl_2$ or $Br_2$
    c. $CH_3Br$ or $CH_3Cl$

22. How much heat is needed to convert 1.0 kg of ice at 0 °C completely into steam at 100 °C? For water, $\Delta H_{vap}$ = 40.79 kJ/mol, $\Delta H_{fus}$ = 6.01 kJ/mol, and the specific heat of water is 4.184 J/g°C.

23. The molar heats of fusion and vaporization of potassium are 2.4 and 79.1 kJ/mol, respectively. Estimate the molar heat of sublimation of potassium metal.

24. Indicate whether each of the following processes is exothermic or endothermic.
    a. melting of ice
    b. evaporation of ethanol
    c. condensation of steam
    d. freezing of water
    e. sublimation of iodine

25. Consider a steel cylinder containing 20 kg of $CO_2$ in the form of liquid and gas. The valve is opened and $CO_2$ gas flows rapidly from the cylinder. Soon, however, the flow slows, and a coating of ice frost forms on the outside of the cylinder. When the valve is closed the cylinder is weighed and is found to contain 14 kg of $CO_2$.

a. Why is the flow faster at first?

b. Why does the flow almost stop before the cylinder is empty?

c. Why does the outside become icy?

26. On top of Pikes Peak water boils at 86 °C. Why do liquids boil at lower temperatures in the mountains as compared to sea level?

27. The viscosities of liquids generally decrease with increasing temperature. Water has the following viscosities in units of N • s/m²; 0.0018 at 0 °C; 0.0010 at 20 °C; 0.0005 at 55 °C; and 0.0003 at 100 °C. Interpret this trend on a molecular basis.

28. Why does your skin feel cool when water evaporates from it?  Why will you feel warmer on a more humid summer day than on a less humid day?

## PRACTICE TEST QUESTIONS
See notes on taking practice test in the Preface

1.  Identify the types of intermolecular forces for each of the following substances:
    a. $H_2O_2$      b. $H_2S$
    c. $SF_6$      d. $NOCl$
    e. $NH_3$

2.  The enthalpies of vaporization of $H_2S$ and $H_2Se$ are 18.7 and 23.2 kJ/mol, respectively. Why is the $\Delta H_{vap}$ of water (40.6 kJ/mol) so much higher than the values for $H_2S$ and $H_2Se$?

3.  Which member of each pair has the stronger intermolecular forces of attraction?
    a. $CO_2$ or $SO_2$      b. $F_2$ or $Br_2$
    c. $H_2O$ or $H_2S$      d. $HCl$ or $C_2H_6$

4.  In the laboratory, glassware is considered clean when water wets the glass in a continuous film. If droplets of water stand on the glass surface, the glass is considered dirty. Discuss the basis of this cleanliness test. Hint: Usually dirty glass has oils and grease as the source of dirt.

5.  The meniscus for mercury in a glass tube is concave downward. Explain.

6.  At what angle would a first-order reflection be observed in the diffraction of 0.090 nm X rays by a set of crystal planes for which d = 0.500 nm?

7.  Calcium oxide like NaCl crystallizes in a face-centered cubic cell, see figure 11.27 in the textbook.
    a.  How many $Ca^{2+}$ and $O^{2-}$ ions are in each unit cell?
    b.  Given the ionic radius of $Ca^{2+}$ (99 pm) and $O^{2-}$ (140 pm), what is the length of an edge of the unit cell?
    c.  Calculate the density of CaO.

8.  From the following data, calculate Avogadro's number. Potassium crystallizes in a body-centered cubic lattice with an edge of 530 pm. The density of K is 0.86 g/cm³.

9.  Aluminum crystallizes in a face-centered cubic lattice with the length of an edge equal to 405 pm. Assume that a face-centered atom touches each of the surrounding corner atoms. Calculate the length of a diagonal of a face.

10. Pick the substance in each pair that has the higher melting point.
    a. $ICl$ or $KCl$
    b. $K$ or $Fe$
    c. $Fe$ or $Cl_2$
    d. $SiO_2$ (quartz) or $SiF_4$
    e. $CO_2$ or $SiO_2$

11. Distinguish between the boiling point of a liquid and the normal boiling point of a liquid.

12. The vapor pressure of mercury is 17.3 mmHg at 200 °C. What is its vapor pressure at 300 °C if its molar heat of vaporization is 59.0 kJ/mol?

13. Use the following data to draw a rough phase diagram for methane: triple point −183.0 °C; normal melting point −182.5 °C; normal boiling point −161.5 °C; critical point −83.0 °C; critical pressure 45.6 atm.
    a. What is the stable state of methane at −160 °C and 2.0 atm pressure?
    b. At a constant temperature of −182 °C the pressure on a sample of methane is gradually increased from 1.0 mmHg to 1.0 atm. What phases would be observed, and in what order would they occur?

14. Given the following data for $N_2$:

| Normal melting point: | −210 °C | Specific heat of liquid: | 2.0 J/g·· °C |
|---|---|---|---|
| Normal boiling point: | −196 °C | Specific heat of gas: | 1.0 J/g·· °C |
| Heat of fusion: | 25 J/g | Specific heat of solid: | 1.6 J/g·· °C |
| Heat of vaporization: | 200 J/g | | |

How much energy in kilojoules is required to convert 1000 g of $N_2$ at −206 °C to $N_2$ gas at 20 °C?

15. In the following pairs indicate which member would have the higher heat of vaporization?
    a. He or Kr
    b. $CH_3OH$ or $CH_3OCH_3$

16. Which one of the following compounds will exhibit hydrogen bonding?
    a. $CH_4$        b. HBr
    c. $CH_3OH$      d. $CCl_4$

# Chapter 12. Physical Properties of Solutions

**The Solution Process (Sections 12.1 – 12.2)**
**Concentration Units (Section 12.3)**
**Temperature and Pressure Effects on Solubility (Sections 12.4 – 12.5)**
**Colligative Properties (Sections 12.6 – 12.7)**
**Colloids (Section 12.8)**


## SUMMARY

### The Solution Process (Sections 12.1 – 12.2)

***Some Definitions.***  The main subject of this chapter concerns the formation and properties of liquid solutions. Recall from chapter 4 that a solution is a homogeneous mixture of two or more substances. The component in greater quantity is called the solvent, and the component in lesser amount is called the solute. In aqueous solutions, water is the solvent. Solutes can be liquids, solids, or gases. Several terms are used to describe the degree to which a solute will dissolve in a solvent:

1.  When two liquids are completely soluble in each other in all proportions, they are said to be miscible. For example, ethanol and water are miscible. If the liquids do not mix, they are said to be immiscible. Oil and water, for example, are immiscible.
2.  A solution that contains the maximum concentration of a solute at a given temperature is a saturated solution. For example, the solubility of NaCl at 25°C is 6.1 moles per liter of water. An unsaturated solution has a concentration of solute that is less than the maximum concentration. Supersaturated solutions have a concentration of solute that is greater than that of a saturated solution.

***The Solution Process.***  Dissolving is a process that takes place at the molecular level and can be discussed in molecular terms. When one substance dissolves in another, the particles of the solute disperse uniformly throughout the solvent. The solute particles occupy positions that are normally taken by solvent molecules. The ease with which a solute particle may replace a solvent molecule depends on the relative strengths of three types of interactions:

-   Solvent-solvent interaction
-   Solute-solute interaction
-   Solvent-solute interaction

Imagine the solution process as taking place in three steps as shown in Figure 12.1. Step 1 is the separation of solvent molecules. Step 2 is the separation of solute molecules. These steps require inputs of energy to overcome attractive intermolecular forces. Step 3 is the mixing of solvent and solute molecules; it is exothermic. According to Hess's law (see Section 6.6 of the text), the heat of solution is given by the sum of the enthalpies of the three steps:

$$\Delta H_{soln} = \Delta H_1 + \Delta H_2 + \Delta H_3$$

The solute will be soluble in the solvent if the solute-solvent attraction is stronger than the solvent-solvent attraction and solute-solute attraction. Such a solution process is exothermic. Only a relatively small amount of the solute will be dissolved if the solute-solvent interaction is weaker than the solvent-solvent and solute-solute interaction; then the solution process will be endothermic. Solvation is the process in which a solute particle (an ion or molecule) is surrounded by solvent molecules due to strong solute-solvent attractive forces. When water is the solvent the process is called hydration.

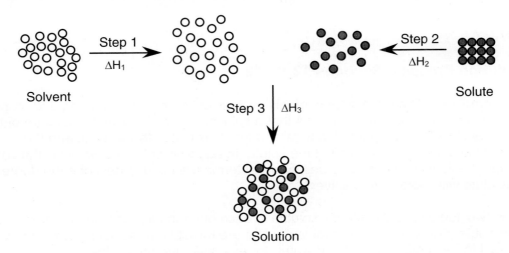

**Figure 12.1.** A molecular view of the solution process. Think of the solute molecules and solvent molecules first being spread apart, and then being mixed together. The relative strength of forces holding solvent molecules together $\Delta H_1$, holding solute particles together $\Delta H_2$, and the forces between solvent and solute molecules $\Delta H_3$ in the solution are important in determining the solubility of a solute in a solvent.

*A General Solubility Rule.* The most general solubility rule is that "like dissolves like." In this rule, the term "like" refers to molecular polarity. "Like dissolves like" means that substances of like, or similar polarity, will mix to form solutions. And substances of different polarity will be immiscible, or will tend only slightly to form solutions. The rule predicts that two polar substances will form a solution, and two nonpolar substances will form a solution, but a polar substance and a nonpolar substance will tend not to mix. For example, water and oil, being polar and nonpolar substances, respectively, are immiscible. Water and ethanol, both being polar, are miscible. Oil will dissolve in carbon tetrachloride because both substances are nonpolar. Many ionic compounds tend to be quite soluble in water (a polar solvent). However, solubility of an ionic compound in water must be determined experimentally. Refer to the solubility rules covered in chapter 4.

Examples
12.1, 12.2

Exercises
1 – 7

## Concentration Units (Section 12.3)

*Concentration.* The term concentration refers to how much of one component of a solution is present in a given amount of solution. In chapter 4 the concentration unit molarity was introduced:

$$molarity = \frac{moles\ of\ solute}{liters\ of\ solution}$$

Three new concentration units are introduced in this chapter.

The percent by mass of solute

$$\text{Percent by mass of solute} = \frac{\text{mass of solute}}{\text{mass solute} + \text{mass solvent}} \times 100\%$$

$$= \frac{\text{mass of solute}}{\text{mass of solution}} \times 100\%$$

The mole fraction of a component of the solution

$$\text{mole fraction of component A} = X_A = \frac{\text{moles of A}}{\text{sum of moles of all compounds}}$$

The molality of a solute in a solution

$$\text{molality} = \frac{\text{moles of solute}}{\text{mass of solvent(kg)}}$$

| Examples 12.3, 12.4 |
| Exercises 12 – 18 |

Notice that molality is the only one of the mentioned concentration units which has the amount of *solvent* (and not *solution*) in the denominator. This is a essential to keep in mind.

It will be very important to be able to convert between the different concentration units. Different situations involve different units. The conversion of these units involves dimensional analysis using the conversion tools you have learned. **However, before doing dimensional analysis, separate the numerator and denominator of the given unit and work with each separately.** Example 12.4 demonstrates this process.

## Temperature and Pressure Effects on Solubility (Sections 12.4 – 12.5)

***Temperature Effect on the Solubility of Solids.*** Temperature changes strongly affect the solublilty of most solid solutes. The effects of temperature on the solubilities of some common salts is shown in Figure 12.3 (text). In most cases the solubility of a solid in water increases with increasing temperature. However, this is not always true. Table 12.1 lists several solids whose solubility *decreases* with increasing temperature. In general, it is not practical to predict just how the temperature will affect the solubility of a given substance. Temperature effects must be determined experimentally.

**TABLE 12.1** Compounds with Solubilities that Decrease with Increasing Temperature

| | |
|---|---|
| $CaSO_4$ | $Ca(OH)_2$ |
| $Na_2SO_4 \cdot 10H_2O$ | $Ca(C_2H_3O_2)_2$ |
| $Ce_2(SO_4)_3$ | $Li_2SO_4$ |

***Pressure and the Solubility of Gases.*** The solubility of gases in liquids is directly proportional to the pressure of the gas being dissolved. Henry's law relates gas concentration c, in moles per liter, to the gas pressure p in atmospheres.

$$c = kP$$

where k is a constant for each gas and has units of mol/L$^\bullet$ atm. The greater the value of k, the greater the solubility of the gas. A few of Henry's law constants for gases dissolved in water at 25 °C are listed in Table 12.2

**TABLE 12.2** Some Henry's Law Constants at 25 °C

| Gas | k(mol/L$^\bullet$ atm) |
| --- | --- |
| $O_2$ | $1.3 \times 10^{-3}$ |
| $N_2$ | $6.8 \times 10^{-4}$ |
| $CO_2$ | $3.4 \times 10^{-2}$ |

It is common, in a "Henry's Law" problem to be given a concentration for a corresponding pressure. This information can be used in order to find the Henry's law constant. Once k is determined, it can be used to convert between pressure of a gas and its concentration in solution. Henry's Law can be rearranged to give

$$\frac{c}{P} = k \quad \text{and therfore} \quad \frac{c_1}{P_1} = \frac{c_2}{P_2}$$

***Gas Solubility and Temperature.*** A common observation is that a glass of cold water, when warmed to room temperature, shows the formation of many small air bubbles. This phenomenon results in part from the decreased solubility of gases in water with increasing temperature.

   The solubility of an unreactive gaseous solute is due to intermolecular attractive forces between solute molecules and solvent molecules. As the temperature of such a solution is increased, more solute molecules attain sufficient kinetic energy to break away from these attractive forces and enter the gas phase, and so the solubility decreases.

> Example
> 12.5
>
> Exercises
> 13 – 15

## Colligative Properties (Sections 12.6 – 12.7)

***Colligative Properties.*** The properties of solutions that depend on the number of solute particles in solution are called colligative properties. The four colligative properties of interest here are:

1.  Vapor-pressure lowering, $\Delta P$, is given by:

    $$\Delta P = X_2 P_1^\circ$$

> Example
> 12.6
>
> Exercise
> 16

where $X_2$ is the mole fraction of the solute, and $P_1^\circ$ is the vapor pressure of the pure solvent. Recall that the vapor pressure of a liquid is the pressure exerted by a vapor in equilibrium, with its liquid phase. For a solution containing a nonvolatile solute, the vapor pressure due to the solvent is less than it is for the pure solvent. The amount of lowering of the vapor pressure can be seen to depend on $X_2$, the mole fraction of the solute. An alternate form of the equation is $P_1 = X_1 P_1^\circ$ where $P_1$ is the pressure above the solution and $X_1$ is the mole fraction of solvent.

2. Boiling-point elevation: the boiling point of a solution is higher than that of pure solvent because the vapor pressure of a solution is always less than the vapor pressure of pure solvent. Thus, a solution must be hotter than a pure solvent, if both vapor pressures are to be 1 atm. Figure 12.10 in the text shows the effect that lowering the vapor pressure has on the boiling point. The boiling-point elevation $\Delta T_b$ of a solution of molality $m$ is given by:

$$\Delta T_b = K_b m$$

where $K_b$ is a constant called the molal boiling-point elevation constant and has the units °C/$m$. There is a $K_b$ for each solvent, and Table 12.2 (text) lists $K_b$ values for five solvents. The magnitude of $\Delta T_b$, the increase in boiling point of the solution over the boiling point of the pure solvent, is proportional to the solute concentration $m$. The preceding equation is independent of the solute, but requires that the solute be nonvolatile.

> Example
> 12.7
>
> Exercises
> 17 - 18

3. Freezing-point depression, $\Delta T_f$, is given by:

$$\Delta T_f = K_f m$$

where $K_f$ is the molal freezing-point depression constant and has the units °C/$m$. As is the case of $K_b$, $K_f$ is different for each solvent, and its value must be found in a table such as Table 12.2 (textbook). The equation applies to all solutes, even volatile ones, because freezing points are rather insensitive to vapor pressure.

4. The osmotic pressure, $\pi$, is given by:

$$\pi = MRT$$

where $M$ is the molar concentration, R is the ideal gas constant (0.0821 L·atm/K·mol), and T is the absolute temperature.

> To help memorize the equation, notice it's similarity to PV=nRT, solving for P gives P=(n/V)RT.

Osmosis is the net flow of solvent molecules through a semipermeable membrane from a more dilute solution to a more concentrated solution. Note that in this process the solvent flows from a region of greater *solvent* concentration to a region of lesser solvent concentration. The osmotic pressure of a solution is the pressure required to stop osmosis.

> Example
> 12.8
>
> Exercise
> 19

**Molar Mass Determination.** Colligative properties provide a means to determine the molar mass of the solute. Keep in mind that molar mass has units of grams per mole. If given the mass of the solute, the moles will need to be determined. (Moles is embedded in molality – in the case of freezing-point depression and boiling-point elevation, and is embedded in molarity in the case of osmotic pressure.)

For instance, the freezing-point depression equation is:

$$\Delta T_f = K_f\, m = K_f \times \frac{\text{mol solute}}{\text{kg solvent}}$$

Since $\Delta T_f$ can be measured and $K_f$ is known, then the molality, $m$, of the solution can be calculated. Recall that molality is

$$\text{molality} = \frac{\text{mol solute}}{\text{kg solvent}} \quad \text{and therefore:}$$

$$\frac{\text{mole solute}}{\text{kg solvent}} \times \text{kg solvent} = \text{mol solute}$$

Now that you know 1) the grams of solute (given in the problem) and 2) the moles of solute (obtained by the equation above), molar mass can be determined:

$$\text{Molar mass} = \frac{\text{grams solute}}{\text{mol solute}}$$

Freezing-point depression is more sensitive to the number of moles of solute than the boiling-point elevation because for the same solvent, $K_f$ is larger then $K_b$. The most sensitive method for the determination of the molar mass of a solute is to measure the osmotic pressure. The equation for osmotic pressure then becomes useful in the following form:

$$\pi = MRT$$

Solve for molarity. $\text{Molarity} = \dfrac{\text{mol solute}}{\text{L of solution}}$ and therefore:

$$\frac{\text{mole solute}}{\text{L of solution}} \times \text{L of solution} = \text{mol solute}$$

Once again you know 1) the grams of solute and 2) the moles of solute and molar mass can be determined.

Example 12.9

Exercises 20, 21

***Colligative Properties of Electrolyte Solutions.*** For the same solute concentrations, solutions of electrolytes always have more pronounced colligative effects than those of nonelectrolyte solutions. For a 0.1 $m$ methanol solution, $\Delta T_f$ is the same as the calculated value. For 0.1 $m$ NaCl, $\Delta T_f$ is 1.9 times greater than calculated. And for $Na_2SO_4$, $\Delta T_f$ is 2.7 times greater than the calculated value. Table 12.3 compares values of $\Delta T_f$ (obs)/$\Delta T_f$ (calc) for several solutes. This ratio is always greater than one for solutions of electrolytes. Since colligative properties depend on the number of solute particles, the most likely explanation of this behavior is that NaCl and $Na_2SO_4$ dissociate into ions and this produces a greater concentration of solute particles. Therefore, the effective concentration of solute particles in 0.10 $m$ NaCl is essentially 2 × 0.10 $m$, and in $Na_2SO_4$ it is almost 3 × 0.10 $m$. The freezing-point depression depends on the total concentration of all solute particles. Each ion acts as an independent particle.

**TABLE 12.3** van't Hoff Factor $i$ for Several Solutes

| Solution | $\Delta T_f$ | $\Delta T_f$ (calc) | $\Delta T_f$(obs)/$\Delta T_f$ (calc) |
|---|---|---|---|
| 0.10 $m$ $CH_3OH$ | 0.186 °C | 0.186 °C | 1.0 |
| 0.10 $m$ NaCl | 0.353 °C | 0.186 °C | 1.9 |
| 0.10 $m$ $Na_2SO_4$ | 0.502 °C | 0.186 °C | 2.7 |

The ratio $\Delta T_f(obs)/\Delta T_f(calc)$ is called the van't Hoff factor, $i$. For 1:1 electrolytes, such as NaCl, KCl, and $MgSO_4$, $i \sim 2$; for 2:1 electrolytes, such as $MgCl_2$ and $K_2SO_4$, $i$ is a little less than 3. The deviation from a whole number results from the temporary formation of some ion pairs such as "NaCl" in the solution.

| Example 12.10 |
| --- |
| Exercises 20 – 23 |

## Colloids (Section 12.8)

Solutions are homogeneous mixtures and are the topic of much of this chapter. Somewhere between homogeneous mixtures and heterogeneous mixtures, lies the topic of colloids. When the particles in the mixture are on the order of $1 \times 10^3 - 1 \times 10^6$ pm, the mixture is no longer called a solution. It is a colloid. The larger particle size causes the solution to become cloudy. The cloudy appearance may be very subtle and only observed when a laser light is passed through the mixture. If a beam of light (say a laser light) is passed through a solution, you cannot see the beam. However, if the light is passed through the colloid, the beam is observed. This is known as the Tyndall effect. Colloidal suspensions are not only liquid, but can be made up of various states. Examples include: solid dispersed in a gas (smoke); gas dispersed in a liquid (foam); liquid dispersed in a liquid (emulsion). For a more complete list of colloid types see Table 12.4 in the textbook.

## GLOSSARY LIST

| | | |
| --- | --- | --- |
| saturated solution | Henry's Law | osmosis |
| unsaturated solution | fractional distillation | semipermeable membrane |
| supersaturated solution | colligative properties | osmotic pressure |
| solvation | Raoult's law | van't Hoff factor |
| miscible | volatile | ideal solution |
| molality | nonvolatile | ion pair |
| percent by mass | | colloid |
| crystallization | boiling-point elevation | hydrophobic |
| fractional crystallization | freezing-point depression | hydrophilic |

## EQUATIONS

| Algebraic Equation | English Translation |
|---|---|
| $\text{molality} = \dfrac{\text{moles of solute}}{\text{mass of solvent(kg)}}$ | Molality is a measure of the number of moles of solute per kg of solvent. |
| $c = kP$ | Concentration of a gas dissolved in solution is proportional to the pressure above the solution. |
| $P_1 = X_1 P_1^\circ$ | Raoult's law (1) |
| $\Delta P = X_2 P_1^\circ$ | Raoult's law (2) |
| $\Delta T_b = K_b m$ | boiling-point elevation as a function of molality |
| $\Delta T_f = K_f\, m$ | freezing-point depression as a function of molality |
| $\pi = MRT$ | osmotic pressure as a function of molarity and temperature |

## WORKED EXAMPLES

---

**EXAMPLE 12.1  Solubility**

Which would be a better solvent for molecular $I_2(s)$:  $CCl_4$ or $H_2O$?

• **Solution**

Using the "like dissolves like" rule, first we identify $I_2$ as a nonpolar molecule. Therefore, it will be more soluble in the nonpolar solvent $CCl_4$ than in the polar solvent $H_2O$.

---

**EXAMPLE 12.2  Solubility**

In which solvent will NaBr be more soluble, benzene or water?

• **Solution**

Benzene is a nonpolar solvent and water is a polar solvent. Since NaBr is an ionic compound, it cannot dissolve in nonpolar benzene, but is very soluble in water.

**Work EXERCISES & PROBLEMS: 1 – 7**

---

**EXAMPLE 12.3  Concentration Units**

The dehydrated form of Epsom salts is magnesium sulfate.
a.  What is the percent $MgSO_4$ by mass in a solution made from 16.0 g $MgSO_4$ and 100 mL of $H_2O$ at 25 °C?  The density of water at 25 °C is 0.997 g/mL.
b.  What is the mole fraction of each component?
c.  Calculate the molality of the solution.

**• Solution**

a.  Write the equation for percent by mass.

$$\text{percent MgSO}_4 = \frac{\text{mass MgSO}_4}{\text{mass MgSO}_4 + \text{mass water}} \times 100\%$$

$$\text{mass H}_2\text{O} = 100 \text{ mL} \times \frac{0.997 \text{ g}}{1 \text{ mL}} = 99.7 \text{ g H}_2\text{O}$$

$$\text{percent MgSO}_4 = \frac{16.0 \text{ g}}{16.0 \text{ g} + 99.7 \text{ g}} \times 100\% = \frac{16.0 \text{ g}}{115.7 \text{ g}} \times 100\%$$

$$\text{percent MgSO}_4 = 13.8\%$$

b.  Write the formula for the mole fraction of $MgSO_4$.

$$X_{\text{MgSO}_4} = \frac{\text{mol MgSO}_4}{\text{mol MgSO}_4 + \text{mol H}_2\text{O}}$$

Convert the masses in part (a) into moles to substitute into the equation.

$$16.0 \text{ g MgSO}_4 \times \frac{1 \text{ mol MgSO}_4}{120.4 \text{ g MgSO}_4} = 0.133 \text{ mol MgSO}_4$$

$$99.7 \text{ g H}_2\text{O} \times \frac{1 \text{ mol H}_2\text{O}}{18.0 \text{ g H}_2\text{O}} = 5.54 \text{ mol H}_2\text{O}$$

The mole fractions of $MgSO_4$ and $H_2O$ are

$$X_{\text{MgSO}_4} = \frac{0.133 \text{ mol}}{0.133 \text{ mol} + 5.54 \text{ mol}} = 0.0235 \qquad X_{\text{H}_2\text{O}} = \frac{5.54 \text{ mol}}{5.67 \text{ mol}} = 0.977$$

Notice that the sum of the two mole fractions is 1.000. The sum of the mole fractions of all solution components is always 1.00.

c.  Write the formula for molality:

$$\text{molality} = \frac{\text{moles of MgSO}_4}{\text{kilograms of H}_2\text{O}}$$

Substitute into this equation, the quantities previously calculated in parts (a) and (b).

$$\text{molality} = \frac{0.133 \text{ mol MgSO}_4}{99.7 \text{ g}} \times \frac{10^3 \text{ g}}{1 \text{ kg}} = 1.33 \; m$$

**Work EXERCISES & PROBLEMS: 8, 9**

**EXAMPLE 12.4  Molality and Molarity of a Solution, Conversions Between Units**

Concentrated hydrochloric acid is 36.5 percent HCl by mass. Its density is 1.18 g/mL.
Calculate:   a.  The molality of HCl   b.  The molarity of HCl

• **Solution**

a.  Write the formula for molality.

$$\text{molality} = \frac{\text{mol HCl}}{\text{kg H}_2\text{O}}$$

Next, find the number of moles of HCl per kilogram of $H_2O$. We take 100 g of solution, and determine how many moles of HCl and how many kilograms of the solvent it contains. A solution that is 36.5 percent HCl by mass corresponds to 36.5 g HCl/100 g solution. Since 100 g of solution contains 36.5 g HCl, the difference 100 g – 36.5 g must equal the mass of water which is 63.5 g. We have two ratios:

$$\frac{36.5 \text{ g HCl}}{100 \text{ g soln}} \quad \text{and} \quad \frac{36.5 \text{ g HCl}}{63.5 \text{ g H}_2\text{O}}$$

The moles of HCl is given by:

$$\text{moles HCl} = 36.5 \text{ g HCl} \times \frac{1 \text{ mol HCl}}{36.5 \text{ g HCl}} = 1.00 \text{ mol HCl}$$

The kilograms of water is given by:

$$\text{kg H}_2\text{O} = 63.5 \text{ g H}_2\text{O} \times \frac{1 \text{ kg}}{1 \times 10^3 \text{ g}} = 0.0635 \text{ kg H}_2\text{O}$$

Then we calculate molality:

$$\text{molality} = \frac{1.00 \text{ mol HCl}}{0.0635 \text{ kg H}_2\text{O}} = 15.7 \; m$$

b.  Write the formula for molarity:

$$\text{molarity} = \frac{\text{moles HCl}}{\text{liters soln}}$$

Find the number of moles of HCl and liters of solution that are present in 100 g of solution.

$$\frac{36.5 \text{ g HCl}}{100 \text{ g soln}}$$

Convert 100 g of solution to volume of solution using the density.  (Note: 36.5 g HCl = 1.00 mol.)

$$\text{Volume of soln} = (100 \text{ g soln}) \times \frac{1 \text{ mL}}{1.18 \text{ g soln}} \times \frac{10^{-3}\text{L}}{1 \text{ mL}} = 0.0847 \text{ L}$$

$$\text{molarity} = \frac{1.00 \text{ mol HCl}}{0.0847 \text{ L soln}} = 11.8 \; M$$

**Work EXERCISES & PROBLEMS: 10 - 12**

---

**EXAMPLE 12.5  Henry's law**

Oxygen at a pressure of 1.00 atm, the concentration of dissolved oxygen is $1.28 \times 10^{-3}$ M. What is the concentration of $O_2$ in air-saturated water at 25 °C and atmospheric pressure of 645 mmHg?  Assume the mole fraction of oxygen in air is 0.209.

**• Solution**

Henry's law states that the concentration of dissolved $O_2$ ($C_{O_2}$) is proportional to its partial pressure ($P_{O_2}$) in atm.

$$C_{O_2} = k\, P_{O_2}$$

Use the information of oxygen a 1 atm to determine k.

$$1.28 \times 10^{-3} \text{ mol/L} = k\,(1.00 \text{ atm})$$
$$1.28 \times 10^{-3} \text{ mol/L} \cdot \text{atm} = k$$

The partial pressure of $O_2$ in air is found by using Dalton's law of partial pressures.

$$P_{O_2} = X_{O_2} P_T = 0.209\,(645 \text{ mmHg}) \times \frac{1 \text{ atm}}{760 \text{ mmHg}} = 0.177 \text{ atm}$$

The concentration of dissolved oxygen is:

$$C_{O_2} = k P_{O_2} = (1.28 \times 10^{-3} \text{ mol/L} \cdot \text{atm})\,(0.177 \text{ atm})$$

$$= 2.27 \times 10^{-4} \text{ mol/L}$$

**Work EXERCISES & PROBLEMS 13 – 15**

---

**EXAMPLE 12.6  Vapor-Pressure Lowering**

Calculate the vapor pressure of an aqueous solution at 30 °C made from 100 g of sucrose ($C_{12}H_{22}O_{11}$) and 100 g of water. The vapor pressure of water at 30 °C is 31.8 mmHg.

**• Solution**

Sucrose is a nonvolatile solute, so the vapor pressure over the solution will be due to $H_2O$ molecules. The problem can be worked in two ways.

a.  The vapor-pressure lowering is proportional to the mole fraction of sucrose $X_2$, and $P_2^{\circ}$, the vapor pressure of pure water at 30 °C.

$$\Delta P = X_2 P_1^{\circ}$$

First, we calculate the mole fraction of sucrose, $X_2$.

$$X_2 = \frac{n_2}{n_1 + n_2}$$

$$n_2 = 100 \text{ g} \times \frac{1 \text{ mol}}{342 \text{ g}} = 0.292 \text{ mol}$$

$$n_1 = 100 \text{ g} \times \frac{1 \text{ mol}}{18.0 \text{ g}} = 5.55 \text{ mol}$$

$$X_2 = \frac{0.292}{0.292 + 5.55} = \frac{0.292}{5.84}$$

$$X_2 = 0.0500$$

The vapor pressure lowering is:

$$\Delta P = (0.0500)(31.8 \text{ mmHg}) = 1.59 \text{ mmHg}$$

The vapor pressure, $P_1$, is $P_1^\circ - \Delta P$.

$$P_1 = 31.8 - 1.59$$
$$= 30.2 \text{ mmHg}$$

b.   Alternatively Raoult's law can be used to calculate the vapor pressure of the solvent:

$$P_1 = X_1 P_1^\circ$$

where $P_1^\circ$ and $P_1$ are the vapor pressures of pure solvent and of the solvent in solution, respectively. From part a, $X_2 = 0.0500$, therefore $X_1 = 1.00 - X_2 = 0.95$, and

$$P_1 = (0.950)(31.8 \text{ mmHg})$$
$$= 30.2 \text{ mmHg}$$

**Work EXERCISES & PROBLEMS: 16**

---

## EXAMPLE 12.7  Boiling-Point Elevation

What is the boiling point of an "antifreeze/coolant" solution made from a 50-50 mixture (by volume) of ethylene glycol, $C_2H_6O_2$ (density 1.11 g/mL), and water?

**• Solution**

The boiling point depends on the molality of the 50-50 mixture. $\Delta T = K_b m$. For simplicity, assume 100 mL of the solution. Then, using the density of water, the mass of 50 mL (50 vol %) of $H_2O$ is 50 g. The mass of 50 mL (50 vol %) of ethylene glycol, using the density given above is 55.5 g.

$$m = \frac{\text{mol } C_2H_6O_2}{\text{kg } H_2O}$$

$$\text{mol } C_2H_6O_2 = 55.5 \text{ g} \times \frac{1 \text{ mol}}{62.0 \text{ g}} = 0.895 \text{ mol}$$

$$m = \frac{0.895 \text{ mol } C_2H_6O_2}{0.050 \text{ kg}} = 17.9 \text{ mol/kg} = 17.9 \, m$$

$$\Delta T_b = K_b m = (0.52 \text{ °C}/m) \times 17.9 \, m = 9.3 \text{ °C}$$

Therefore the boiling point of the solution is 9.3 °C above the normal boiling point of water. Boiling point = 109.3 °C.

**Work EXERCISES & PROBLEMS: 17, 18**

**EXAMPLE 12.8 Osmotic Pressure**

A solution of glucose ($C_6H_{12}O_6$) is made by dissolving 15.0 grams of glucose in enough water to make 100.0 mL of solution. What is the osmotic pressure of the solution at 25°C?

- **Solution**

Osmotic pressure is determined according to the equation: $\Pi$=MRT. Therefore we will first determine the molarity of the solution:

$$M = \frac{\text{mol of } C_6H_{12}O_6}{\text{L of solution}} = \frac{15.0\,g\,C_6H_{12}O_6 \times \dfrac{1\,mol\,C_6H_{12}O_6}{180.0\,g\,C_6H_{12}O_6}}{100.0\,mL \times \dfrac{1L}{1000\,mL}} = 0.833\,mol/L$$

Inserting the molarity into the osmotic pressure equation gives:

$$\Pi = 0.833\,\frac{mol}{L} \times 0.0821\,\frac{L \cdot atm}{mol \cdot K} \times 298\,K = 20.4\,atm$$

**Work EXERCISES & PROBLEMS: 19**

---

**EXAMPLE 12.9  Finding the Molar Mass of a Solute**

Benzene has a normal freezing point of 5.51 °C. The addition of 1.25 g of an unknown compound to 85.0 g of benzene produces a solution with a freezing point of 4.52 °C. What is the molecular mass of the unknown compound?

• **Solution**

We need to find the number of moles in 1.25 g of unknown compound X. The freezing-point depression is proportional to the number of moles of X per kilogram of solvent.

The freezing-point depression is:

$$\Delta T_f = 5.51\,°C - 4.52\,°C$$
$$= 0.99\,°C$$

We can calculate the molality of the solution because $\Delta T_f$ is given above.

$$m = \frac{\Delta T_f}{K_f}$$

where $K_f$ is the freezing-point depression constant for benzene (Table 12.2 textbook).

$$m = \frac{0.99\,°C}{5.12\,°C/m} = 0.19\,m$$

This means that there are 0.19 mol of X per kg of benzene. The number of moles of X in 0.085 kg of the solvent benzene (the given amount) is:

$$0.085 \text{ kg} \times \frac{0.19 \text{ mol X}}{1 \text{ kg benzene}} = 1.6 \times 10^{-2} \text{ mol X}$$

Therefore, 1.25 g X = $1.6 \times 10^{-2}$ mol X, and

$$\text{molar mass of X} = \frac{1.25 \text{ g}}{1.6 \times 10^{-2} \text{ mol}} = 78 \text{ g/mol}$$

**Work EXERCISES & PROBLEMS: 20, 21**

**EXAMPLE 12.10  Colligative Properties of Electrolytes and Nonelectrolytes**

List the following aqueous solutions in the order of increasing boiling points:  0.10 *m* glucose $(C_6H_{12}O_6)$; 0.10 *m* $Ca(NO_3)_2$; 0.10 *m* NaCl.

**• Solution**

Ethanol is a nonelectrolyte, whereas NaCl and $Ca(NO_3)_2$ are electrolytes. NaCl dissociates into 2 ions per formula unit, and $Ca(NO_3)_2$ dissociates into three ions per formula unit. The effective molalities are approximately:

> glucose 0.10 *m*
> NaCl ≅ 0.20 *m*
> $Ca(NO_3)_2$ ≅ 0.30 *m*

Therefore, the boiling-point elevation and the boiling point are greatest for a solution of $Ca(NO_3)_2$, second highest for NaCl, and lowest for glucose.

**• Comment**

If you included the van't Hoff factors, the effective molalities would be affected slightly, but not enough to change the predicted results.

**Work EXERCISES & PROBLEMS: 22, 23**
**Work the rest of the EXERCISES & PROBLEMS**


## EXERCISES & PROBLEMS

1.  Isopropyl alcohol and water dissolve in each other regardless of the proportions of each. What term describes the solubilities of these liquids in each other?

2.  Explain why hexane $(C_6H_{14})$ even though a liquid, is not miscible with water.

3.  Indicate whether each compound listed is soluble or insoluble in water.
    a. $CH_3OH$    b. LiBr    c. $C_8H_{18}$    d. $CCl_4$    e. $BaCl_2$    f. $HOCH_2CH_2OH$

4.  Explain why ammonia gas, $NH_3$, is very soluble in water, but not in hexane, $C_6H_{14}$.

5.  What is the difference between solvation and hydration?

6.  Predict which substance of the following pairs, will be more soluble in water.
    a. NaCl(s) or $I_2$(s)

    b. $CH_4(g)$ or $NH_3(g)$
    c. $CH_3OH(\ell)$ or $C_6H_6$(benzene)

7. For each of the following pairs, predict which substance will be more soluble in $CCl_4(\ell)$.
    a. $I_2(s)$ or $KBr(s)$
    b. $CS_2(\ell)$ or $CH_3OH(\ell)$
    c. $CO_2(g)$ or $HCl(g)$

8. Calculate the percent by mass of the solute in the following aqueous solutions.
    a. 6.50 g NaCl in 75.2 g of water
    b. 27.2 g ethanol in 250 g of solution
    c. 2.0 g $I_2$ in 125 g methanol

9. Calculate the molality of each of the following solutions.
    a. 6.50 g NaCl in 75.2 g of water
    b. 27.5 g glucose ($C_6H_{12}O_6$) in 425 g of water

10. Calculate the molarity of a 2.44 $m$ NaCl solution given that its density is 1.089 g/mL.

11. Calculate the percent $AgNO_3$ by mass in a 0.650 $m$ $AgNO_3$ solution.

12. Calculate the molality of a 5.5% $AgNO_3$ solution.

13. The solubility of $KNO_3$ at 70 °C is 135 g per 100 g of water. At 10 °C the solubility is 21 g per 100 g of water. What mass of $KNO_3$ will crystallize out of solution if exactly 100 g of its saturated solution at 70 °C is cooled to 10 °C?

14. As temperature increases, the solubility of all gases in water _____ (increases/decreases).

15. The Henry's law constant for argon is $1.5 \times 10^{-3}$ mol/L• atm at 20 °C. Calculate the solubility of argon in water at 20 °C and 7.6 mmHg.

16. Calculate the vapor pressure at 30°C above an aqueous solution which is 10.5% by mass glucose ($C_6H_{12}O_6$). The vapor pressure of pure water at 30 °C is 31.8 mmHg.

17. What is the freezing point of a solution made from 1.00 g of $C_6H_{12}O_6$ (glucose) and 100.0 g of $H_2O$? $K_f$ ($H_2O$) = 1.86 °C/$m$.

18. Calculate the freezing point of an aqueous solution that boils at 100.5 °C. $K_b(H_2O)$ = 0.52 °C/$m$.

19. Calculate the *approximate* osmotic pressure at 25 °C of an aqueous solution that has a freezing point of –1.5 °C.

20. Calculate the molar mass of naphthalene given that a solution of 2.11 g of naphthalene in 100 g of benzene has a freezing-point depression of 0.85 °C. $K_f$(benzene) = 5.12 °C/$m$.

21. A solution contains 1.00 g of a compound (a nonelectrolyte) dissolved in 100.0 g of water. The freezing point of the solution is –0.103 °C. What is the molar mass of the compound?

22. How many solute particles does each basic unit of the following compounds give in aqueous solution?
    a. $C_2H_6O_2$ (ethylene glycol)    b. $(H_2N)_2CO$ (urea)    c. HBr    d. $(NH_4)_3PO_4$    e. NaOH
    f. $Ca(OH)_2$

23. Arrange the following aqueous solutions in order of decreasing freezing point.
    0.15 *m* $MgCl_2$,  0.20 *m* $C_6H_{12}O_6$,  0.20 *m* NaCl,  0.15 *m* KI.

24. Which of following aqueous solutions has the higher boiling point?
    0.10 *m* HCl  or  0.10 *m* acetic acid

25. List three things that happen when a salt dissolves in water.  Give the sign of $\Delta H$ for each step and identify the forces involved in each.

26. Many salts with $\Delta H_{soln}$ near zero still dissolve in appreciable amounts. Explain.

27. When a "seed" crystal is placed in a supersaturated solution at constant temperature, precipitation is induced. When precipitation stops, is the solution saturated, unsaturated, or supersaturated?

28. What are ion pairs?  What effect does ion-pair formation have on the osmotic pressure?

## PRACTICE TEST QUESTIONS
See notes on taking practice test in the Preface

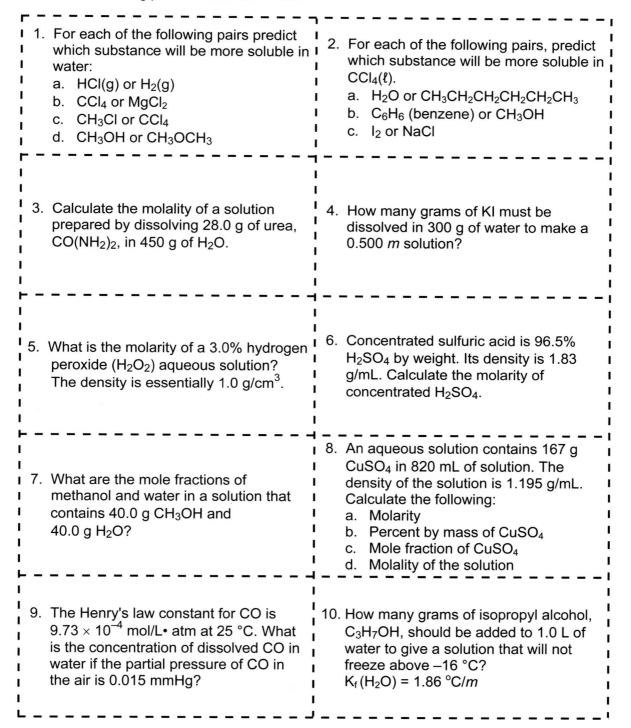

1. For each of the following pairs predict which substance will be more soluble in water:
   a. HCl(g) or H₂(g)
   b. CCl₄ or MgCl₂
   c. CH₃Cl or CCl₄
   d. CH₃OH or CH₃OCH₃

2. For each of the following pairs, predict which substance will be more soluble in CCl₄(ℓ).
   a. H₂O or CH₃CH₂CH₂CH₂CH₂CH₃
   b. C₆H₆ (benzene) or CH₃OH
   c. I₂ or NaCl

3. Calculate the molality of a solution prepared by dissolving 28.0 g of urea, CO(NH₂)₂, in 450 g of H₂O.

4. How many grams of KI must be dissolved in 300 g of water to make a 0.500 *m* solution?

5. What is the molarity of a 3.0% hydrogen peroxide (H₂O₂) aqueous solution? The density is essentially 1.0 g/cm³.

6. Concentrated sulfuric acid is 96.5% H₂SO₄ by weight. Its density is 1.83 g/mL. Calculate the molarity of concentrated H₂SO₄.

7. What are the mole fractions of methanol and water in a solution that contains 40.0 g CH₃OH and 40.0 g H₂O?

8. An aqueous solution contains 167 g CuSO₄ in 820 mL of solution. The density of the solution is 1.195 g/mL. Calculate the following:
   a. Molarity
   b. Percent by mass of CuSO₄
   c. Mole fraction of CuSO₄
   d. Molality of the solution

9. The Henry's law constant for CO is 9.73 × 10⁻⁴ mol/L• atm at 25 °C. What is the concentration of dissolved CO in water if the partial pressure of CO in the air is 0.015 mmHg?

10. How many grams of isopropyl alcohol, C₃H₇OH, should be added to 1.0 L of water to give a solution that will not freeze above −16 °C? Kf (H₂O) = 1.86 °C/m

12        12

12        12

12        12

12        12

12        12

11. What is the boiling point of an aqueous solution of a nonvolatile solute that freezes at –3.0 °C?
$K_f (H_2O) = 1.86 \, °C/m$
$K_b (H_2O) = 0.52 \, °C/m$

12. When 46 g of glucose (a nonelectrolyte) is dissolved in 500 g of $H_2O$, the solution has a freezing point of –0.94 °C. What is the molar mass of glucose? $K_f (H_2O) = 1.86 \, °C/m$.

13. 7.85 g of a compound with an unknown formula is dissolved in 300 g of benzene. The freezing point of the solution is 2.10 °C below that of pure benzene. What is the molar mass of the compound?

14. The dart poison in root extracts used by the Peruvian Indians is called curare. It is a nonelectrolyte. The osmotic pressure at 20 °C of an aqueous solution containing 0.200 g of curare in 100 mL is 56.2 mmHg. Calculate the molar mass of curare.

15. What is the concentration of a NaCl solution with an osmotic pressure of 10 atm at 25 °C?

16. The walls of red blood cells are semipermeable membranes, and the solution of NaCl within those walls exerts an osmotic pressure of 7.82 atm at 37 °C. What concentration of NaCl must a surrounding solution have so that this pressure is balanced and cell rupture (hemolysis) is prevented?

17. Arrange the following aqueous solutions in order of increasing freezing points, lowest to highest.
0.100 m ethanol
0.050 m $Ca(NO_3)_2$
0.100 m NaBr
0.050 m HCl

18. Calculate the value of i for a 1:1 electrolyte, if a 1.0 $m$ aqueous solution of the electrolyte freezes at –3.28 °C.
$K_f (H_2O) = 1.86 \, °C/m$

19. In the course of research a chemist isolated a new compound. An elemental analysis gave the following: C 50.7%, H 4.25%, O 45.1%. When 5.01 g of the compound was dissolved in exactly 100 g of water, it produced a solution with a freezing point of –0.74 °C. What is the molecular formula of the compound?
$K_f (H_2O) = 1.86°C/m$.

# Chapter 13. Chemical Kinetics

The Rate of Reaction (Section 13.1)
Rate Laws (Section 13.2)
The Relation between Reactant Concentration and Time (Section 13.3)
Activation Energy and Temperature Dependence of Reaction Rates (Section 13.4)
Reaction Mechanisms (Section 13.5)
Catalysis (Section 13.6)


## SUMMARY

### The Rate of Reaction (Section 13.1)

***Expressing the Rate of Reaction.*** Chemical kinetics is the area of chemistry concerned with the study of the rates of chemical reactions. The purposes of kinetic studies are to find the factors that affect reaction rates and determine the reaction mechanism. Knowledge of the factors that affect reaction rates enables chemists to control rates. Finding the reaction mechanism means we can identify the intermediate steps by which reactants are converted into products.

A rate is a change in some quantity with time. The rate of population growth is the change in population per change in time. The change in a quantity such as population is always equal to the difference, population (after) minus population (before). The symbol for "the change in" is $\Delta$.

$$\Delta(\text{population}) = \text{population}_{final} - \text{population}_{initial}$$

The change in time is some appropriate time interval, $\Delta t$, where

$$\Delta t = t_{final} - t_{initial}$$

The rate of population change is:

$$\text{rate} = \frac{\Delta(\text{population})}{\Delta t}$$

As a chemical reaction proceeds, the concentrations of reactants and products change with time. For instance, as the reaction $A + B \rightarrow C$ progresses, the concentration of C increases. The rate is expressed as the change in the molar concentration of C, $\Delta[C]$, during the time interval $\Delta t$.

$$\text{rate} = \frac{\Delta[C]}{\Delta t}$$

For a specific reaction we need to take into account the stoichiometry; that is, we need the balanced equation. For example, let's express the rate of the following reaction in terms of the concentrations of the individual reactants and products.

$$2 NO(g) + O_2(g) \rightarrow 2 NO_2(g)$$

These concentrations can be monitored experimentally as a function of time. Notice from the balanced equation, that 2 mol of NO reacts with 1 mol of $O_2$; therefore, the concentration of NO will decrease twice as fast as that of $O_2$.

$$\frac{\Delta[NO]}{\Delta t} = \frac{2\Delta[O_2]}{\Delta t}$$

Since the rates of change of individual reactants and products may differ, the convention is to make the reaction rate the same no matter which reactant or product is used to calculate it. First, we divide each concentration change by the coefficient from the balanced equation

$$rate = -\frac{\Delta[NO]}{2\,\Delta t} = -\frac{\Delta[O_2]}{\Delta t} = \frac{\Delta[NO_2]}{2\,\Delta t}$$

Second, a negative sign is inserted before terms involving reactants. The change in NO concentration, $\Delta[NO]$, is negative because the concentration of NO *decreases* with time. Inserting a negative sign in the expression makes the rate of reaction a positive quantity.

For a general equation:

$$aA + bB \rightarrow cC$$

the rate can be expressed in terms of any individual reactant or product.

$$rate = -\frac{\Delta[A]}{a\,\Delta t} = -\frac{\Delta[B]}{b\,\Delta t} = \frac{\Delta[C]}{c\,\Delta t}$$

No matter which reactant or product we use, the reaction rate will be positive and have the same value.

***Calculating an Average Rate.*** The average rate of reaction over any time interval is equal to the change in the concentration of a reactant $\Delta[A]$, or of a product $\Delta[C]$, divided by the time interval, $\Delta t$, during which the change occurred.

| Examples |
| 13.1 – 13.3 |
| |
| Exercises |
| 1 – 3 |

$$average\ rate = -\frac{change\ in\ the\ concentration\ of\ A}{length\ of\ time\ interval}$$

$$average\ rate = -\frac{\Delta[A]}{\Delta t} = -\frac{[A]_{final} - [A]_{initial}}{t_{final} - t_{initial}}$$

## Rate Laws (Section 13.2)

***Effect of Concentration.*** The rate of a reaction is proportional to the reactant concentrations. For the reaction,

$$NO + \tfrac{1}{2}O_2 \rightarrow NO_2$$

the rate is proportional to the concentrations of NO and $O_2$.

The *rate law* (or *rate equation*) for the reaction is:

rate = $k[NO]^x[O_2]^y$

The proportionality constant k is called the rate constant. The value of k depends on the reaction and the temperature. The values of x and y are often 1 or 2. However, other values (including fractions) are possible.

The exponents x and y determine how strongly the concentrations affect the reaction rate. The exponent x is called the order with respect to NO, and y is the order with respect to $O_2$. The sum (x + y) is the overall order.

---
*The values of x and y must be determined from experiment, and cannot be derived by any other means.*

---

We will discuss how to determine the order of reaction in the next section. For now we will just use the results. For the NO reaction with $O_2$ experiment, x = 2 and y = 1. Therefore, the rate law for this reaction is:

rate = $k[NO]^2[O_2]$

This reaction is second-order in nitric oxide, and first-order in oxygen. It is third-order overall.

The fact that the reaction is first-order in $O_2$ means that the rate is directly proportional to the $O_2$ concentration. If $[O_2]$ doubles or triples, the rate will double or triple also. We can show this mathematically. Consider two experiments. In experiment 1 the concentration of $O_2$ is c. In experiment 2 the concentration of $O_2$ is doubled from c to 2c. If the concentration of NO is the same in both experiments, it will have no effect on the rate. Use of the rate law allows us to write the ratio of the two rates:

$$\frac{\text{rate (expt 2)}}{\text{rate (expt 1)}} = \frac{k[NO]^2(2c)}{k[NO]^2(c)} = 2$$

As discussed, we see that doubling the concentration of a reactant that is first-order will cause the rate to double.

If the concentration of $O_2$ is held constant in two experiments and the concentration of NO doubles (from c to 2c), the rate law predicts that the rate will quadruple.

$$\frac{\text{rate (expt 2)}}{\text{rate (expt 1)}} = \frac{k[O_2](2c)^2}{k[O_2](c)^2} = 2^2 = 4$$

The fact that the reaction is second-order in NO means that the rate is proportional to the square of the concentration of NO. Doubling or tripling of [NO] causes the rate to increase four- or nine-fold, respectively.

In general, if the concentration of one reactant is doubled while the other reactant concentration is unchanged, *and* the rate is:

1. *unchanged,* the order of the reaction is *zero-order* with respect to the changing reactant.
2. *doubled,* the order of the reaction is *first-order* with respect to the changing reactant.
3. *quadrupled,* the order of the reaction is *second-order* with respect to the changing reactant.

***The Isolation Method.*** One procedure used to determine the rate law for a reaction involves the isolation method. In this method, the concentration of all but one reactant is fixed, and the rate of reaction is measured as a function of the concentration of the one reactant whose concentration is varied. Any variation in the rate is due to the variation of this reactant's concentration. In practice, the experimenter observes the dependence of the initial rate on the concentration of the reactant.

To determine the order with respect to A in the following chemical reaction,

$$2A + B \rightarrow C$$

the initial rate would be measured in several experiments in which the concentration of A is varied and the concentration of B is held constant. To determine the order with respect to B, the concentration of A must be held constant and the concentration of B is varied in several experiments. The application of this method is illustrated in Example 13.5.

> Examples
> 13.4, 13.5
>
> Exercises
> 4 – 5

## The Relation Between Reactant Concentration and Time (Section 13.3)

***First-Order Reactions.*** One of the most widely encountered kinetic forms is the first-order rate equation. In this case, the exponent of [A] in the rate law is 1.

$$A \rightarrow products$$

$$rate = -\frac{\Delta[A]}{\Delta t} = k[A]$$

For a first-order reaction, the unit of the rate constant is reciprocal time, $1/t$. Convenient units are $1/s$, $1/h$, etc.

The equation that relates the concentration of A remaining to the time since the reaction started is:

$$\ln \frac{[A]_t}{[A]_0} = -kt$$

This is a very useful equation called the integrated first-order equation. Here $[A]_0$ is the concentration of A at time = 0, and $[A]_t$ is the concentration of A at time = t. The rate constant k is the first-order rate constant. The concentration $[A]_t$ decreases as the time increases. This equation allows the calculation of the rate constant k when $[A]_0$ is known, and $[A]_t$ is measured at time t. Also, once k is known, $[A]_t$ can be calculated for any future time.

To determine whether a reaction is first-order, we rearranged the first-order equation into the form:

$$\ln [A]_t = -kt + \ln [A]_0$$

corresponding to the linear equation

$$y = mx + b$$

Here, m is the slope of the line and b is the intercept on the y axis. Comparing the last two equations, we can equate y and x to experimental quantities.

$$y = \ln [A]_t \quad \text{and} \quad x = t$$

Therefore, the intercept $b = \ln [A]_0$, and the slope of the line $m = -k$. Thus, a plot of $\ln [A]_t$ versus t for a first-order reaction gives a straight line with a slope of $-k$ as shown in Figure 13.1 below. If a plot of ln [A] versus t yields a curved line, rather than a straight line, the reaction is not a first-order reaction. This graphical procedure is the method used by most chemists to determine whether or not a given reaction is first order.

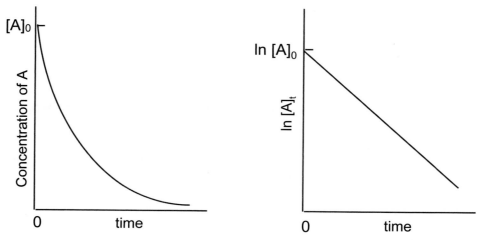

**Figure 13.1** A plot of $[A]_t$ versus time for a first-order reaction gives a curved line. A plot of $\ln [A]_t$ versus time gives a straight line for a first-order reaction.

**Half-life.** The half-life of a reaction, $t_{1/2}$, is a useful concept. For a first-order reaction, the half-life is given by:

$$t_{1/2} = \frac{0.693}{k}$$

where 0.693 is a constant and k is the rate constant. Knowledge of the half-life allows the calculation of the rate constant k. The half-life of a reaction is the time required for the concentration of a reactant to decrease to half of the initial value. After one half-life, the ratio $[A]_t/[A]_0$ is equal to 0.5.

If the reaction continues, then $[A]_t$ will drop by 1/2 again during the second half-life period as shown in Figure 13.2. After two half-life periods the fraction of the original concentration of A remaining, $[A]_t/[A]_0$, will be 1/2 of the concentration remaining after the first half-life, so $[A]_t/[A]_0 = 0.5 \times 1/2 = 0.25$.

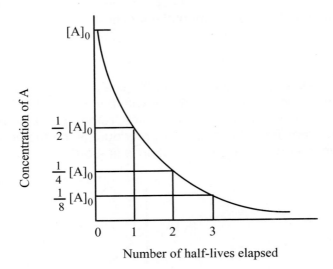

**Figure 13.2** A plot of $[A]_t$ versus time for a first-order reaction gives a curved (exponential) line. Over each half-life period, $[A]_t$ drops in half.

***Second-Order Reactions.*** In a second-order reaction, the rate is proportional either (1) to the square of the concentration of one reactant,

A → product

$$\text{rate} = -\frac{\Delta[A]}{\Delta t} = k[A]^2$$

or (2) to the product of the concentrations of two reactants, each raised to the first power:

A + B → product

$$\text{rate} = -\frac{\Delta[A]}{\Delta t} = k[A][B]$$

This reaction is first-order in A and first-order in B, and so it is second-order overall. For a second-order reaction, the rate constant has units 1/(molarity × time) which is $1/M{\cdot}s$.

An important equation called the integrated second-order equation is:

$$\frac{1}{[A]_t} = \frac{1}{[A]_0} + kt$$

where $[A]_0$, $[A]_t$, and t have their usual meaning, and k is the second-order rate constant. This equation applies to the second-order reaction A → products. This equation allows the calculation of the concentration of A at any time, t, after the reaction has begun. Alternatively, if $[A]_0$, $[A]_t$, and t are known, the rate constant can be calculated. Example 13.7 in the text shows a calculation example using this equation.

The half-life for a second-order reaction is given by:

$$t_{1/2} = \frac{1}{k[A]_0}$$

Here, we see that the half-life is inversely proportional to the initial concentration, $[A]_0$. This situation is different from that for a first-order reaction, where $t_{1/2}$ is independent of $[A]_0$.

As we did with first-order kinetics, we can manipulate the equation to make a linear plot. The manipulation is quite simple. The equation is simply rearranged to give

$$\frac{1}{[A]_t} = kt + \frac{1}{[A]_0}$$ and is then in the form y = mx + b. A plot of $\frac{1}{[A]_t}$ (on the y axis) vs. t (on the x axis) gives a straight line with slope of k.

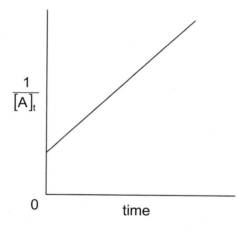

**Figure13.3** A plot of 1/[A] vs. time gives a straight line for a second-order reaction. The slope of the line equals the rate constant, k.

***Zero-Order Reactions.*** A zero-order reaction is one in which the rate does not depend on the concentrations of reactants.

A → products

$$\text{rate} = \frac{-[A]}{\Delta t} = k[A]^0 = k$$

> Examples
> 13.6, 13.7
>
> Exercises
> 6 – 11

A plot of [A] versus time is a straight line for a zero-order reaction. The rate does not slow down as the reactant is used up.

***Summary.*** The integrated rate equations and graphical methods allow us to distinguish between the various overall orders of reaction.

> 1.  A reaction is first-order when a plot of ln $[A]_t$ versus t is a straight line. A plot of $[A]_t$ versus time will be curved.
> 2.  A reaction is second-order when a plot of 1/$[A]_t$ versus t is a straight line, and ln $[A]_t$ versus t is curved.
> 3.  A reaction is zero-order when a plot of $[A]_t$ versus t is a straight line.

> 4.  The half-life equations also provide a way to distinguish between first- and second-order reactions. The half-life of a first-order reaction is independent of the starting concentration, whereas the half-life of a second-order reaction is inversely proportional to the initial concentration.

## Activation Energy and Temperature Dependence of Reaction Rates (Section 13.4)

***Collision Theory.*** The collision theory of chemical reactions provides a general explanation of how reaction rates are affected by reactant concentrations and temperature. The basic ideas of the theory are:

1.  In order for atoms, molecules, or ions to react, they must first collide with each other. The rate of reaction is proportional to the rate of collisions, called the collision frequency. The more concentrated the reactants, the greater the collision frequency, and the reaction rate.
2.  For molecules to react, they must come together in the proper orientation. Molecules can be complex and often it is just one atom in a molecule that reacts upon collision with another molecule.
3.  When reactant molecules collide, they must possess a minimum amount of kinetic energy, in order for an effective collision—a reaction—to occur. Without this necessary energy two molecules will just bump each other and bounce back without reacting. The minimum amount of energy required to initiate a chemical reaction is called the activation energy.

***Effect of Temperature.*** The temperature of a reaction system is an important variable because of its strong effect on reaction rates. As a rough rule, reaction rates approximately double with a 10 °C rise in temperature. In general, the rate equation is rate = $k[A]^x[B]^y$. Since the concentrations [A] and [B] are unaffected by temperature, it is the rate constant that changes with temperature. In 1889, S. Arrhenius found that a plot of the natural logarithm of the rate constant (ln k) versus the reciprocal of the absolute temperature 1/T gave a straight line. See Figure 13.18 (text). Arrhenius identified the slope of the line as being related to an energy term:

$$m = -\frac{E_a}{R}$$

where R is the ideal gas constant (in units of joules) and $E_a$ is the activation energy. The logarithmic form of the Arrhenius equation is:

$$\ln k = -\frac{E_a}{RT} + \ln A$$

where ln A is the intercept. Both the Arrhenius A and $E_a$ are constants for a particular reaction. Taking the antilog of both sides gives the Arrhenius equation, which relates the rate constant to the temperature.

$$k = A\, e^{-E_a/RT}$$

***The Meaning of A and $E_a$.*** The parameter A is called the frequency factor. It is related to the frequency of molecular collisions, and the fraction of collisions that have the correct orientation. As discussed above, the collision frequency is important because molecules must collide in order to react.

The activation energy, $E_a$, is related to the formation of the activated complex. The activated complex is the high-energy intermediate species that dissociates into the products. Figure 13.3 shows that the *activation energy* is the difference in energy between the activated complex and the reactants. The activation energy is provided by the kinetic energy of rapidly moving molecules during collisions. Reactants must "get over the barrier" before they become products. The factor $e^{-E_a/RT}$ that appears in the Arrhenius equation is the fraction of molecules with energies equal to or greater than the activation energy. This factor changes significantly with temperature. As temperature increases, a greater fraction of molecules have energy equal to or greater than the activation energy.

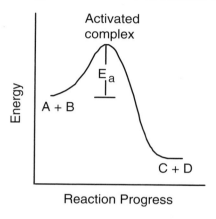

**Figure 13.4** Molecules of A and B with average energies must first acquire an energy $E_a$ before they can react.

From the Arrhenius equation we can point out that reactions for which $E_a$ is large will be much slower than those for which $E_a$ is small. As $E_a$ increases, the negative exponent increases, and so k decreases.

A convenient equation that can be used to calculate the activation energy is:

$$\ln\left(\frac{k_1}{k_2}\right) = \frac{E_a}{R}\left(\frac{T_1 - T_2}{T_1 T_2}\right)$$

Examples
13.8, 13.9

Exercises
12 – 15

where $k_1$ is the rate constant at temperature $T_1$, and $k_2$ is the rate constant at temperature $T_2$. Use of this equation requires that rate constants $k_1$ and $k_2$ be measured at two temperatures $T_1$ and $T_2$, respectively.

## Reaction Mechanisms (Section 13.5)

***Elementary Steps.*** The purpose of a reaction mechanism is to indicate how the reactants are converted into products. A reaction mechanism consists of a set of so-called elementary steps. A mechanism indicates which molecules must collide with each other and in what sequence. Each elementary step marks the progress of the conversion of molecules of reactants into products. A mechanism indicates the intermediates that are formed and consumed along the way, as well as, which steps are fast and which are slow. This information is not supplied by the balanced chemical equation, because it is intended only to indicate the number of moles of one reactant that are consumed per mole of the other reactant.

Elementary steps have a property called molecularity, which pertains to the number of molecules that must collide in a single step. An elementary step that involves three molecules is called termolecular. A step that involves two molecules is bimolecular, and a step in which one molecule decomposes or rearranges is called unimolecular. Thus, the equation:

$$A \rightarrow B$$

represents a unimolecular reaction in which a single molecule of A reacts to form a single molecule of B. The following equations represents a bimolecular reaction.

$$A + B \rightarrow C \quad \text{or} \quad A + B \rightarrow C + D$$

That is one in which a molecule of A collides with a molecule of B, and as a result, a molecule of C is formed or the molecules C and D are formed.

***Reaction Order for Elementary Steps.*** The rate of a unimolecular elementary step

$$A \rightarrow B$$

will depend on the number of molecules of A per unit volume of the container. Therefore, unimolecular reactions follow first-order rate laws.

$$\text{rate} = k[A]$$

The rate of a bimolecular step ($A + B \rightarrow C$) will depend on the number of molecules of A per unit volume, times the number of molecules of B per unit volume.

$$\text{rate} = k[A][B]$$

Therefore, bimolecular steps follow second-order rate laws.

This also can be understood by realizing that A and B must collide with each other if they are to react, and so the rate of a bimolecular process depends on the rate of collisions of A and B. Doubling the concentration of A will double the probability of collisions between A and B. Similarly, doubling the concentration of B will also double the rate of collisions of A and B. See Figure 13.16 in the textbook. Reasoning further along these lines, we conclude that termolecular reactions are third-order.

***Rate-Determining Step.*** It often turns out that one of the elementary steps in a mechanism is much slower than all the rest. This step determines the overall rate of reaction just as the slowest person ahead of you in a cafeteria line determines how fast you and others move through the line. The slowest step in a sequence of elementary steps is called the rate-determining step. This allows us to say that the overall law predicted by a mechanism will be the one corresponding to the rate-determining step.

***Testing a Mechanism.*** The sequence of events in a kinetic study of a reaction that leads to a proposed mechanism is as follows:

1. Determine the rate law experimentally.
2. Propose a mechanism for the reaction.
3. Test the mechanism.

To test a proposed mechanism, assume one step to be the rate-determining step. Then, establish the rate law for that step. This gives the rate law predicted by

Example
13.10

Exercises
16 - 18

the mechanism. If the mechanism is adequate, when the predicted rate law is compared with the experimental rate law, the two will match. On the other hand, if the predicted and experimental rate laws do not match, the mechanism has been proved inadequate. Only those mechanisms consistent with all the data can be considered adequate.

## Catalysis (Section 13.6)

***Mechanism of a Catalyzed Reaction.*** A catalyst is a substance that increases the rate of a chemical reaction without being consumed in the reaction. For this reason a catalyst does not appear in the stoichiometric equation. Chromium(III) oxide, $Cr_2O_3$, is the catalyst present in catalytic converters in American automobiles. This compound accelerates the reaction of carbon monoxide with oxygen. By converting CO to $CO_2$ in the exhaust stream, CO emissions are reduced.

$$CO(g) + \frac{1}{2}O_2(g) \xrightarrow{Cr_2O_3} CO_2(g)$$

The lifetime of catalytic converters is quite long; the $Cr_2O_3$ and other catalysts do not need to be replaced.

Catalysts accelerate reaction rates by providing a new reaction pathway (mechanism) which has a lower activation energy. See Figure 13.23 (textbook). In the catalytic converter, both CO and $O_2$ are chemically adsorbed on the catalyst's surface. Adsorption of $O_2$ by $Cr_2O_3$ weakens the O—O bond enough so that oxygen atoms can react with adsorbed CO. The reaction path involving a weakened O—O bond has a significantly lower activation energy than that of the reaction occurring purely in the gas phase. Recall that as the activation energy is lowered, the rate constant for a reaction increases. After $CO_2$ is formed, it desorbs from the surface, leaving the $Cr_2O_3$ catalyst chemically unaltered. The catalyst is not consumed in the reaction. Catalysts are regenerated in one of the last steps of the mechanism. The textbook describes three types of catalysts, which we will review here.

***Heterogeneous Catalysis.*** In heterogeneous catalysis, the reactants are in one phase and the catalyst is in another phase. The catalyst is usually a solid, and the reactants are gases or liquids. Ordinarily, the site of the reaction is the surface of the solid catalyst. Many industrially important reactions involve gases and are catalyzed by solid surfaces. For example, the Haber process for the synthesis of ammonia is catalyzed by iron plus a few percent of the oxides of potassium and aluminum. The famous Ziegler-Natta catalyst for polymerizing ethylene gas ($C_2H_4$) into a polyethylene polymer contains triethylaluminum and titanium or vanadium salts.

***Homogeneous Catalysis.*** In homogeneous catalysis, the reactants, products, and catalysts are all in the same phase, which is usually the gas or the liquid phase. Many reactions are catalyzed by acids. The decomposition of formic acid ($HCO_2H$) is an example.

$$HCO_2H \rightarrow H_2O + CO$$

Formic acid is normally stable and lasts a long time on the shelf. However, when sulfuric acid is added, bubbles of carbon monoxide gas can be observed immediately. The hydrogen ions from the sulfuric acid initiate a new reaction path. The mechanism is

| Example 13.11 |
| :--- |
| Exercises 19, 20 |

$$HCO_2H + H^+ \rightarrow HCO_2H_2^+$$

$$HCO_2H_2^+ \rightarrow H_2O + HCO^+$$

$$HCO^+ \rightarrow CO + H^+$$

$$\overline{HCO_2H \rightarrow CO + H_2O}$$

Note that the $H^+$ ion is consumed in the first step, but, as for all catalysts, $H^+$ is regenerated. The net reaction is the sum of the three steps. The intermediate $HCO_2H_2^+$ cancels out as does the catalyst.

***Enzyme Catalysis.***  Nearly all chemical reactions that occur in living organisms require catalysts. Enzymes are biological catalysts. Most enzymes are protein molecules with molecular masses well over 10,000 amu. All enzyme molecules are highly specific with respect to the reactions they catalyze. That is, they can only affect the reaction rates of a few specific reactant molecules. Typically, an enzyme catalyzes a single reaction, or a group of closely related reactions. The molecule on which an enzyme acts is called a substrate.

The simplest mechanism which explains enzyme activity, and is consistent with these trends is one in which the substrate S and the enzyme E form an enzyme-substrate complex, ES. This complex has a lower energy than the activated complex without the enzyme. The enzyme-substrate complex can either dissociate back into E and S, or break apart into the products P, and the regenerated enzyme E. The simplest mechanism has the two steps shown below.

$$E + S \rightleftharpoons ES$$

$$ES \rightarrow E + P$$

## GLOSSARY LIST

| | | |
|---|---|---|
| chemical kinetics | activation energy | unimolecular reaction |
| reaction rate | activated complex | bimolecular reaction |
| rate constant | catalyst | termolecular reaction |
| rate law | reaction mechanism | |
| reaction order | intermediate | |
| first-order reaction | rate-determining step | |
| half-life | molecularity | |
| second-order reaction | elementary step | enzymes |

## EQUATIONS

| Algebraic Equation | English Translation |
|---|---|
| $\text{rate} = k[A]^x[B]^y$ | rate law expression |
| $\ln\dfrac{[A]_t}{[A]_0} = -kt$ | relationship between concentration and time for a 1st order reaction |
| $\ln[A]_t = -kt + \ln[A]_0$ | linear form for relationship between concentration and time for a 1st order reaction |
| $t_{\frac{1}{2}} = \dfrac{0.693}{k}$ | half-life for first-order reaction |
| $\dfrac{1}{[A]_t} = \dfrac{1}{[A]_0} + kt$ | relationship between concentration and time for a 2nd order reaction |
| $\ln\left(\dfrac{k_1}{k_2}\right) = \dfrac{E_a}{R}\left(\dfrac{T_1 - T_2}{T_1 T_2}\right)$ | relationships of rate constants at two different temperatures |

## WORKED EXAMPLES

**EXAMPLE 13.1  Expressing the Rate of Reaction**

Write expressions for the rate of the following reaction in terms of each of the reactants and products.

$$2\,N_2O_5(g) \rightarrow 4\,NO_2(g) + O_2(g)$$

**• Solution**

Recall that the rate is defined as the change in concentration of a reactant or product with time. Each "change-in-concentration" term is divided by the corresponding stoichiometric coefficient. Terms involving reactants are preceded by a minus sign. Therefore, the rate is expressed as:

$$\text{rate} = -\frac{\Delta[N_2O_5]}{2\,\Delta t} = \frac{\Delta[NO_2]}{4\,\Delta t} = \frac{\Delta[O_2]}{\Delta t}$$

## EXAMPLE 13.2  Rate of Reaction

Oxygen gas is formed by the decomposition of nitric oxide:

$$2\,NO(g) \rightarrow O_2(g) + N_2(g)$$

If the rate of formation of $O_2$ is 0.054 M/s, what is the rate of change of NO concentration?

• **Solution**

From the stoichiometry of the reaction, 2 mol NO react for each mole of $O_2$ that forms.

$$\text{rate} = -\frac{\Delta[NO]}{2\,\Delta t} = \frac{\Delta[O_2]}{\Delta t}$$

$$\frac{\Delta[NO]}{\Delta t} = -\frac{2\,\Delta[O_2]}{\Delta t} = -2(0.054\ \text{M/s})$$

$$= -0.11\ \text{M/s}$$

## EXAMPLE 13.3  Calculation of the Average Rate

Experimental data for the hypothetical reaction:

$$A \rightarrow 2B$$

are listed in the following table.

| Time (s) | [A] (mol/L) |
|----------|-------------|
| 0.00 | 1.000 |
| 10.0 | 0.891 |
| 20.0 | 0.794 |
| 30.0 | 0.707 |
| 40.0 | 0.630 |

a.  Calculate the average rates of change of [A], and the average reaction rates for the two time intervals from 0 to 10 s and from 30 to 40 s.
b.  Why does the rate decrease from one time interval to the next?

• **Solution**

a.  The average rate of change of [A] is given by:

$$\frac{\Delta[A]}{\Delta t} = \frac{[A]_2 - [A]_1}{t_2 - t_1}$$

For the time interval 0 − 10 s, we get:

$$\frac{\Delta[A]}{\Delta t} = \frac{0.891\ \text{mol/L} - 1.00\ \text{mol/L}}{10.0\ \text{s} - 0\ \text{s}} = \frac{-0.109\ \text{mol/L}}{10.0\ \text{s}} = -0.0109\ \text{mol/L} \cdot \text{s}$$

Since the average reaction rate $= -\dfrac{\Delta[A]}{\Delta t}$ , then:

average reaction rate $= - (- 0.0109 \text{ mol/L·s}) = 0.0109 \text{ mol/L·s}$

For the time interval $30 - 40$ s, the rate of change of [A] is:

$$\frac{\Delta[A]}{\Delta t} = \frac{0.630 \text{ mol/L} - 0.707 \text{ mol/L}}{40.0 \text{ s} - 30.0 \text{ s}} = \frac{-0.77 \text{ mol/L}}{10.0 \text{ s}} = -0.0077 \text{ mol/L·s}$$

And the average reaction rate is:

$$\text{average rate} = -\frac{\Delta[A]}{\Delta t} = 0.0077 \text{ mol/L·s}$$

b.  The reaction rate decreases as the total reaction time increases because the rate is proportional to the concentration of the reactant A, and the concentration of A decreases as the time of reaction increases.

**Work EXERCISES & PROBLEMS:  1 – 3**

---

**EXAMPLE 13.4  Concentration Effect on the Rate**

The reaction, A + 2B → products, was found to have the rate law:  rate $= k[A][B]^3$. By what factor will the rate of reaction increase if the concentration of B is increased from c to 3c, while the concentration of A is held constant?

**• Solution**

Write a ratio of the rate law expressions for the two different concentrations of B.

$$\frac{\text{rate (expt 2)}}{\text{rate (expt 1)}} = \frac{k[A][3c]^3}{k[A][c]^3} = \frac{3^3 c^3}{c^3} = 27$$

The rate will increase 27-fold when [B] is increased three-fold.

---

**EXAMPLE 13.5  Finding the Rate Law**

The following rate data were collected for the reaction:

$$2 \text{ NO} + 2 \text{ H}_2 \rightarrow \text{N}_2 + 2 \text{ H}_2\text{O}$$

| Experiment | $[NO]_0$ (M) | $[H_2]_0$ (M) | $\Delta[N_2]/\Delta t$ (M/h) |
|---|---|---|---|
| 1 | 0.60 | 0.15 | 0.076 |
| 2 | 0.60 | 0.30 | 0.15 |
| 3 | 0.60 | 0.60 | 0.30 |
| 4 | 1.20 | 0.60 | 1.21 |

a.  Determine the rate law.
b.  Calculate the rate constant.

• **Solution**

a.  We want to determine the exponents in the equation

rate = k[NO]$^x$[H$_2$]$^y$

　　To determine the order with respect to H$_2$, first find two experiments in which [NO] is held constant. This can be done by comparing the data of experiments 1 and 2. When the concentration of H$_2$ is doubled, the reaction rate doubles. Thus, the reaction is first-order in H$_2$. When the NO concentration is doubled (experiments 3 and 4) the reaction rate quadruples. Therefore, the reaction is second-order in NO.

　　The rate law is:

rate = k[NO]$^2$[H$_2$]

b.  Rearrange the rate law from part (a).

$$k = \frac{\text{rate}}{[NO]^2[H_2]}$$

Then, substitute data from any one of the experiments. Using experiment 1:

$$k = \frac{0.076\ M/h}{(0.60\ M)^2(0.15\ M)} = 1.4/M^2\ h$$

• **Comment**

You should get the same value of k from all four experiments. Note that the units of k are those of a third-order rate constant. Take a closer look at proving that x = 2. Write the ratio of rate laws for experiments 4 and 3.

$$\frac{\text{rate}_4}{\text{rate}_3} = \frac{k[NO]^x[H_2]}{k[NO]^x[H_2]}$$

$$\frac{1.21}{0.30} = \frac{k(1.20)^x(0.60)}{k(0.60)^x(0.60)} = \left(\frac{1.20}{0.60}\right)^x$$

4.0 = 2.0$^x$
Therefore, x = 2

**Work EXERCISES & PROBLEMS: 4, 5**

**EXAMPLE 13.6 First-Order Reaction**

Methyl isocyanide undergoes a rearrangement to form methyl cyanide that follows first-order kinetics.

$$CH_3NC(g) \rightarrow CH_3CN(g)$$

The reaction was studied at 199 °C. The initial concentration of $CH_3NC$ was 0.0258 mol/L, and after 11.4 min analysis showed the concentration of product was $1.30 \times 10^{-3}$ mol/L.
a.  What is the first-order rate constant?
b.  What is the half-life of methyl isocyanide?
c.  How long will it take for 90 percent of the $CH_3NC$ to react?

**• Solution**

a.  This problem illustrates the use of the integrated first-order rate equation

$$\ln \frac{[CH_3NC]}{[CH_3NC]_0} = -kt$$

Where k is the rate constant and $[CH_3NC]$ is the reactant concentration at time t. The initial concentration is $[CH_3NC]_0 = 0.0258$ *M*. After 11.4 min the product concentration is $1.30 \times 10^{-3}$ *M*. This implies that the concentration of $CH_3NC$ remaining unreacted is $0.0258 - 0.0013 = 0.0245$ *M*.

Substitution gives:

$$\ln \frac{0.0245\,M}{0.0258\,M} = -k\,(11.4 \text{ min})$$

$$\ln 0.950 = -k\,(11.4 \text{ min})$$

$$k = 4.54 \times 10^{-3}/\text{min}$$

b.  For a first-order reaction the half-life equation is:

$$t_{1/2} = \frac{0.693}{k} = \frac{0.693}{4.54 \times 10^{-3}/\text{min}} = 153 \text{ min}$$

c.  If 90 percent of the initial $CH_3NC$ is consumed, then 10 percent remains. Therefore:

$$[CH_3NC] = 0.10\,[CH_3NC]_0$$

Substitution into the rate equation gives:

$$\ln \frac{0.10[CH_3NC]_0}{[CH_3NC]_0} = -(4.54 \times 10^{-3}/\text{min})t$$

Solving for t, we get:

$$t = -\frac{\ln 0.10}{4.54 \times 10^{-3}/\text{min}} = \frac{2.303}{4.54 \times 10^{-3}/\text{min}} = 507 \text{ min}$$

## EXAMPLE 13.7 First-Order Reaction

At 230 °C the rate constant for methyl isocyanide isomerization is $9.25 \times 10^{-4}$/s.

$$CH_3NC \rightarrow CH_3CN$$

a. What fraction of the original isocyanide will remain after 60.0 min?
b. What is the half-life of methyl isocyanide at this temperature?

### • Solution

a. Again, applying the first-order equation:

$$\ln \frac{[CH_3NC]}{[CH_3NC]_0} = -kt$$

The fraction of methyl isocyanide remaining is given by the fraction $\dfrac{[CH_3NC]}{[CH_3NC]_0}$.

Plug in the rate constant and the time, and solve for the fraction remaiming:

$$\ln \frac{[CH_3NC]}{[CH_3NC]_0} = -9.25 \times 10^{-4}/s \ (60 \ min) \times \frac{60 \ s}{1 \ min} = -3.33$$

Taking the antilog of both sides gives:

$$\frac{[CH_3NC]}{[CH_3NC]_0} = e^{-3.33} = 0.0358$$

b. The half-life is

$$t_{1/2} = \frac{0.693}{9.25 \times 10^{-4}/s}$$

$$= 749 \ s \ or \ 12.5 \ min$$

### • Comment

Note the much shorter half-life at 230 °C than at 199 °C (Example 13.6).

**Work EXERCISES & PROBLEMS: 6 - 11**

## EXAMPLE 13.8 Calculating Activation Energy

For the reaction,

$$NO + O_3 \rightarrow NO_2 + O_2$$

the following rate constants have been obtained:

| Temperature, °C | $k(1/M \cdot s)$ |
|---|---|
| 10.0 | $9.30 \times 10^6$ |
| 30.0 | $1.25 \times 10^7$ |

Calculate the activation energy for this reaction.

---

**• Solution**

Here we are given two rate constants corresponding to two temperatures. The activation energy ($E_a$) can be calculated by substituting into the equation given earlier that shows the temperature dependence of the rate constant.

$$\ln\left(\frac{k_1}{k_2}\right) = \frac{E_a}{R}\left(\frac{T_1 - T_2}{T_1 T_2}\right)$$

Let    $k_1 = 9.3 \times 10^6/M{\cdot}s$   at $T_1 = 273 + 10.0 = 283$ K
          $k_2 = 1.25 \times 10^7/M{\cdot}s$   at $T_2 = 273 + 30.0 = 303$ K

Recall $R = 8.31$ J/mol·K. Substitute into the preceding equation to solve for $E_a$:

$$\ln\left(\frac{9.3 \times 10^6/M \cdot s}{1.25 \times 10^7/M \cdot s}\right) = \frac{E_a}{8.314\,\text{J/mol}\cdot\text{K}}\left(\frac{283\,\text{K} - 303\,\text{K}}{(283\,\text{K})(303\,\text{K})}\right)$$

Solving for $E_a$ yields:

$$(\ln 0.744)(8.314\,\text{J/mol}\cdot\text{K}) = E_a\left(\frac{-20\,\text{K}}{85749\,\text{K}^2}\right)$$

$$(-0.296)(8.314\,\text{J/mol K}) = E_a(-2.33 \times 10^{-4}/\text{K})$$

$$E_a = 1.06 \times 10^4\,\text{J/mol} \quad (\text{or } 10.6\,\text{kJ/mol})$$

---

**EXAMPLE 13.9   Reaction Energy Profile**

Draw a reaction energy profile for the following endothermic reaction:

$$2\,HI(g) \rightarrow H_2(g) + I_2(g) \qquad \Delta H^\circ_{rxn} = 12.5\,\text{kJ}$$

Given the activation energy, $E_a = 185$ kJ/mol. Label the activation energy and the activated complex. What is the activation energy for the reverse reaction?

**• Solution**

The reaction is endothermic so the products are at a higher level than the reactants.

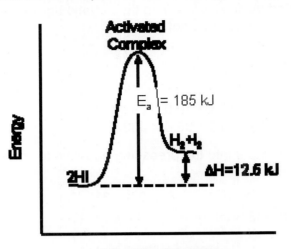

Using the diagram we can see that the activation energy for the reverse reaction is 185 kJ – 12.5 kJ = 173 kJ.

**Work EXERCISES & PROBLEMS: 12 – 15**

**EXAMPLE 13.10  Reaction Mechanism**

The rate law for the following reaction has been experimentally determined to be third-order:

$$2\,NO(g) + O_2(g) \rightarrow 2\,NO_2(g)\quad rate = k[NO]^2[O_2]$$

Which of the two proposed mechanisms that follow is more satisfactory?

a.

$$NO + NO \xrightarrow{k_1} N_2O_2 \qquad slow$$

$$N_2O_2 + O_2 \xrightarrow{k_2} 2NO_2 \qquad fast$$

b.

$$NO + NO \underset{k_{-1}}{\overset{k_1}{\rightleftharpoons}} N_2O_2 \qquad fast$$

$$N_2O_2 + O_2 \xrightarrow{k_2} 2NO_2 \qquad slow$$

**• Solution**

Testing mechanism (a) first, we note that the first step is the rate-determining step. The rate law for this bimolecular step is:

$$rate = k_1[NO]^2$$

This mechanism predicts a rate law that is second-order in NO concentration and zero-order in $O_2$. By comparison, the experimental rate law is second-order in NO and firs- order in $O_2$. The predicted rate law and the experimental rate law do not match, and so mechanism (a) is not satisfactory.

Testing mechanism (b), we find that the rate-determining step is the second step. Since it is bimolecular, its rate law should be:

$$rate = k_2[N_2O_2][O_2]$$

It is not possible to compare this predicted rate law directly with the experimental rate law because of the unique term, which is the concentration of $N_2O_2$. In this mechanism, $N_2O_2$ is an intermediate. Being formed in step 1 and consumed in step 2, its concentration is always small and usually not measurable.

The way around this situation is to find a mathematical substitution for $[N_2O_2]$. Note that step 1 is reversible and equilibrium can be established. This means that the rates of the forward and reverse reactions are equal:

$$k_1[NO]^2 = k_{-1}[N_2O_2]$$

where $k_1$ is the rate constant for the forward reaction, and $k_{-1}$ for the reverse.

Rearranging terms gives us:

$$[N_2O_2] = \frac{k_1}{k_{-1}}[NO]^2$$

We now have an expression for $[N_2O_2]$, which we can substitute into equation II. This yields:

$$\text{rate} = \frac{k_2k_1}{k_{-1}}[NO]^2[O_2]$$

Whenever a reaction step is reversible, we can use the equality of the forward and reverse rates to form an expression to substitute for the concentration of an intermediate.

Now we compare this rate law with the experimental rate law, which is

$$\text{rate} = k[NO]^2[O_2]$$

We see that this mechanism predicts that the reaction will be second-order in NO and first-order in $O_2$, just as observed. Also, the collection of constants $k_2k_1/k_{-1}$ will equal the rate constant.

$$k = \frac{k_2k_1}{k_1}$$

Therefore, the second mechanism predicts a rate law that matches the experimental order of reaction.

**Work EXERCISES & PROBLEMS 16 – 18**

---

**EXAMPLE 13.11  Intermediates and Catalysts**

Given the following mechanism for the decomposition of ozone in the stratosphere, identify the intermediate and the catalyst:

$$O_3 + Cl \rightarrow O_2 + ClO$$
$$ClO + O \rightarrow Cl + O_2$$

**• Solution**

Adding the two steps gives the overall reaction

$$O_3 + O \rightarrow 2\,O_2$$

An intermediate is formed in an early step and removed in a later step and does not appear in the overall equation; therefore ClO is the intermediate. A catalyst is consumed in an early step and is later regenerated. Atomic chlorine, Cl, is the catalyst.

**Work EXERCISES & PROBLEMS 19, 20**
**Work the rest of the EXERCISES & PROBLEMS**

## EXERCISES & PROBLEMS

1. Write expressions for the rate of reaction in terms of each of the reactants and products.

   $N_2(g) + 3H_2(g) \rightarrow 2NH_3(g)$

2. Thiosulfate ion is oxidized by iodine in aqueous solution according to the equation:

   $2S_2O_3^{2-}(aq) + I_2(aq) \rightarrow S_4O_6^{2-}(aq) + 2I^-(aq)$

   If 0.025 mol of $S_2O_3^{2-}$ is consumed in 0.50 L solution per minute:

   a. Calculate the rate of removal of $S_2O_3^{2-}$ in *M*/s.

   b. What is the rate of removal of $I_2$?

3. $N_2O_5$ is an unstable compound that decomposes according to the following equation.:

   $2N_2O_5 \rightarrow 4NO_2 + O_2$

   The following data was obtained at 50 °C.

   | [N₂O₅] (*M*) | Time (s) |
   |--------------|----------|
   | 1.00 | 0 |
   | 0.88 | 200 |
   | 0.78 | 400 |
   | 0.69 | 600 |
   | 0.61 | 800 |
   | 0.54 | 1000 |
   | 0.48 | 1200 |
   | 0.43 | 1400 |

   a. What is the average rate of $N_2O_5$ disappearance in the time interval 200-400 s?
   b. What is the average rate of $N_2O_5$ disappearance in the time interval 800-1000 s?
   c. What is the rate of $O_2$ production in the time interval 800-1000 s?

4. The rate law for the reaction $2A + B \rightarrow C$ was found to be rate = $k[A][B]^2$. If the concentration of B is tripled and the concentration of A is unchanged, by how many times will the reaction rate increase?

5. Use the following data to determine (a) the rate law and (b) the rate constant for the reaction.

   | | 2A + B → C | | |
   |------|-------|-------|-----------|
   | Expt | [A]₀ | [B]₀ | Rate (*M*/s) |
   | 1 | 0.25 | 0.10 | 0.012 |
   | 2 | 0.25 | 0.20 | 0.048 |
   | 3 | 0.50 | 0.10 | 0.024 |

6. A certain first-order reaction A → B is 40% complete (40% of the reactant is used up) in 75 s. What is (a) the rate constant and (b) the half-life of this reaction?

7. a. Write the integrated rate equation for a first-order reaction.
   b. How can the equation be plotted to give a straight line? Define x, y, and the slope.
   c. What information do you need in order to calculate the first-order rate constant?

8. The half-life of a certain first-order reaction is 112 min. What percent of the initial concentration of reactant will remain after 89 min?

9. Consider the reaction in Exercise 3 above. The reaction obeys first-order kinetics. For the reaction:

   $2 N_2O_5 \rightarrow 4 NO_2 + O_2$

   a. Use the data to graphically calculate the rate constant.
   b. How long will it take for the concentration of $N_2O_5$ to fall to 0.25 *M*?
   c. What is the half-life?

10. What are the units of a first-order and a second-order rate constant?

11. A decomposition reaction has a rate constant of 0.12/y at a certain temperature.
    a. What is the half-life of the reaction at the same temperature?
    b. How long will it take for the concentration of the reactant to reach 20% of its initial value?

12. Calculate the activation energy for a reaction given that the rate constant is $4.60 \times 10^{-4}$/s at 350 °C and $1.87 \times 10^{-4}$/s at 320 °C.

13. The rate constant for a first-order reaction is $4.60 \times 10^{-4}$/s at 250 °C. If the activation energy is 100.0 kJ/mol, calculate the rate constant at 300 °C.

14. A substance decomposes according to first-order kinetics, the rate constants at various temperatures being as follows:

    | Temp ( °C) | k (1/s) |
    |------------|---------|
    | 15.0 | $4.41 \times 10^{-6}$ |
    | 25.0 | $1.80 \times 10^{-5}$ |
    | 30.0 | $2.44 \times 10^{-5}$ |

    a. If one were to make an Arrhenius plot what would its slope be equal to according to the Arrhenius theory?
    b. Estimate the numerical value of the slope by using the first and third points.
    c. Calculate the activation energy from your value for the slope.

15. The reaction $2 NOCl \rightarrow 2 NO + Cl_2$ has an $E_a$ of 102.0 kJ/mol and a $\Delta H_{rxn}$ of 75.5 kJ/mol. Sketch the reaction energy diagram and determine the activation energy for the reverse reaction.

16. Which of the following species can be isolated from a chemical reaction?
    product, activated complex, intermediate

17. Carbon monoxide can be converted to carbon dioxide by the following overall reaction.

    $NO_2(g) + CO(g) \rightarrow NO(g) + CO_2(g)$

    The experimentally determined rate law is rate = k $[NO_2]^2$. The suggested mechanism involves two bimolecular elementary steps.

    $NO_2(g) + NO_2(g) \rightleftharpoons NO_3(g) + NO(g)$   step 1
    $NO_3(g) + CO(g) \rightarrow NO_2(g) + CO_2(g)$     step 2

    a. What is the rate law for each step?
    b. Derive the predicted rate laws when step 1 is rate determining, and when step 2 is rate determining.

18. The reaction of nitric oxide and chlorine has the following rate law as determined by experiment:

    $2 NO(g) + Cl_2(g) \rightarrow 2 NOCl(g)$     rate = k [NO] [$Cl_2$]

    Is the following proposed mechanism consistent with the experimental rate law?
    $NO(g) + Cl_2(g) \rightarrow NOCl_2(g)$        slow
    $NO(g) + NOCl_2(g) \rightarrow 2 NOCl(g)$   fast

19. Define a catalyst. Why is a catalyst not consumed in the reaction?

20. Identify the catalyst, any intermediates, and the overall equation for the following mechanism.

    $O_3 + NO \rightarrow O_2 + NO_2$
    $NO_2 + O \rightarrow NO + O_2$

21. Explain why the rate of reaction increases rapidly with increasing temperature.

22. Cooking an egg involves the denaturation of a protein called albumin from a soluble globular protein to an insoluble fibrous protein. This reaction has an activation energy of roughly 75 kJ/mol. At sea level it takes three minutes to prepare a "soft-boiled egg." However on top of Pikes Peak at 4452 m elevation it takes 7.5 minutes to cook a "soft-boiled egg."  Explain why its takes longer to cook an egg on the top of mountains.

23. Describe the difference between an overall reaction (balanced chemical equation) and an elementary reaction.

## PRACTICE TEST QUESTIONS
See notes on taking practice test in the Preface

1. The following experimental data were obtained for the reaction $2A + B \rightarrow$ products. What is the rate law for this reaction?

| Exp. | $[A]_o$ (M) | $[B]_o$ (M) | Rate (M/s) |
|------|-------------|-------------|------------|
| 1 | 0.80 | 0.20 | $5.5 \times 10^{-3}$ |
| 2 | 0.40 | 0.20 | $5.6 \times 10^{-3}$ |
| 3 | 0.80 | 0.40 | $2.23 \times 10^{-2}$ |

2. For the reaction

$$30\ CH_3OH + B_{10}H_{14} \rightarrow 10\ B(OCH_3)_3 + 22\ H_2$$

express the rate in terms of the change in concentration with time for each of the reactants and each of the products.

3. The hydrolysis of sucrose ($C_{12}H_{22}O_{11}$) yields the simple sugars, glucose ($C_6H_{12}O_6$) and fructose ($C_6H_{12}O_6$), which just happen to be isomers.
$$C_{12}H_{22}O_{11} + H_2O \rightarrow C_6H_{12}O_6 + C_6H_{12}O_6$$
The rate follows the rate equation, rate = $k[C_{12}H_{22}O_{11}]$. At 27 °C the rate constant is $2.1 \times 10^{-6}$ /s.
   a. Starting with the sucrose solution with a concentration of 0.10 M at 27 °C, what would the concentration of sucrose be 24 hours later? (The solution is kept at 27 °C).
   b. What is the half-life of sucrose in seconds at 27 °C?

4. $N_2O_5$ decomposes according to first-order kinetics:
$$2\ N_2O_5(g) \rightarrow 4\ NO_2(g) + O_2(g)$$
   a. At a certain temperature 20.0% of the initial $N_2O_5$ decomposes in 2.10 h. Determine the rate constant.
   b. What fraction of the initial $N_2O_5$ will remain after 13.0 h?
   c. What is the half-life?
   d. If the initial concentration of $N_2O_5$ was 0.222 M, then what concentration remains after 24 h?

5. The following data were obtained on the rate of disintegration of a pesticide in soil at 30 °C.

| Time (days) | 11 | 60 | 93 |
|-------------|----|----|----|
| % pest. remaining | 96 | 80 | 71 |

   a. What is the order of reaction with respect to pesticide concentration?
   b. What is the value of the rate constant?

6. a. It takes 30.0 min for the concentration of a reactant in a second-order reaction to drop from 0.40 M to 0.30 M. What is the value of the rate constant for this reaction?
   b. How long will it take for the concentration to drop from 0.40 M to 0.20 M?

7. The reaction $A + 2B \rightarrow C + D$ has for its experimental rate law rate = $k[A]^2[B]$. By what factor will the rate increase if the concentration of A is doubled and the concentration of B is tripled?

8. The rate of the reaction of hydrogen with iodine

$$H_2 + I_2 \rightarrow 2HI$$

has rate constants of $1.41 \times 10^{-5}$ /M·s at 393 °C and $1.40 \times 10^{-4}$ /M·s at 443 °C. Calculate the activation energy for this reaction.

9. At 300 K the rate constant is $1.5 \times 10^{-5}$ /M·s for the reaction:

$$2\,NOCl \rightarrow 2\,NO + Cl_2$$

The activation energy is 90.2 kJ/mol. Calculate the value of the rate constant at 310 K. By what factor did the rate constant increase?

10. The activation energy for the reaction

$$CO + NO_2 \rightarrow CO_2 + NO$$

is 116 kJ/mol. How many times greater is the rate constant for this reaction at 250 °C than it is at 200 °C?

11. If the reaction $2\,HI \rightarrow H_2 + I_2$ has an activation energy of 190 kJ/mol and a $\Delta H = 10$ kJ, what is the activation energy for the reverse reaction?

$$H_2 + I_2 \rightarrow 2\,HI$$

12. The rate law for the net reaction

$$2\,NO + O_2 \rightarrow 2\,NO_2$$

is rate = $k[NO_2]^2[O_2]$. Could the mechanism of this reaction be a single termolecular process?

13. The reaction between nitrite ion and oxygen gas, in aqueous solution is:

$$2\,NO_2^-\,(aq) + O_2(aq) \rightarrow 2\,NO_3^-\,(aq)$$

and proceeds at a rate that is first-order in $[NO_2^-]$ and zero-order in $[O_2]$. A mechanism has been proposed:

$$NO_2^- + O_2 \rightarrow NO_4^- \qquad \text{slow}$$
$$NO_4^- + NO_2^- \rightarrow 2\,NO_3^- \qquad \text{fast}$$

Show that this mechanism is consistent with the experimental rate law.

14. The rate law for the replacement of $H_2O$ by $NH_3$ in the reaction of a complex ion:

$$Ni(H_2O)_6^{2+}\,(aq) + NH_3(aq) \rightarrow Ni(H_2O)_5(NH_3)^{2+}(aq) + H_2O(\ell)$$

is first-order in $Ni(H_2O)_6^{2+}$, and zero-order in $NH_3$.

rate = $k[Ni(H_2O)_6^{2+}]$

Show that the following mechanism is consistent with the experimental rate law.

$$Ni(H_2O)_6^{2+} \rightarrow Ni(H_2O)_5^{2+} + H_2O \qquad \text{slow}$$
$$Ni(H_2O)_5^{2+} + NH_3 \rightarrow Ni(H_2O)_5(NH_3)^{2+} \qquad \text{fast}$$

15. The rate law for the net reaction

$$H_2(g) + I_2(g) \rightarrow 2\,HI(g)$$

is rate = $k[H_2][I_2]$. A possible mechanism involves a bimolecular elementary step:

$$H_2 + I_2 \rightarrow 2\,HI$$

A second possibility has also been proposed:

$$I_2 \rightarrow 2\,I \qquad \text{fast}$$
$$2\,I + H_2 \rightarrow 2\,HI \qquad \text{slow}$$

Show that both mechanisms are consistent with the experimental rate law.

16. The following statements are sometimes made with reference to catalysts. Explain the fallacy in each one of them.
    a. A catalyst is a substance that accelerates the rate of a chemical reaction but does not take part in the reaction.
    b. A catalyst may increase the rate of a reaction going in one direction without increasing the rate of the reaction going in the reverse direction.

17. a. Identify the catalyst in the following mechanism.
    b. What is the overall reaction?
    c. Is this reaction an example of homogeneous or heterogeneous catalysis?

$$H_2O_2(aq) + Br_2(aq) \rightarrow 2H^+(aq) + O_2(aq) + 2Br^-(aq)$$
$$H_2O_2(aq) + 2H^+(aq) + 2Br^-(aq) \rightarrow Br_2(aq) + 2H_2O(\ell)$$

# 13

# 13

# Chapter 14. Chemical Equilibrium

**Chemical Equilibrium and the Equilibrium Constant (Section 14.1)**
**Writing Equilibrium Constant Expressions (Section 14.2)**
**Calculating Equilibrium Constants (Section 14.4)**
**What the Equilibrium Constant Tells Us (Section 14.4)**
**Factors Affecting Chemical Equilibrium (Section 14.5)**

## SUMMARY

### Chemical Equilibrium and the Equilibrium Constant (Section 14.1)

***Reversible Chemical Reactions.*** In this chapter we will discuss how equilibrium is established in several types of chemical reaction systems. We will define the equilibrium constant and show how to calculate equilibrium concentrations of reactants and products. A state of chemical equilibrium exists when the concentrations of reactants and products are observed to remain constant with time. When a mixture of $SO_2$ and $O_2$, for instance, is introduced into a closed reaction vessel at a temperature of 700 K, a reaction that produces $SO_3$ occurs:

$$2 SO_2 + O_2 \rightarrow 2 SO_3$$

When a specific concentration of $SO_3$ is reached, no additional $SO_3$ is formed even though some $SO_2$ and $O_2$ remain. From that time on, the concentrations of $SO_2$, $O_2$, and $SO_3$ stay constant, and we say the system has reached chemical equilibrium.

The constant concentrations are the result of a reversible chemical reaction. In the reverse reaction some $SO_3$ decomposes back into $SO_2$ and $O_2$.

$$2 SO_3 \rightarrow 2 SO_2 + O_2$$

When the rates of the forward reaction and reverse reactions are the same, no *net* chemical change occurs and a state of chemical equilibrium exists. The equilibrium state is referred to as dynamic, because of the continual conversion of reactants into products, and of products into reactants at the molecular level. A reversible reaction is represented by the opposing arrows in the chemical equation:

$$2 SO_2 + O_2 \rightleftharpoons 2 SO_3$$

At 700 K all three species exist together in the reaction vessel. The three compounds are said to be in equilibrium. Equilibrium can be reached by the reverse reaction as well. That is, if only $SO_3$ is added to the reaction vessel at 700 K, it will form an equilibrium amount of $SO_2$ and $O_2$.

***The Equilibrium Constant.*** In quantitative terms, the equilibrium state is described by the equilibrium constant expression, which is also called the mass action expression.

$$K_c = \frac{[SO_3]^2}{[SO_2]^2[O_2]}$$

Here, the brackets [ ] indicate concentration in moles per liter. In this expression, note that the concentration of the product is in the numerator, and the concentrations of the reactants appear in the denominator. Since there are two reactants, their concentrations are multiplied. Also, the concentration of each component is raised to a power equal to its coefficient in the balanced equation. The value of this expression is called the equilibrium constant.

The equilibrium constant for a reaction can be determined by measuring the concentrations of all components at equilibrium and substituting these values into the $K_c$ expression. The equilibrium concentrations of $SO_2$, $O_2$, and $SO_3$ are related to each other by the equilibrium constant expression, which, at 700 K, has the following value.

$$K_c = \frac{[SO_3]^2}{[SO_2]^2[O_2]} = 4.3 \times 10^6$$

## Writing Equilibrium Constant Expressions (Section 14.2)

***The Form of $K_c$ and the Equilibrium Equation.*** In general the, equilibrium constant expression has the form of a ratio of product concentrations over reactant concentrations at equilibrium. For the general equation

$$aA + bB \rightleftharpoons cC + dD$$

where a, b, c, and d are the stoichiometric coefficients for the balanced equation, The equilibrium constant expression is:

$$K_c = \frac{[C]^c[D]^d}{[A]^a[B]^b}$$

In an equilibrium constant expression, each concentration is raised to a power equal to its stoichiometric coefficient. The numerator is found by multiplying together the equilibrium concentrations of the products, each raised to an exponent equal to its stoichiometric coefficient. The denominator is found by multiplying together the equilibrium concentrations of the reactants, each raised to an exponent equal to its stoichiometric coefficient.

The equilibrium constant expression and its value depend on how the equation is balanced. An equation can be balanced with more than one set of coefficients, as shown below:

$$2 SO_2 + O_2 \rightleftharpoons 2 SO_3$$
$$SO_2 + \tfrac{1}{2}O_2 \rightleftharpoons SO_3$$

How does this affect the equilibrium constant? For the latter equation, the equilibrium constant expression is written:

$$K_c' = \frac{[SO_3]}{[SO_2][O_2]^{1/2}}$$

where the prime (') is just to help us keep track of the constant to which we are referring. Note that $K_c'$ is the square root of $K_c$ from the previous section.

$$K_c = \frac{[SO_3]^2}{[SO_2]^2[O_2]}$$

$$K_c' = \frac{[SO_3]}{[SO_2][O_2]^{1/2}}$$

Therefore, by inspection:

$$K_c' = \sqrt{K_c}$$

The value of $K_c'$ is:

$$K_c' = \sqrt{4.3 \times 10^6} = 2.1 \times 10^3$$

The equilibrium constant expression for the reaction written in the reverse direction

$$2\,SO_3 \rightleftharpoons 2\,SO_2 + O_2$$

Is:

$$K_c'' = \frac{[SO_2]^2[O_2]}{[SO_3]^2}$$

Compare this to $K_c$ for the original forward reaction:

$$K_c = \frac{[SO_3]^2}{[SO_2]^2[O_2]}$$

By inspection, you can tell that $K_c''$ is the reciprocal of $K_c$ for the forward reaction.

$$K_c'' = \frac{[SO_3]^2[O_2]}{[SO_3]^2} = \frac{1}{K_c}$$

Therefore, the value of $K_c''$ is

$$K_c'' = \frac{1}{4.3 \times 10^6} = 2.3 \times 10^{-7}$$

Always use the $K_c$ expression and value that are consistent with the way in which the equation is written and balanced.

***$K_p$ and $K_c$.*** For reactions involving gases, the equilibrium constant expression can be written in terms of the partial pressures, $P_i$, of each gaseous component. For the equation:

$$2\,SO_2 + O_2 \rightleftharpoons 2\,SO_3$$

$$K_p = \frac{P_{SO_3}^2}{P_{SO_2}^2 P_{O_2}}$$

The pressure, in atmospheres, of each gas at equilibrium is raised to a power corresponding to its coefficient in the balanced equation. The subscript p in $K_p$ indicates that the equilibrium constant has a value that was calculated using equilibrium partial pressures in atmospheres, rather than units of moles per liter.

The values of $K_p$ and $K_c$ are related. Recall that for an ideal gas, $PV = nRT$, and so the pressure of an ideal gas is proportional to its concentration:

$$P = \frac{n}{V}RT$$

where $n/V$ = moles per liter. Substitution of $(n/V)RT$ for the pressure of each gas, gives the equation relating $K_p$ and $K_c$ that was derived in the text,

$$K_p = K_c(RT)^{\Delta n}$$

where R is the ideal gas constant (0.0821 L·atm/K·mol) and $\Delta n$ is the change in the number of moles of gas, $\Delta n = n(\text{gas products}) - n(\text{gas reactants})$. For the preceding reaction $\Delta n = -1$, so at 700 K, $K_p$ is given by

$$K_p = K_c(RT)^{-1} = \frac{K_c}{RT} = \frac{4.3 \times 10^6}{(0.0821)(700)} = 7.5 \times 10^4$$

In general, $K_p \neq K_c$ except in the case when $\Delta n = 0$.

***Heterogeneous Equilibria.*** Whenever a reaction involves reactants and products that exist in different phases, it is called a heterogeneous reaction, and in a closed container a heterogeneous equilibrium will result. For example, when steam is brought into contact with charcoal the following equilibrium is established in a closed reaction vessel.

$$C(s) + H_2O(g) \rightleftharpoons H_2(g) + CO(g)$$

The equilibrium constant expression for this reaction will *not* be the same as for a homogeneous reaction.

The usual expression for the constant would be

$$K = \frac{[H_2][CO]}{[C][H_2O]}$$

However, the concentration of a pure solid is itself a constant, and is not changed by the addition to, or removal of, some of that solid so long as some solid remains throughout the reaction. Remember that solids have their own volume, and do not fill their containers as gases do. The concentration (mol/L) of a solid such as charcoal depends only on its density. This means that the equilibrium constant expression can be written:

$$K_c = K[C] = \frac{[H_2][CO]}{[H_2O]}$$

where the constant concentration of the pure solid [C] has been combined with the equilibrium constant. Thus, the equilibrium constant expression does not contain the solid. In general, concentrations of solids and pure liquids do not appear in equilibrium constant expressions. In this reaction, equilibrium will be maintained as long as some C(s) is present. The amount of solid carbon present does not affect the position of equilibrium.

***Multiple Equilibria.*** When the product molecules of one equilibrium reaction become reactants in a second equilibrium process, we have an example of multiple equilibria. In such a case, the overall reaction is the sum of the two individual reactions.

For instance, consider the following reactions at 700 °C:

$$NO_2 \rightleftharpoons NO + \tfrac{1}{2}O_2 \quad K_1 = 0.012$$

followed by

$$SO_2 + \tfrac{1}{2}O_2 \rightleftharpoons SO_3 \quad K_2 = 20$$

The overall reaction is the sum of the two steps:

$$NO_2 + SO_2 \rightleftharpoons NO + SO_3 \quad K_c = ?$$

The equilibrium constant value for the overall reaction is given by *the **product** of the equilibrium constants of the individual reactions.*

$$K_c = K_1 K_2 = (0.012)(20) = 0.24$$

| |
|---|
| **Guidelines for Writing Equilibrium Constant Expressions** |

- The concentrations of the reacting species in the solution phase are expressed in mol/L. In the gaseous phase, the concentrations can be expressed in mol/L or in atm. The constant $K_c$ is related to $K_p$ by a simple equation.
- The concentrations of pure solids, pure liquids, and solvents are constants, and do not appear in equilibrium constant expressions.
- The equilibrium constant ($K_c$ or $K_p$) is a dimensionless quantity.
- In stating a value for the equilibrium constant, we must specify the balanced equation and the temperature.
- If a reaction can be expressed as the sum of two or more reactions, the equilibrium constant for the overall reaction is given by the product of the equilibrium constants of the individual reactions.

Examples
14.1 – 14.5

Exercises
1 - 4

## Calculating Equilibrium Constants (Section 14.4)

*Equilibrium Concentrations.* The equilibrium constant for a reaction can be determined by measuring the concentrations of all components at equilibrium and substituting these values into the $K_c$ expression. Two situations arise. In one, all of the equilibrium concentrations are given and a straight forward calculation yields the equilibrium constant. In the second, the initial concentrations of all reactants are given along with the equilibrium concentration of only one product or one reactant. You need to determine the other equilibrium concentrations first, and then $K_c$.

Examples
14.6, 14.7

Exercises
5 – 8

## What the Equilibrium Constant Tells Us (Section 14.4)

*Predicting the Direction of Reaction.* Equilibrium constants provide useful information about chemical reaction systems. In this section, we will use them to predict the direction a reaction will proceed to establish equilibrium. We will also use them to determine the extent of reaction.

The reaction quotient, $Q_c$, is a useful aid in predicting whether or not a reaction system is at equilibrium. Take, for instance, the reaction,

$$2\,SO_2 + O_2 \rightleftharpoons 2\,SO_3$$

The reaction quotient is:

$$Q_c = \frac{[SO_3]_0^2}{[SO_2]_0^2[O_2]_0}$$

You will notice that Q has the same algebraic form of the concentrations terms as does $K_c$. The difference is that these concentrations substituted into this expression are not necessarily equilibrium concentrations. We will call them initial concentrations, $[\ ]_0$. When a set of initial concentrations is substituted into the reaction quotient, $Q_c$ takes on a certain value. In order to predict whether the system is at equilibrium at this point or not, the magnitude of $Q_c$ must be compared with that of $K_c$.

- When $Q_c = K_c$, the reaction is at equilibrium, and no net reaction will occur.
- When $Q_c > K_c$, the system is not at equilibrium, and a net reaction will occur in the reverse direction until $Q_c = K_c$.
- When $Q_c < K_c$, the system is not at equilibrium, and a net reaction will occur in the forward direction until $Q_c = K_c$.

Example
14.8

Exercises
9 – 10

**The Magnitude of the Equilibrium Constant.** The magnitude of the equilibrium constant is related to the degree of conversion of reactants to products before chemical equilibrium is reached. Since $K_c$ is proportional to the concentrations of products divided by the concentrations of reactants present at equilibrium, the value of $K_c$ tells us the relative quantities of reactants and products present at equilibrium. When $K_c \gg 1$ more products are present than reactants at equilibrium. Conversely, when $K_c \ll 1$ more reactants are present at equilibrium than products. In this case equilibrium is reached before appreciable concentrations of products build up. The larger the value of $K_c$, the greater the extent of reaction before equilibrium is reached. In general, we can say:

Example
14.9

- When $K_c \gg 1$, the forward reaction will go nearly to completion. The equilibrium will lie to the right and favor the products.

- When $K_c \ll 1$, the reaction will not go forward to an appreciable extent. The equilibrium will lie to the left and favor the reactants.

**Calculation of Equilibrium Concentrations.** Not only can we estimate the extent of reaction from the $K_c$ value, but you can also calculate the expected concentrations of reactants and products at equilibrium from a knowledge of the initial concentrations and the $K_c$ value. See Example 14.10 for this important type of calculation. In these types of problems, it will be very helpful to use the following approach.

1. Express the equilibrium concentrations of all species in terms of the initial concentrations and an unknown x, which represents *the change in concentration*, (ICE table in examples.)
2. Substitute the equilibrium concentrations derived in part 1 into the equilibrium constant expression, and solve for x. The equilibrium concentration is given by:

   equilibrium concentration = initial concentration ± the change due to reaction
   equilibrium concentration = initial concentration ± x

Example
14.10

Exercises
11 – 13

   where the + sign is used for the substances which are formed, and the – sign for substances which are consumed.

3. Use x to calculate the equilibrium concentrations of all species.

## Factors Affecting Chemical Equilibrium (Section 14.5)

***Le Chatelier's Principle.*** When a reaction reaches a state of chemical equilibrium under a particular set of conditions, then no further change in the concentrations of reactants and products occurs. If a change is made in the conditions under which the system is at equilibrium, chemical change will occur in such a way as to establish a new equilibrium. The factors that can influence equilibrium are change in concentration of a reactant or product, change in pressure (or volume), and change in temperature.
　　What effect does a change in one of these factors have on the extent of reaction? This question can be answered qualitatively by using Le Châtelier's principle: When a stress (or change) is applied to a system in a state of dynamic equilibrium, the system will, if possible, shift to a new position of equilibrium in which the stress is partially offset. To interpret this statement consider, for example, the reaction:

$$2 NO_2(g) \rightleftharpoons N_2O_4(g)$$

First, the reaction must be at equilibrium. Let's add more $N_2O_4$ as an illustration of a change in concentration. The concentration of $N_2O_4$ increases, and the equilibrium is disturbed. The system will respond by using up part of the additional $N_2O_4$ and forming $NO_2$. In this case, a net reverse reaction will partially offset the increased $N_2O_4$ concentration. The net reverse reaction brings the system to a new state of equilibrium. When equilibrium is reestablished, more $NO_2$ will be present than there was before the $N_2O_4$ was added. Thus, the position of equilibrium has shifted to the left.
　　Le Châtelier's principle will predict the direction of the net reaction or "shift in equilibrium" that brings the system to a new equilibrium. In this case, the stress of adding more $N_2O_4$ was partially offset by a net reverse reaction that consumed some of the additional $N_2O_4$. The key to the use of Le Châtelier's principle is to recognize which net reaction, forward or reverse, will partially offset the change in conditions.

***Changes in Volume and Pressure.*** The pressure of a system of gases in chemical equilibrium can be increased by decreasing the available volume. This change causes the concentration of all components to increase. The stress will be partially offset by a net reaction that will lower the total concentration of gas molecules. Consider our previous reaction:

$$2 NO_2(g) \rightleftharpoons N_2O_4(g)$$

When the molecules of both gases are compressed into a smaller volume, their total concentration increases (this is the stress). A net forward reaction (shift to the right) will bring the system to a new state of equilibrium, in which the total concentration of molecules will be lowered somewhat. This partially offsets the initial stress on the system. Notice that when 2 moles of $NO_2$ react, only 1 mole of $N_2O_4$ is formed. When equilibrium is reestablished, more moles of $N_2O_4$ and fewer moles of $NO_2$ will be present than before the pressure increase occurred. In this case, the equilibrium has shifted to the right. In general, an increase in pressure by decreasing the volume will result in a net reaction that decreases the total concentration of gas molecules. A special case arises when the total number of moles of gaseous products and of gaseous reactants are equal in the balanced equation. In this case, no shift in equilibrium will occur.

***Changes in Temperature.*** If the temperature of a system is changed, a change in the value of $K_c$ occurs. An increase in temperature always shifts the equilibrium in the direction of the endothermic reaction, while a temperature decrease shifts the equilibrium in the direction of the exothermic reaction. Therefore, for endothermic reactions the value of $K_c$ increases with increasing temperature, and for exothermic reactions the value of $K_c$ decreases with increasing temperature. In the case of our example reaction, $K_c$ will decrease as the temperature is increased, because the equilibrium will shift in the direction of the endothermic reaction, that is, in the reverse direction.

$$2\ NO_2(g) \rightleftharpoons N_2O_4(g) \qquad \Delta H^\circ_{rxn} = -58.0\ kJ/mol$$

We can explain this in terms of Le Chatelier's principle. As heat is added to the system, it represents a stress on the equilibrium. The equilibrium will shift in the direction that will consume some of the added heat. This partially offsets the stress. In this reaction, the equilibrium shifts to the left, and $K_c$ decreases.

Remember, of these three types of changes; concentration, pressure, and temperature, only changes in temperature will actually alter the $K_c$ value.

There are two other factors related to chemical reactions that do not affect the position of equilibrium. The first of these is the presence of a catalyst. Catalysts increase the rate at which equilibrium is reached by lowering the activation energy barrier reduction in $K_c$. Of course, this affects both the forward reaction and the reverse reaction as well. The net result is that changes in the concentration of a catalyst will not affect the equilibrium concentrations of reactants and products, nor will they change the value of $K_c$. Catalyst concentration changes affect only the amount of time required for the chemical system to reach an equilibrium state.

The second of the two factors is the addition of an inert gas to a system at equilibrium. This will cause an increase in the total pressure within the reactor vessel. However, none of the partial pressures of reactants or products are changed, and so the equilibrium is not upset, and no shifting is needed to bring the system back to equilibrium.

| Example |
| 14.11 |
| Exercises |
| 14 - 17 |

## GLOSSARY LIST

| | | |
|---|---|---|
| chemical equilibrium | law of mass action | reaction quotient |
| physical equilibrium | homogeneous equilibrium | Le Châtelier's principle |
| equilibrium constant | heterogeneous equilibrium | |

## EQUATIONS

| Algebraic Equation | English Translation |
|---|---|
| $K = \dfrac{[C]^c[D]^d}{[A]^a[B]^b}$ | equilibrium constant expression |
| $K_p = K_c(RT)^{\Delta n}$ | conversion between $K_p$ and $K_c$ |

## WORKED EXAMPLES

**EXAMPLE 14.1  The Equilibrium Constant Expression**

Write the equilibrium constant expressions for the following reversible reactions:

a.  $4\,NH_3(g) + 5\,O_2(g) \rightleftharpoons 4\,NO(g) + 6\,H_2O(g)$
b.  $BaO(s) + CO_2(g) \rightleftharpoons BaCO_3(s)$

**• Solution**

Remember that the equilibrium constant expression has the concentrations of the products in the numerator and those of the reactants in the denominator. Raise each concentration to a power equal to the coefficient of that substance in the balanced equation. Include concentration terms for gaseous components only; leave out the concentrations of pure liquids and solids as they are not included in the equilibrium constant expression. The expressions are:

a.  $K_c = \dfrac{[NO]^4[H_2O]^6}{[NH_3]^4[O_2]^5}$    b.  $K_c = \dfrac{1}{[CO_2]}$

**Work EXERCISES & PROBLEMS: 1**

---

**EXAMPLE 14.2  The Equilibrium Constant for a Reverse Reaction**

The equilibrium constant for the reaction,

$$H_2(g) + I_2(g) \rightleftharpoons 2\,HI(g)$$

at 400 °C is:

$$K_c = \frac{[HI]^2}{[H_2][I_2]} = 64$$

What is the equilibrium constant value for the reverse reaction?

$$2\,HI(g) \rightleftharpoons H_2(g) + I_2(g)$$

**• Solution**

Look to see if the reactions are related in some way. Note that the equilibrium constant for the second reaction is the reciprocal of the equilibrium expression for the forward reaction.

$$K_c' = \frac{[H_2][I_2]}{[HI]^2} = \frac{1}{K_c}$$

$$K_c' = \frac{1}{K_c} = \frac{1}{64} = 0.016$$

**Work EXERCISES & PROBLEMS: 2**

---

**EXAMPLE 14.3  $K_p$ and $K_c$ Values**

What are the values of $K_p$ and $K_c$ at 1000 °C for the reaction,

$$CaCO_3(s) \rightleftharpoons CaO(s) + CO_2(g)$$

if the pressure of $CO_2$ in equilibrium with $CaCO_3$ and CaO is 3.87 atm?

**• Solution**

Enough information is given to find $K_p$ first. Writing the $K_p$ expression for this heterogeneous reaction:

$$K_p = P_{CO_2} = 3.87$$

Then to get $K_c$, rearrange the equation,

$$K_p = K_c(RT)^{\Delta n}$$

where $\Delta n$, the change in the number of moles of gas in the reaction is +1.

$$K_c = [CO_2] = \frac{K_p}{(RT)^{\Delta n}} = \frac{3.87}{[(0.0821)(1273)]^1} = 0.0370$$

**EXAMPLE 14.4 $K_p$ and Partial Pressures at Equilibrium**

At 400 °C, $K_c = 64$ for the reaction,

$$H_2(g) + I_2(g) \rightleftharpoons 2HI(g)$$

a. What is the value of $K_p$ for this reaction?
b. If, at equilibrium, the partial pressures of $H_2$ and $I_2$ in a container are 0.20 atm and 0.50 atm, respectively, what is the partial pressure of HI in the mixture?

**• Solution**

a. The equation relating $K_p$ to $K_c$ is:

$$K_p = K_c(RT)^{\Delta n}$$

Here, the change in the number of moles of gas $\Delta n$ is:

$$\Delta n = 2 \text{ mol HI} - (1 \text{ mol } H_2 + 1 \text{ mol } I_2) = 0$$

Since, $\Delta n = 0$, $K_p$ and $K_c$ are the same.

$$K_p = K_c(RT)^0 = K_c$$

$$K_p = K_c = 64$$

b. Writing the equilibrium constant expression:

$$K_p = \frac{P_{HI}^2}{P_{H_2} P_{I_2}} = 64$$

and substituting the given pressures:

$$\frac{P_{HI}^2}{(0.20)(0.50)} = 64$$

the partial pressure of HI is

$$P_{HI} = \sqrt{(0.20)(0.50)(64)}$$

$$P_{HI} = 2.5 \text{ atm}$$

**Work EXERCISES & PROBLEMS: 3**

## EXAMPLE 14.5 Multiple Equilibria

Consider the following equilibra:

$$A(g) + 2B(g) \rightleftharpoons 2C(g) \qquad K'_c = 10 \qquad \text{(equation 1)}$$

$$A(g) + B(g) \rightleftharpoons D(g) \qquad K''_c = 0.10 \qquad \text{(equation 2)}$$

What is the value of $K_c$ for the reaction :

$$2D(g) \rightleftharpoons A(g) + 2C(g) \qquad \text{(equation 3)}$$

• **Solution**

To create the reaction equation 3 from the first two reactions, equation 2 must be doubled and flipped. The equation 1 and 2 can be added.

$$A(g) + 2B(g) \rightleftharpoons 2C(g) \qquad K'_c = 10$$

$$2D(g) \rightleftharpoons 2A(g) + 2B(g) \qquad K'''_c = (1/0.10)^2 \qquad \text{(equation 4)}$$

_____

$$2D(g) \rightleftharpoons A(g) + 2C(g) \qquad K_c = 10 \times 100 = 1000$$

• **Comment:**

Doubling a reaction equation squares the value of $K''_c$ and when a reaction is flipped, take the reciprocal of K. This is demonstrated in equation 4.

When equations are added to give a third, the K's are multiplied together.

**Work EXERCISES & PROBLEMS: 4**

## EXAMPLE 14.6  Calculating an Equilibrium Constant

Ammonia is synthesized from hydrogen and nitrogen according to the equation:

$$3 H_2(g) + N_2(g) \rightleftharpoons 2 NH_3(g)$$

An equilibrium mixture at a given temperature was analyzed and the following concentrations were found:  0.31 mol $N_2$/L; 0.90 mol $H_2$/L; and 1.4 mol $NH_3$/L. What is the equilibrium constant value?

• **Solution**

When the concentrations of each component of a chemical system at equilibrium are known, the value of $K_c$ can be determined readily by substituting these concentrations into the equilibrium constant expression.

$$K_c = \frac{[NH_3]^2}{[H_2]^3[N_2]}$$

$$K_c = \frac{(1.4)^2}{(0.90)^3(0.31)} = 8.7$$

**Work EXERCISES & PROBLEMS: 5**

---

**EXAMPLE 14.7  Calculating an Equilibrium Constant**

When 3.0 mol of $I_2$ and 4.0 mol of $Br_2$ are placed in a 2.0-L reaction chamber at 150 °C, the following reaction occurs until equilibrium is reached:

$$I_2(g) + Br_2(g) \rightleftharpoons 2\ IBr(g)$$

Chemical analysis then shows that the reactor contains 3.2 mol of IBr. What is the equilibrium constant $K_c$ for the reaction?

**• Solution**

We need to know the equilibrium concentrations of $I_2$, $Br_2$, and IBr. These, when substituted into the $K_c$ expression, will give the value of $K_c$.

Determine the molarity of $I_2$ and $Br_2$ initially put in the reaction chamber. Initially, the amounts of each component in the reactor are 3.0 mol of $I_2$/ 2.0 L = 1.5 $M$ $I_2$ and 4.0 mol of $Br_2$/2.0 L = 2.0 $M$ $Br_2$ and zero molar of IBr. The molarity of $I_2$ remaining at equilibrium is given by the initial molarity of $I_2$ minus the molarity of $I_2$ reacted. This is also true for $Br_2$. The information that 3.2 mol IBr/2.0 L = 1.6 $M$ IBr  formed tells us that 0.8 $M$ $I_2$ and 0.8 $M$ $Br_2$ must have reacted. We know this because the balanced equation states that 2 mol IBr are formed whenever 1 mol $I_2$ and 1 mol of $Br_2$ react.  This is best illustrated using an ICE table:

|   | $I_2$ | + | $Br_2$ | $\rightleftharpoons$ | 2 IBr |
|---|---|---|---|---|---|
| I | 1.5 *M* | | 2.0 *M* | | 0 *M* |
| C | - 0.8 *M* | | - 0.8 *M* | | + 1.6 *M* |
| E | 0.7 *M* | | 1.2 *M* | | 1.6 *M* |

Substituting the equilibrium concentrations into the equilibrium constant expression gives the $K_c$ value:

$$K_c = \frac{[IBr]^2}{[I_2][Br_2]} = \frac{(1.6)^2}{(0.7)(1.2)} =$$

$$K_c = 3.0$$

### • Comment

The key to finding all the equilibrium concentrations was that the changes in $I_2$ and $Br_2$ could be related to the change in IBr through the balanced chemical equation. Keep in mind that the equilibrium concentration equals the initial concentration plus the change in concentration due to reaction to reach equilibrium. The change in concentration is positive for a product, and negative for a reactant.

**Work EXERCISES & PROBLEMS: 6 - 8**

---

### EXAMPLE 14.8  Predicting Direction of Reaction

At a certain temperature the reaction,

$$CO(g) + Cl_2(g) \rightleftharpoons COCl_2(g)$$

has an equilibrium constant $K_c$ = 13.8. Is the following mixture an equilibrium mixture?  If not, in which direction (forward or reverse) will reaction occur to reach equilibrium?

$$[CO]_0 = 2.5 \ M; \ [Cl_2]_0 = 1.2 \ M; \ \text{and} \ [COCl_2]_0 = 5.0 \ M$$

### • Solution

Recall that for the system to be at equilibrium $Q_c = K_c$. Substitute the given concentrations into the reaction quotient for the reaction, and determine $Q_c$.

$$Q_c = \frac{[COCl_2]_0}{[CO]_0[Cl_2]_0} = \frac{(5.0)}{(2.5)(1.2)} = 1.7$$

Compare $Q_c$ to $K_c$. Since $Q_c < K_c$ the reaction mixture is not an equilibrium mixture. The product concentrations are too low and a net forward reaction will occur until $Q_c = K_c$, and the reaction reaches equilibrium.

**Work EXERCISES & PROBLEMS: 9, 10**

---

### EXAMPLE 14.9  Extent of Reaction

Arrange the following reactions in order of their increasing tendency to proceed toward completion (least extent → greatest extent).

a.  $CO + Cl_2 \rightleftharpoons COCl_2$       $K_c = 13.8$
b.  $N_2O_4 \rightleftharpoons 2 \ NO_2$       $K_c = 2.1 \times 10^{-4}$
c.  $2 \ NOCl \rightleftharpoons 2 \ NO + Cl_2$       $K_c = 4.7 \times 10^{-4}$

### • Solution

The larger the value of $K_c$, the more products there are at equilibrium compared to reactants, and the farther a reaction will proceed toward completion (the greater the extent of reaction). Therefore,  b < c < a.

**EXAMPLE 14.10  Calculating the Equilibrium Concentrations**

At 400 °C, $K_c$ = 64 for the equilibrium,

$$H_2(g) + I_2(g) \rightleftharpoons 2\,HI(g)$$

If 1.00 mol $H_2$ and 2.00 mol $I_2$ are introduced into an empty 0.50-L reaction vessel, find the equilibrium concentrations of all components at 400 °C.

• **Solution**

The equilibrium constant expression relates the concentrations $H_2$, $I_2$, and HI at equilibrium.

$$K_c = \frac{[HI]^2}{[H_2][I_2]} = 64$$

First we need expressions for the equilibrium concentrations of $H_2$, $I_2$, and HI. Begin by tabulating the initial concentrations.

| Concentration | $H_2$ | + | $I_2$ | $\rightleftharpoons$ | 2 HI |
|---|---|---|---|---|---|
| Initial | 1.00 mol/0.50 L | | 2.00 mol/0.50 L | | 0 |
| Change | — | | — | | — |
| Equilibrium | — | | — | | — |

Since the answer involves three unknowns, we will relate the concentrations to each other by introducing the variable x. Recall, that the equilibrium concentration = initial concentration ± change in concentration. Let x = the change in concentration of $H_2$. That is, let x = the number of moles of $H_2$ reacted per liter. From the coefficients of the balanced equation, we can tell that if the change in $H_2$ is –x, then the change in $I_2$ must also be –x, and the change in HI must be +2x.

The next step is to complete the table in units of molarity:

| Concentration | $H_2$ | + | $I_2$ | $\rightleftharpoons$ | 2HI |
|---|---|---|---|---|---|
| Initial (*M*) | 2.0 | | 4.0 | | 0 |
| Change (*M*) | –x | | –x | | 2x |
| Equilibrium (*M*) | (2.0 – x) | | (4.0 – x) | | 2x |

Now, substitute the equilibrium concentrations from the table into the $K_c$ expression,

$$K_c = \frac{(2x)^2}{(2.0-x)(4.0-x)} = 64$$

and solve for x.

$$\frac{(2x)^2}{x^2 - 6.0x + 8.0} = 64$$

Rearranging, we get:

$$4x^2 = 64x^2 - 384x + 512$$

and grouping yields a quadratic equation:

$$60x^2 - 384x + 512 = 0$$

We will use the general method of solving a quadratic equation of the form:

$$ax^2 + bx + c = 0$$

The root x is given by,

$$x = \frac{-b \pm \sqrt{b^2 - 4ac}}{2a}$$

In this case, a = 60, b = −384, and c = 512. Therefore,

$$x = \frac{-(-384) \pm \sqrt{(-384)^2 - 4(60)(512)}}{2(60)} = \frac{384 \pm \sqrt{2.5 \times 10^4}}{120}$$

$$x = \frac{384 \pm 158}{120} = \text{ 1.9 and 4.5 mol/L}$$

Recall that  x = the number of moles of $H_2$ (or $I_2$) reacted per liter. Of the two answers (roots), only 1.9 is reasonable, because the value 4.5 *M* would mean that more $H_2$ (or $I_2$) reacted than was present at the start. This would result in a negative equilibrium concentration, which is physically meaningless. We therefore use the root x = 1.9 *M* to calculate the equilibrium concentrations:

$$[H_2] = 2.0 - x = 2.0\ M - 1.9\ M = 0.1\ M$$

$$[I_2] = 4.0 - x = 4.0\ M - 1.9\ M = 2.1\ M$$

$$[HI] = 2x = 2(1.9\ M) = 3.8\ M$$

The results can be checked by plugging these concentrations back into the $K_c$ expression to see if $K_c = 64$:

$$K_c = \frac{[HI]^2}{[H_2][I_2]} = \frac{(3.8)^2}{(0.1)(2.1)} = 68$$

Thus, the concentrations we have calculated are correct. The difference between 64 and 68 results from rounding off to maintain the correct number of significant figures. Therefore, our result is correct only to the number of signficant figures given in the problem.

**Work EXERCISES & PROBLEMS: 11-13**

---

**EXAMPLE 14.11 Changing Conditions Affecting Equilibrium**

For the reaction at equilibrium,

$$2\,NaHCO_3(s) \rightleftharpoons Na_2CO_3(s) + H_2O(g) + CO_2(g) \qquad \Delta H^\circ_{rxn} = 128\ kJ/mol$$

state the effects (increase, decrease, no change) of the following stresses on the number of moles of sodium carbonate, $Na_2CO_3$, at equilibrium in a closed container. Note that $Na_2CO_3$ is a solid (this is a heterogeneous equation); its concentration will remain constant, but its amount can change.

a. removing $CO_2(g)$.
b. adding $H_2O(g)$.
c. raising the temperature.
d. adding $NaHCO_3(s)$.

**• Solution**

Apply Le Châtelier's principle.

a. If $CO_2$ concentration is lowered, the system will react in such a way as to offset the change. That is, a shift to the right will replace some of the missing $CO_2$. The number of moles of $Na_2CO_3$ increases.
b. Addition of $H_2O(g)$ exerts a stress on the equilibrium that is partially offset by a shift in the equilibrium to the left (net reverse reaction). This consumes $Na_2CO_3$ as well as some of the extra $H_2O$. The number of moles of $Na_2CO_3$ decreases.
c. An increase in temperature will increase the $K_c$ value of an endothermic reaction. There is a shift to the right, and more $Na_2CO_3$ is formed.
d. The position of a heterogeneous equilibrium does not depend on the amounts of pure solids or liquids present. The same equilibrium is reached whether the system contains 1 g of $NaHCO_3(s)$ or 10 g of $NaHCO_3$. No shift in the equilibrium occurs. No change in the amount of $Na_2CO_3$ occurs.

**Work EXERCISES & PROBLEMS: 14 – 17**
**Work the rest of the EXERCISES & PROBLEMS**

## EXERCISES & PROBLEMS

1. Write the equilibrium constant expressions ($K_c$) for the following equations.

   a. $2\,HgO(s) \rightleftharpoons 2\,Hg(\ell) + O_2(g)$

   b. $Ni(s) + 4\,CO(g) \rightleftharpoons Ni(CO)_4(g)$

   c. $2\,NO(g) + Br_2(g) \rightleftharpoons 2\,NOBr(g)$

2.  For the reaction

    $H_2(g) + Br_2(g) \rightleftharpoons 2HBr(g)$

    $K_p = 7.1 \times 10^4$ at 700 K. What is the value of $K_p$ for the following reactions at the same temperature?

    a. $2\ HBr(g) \rightleftharpoons H_2(g) + Br_2(g)$

    b. $\frac{1}{2} H_2(g) + \frac{1}{2} Br_2(g) \rightleftharpoons HBr(g)$

3.  $K_p$ for the decomposition of ammonium chloride at 427 °C is

    $NH_4Cl(s) \rightleftharpoons NH_3(g) + HCl(g) \quad K_p = 4.8$

    Calculate $K_c$ for this reaction.

4.  The following equilibrium constants were determined at 1123 K:

    | | |
    |---|---|
    | $C(s) + CO_2(g) \rightleftharpoons 2\ CO(g)$ | $K_c = 1.4 \times 10^{12}$ |
    | $CO(g) + Cl_2(g) \rightleftharpoons COCl_2(g)$ | $K_c = 5.5 \times 10^{-1}$ |

    Write the equilibrium constant expression $K_c$ and calculate the equilibrium constant at 1123 K for the following reaction:

    $C(s) + CO_2(g) + 2\ Cl_2(g) \rightleftharpoons 2\ COCl_2(g)$

5.  A sample of nitrosyl bromide was heated to 100 °C in a 10.0-L container in order to partially decompose it.

    $2\ NOBr(g) \rightleftharpoons 2\ NO(g) + Br_2(g)$

    At equilibrium, the container was found to contain 0.0585 mol of NOBr, 0.105 mol of NO, and 0.0524 mol of $Br_2$. Calculate the value of $K_c$.

6.  1.25 mol NOCl was placed in a 2.50-L reaction chamber at 427 °C. After equilibrium was reached, 1.10 mol NOCl remained. Calculate the equilibrium constant $K_c$ for the reaction.

    $2\ NOCl(g) \rightleftharpoons 2\ NO(g) + Cl_2(g)$

7.  The brown gas $NO_2$ and the colorless gas $N_2O_4$ exist in equilibrium.

    $N_2O_4(g) \rightleftharpoons 2\ NO_2(g)$

    0.625 mol of $N_2O_4$ was introduced into a 5.00-L vessel and was allowed to decompose until it reached equilibrium with $NO_2$. The concentration of $N_2O_4$ at equilibrium was 0.0750 *M*. Calculate $K_c$ for the reaction.

8. 5.0 mol of ammonia were introduced into a 5.0-L reaction chamber in which it partially dissociated at high temperatures.

   $$2 NH_3(g) \rightleftharpoons 3 H_2(g) + N_2(g)$$

   At equilibrium, at a particular temperature, 80.0% of the ammonia had reacted. Calculate $K_c$ for the reaction.

9. The reaction,

   $PCl_5(g) \rightleftharpoons PCl_3(g) + Cl_2(g)$ has the equilibrium constant value $K_c = 0.24$ at 300 °C.

   a. Is the following reaction mixture at equilibrium?

   $[PCl_5] = 5.0$ mol/L; $[PCl_3] = 2.5$ mol/L; $[Cl_2] = 1.9$ mol/L

   b. If not, predict the direction in which the system will react to reach equilibrium.

10. At 700 K, the reaction,

    $$2 SO_2(g) + O_2(g) \rightleftharpoons 2 SO_3(g)$$

    has an equilibrium constant $K_c = 4.3 \times 10^6$.

    a. Is a mixture with the following concentrations at equilibrium?

    $[SO_2] = 0.10\ M$; $[SO_3] = 10\ M$; $[O_2] = 0.10\ M$

    b. If not at equilibrium, predict the direction in which a net reaction will occur to reach equilibrium.

11. The decomposition of NOBr is represented by the equation:

    $$2 NOBr(g) \rightleftharpoons 2 NO(g) + Br_2(g) \qquad K_c = 0.0169$$

    At equilibrium the concentrations of NO and $Br_2$ are $1.05 \times 10^{-2}\ M$ and $5.24 \times 10^{-3}\ M$, respectively. What is the concentration of NOBr?

12. At 400 °C, the equilibrium constant for the reaction,

    $$H_2(g) + I_2(g) \rightleftharpoons 2HI(g)$$

    is 64. A mixture of 0.250 mol $H_2$ and 0.250 mol $I_2$ was introduced into an empty 0.75-L reaction vessel at 400 °C, find the equilibrium concentrations of all components.

13. For the equilibrium,

    $$N_2O_4(g) \rightleftharpoons 2 NO_2(g) \qquad K_c = 0.36 \text{ at } 100 \text{ °C}$$

    a sample of 0.25 mol $N_2O_4$ is allowed to dissociate and come to equilibrium in a 1.5-L flask at 100 °C. What are the equilibrium concentrations of $NO_2$ and $N_2O_4$?

14. Copper can be extracted from its ores by heating $Cu_2S$ in air.

$$Cu_2S(s) + O_2(g) \rightleftharpoons 2\,Cu(s) + SO_2(g) \qquad \Delta H^{\circ}_{rxn} = -250 \text{ kJ/mol}$$

Predict the direction of the shift of the equilibrium position in response to each of the following changes in conditions. If no shift occurs, say so.
a. Adding more $O_2(g)$.
b. Compressing the vessel volume in half.
c. Raising the temperature.
d. Adding more $SO_2(g)$.
e. Adding more $Cu(s)$.
f. Adding a catalyst.

15. Consider the following reaction at equilibrium:

$$2\,SO_2(g) + O_2(g) \rightleftharpoons 2\,SO_3(g)$$

If the volume of the container is decreased at constant temperature, state which way the equilibrium will shift to reach a new equilibrium: left, right, or no change.

16. The following reaction is at equilibrium. Describe what change in the concentration of $NO(g)$, if any, will occur if bromine is added to the reaction chamber.

$$2\,NOBr(g) \rightleftharpoons 2\,NO(g) + Br_2(g)$$

17. The following reaction is exothermic:

$$2\,SO_2(g) + O_2(g) \rightleftharpoons 2\,SO_3(g)$$

Describe what will happen to the concentration of $SO_2$ when the temperature is increased.

18. Consider the following reaction at 400 °C:

$$H_2(g) + I_2(g) \rightleftharpoons 2HI(g)$$

$H_2$ and $I_2$ were placed into a flask and allowed to react until equilibrium was reached. Then a very small amount of $H^{131}I$ was added. $^{131}I$ is an isotope of iodine that is radioactive. Will radioactive $^{131}I$ atoms stay attached to H atoms in HI, or will some or all of them find their way into the $I_2$ molecules forming $I^{131}I$? Explain your answer.

19. Consider the following reaction at 400 °C:

$$H_2O(g) + CO(g) \rightleftharpoons H_2(g) + CO_2(g)$$

$H_2O$, $CO(g)$, $H_2(g)$, and $CO_2(g)$ were put into a flask so that the composition corresponded to an equilibrium mixture. A lab technician added an iron catalyst to the mixture, but was surprised when no additional $H_2(g)$ and $CO_2(g)$ were formed, even after many days. Explain why the technician should not have been surprised.

## PRACTICE TEST QUESTIONS
See notes on taking practice test in the Preface

1. Write the equilibrium constant expressions for the following reactions:
   a. $4 NH_3(g) + 3 O_2(g) \rightleftharpoons$
   $\qquad\qquad 2 N_2(g) + 6 H_2O(g)$
   b. $2 N_2O(g) + 3 O_2(g) \rightleftharpoons 2 N_2O_4(g)$
   c. $2 ClO_2(g) + F_2(g) \rightleftharpoons 2 FClO_2(g)$
   d. $H_2(g) + Br_2(\ell) \rightleftharpoons 2 HBr(g)$
   e. $C(s) + CO_2(g) \rightleftharpoons 2 CO(g)$
   f. $CuO(s) + H_2(g) \rightleftharpoons Cu(s) + H_2O(g)$

2. The equilibrium constant $K_c$ for the reaction:
   $$Ni(s) + 4 CO(g) \rightleftharpoons Ni(CO)_4(g)$$
   is $5.0 \times 10^4$ at 25 °C. What is value of the equilibrium constant for the reaction?

   $$Ni(CO)_4(g) \rightleftharpoons Ni(s) + 4 CO(g)$$

3. The decomposition of HI(g) is represented by the equation
   $$2 HI(g) \rightleftharpoons H_2(g) + I_2(g) \quad K_c = 64$$
   If the equilibrium concentrations of $H_2$ and $I_2$ at 400 °C are found to be $[H_2] = 4.2 \times 10^{-4}$ M and $[I_2] = 1.9 \times 10^{-3}$ M, what is the equilibrium concentration of HI?

4. For the reaction
   $$CH_4(g) + 2 H_2S(g) \rightleftharpoons CS_2(g) + 4 H_2(g)$$
   $$K_p = 2.05 \times 10^9 \text{ at 25 °C.}$$

   Calculate $K_p$ and $K_c$, at this temperature, for
   $$2 H_2(g) + \tfrac{1}{2} CS_2(g) \rightleftharpoons H_2S(g) + \tfrac{1}{2} CH_4(g)$$

5. A 1.00-L vessel initially contains 0.777 moles of $SO_3(g)$ at 1100 K. What is the value of $K_c$ for the following reaction, if 0.520 moles of $SO_3$ remain at equilibrium?

   $$2 SO_3(g) \rightleftharpoons 2 SO_2(g) + O_2(g)$$

6. Initially, a 1.0-L vessel contains 10.0 moles of NO and 6.0 moles of $O_2$ at a certain temperature. They react until equilibrium is established.
   $$2 NO(g) + O_2(g) \rightleftharpoons 2 NO_2(g)$$
   At equilibrium, the vessel contains 8.8 moles of $NO_2$. Determine the value of $K_c$ at this temperature.

7. Given the reaction:
   $$N_2 + O_2 \rightleftharpoons 2 NO$$
   $$K_c = 2.5 \times 10^{-3} \text{ at 2130 °C}$$
   Decide whether the following mixture is at equilibrium, or if a net forward or reverse reaction will occur. [NO] = 0.005; $[O_2]$ = 0.25; $[N_2]$ = 0.020 mol/L.

8. Hydrogen iodide decomposes according to the equation:
   $$2 HI(g) \rightleftharpoons H_2(g) + I_2(g)$$
   $$K_c = 0.0156 \text{ at 400 °C}$$
   A 0.55 mol sample of HI was injected into a 2.0-L vessel held at 400 °C. Calculate the concentration of HI at equilibrium.

9. 2.00 mol of NOCl was placed in a 2.00-L reaction vessel at 400 °C. After equilibrium was established, it was found that 24 percent of the NOCl had dissociated according to the equation:
   $$2 NOCl(g) \rightleftharpoons 2 NO(g) + Cl_2(g)$$
   Calculate the equilibrium constant $K_c$ for the reaction.

10. For the reaction,
    $$N_2(g) + O_2(g) \rightleftharpoons 2 NO(g)$$
    $$K_p = 3.80 \times 10^{-4} \text{ at 2000 °C.}$$
    What equilibrium pressures of $N_2$, $O_2$, and NO will result when a 10-L reactor vessel is filled with 2.00 atm of $N_2$ and 0.400 atm of $O_2$ and the reaction is allowed to come to equilibrium?

14 14

14 14

14 14

14 14

14 14

11. For the equilibrium,

$$N_2O_4(g) \rightleftharpoons 2\,NO_2(g)$$
$$K_c = 0.36 \text{ at } 100\,°C$$

Suppose 30.0 g $N_2O_4$ is placed in a 2.5-L flask at 100 °C. (a) Calculate the number of moles of $NO_2$ present at equilibrium and (b) the percentage of the original $N_2O_4$ that is dissociated.

12. Arrange the following reactions in their increasing tendency to proceed toward completion:

a. $2\,HF \rightleftharpoons H_2F_2$ $\qquad K_c = 1 \times 10^{-13}$

b. $2\,H_2 + O_2 \rightleftharpoons 2\,H_2O$ $\qquad K_c = 3 \times 10^{81}$

c. $2\,NOCl \rightleftharpoons 2\,NO + Cl_2$ $\quad K_c = 4.7 \times 10^{-4}$

13. What changes in the equilibrium composition of the reaction,

$$2\,SO_2(g) + O_2(g) \rightleftharpoons 2\,SO_3(g) \qquad \Delta H^\circ_{rxn} = -197 \text{ kJ/mol}$$

will occur if it experiences the following stresses?
a. The partial pressure of $SO_3(g)$ is increased.
b. Inert Ar gas is added.
c. The temperature of the system is decreased.
d. The total pressure of the system is increased by reducing the available volume.
e. The partial pressure of $O_2(g)$ is decreased.

14. For the chemical equilibrium,

$$PCl_5(g) \rightleftharpoons PCl_3(g) + Cl_2(g) \qquad \Delta H^\circ_{rxn} = 92.9 \text{ kJ/mol}$$

a. What is the effect on $K_c$ of lowering the temperature?
b. What is the effect on the equilibrium concentration of $PCl_3$ of adding $Cl_2$?
c. What is the effect on the equilibrium concentrations of compressing the mixture to a smaller volume?
d. What is the effect on the equilibrium pressure of $Cl_2$ of removing $PCl_3$?

15. Consider the equilibrium,

$$SO_2(g) + NO_2(g) \rightleftharpoons NO(g) + SO_3(g)$$

where $K_c = 85.0$ at 460 °C.
The following mixture was prepared in a reactor at 460 °C.
$[SO_2] = 0.0200\,M$; $[NO_2] = 0.0200\,M$; $[NO] = 0.100\,M$; $[SO_3] = 0.100\,M$

What will the concentrations of the four gases be when equilibrium is reached?

16. For the decomposition of calcium carbonate,

$$CaCO_3(s) \rightleftharpoons CaO(s) + CO_2(g)$$
$$\Delta H^\circ_{rxn} = 175 \text{ kJ/mol}$$

how will the amount (not concentration) of $CaCO_3(s)$ change with the following stresses?
a. $CO_2(g)$ is removed.
b. $CaO(s)$ is added.
c. The temperature is raised.
d. The volume of the container is decreased.

17. In a closed container at 1900 K nitrogen reacts with oxygen to yield NO:

$N_2(g) + O_2(g) \rightleftharpoons 2\,NO(g)$ $K_p = 2.3 \times 10^{-4}$

However, NO(g) quickly reacts with oxygen to produce $NO_2(g)$:

$2\,NO(g) + O_2(g) \rightleftharpoons 2\,NO_2(g)$
$$K_p = 1.3 \times 10^{-4}$$

Calculate $K_p$ and $K_c$ at 1900 K for the net reaction.

$N_2(g) + 2\,O_2(g) \rightleftharpoons 2\,NO_2(g)$

14

14

14

14

14

14

14

# Chapter 15.  Acids & Bases

**Brønsted Acids and Bases (Section 15.1)**
**Acid-Base Properties of Water and the pH Scale (Sections 15.2 – 15.3)**
**Strengths of Acids and Bases (Section 15.4)**
**Weak Acids, Weak Bases, and Ionization Constants (Sections 15.5 – 15.7)**
**Diprotic and Polyprotic Acids (Section 15.8)**
**Molecular Structure and Strengths of Acids (Section 15.9)**
**Hydrolysis and Acid-Base Properties of Salts (Section 15.10)**
**Lewis Acids and Bases (Section 15.12)**

## SUMMARY

### Brønsted Acids and Bases (Section 15.1)

In this chapter, we will develop the Brønsted definition of acids and bases previously discussed in chapter 4. The concept of equilibrium will be applied to acid-base reactions and the pH scale will be introduced. Later in the chapter, the structural features of molecules that make them acids or bases will be discussed.

***Brønsted Acids and Bases.***  According to the Brønsted acid-base theory, an acid is defined as a substance that can donate a proton to another substance. (A proton is a $H^+$ ion.) A base is a substance that can accept a proton from another substance. For example, when HBr gas dissolves in water, it donates a proton to the solvent. Therefore, HBr is a Brønsted acid.

$$HBr(aq) + H_2O(\ell) \rightarrow H_3O^+(aq) + Br^-(aq)$$
$$\text{acid} \qquad\quad \text{base}$$

The water molecule is the proton acceptor. Therefore, water is a Brønsted base in this reaction. All acid-base reactions, according to Brønsted, involve the transfer of a proton. When a water molecule accepts a proton, a hydronium ion ($H_3O^+$) is formed.

In the above example, HBr and $Br^-$ are related chemical species. A conjugate acid-base pair consists of two species that differ from each other by the presence of one $H^+$ unit. HBr and $Br^-$ ions are a conjugate acid-base pair. Removing a proton ($H^+$) from the acid (HBr) gives its conjugate base ($Br^-$). $H_3O^+$ and $H_2O$ are also a conjugate acid-base pair.

Substances that can behave as an acid in one proton transfer reaction and as a base in another reaction are called *amphoteric*. Water is an amphoteric substance. When ammonia dissolves in water for example, a proton transfer reaction takes place. Only this time water is the acid and ammonia is the base.

$$NH_3(aq) + H_2O(\ell) \rightleftharpoons NH_4^+(aq) + OH^-(aq)$$

The conjugate acid-base pairs are:

$$NH_4^+ / NH_3, \text{ and } H_2O / OH^-$$

Such pairs are usually labeled as shown below in the following equation:

Examples
15.1, 15.2

Exercises
1, 2

$$NH_3(aq) + H_2O(\ell) \rightleftharpoons NH_4^+ (aq) + OH^- (aq)$$
$$\text{base}_1 \qquad \text{acid}_2 \qquad \text{acid}_1 \qquad \text{base}_2$$

The subscript 1 designates one conjugate acid-base pair, and the subscript 2 the other pair. The double arrow indicates that the reaction is reversible. In the reverse reaction $OH^-$ acts as a base and $NH_4^+$ acts as an acid.

## Acid-Base Properties of Water and the pH Scale (Sections 15.2 – 15.3)

*Autoionization and the Ion-Product of Water.* Pure water is a very weak electrolyte and ionizes according to the equation:

$$H_2O(\ell) \rightleftharpoons H^+(aq) + OH^- (aq)$$

According to the Brønsted theory, the reaction is viewed as a proton transfer from one water molecule to another.

$$H_2O(\ell) + H_2O(\ell) \rightleftharpoons H_3O^+(aq) + OH^- (aq)$$
$$\text{acid}_1 \qquad \text{base}_2 \qquad \text{acid}_2 \qquad \text{base}_1$$

Since water can act as both an acid and a base, it is an *amphoteric* substance.

This reaction is reversible and $H_2O$, $H_3O^+$ and $OH^-$ are in equilibrium. In pure water at 25 °C, $[H^+] = [OH^-] = 1.0 \times 10^{-7}$ M. These are low concentrations and tell us that very few $H_2O$ molecules are ionized, and that the equilibrium lies to the left.

At equilibrium, the product of the hydrogen ion concentration and hydroxide ion concentration equals a constant, called the ion-product constant for water, $K_w$.

$$K_w = [H^+][OH^-] = (1.0 \times 10^{-7})(1.0 \times 10^{-7}) = 1.0 \times 10^{-14}$$

Like other equilibrium constants we treat $K_w$ as unitless. The value of $K_w$ applies to all aqueous solutions at 25 °C. When an acid is added to water, the $[H^+]$ increases. Therefore, the $[OH^-]$ must decrease in order for $K_w$ to remain constant. In acidic solutions $[H^+] > [OH^-]$. Similarly, when a base is added to water, and $[OH^-]$ increases, then $[H^+]$ must decrease. In basic solutions $[OH^-] > [H^+]$. Acidic, basic, and neutral solutions are characterized by the following conditions:

| Example 15.3 |
| Exercises 3 – 5 |

| | |
|---|---|
| neutral | $[H^+] = [OH^-]$ |
| acidic | $[H^+] > [OH^-]$ |
| basic | $[H^+] < [OH^-]$ |

The ion product provides a useful relationship for aqueous solutions. If the value of $[H^+]$ is known, then the concentration of $OH^-$ can be calculated. Similarly, the $H^+$ ion concentration can be calculated, if the value of $[OH^-]$ is known. Example 15.3 illustrates this type of calculation. In Table 15.1 on page 332, each row corresponds to a solution with the given $H^+$ and $OH^-$ concentrations. The table covers the entire practical range of concentrations found in aqueous solutions. Note that the product of the two concentrations in all aqueous solutions is $1.0 \times 10^{-14}$.

*The pH Scale.* The concentration of $H^+(aq)$ in a solution can be expressed in terms of the pH scale. The pH of a solution is defined as the negative logarithm of the hydrogen ion concentration.

$$pH = -\log [H^+]$$

Recall that the logarithm of a number is the power to which 10 must be raised in order to equal the number. For example, the logarithm of 100 is 2.0 because raising 10 to the $2^{nd}$ power gives 100.

$$100 = 10^2$$
$$\log 100 = 2$$

The log of a number less than 1 is a negative number.

$$\frac{1}{100} = 0.01 = 10^{-2}$$
$$\log 0.01 = -2$$

First, let's find the pH of a neutral solution. In pure water at 25 °C; $[H^+] = 1 \times 10^{-7}$ *M*. Using the definition of pH given above, take the log of the $H^+$ ion concentration first:

$$pH = -\log (1 \times 10^{-7})$$
$$pH = -(-7.0) = 7.0$$

The pH of a neutral solution is 7.0.

Likewise, for an acidic solution, where for example, the $H^+$ ion concentration is $1 \times 10^{-5}$ *M*, the pH is 5.0.

$$pH = -\log (1 \times 10^{-5}) = -(-5.0) = 5.0$$

All acidic solutions have a pH < 7.0.

When $[H^+]$ is not an exact power of 10, the pH is not a round number. Take the following basic solution, for example, if $[H^+] = 2.5 \times 10^{-9}$ *M*, the pH is

$$pH = -\log (2.5 \times 10^{-9}) = -(-8.60)$$
$$pH = 8.60$$

Note that all basic solutions have a pH > 7.0. The pH values corresponding to selected sets of $H^+(aq)$ and $OH^-(aq)$ concentrations are given in Table 15.1. In terms of pH, solutions that are acidic, basic, and neutral are defined as follows:

| | |
|---|---|
| neutral | pH = 7.0 |
| acidic | pH < 7.0 |
| basic | pH > 7.0 |

***The pOH Scale.*** A scale just like the pH scale has been devised for the hydroxide ion concentration, where

$$pOH = -\log [OH^-]$$

Just as the $H^+$ ion and $OH^-$ ion concentrations are related by the ion-product constant of water, $K_w$, the pH and pOH are also related.

$$K_w = [H^+][OH^-] = 1.0 \times 10^{-14}$$
$$pH + pOH = 14.00$$

The sum of the pH and pOH values of any solution is always 14 at 25 °C. You can see this in Table 15.1. Sum the pH and pOH for each set of $H^+$ and $OH^-$ concentrations, and see what you get.

| Example 15.4 |
| Exercises 6, 7 |

It is also important to notice that a change in pH of one unit corresponds to a 10-fold change in $[H^+]$. As $H^+$ drops from $10^{-8}$ to $10^{-9}$ $M$, for instance, the pH changes from 8 to 9. A change of 2.0 pH units corresponds to a 100-fold change in $H^+$ ion concentration. Never say, "a pH of 2 is twice as acidic as a pH of 4." It is really 100 times more acidic!

**Table 15.1** Relationship of pH and pOH in Aqueous Solutions

| $[H^+]$ | $[OH^-]$ | pH | pOH | Nature of Solution |
|---------|----------|-----|------|--------------------|
| $10^0$ | $10^{-14}$ | 0 | 14 | acidic |
| $10^{-1}$ | $10^{-13}$ | 1 | 13 | acidic |
| $10^{-2}$ | $10^{-12}$ | 2 | 12 | acidic |
| $10^{-3}$ | $10^{-11}$ | 3 | 11 | acidic |
| $10^{-6}$ | $10^{-8}$ | 6 | 8 | acidic |
| $10^{-7}$ | $10^{-7}$ | 7 | 7 | neutral |
| $10^{-8}$ | $10^{-6}$ | 8 | 6 | basic |
| $10^{-11}$ | $10^{-3}$ | 11 | 3 | basic |
| $10^{-12}$ | $10^{-2}$ | 12 | 2 | basic |
| $10^{-13}$ | $10^{-1}$ | 13 | 1 | basic |
| $10^{-14}$ | $10^0$ | 14 | 0 | basic |

| Example 15.5 |
| Exercise 8 |

***Changing pH Values to [H+].*** Given the pH, how do we calculate the $[H^+]$? Rearrange the equation for pH:

$$pH = -\log [H^+]$$

$$\log [H^+] = -pH$$

taking the antilog of both sides:

A different way of looking at antilog:

$$10^{\log[H^+]} = 10^{-pH}$$

| Example 15.6 |
| Exercises 9 – 11 |

gives:

$$antilog (\log [H^+]) = antilog (-pH)$$

$$[H^+] = 10^{-pH}$$

Any electronic calculator with a $10^x$ key will easily make the calculation of $H^+$ ion concentrations from pH values. Just enter $-pH$ for x and push $10^x$.

## Strengths of Acids and Bases (Section 15.4)

***Strong Acids and Bases.*** The stronger the acid, the more completely it ionizes in water, producing $H_3O^+(aq)$ and an anion. Strong acids are strong electrolytes and ionize completely in water. HBr, for example, is a strong acid.

$$HBr(aq) + H_2O(\ell) \rightarrow H_3O^+(aq) + Br^-(aq)$$

A solution of hydrobromic acid consists of $H_3O^+(aq)$ and $Br^-(aq)$ ions. The equilibrium concentration of HBr molecules is zero because all have ionized. The concentration of $H_3O^+$ and $Br^-$ is equal to the initial concentration of the acid.

Like a strong acid, a strong base is one that ionizes completely in aqueous solution. KOH, a commonly used base, is an example. It is purchased as a white solid. It dissolves readily in water to give a solution of $K^+$ and $OH^-$ ions:

$$KOH(aq) \rightarrow K^+(aq) + OH^-(aq)$$

A solution of 0.10 *M* KOH is made by dissolving 0.10 mol of KOH in 1 L of solution. In this solution, $[K^+]$ is 0.10 *M*, $[OH^-]$ is 0.10 *M*, and [KOH] is essentially zero. Of the strong bases, only NaOH and KOH are commonly used in the laboratory.

***The pH of Strong Acid and Strong Base Solutions.*** The pH of a strong acid solution depends on the hydrogen ion concentration. What is the pH of a 0.0052 *M* HBr solution? When hydrobromic acid is added to water, it ionizes completely.

$$HBr(aq) + H_2O(\ell) \rightarrow H_3O^+(aq) + Br^-(aq)$$

| Example 15.7 |
|---|
| Exercises 12, 13 |

For simplicity, we can write the reaction without the $H_2O$ molecules.

$$HBr(aq) \rightarrow H^+(aq) + Br^-(aq)$$

Before any ionization occurs, the HBr concentration is 0.0052 *M*, but after ionization, its concentration is zero. Since each mole of HBr that reacts yields 1 mol of $H^+$ and 1 mol of $Br^-$, the concentrations of $H^+$ and $Br^-$ ions are both 0.0052 *M*. 0.0052 mol HBr reacts to give 0.0052 mol $H^+$ ion and 0.0052 mol $Br^-$ ion per liter of solution.

The pH is given by:

$$pH = -\log(5.2 \times 10^{-3}) = 2.28$$

***Weak Acids and Bases.*** Not all acids are strong proton donors in aqueous solutions. A weak acid is one that is only partially ionized in water. Hydrofluoric acid is a typical weak acid in water:

$$HF(aq) + H_2O(\ell) \rightleftharpoons H_3O^+(aq) + F^-(aq)$$

Since HF is only partially ionized, the concentration of HF molecules is much greater than the concentrations of $H^+(aq)$ and $F^-(aq)$ ions. For example, a 1.0 *M* HF solution has the following concentrations: $[H^+] = [F^-] = 0.026$ *M*, and [HF] = 0.974 *M*. About 97% of the initial HF molecules are *not* ionized. For weak acids, the ionization equilibrium lies to the left. The extent of reaction is small. A number of weak acids are listed in Table 15.2 of the text.

Figure 15.1 compares the relative concentrations of HA, H⁺, and A⁻ in solutions of strong and weak acids.

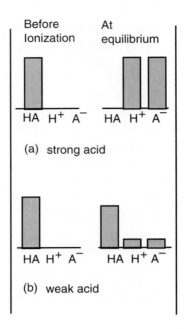

**Figure 15.1** Illustration of the extent of ionization of (a) a strong acid, HA, that ionizes 100%, and (b) a weak acid, HA, that ionizes a few percent. A⁻ represents the anion. The height of the bar represents the concentration of the species present.

Some bases are not ionized completely in aqueous solution.  A weak base ionizes to a limited extent in water. Ammonia is an example of a weak base. In a 0.10 $M$ solution of ammonia, only about 1% of the ammonia molecules are ionized. A dominant reverse reaction means that the equilibrium lies to the left.

$$NH_3(aq) + H_2O(\ell) \rightleftharpoons NH_4^+ (aq) + OH^- (aq)$$
$$\text{base} \qquad \text{acid}$$

***Strength of Conjugate Bases.***  The strength of an acid is related to the strength of its conjugate base. Let's compare the ionization of HCl, a strong acid, and HF, a weak acid.

$$HCl(aq) + H_2O(\ell) \rightarrow H_3O^+(aq) + Cl^- (aq)$$

$$HF(aq) + H_2O(\ell) \rightleftharpoons H_3O^+(aq) + F^- (aq)$$

Chloride ion is the conjugate base of HCl. Here, we see that it must be a very weak base, because it has no tendency to accept a proton from $H_3O^+$ in the reverse reaction. In the ionization of the weak acid HF, the reverse reaction occurs to a much greater extent than for HCl. Fluoride ion accepts a proton from $H_3O^+$. The F⁻ ion is a much stronger base than Cl⁻. The conjugate bases of strong acids have no measurable base strength.

When comparing the strength of different acids, we see that as acid strength increases, the strength of its conjugate base decreases. Table 15.2 in the text lists several important conjugate acid-base pairs, according to their relative strengths. A short version is reproduced here in Table 15.2.

**Table 15.2** Relative Strengths of Conjugate Acid-Base Pairs

| | Acid | Conjugate Base | |
|---|---|---|---|
| ↑ Increasing Acid Strength | $HClO_4$ | $ClO_4^-$ | Increasing Base Strength ↓ |
| | $HCl$ | $Cl^-$ | |
| | $HNO_3$ | $NO_3^-$ | |
| | $H_3O^+$ | $H_2O$ | |
| | $HF$ | $F^-$ | |
| | $HNO_2$ | $NO_2^-$ | |
| | $HCN$ | $CN^-$ | |
| | $H_2O$ | $OH^-$ | |

***Predicting the Direction of Acid-Base Reactions.*** The information in Table 15.2 can be used to predict the direction of an acid-base reaction. When an acid reacts with a base, will the reaction tend to go from left to right and favor the products, or will it tend to go from right to left and favor the reactants? For instance, when HF reacts with water, on which side will the equilibrium lie?

$$HF(aq) + H_2O(\ell) \rightleftharpoons H_3O^+(aq) + F^-(aq)$$

To decide, compare the strength of HF to that of the acid in the reverse reaction, $H_3O^+$. According to Table 15.2, $H_3O^+$ is the stronger acid. Compare the base strength of $H_2O$ to that of the base in the reverse reaction, $F^-$ ion. $F^-$ ion is the stronger base. The reaction can be summarized as:

| | | | | | Examples 15.8, 15.9 |
|---|---|---|---|---|---|

$$HF(aq) + H_2O(\ell) \rightleftharpoons H_3O^+(aq) + F^-(aq)$$

| weaker acid | weaker base | stronger acid | stronger base |
|---|---|---|---|

Examples 15.8, 15.9

Exercises 14 - 15

All proton transfers proceed from the stronger acid and stronger base toward the weaker acid and weaker base. Therefore, in this case the reaction proceeds from right to left and equilibrium lies to the left. There will be more HF at equilibrium than $H_3O^+$ and $F^-$ ion. A simplification is to just compare the strengths of the two bases. The reaction can be viewed as a competition of two bases for the proton. Who wins? The stronger base, of course. $F^-$ ions being stronger accept protons more readily than do $H_2O$ molecules, and so HF predominates over $H_3O^+$ in solution. The weaker acid and weaker base predominate at equilibrium.

## Weak Acids, Weak Bases, and Ionization Constants (Sections 15.5 – 15.7)

*Ionization Constants.* Weak acids and weak bases are ionized only to a small extent in aqueous solution. In both cases, the ionization reaction is reversible and equilibrium is established. For instance, hydrocyanic acid ionizes in water as follows:

$$HCN(aq) \rightleftharpoons H^+(aq) + CN^-(aq)$$

The equilibrium constant for the ionization of a weak acid is called the acid ionization constant, $K_a$. The expression for this constant is:

$$K_a = \frac{[H^+][CN^-]}{[HCN]} = 4.9 \times 10^{-10}$$

The value of $K_a$ is experimentally determined, and Table 15.3 (text) and Table 15.3 below list $K_a$ values for a number of weak acids.

**Table 15.3**  Values of $K_a$ for Some Monoprotic Acids

| Formula | Name | $K_a$ Value |
|---------|------|-------------|
| $HSO_4^-$ | Hydrogen sulfate ion | $1.3 \times 10^{-2}$ |
| HF | Hydrofluoric acid | $7.1 \times 10^{-4}$ |
| $HNO_2$ | Nitrous acid | $4.5 \times 10^{-4}$ |
| HCOOH | Formic acid | $1.7 \times 10^{-4}$ |
| $CH_3COOH$ | Acetic acid | $1.8 \times 10^{-5}$ |
| $HSO_3^-$ | Hydrogen sulfite ion | $6.3 \times 10^{-8}$ |
| HCN | Hydrocyanic acid | $4.9 \times 10^{-10}$ |

Methylamine is an example of a weak base. It ionizes in water as shown in the following equation:

$$CH_3NH_2(aq) + H_2O(\ell) \rightleftharpoons CH_3NH_3^+(aq) + OH^-(aq)$$

The equilibrium constant for the ionization of a weak base is called the base ionization constant, $K_b$.

$$K_b = \frac{[CH_3NH_3^+][OH^-]}{[CH_3NH_2]} = 4.4 \times 10^{-4}$$

Table 15.4 in the text and Table 15.4 on next page list ionization constant values ($K_b$) for a number of weak bases. Note that the concentration of water is not shown in the $K_b$ expression. The concentration of water is not measurably affected by the reaction and is essentially a constant.

**Table 15.4** Values of $K_b$ for Several Weak Bases

| Formula | Name | $K_b$ Value |
|---|---|---|
| $CH_3CH_2NH_2$ | Ethylamine | $5.6 \times 10^{-4}$ |
| $CH_3NH_2$ | Methylamine | $4.4 \times 10^{-4}$ |
| $CN^-$ | Cyanide ion | $2.0 \times 10^{-5}$ |
| $NH_3$ | Ammonia | $1.8 \times 10^{-5}$ |
| $C_5H_5N$ | Pyridine | $1.7 \times 10^{-9}$ |

*Acid and Base Strength.* The $K_a$ value is the quantitative measure of acid strength. The larger the $K_a$, the greater the extent of the ionization reaction and the stronger the acid. Therefore, acetic acid, with $K_a = 1.8 \times 10^{-5}$, is a stronger acid than hydrocyanic acid, with $K_a = 4.9 \times 10^{-10}$. The acids listed in Table 15.3 are arranged in order of decreasing acid strength as you read down in the table. The $K_a$ values for strong acids are never tabulated. Strong acids are essentially 100 percent ionized. It will help you to be able to recognize the following 6 strong acids by their names and formulas, so that you won't confuse them with weak acids.

**Table 15.5** Values of $K_b$ for Several Weak Bases

| Formula | Name |
|---|---|
| $HClO_4$ | Perchloric Acid |
| $HI$ | Hydroiodic Acid |
| $HBr$ | Hydrobromic Acid |
| $HCl$ | Hydrochloric Acid |
| $H_2SO_4$ | Sulfuric Acid |
| $HNO_3$ | Nitric Acid |

*Percent Ionization.* The strength of an acid is also reflected in its percent ionization. The percent ionization is defined as follows:

$$\text{percent ionization} = \frac{\text{concentration of acid ionized}}{\text{initial concentration of acid}} \times 100\%$$

For a weak acid, HA, the concentration of the acid that is ionized is equal to the concentration of $H^+$ ions (or that of $A^-$ ions) at equilibrium.

$$HA(aq) \rightleftharpoons H^+(aq) + A^-(aq)$$

$$\text{percent ionization} = \frac{[H^+]}{[HA]_0} \times 100\%$$

Examples
15.10, 15.11

Exercises
16 - 23

Be sure to work through Examples 15.10 and 15.11 because they illustrate the problem-solving techniques used to calculate $[H^+]$, $[OH^-]$, pH, and percent ionization of weak acids and bases.

*The $K_b$ of a Base is Related to the $K_a$ of Its Conjugate Acid.* The $K_a$ and $K_b$ for *any* conjugate acid-base pair are always related by the equation:

$$K_a K_b = K_w$$

The $K_b$ for a conjugate base of a weak acid is found by rearranging:

$$K_b = \frac{K_w}{K_a}$$

From this equation, you can see that as $K_a$ decreases, $K_b$ increases, the weaker the acid, the stronger its conjugate base – a relationship we have noted before.

Fluoride ion is a base. What is its $K_b$ value?

$$F^-(aq) + H_2O(\ell) \rightleftharpoons HF(aq) + OH^-(aq) \qquad K_b = ?$$

The base ionization constant $K_b$ for $F^-$ is related to the $K_a$ for the conjugate acid HF by:

$$K_b = \frac{K_w}{K_a}$$

Substituting $K_a$ for HF gives:

$$K_b = \frac{K_w}{7.1 \times 10^{-4}} = 1.4 \times 10^{-11}$$

> Examples
> 15.12, 15.13
>
> Exercises
> 24 - 25

## Diprotic and Polyprotic Acids (Section 15.8)

***Stages of Ionization.*** The acids listed in Table 15.3 of the text are monoprotic acids; that is, they produce one proton per acid molecule. Acetic acid ($CH_3COOH$) contains four hydrogen atoms per molecule, but it is still monoprotic, because only one of them is ionized in solution. The H atoms bonded to the C atom do not ionize because the electronegativity difference between C and H is not great enough to produce sufficient positive charge on the hydrogen atoms.

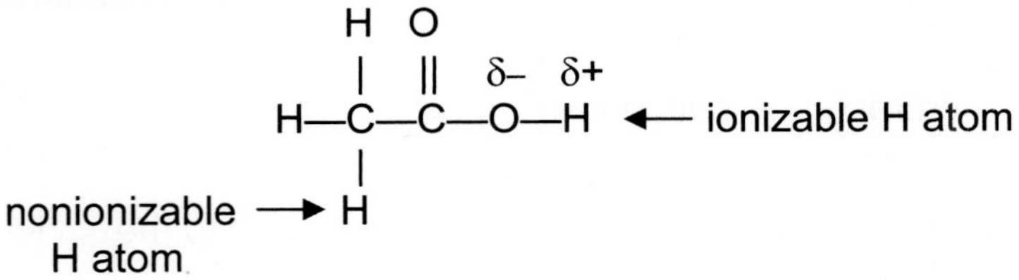

Acids that contain two ionizable H atoms are called diprotic acids. Those with three or more ionizable H atoms are called polyprotic acids. For example, sulfuric acid ($H_2SO_4$) is a diprotic acid, and phosphoric acid ($H_3PO_4$) with its three ionizable H atoms is a polyprotic acid. The ionization reactions occur stepwise; the conjugate base in the first step becomes the acid in the second step of ionization. For oxalic acid, $H_2C_2O_4$:

$$H_2C_2O_4 \rightleftharpoons H^+ + HC_2O_4^- \qquad K_{a1} = 6.5 \times 10^{-2}$$

$$HC_2O_4^- \rightleftharpoons H^+ + C_2O_4^{2-} \qquad K_{a2} = 6.1 \times 10^{-5}$$

This example is typical in that $K_{a1} >> K_{a2}$. It is always more difficult to remove an $H^+$ ion from a negative ion such as $HC_2O_4^-$ than from a neutral molecule such as $H_2C_2O_4$. Table 15.5 of the text lists acid ionization constants for several polyprotic acids.

---

**Here are some important observations concerning polyprotic acids:**
1. The major product species in solution come from the first ionization step.
2. Since all reactants and products (ions and nonionized acid molecules) exist in the same solution, they are in equilibrium with each other. At equilibrium, each species can have only one concentration.
3. The concentration of all species in solution must be consistent with both $K_{a1}$ and $K_{a2}$.

---

Example
15.14

Exercises
26, 27

## Molecular Structure and Strengths of Acids (Section 15.9)

*Binary Acids.* The acid strengths of nonmetal hydrides are related to two features of these molecules. These are the polarity of the X—H bond (X stands for a nonmetal atom), and the strength of the H—X bond. The polarity of a H—X bond increases with the electronegativity of the nonmetal atom. The greater the electronegativity of the nonmetal atom, the more it withdraws electrons away from the hydrogen atom. This facilitates the transfer of $H^+$ to another species. Thus, acid strength increases as the electronegativity of the nonmetal atom increases.

The strength of the X—H bond decreases as the atomic radius of X increases. A hydrogen ion is more easily broken away from a larger atom than a smaller one. This happens because the electron cloud in a large atom is more diffuse. The smaller the nonmetal atom, the more dense the electron cloud. This greater electron density results in a greater attraction for the proton. Therefore, acid strength increases as the atomic radius of the nonmetal atom increases.

When comparing acid strengths of hydrides in a group of the periodic table versus those within a period, one must consider both the electronegativity of the nonmetal atom and the bond energy. Within a series of hydrides of elements in a group the acid strength increases with increasing atomic radius of the nonmetal. For example, note the following hydrides of the Group 6A elements.

$H_2S < H_2Se < H_2Te$
increasing acid strength →
increasing nonmetal radius →
← increasing electronegativity of X

In this series, the electronegativity of X varies only slightly, and so is outweighed by the more significant increases in atomic radius, as shown below.

|  | $H_2S$ | $H_2Se$ | $H_2Te$ |
|---|---|---|---|
| Radius of X (pm) | 104 | 117 | 137 |
| Electronegativity of X | 2.5 | 2.4 | 2.1 |

When reading across a row of the periodic table, the situation is reversed. The importance of bond polarity in determining acid strength outweighs that of atomic radius. For instance, the hydrides of the elements; phosphorus, sulfur, and chlorine show increasing acid strength as shown next. Increasing acid strength parallels the increasing electronegativity on the nonmetal atom.

$PH_3$ < $H_2S$ < HCl
increasing acid strength →
increasing electronegativity of X →
← increasing nonmetal radius

|  | $PH_3$ | $H_2S$ | HCl |
|---|---|---|---|
| Radius of X (pm) | 110 | 104 | 99 |
| Electronegativity of X | 2.1 | 2.5 | 3.0 |

The smaller decrease in atomic radius is insignificant in comparison to the larger change in electronegativity. The trend in acid strength results from significant increases in bond polarity when going across the row.

***Oxoacids.*** For oxoacids with the same structure, but with different central atoms whose elements are in the same group, acid strength increases with increasing electronegativity of the central atom. Thus, acid strength increases in the following series where the central atom is a halogen element:

HOI < HOBr < HOCl

In each molecule, the O—H bond strength is approximately the same. In this series the ability of the halogen atom to withdraw electron density from the O—H bond increases with increasing electronegativity. As the O—H bond in a series of acids becomes more polar, the acid strength increases.
    For oxoacids that have the same central atom, but differing numbers of attached oxygen atoms, the acid strength increases with increasing oxidation number of the central atom. Therefore, $HNO_3$ is a stronger acid than $HNO_2$, for example.

| Examples 15.15, 15.16 |
| --- |
| Exercise 28 |

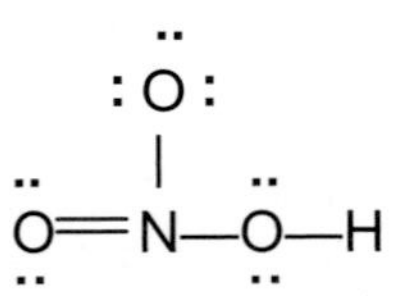

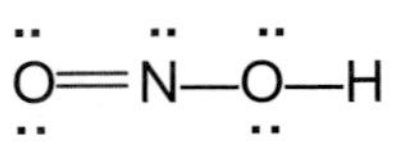

```
      ..
     : O :
      |   ..
  O=N—O—H
  ..      ..
```

```
  ..   ..  ..
  O=N—O—H
  ..   ..  ..
```

The oxygen atoms draw electrons away from the nitrogen atom, making it more positive. The more positive the N atom, the more effective it is in withdrawing electrons from the O—H bond. This increases the polarity of the O—H bond.

***Carboxylic Acids.*** Carboxylic acids are organic acids with have the general form:

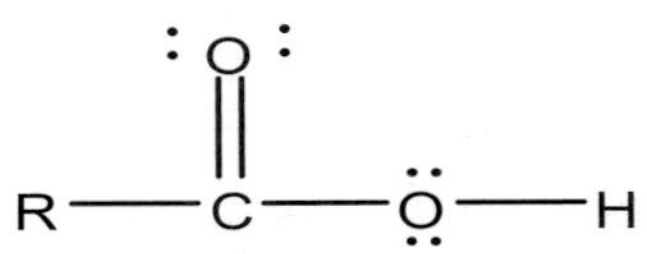

```
        : O :
        ||
        ..
R——C——O——H
        ..
```

where R is the rest of the molecule. The COOH group is the carboxyl group. Acetic acid is an example of a carboxylic acid. In comparing the strength of two carboxylic acids, the carboylic acid with more electronegative atoms will be a stronger acid. This is attributed to the fact that the more electronegative atoms will pull electron density towards themselves away from the hydrogen on the COOH group. This weakens the OH bond and thus the acid is strong.

## Hydrolysis and Acid-Base Properties of Salts (Section 15.10)

*Hydrolysis.* A salt is an ionic compound formed by the reaction of an acid and a base. In dealing with salts, it will be useful to recall that they are *strong* electrolytes and are completely dissociated in water. The term salt hydrolysis refers to the reaction of a cation or an anion, with water after dissociation splits the salt molecule into two parts. Salt hydrolysis usually affects the pH of a solution.

Hydrolysis can be cation hydrolysis if the cation of the salt reacts with water, and/or anion hydrolysis if the anion of the salt reacts with water.

*Anion Hydrolysis.* Anions that cause hydrolysis are Brønsted bases because they can accept a proton from water.

$$A^- + H_2O \rightleftharpoons HA + OH^-$$

Not all anions cause hydrolysis. You must be able to recognize those that do and those that don't. Quite simply, those anions with the ability to undergo hydrolysis are conjugate bases of weak acids. The conjugate base of any weak acid is a weak base and will raise the pH of an aqueous solution. The extent of hydrolysis is proportional to the base ionization constant, $K_b$.

Keep in mind that the conjugate base of a *strong acid* has no tendency to accept protons from $H_2O$. Anions such as $Cl^-$, $NO_3^-$, and $ClO_4^-$ do not cause hydrolysis. They do not have acid-base properties.

*The Conjugate Bases of Weak Acids.* Fluoride ion is the conjugate base of hydrofluoric acid (HF) a weak acid. As a weak base it can accept a proton from water. The resulting $OH^-$ ions make the solution basic.

$$F^- + H_2O \rightleftharpoons HF + OH^- \qquad\qquad K_b = \frac{[HF][OH^-]}{[F^-]}$$

The force that makes this reaction possible is the tendency of the weak acid HF to remain undissociated. Three realms of base strength can be distinguished:

1. The conjugate bases of strong acids ($K_a > 1.0$) are too weak to cause measurable hydrolysis. Thus, $Cl^-$, $Br^-$, and $NO_3^-$ ions do not hydrolyze in aqueous solutions.

   $$Cl^- + H_2O \rightarrow HCl + OH^- \quad \text{(no reaction)}$$

2. The conjugate bases of acids with a $K_a$ between $10^{-1}$ and $10^{-5}$ are weak bases. Thus, $F^-$, $NO_2^-$, $HCOO^-$(formate), and $CH_3COO^-$ (acetate) ions hydrolyze to a small extent and produce low concentrations of $OH^-$ ions in solution.

   $$NO_2^- + H_2O \rightleftharpoons HNO_2 + OH^-$$

3.  The conjugate bases of very weak acids ($K_a < 10^{-5}$) are moderately strong bases. Thus solutions containing $CN^-$ and $HS^-$ ions can be quite basic.

***Cation Hydrolysis.*** It is important to know which metal ions cause hydrolysis and which do not. Those that cause hydrolysis are small, highly charged metal ions, such as $Al^{3+}$, $Cr^{3+}$, $Fe^{3+}$, $Bi^{3+}$, and $Be^{2+}$. These can split $H_2O$ molecules, yielding acidic solutions. In the case of the hydrated $Al^{3+}$ ion, the reaction is

$$Al(H_2O)_6^{3+}(aq) + H_2O(\ell) \rightleftharpoons Al(OH)(H_2O)_5^{2+}(aq) + H_3O^+(aq)$$

The singly charged metal ions associated with strong bases do not cause hydrolysis. The commonly encountered ions $Na^+$, $K^+$, $Ca^{2+}$, and $Mg^{2+}$ do not have a great enough positive charge to attract an $OH^-$ from a $H_2O$ molecule and bond to it. Recall that NaOH and KOH are strong bases. That is, they ionize completely in solution. These ions, $K^+$ and $OH^-$ for instance, have no tendency to stay together in solution.

$$KOH(aq) \rightarrow K^+(aq) + OH^-(aq)$$

The hydrolysis reaction that would bond an $OH^-$ to a $K^+$ ion does not occur!

$$K^+(aq) + H_2O(\ell) \rightarrow KOH(aq) + H^+(aq) \quad \text{(no reaction)}$$

Another acidic cation that you will need to be familiar with is the ammonium ion ($NH_4^+$). Solutions of ammonium ion are acidic because $NH_4^+$ is a Brønsted acid. Recall that $NH_4^+$ is the conjugate acid of the weak base $NH_3$.

$$NH_4^+ + H_2O \rightleftharpoons NH_3 + H_3O^+$$

or simply,

$$NH_4^+ \rightleftharpoons NH_3 + H^+ \quad K_a = 5.6 \times 10^{-10}$$

Table 15.5 lists the various anions and cations we have discussed according to their acid-base properties. You should review the material above, and in the textbook to see why each ion has the property described.

**Table 15.5** Acid-Base Properties of Ions

|  | Acidic | Basic | Neutral |
|---|---|---|---|
| Cations | $Al^{3+}$, $Cr^{3+}$, $Fe^{3+}$ $Bi^{3+}$, $Be^{2+}$, $NH_4^+$ |  | $Na^+$, $K^+$ $Mg^{2+}$, $Ca^{2+}$ |
| Anions |  | $CH_3COO^-$, $F^-$ all conjugate bases of weak acids | $Cl^-$, $Br^-$ all conjugate bases of strong acids |

***Salts That Produce Neutral Solutions.*** When a salt is dissolved in water the solution will acquire a pH consistent with the acid-base properties of the cation and anion making up the salt. In general, salts containing alkali metal ions or alkaline earth metal ions (except $Be^{2+}$) and the conjugate bases of strong acids do not undergo hydrolysis; their solutions are neutral. Consequently, a solution of KCl, for example, is neutral. A $K^+$ ion does not accept an $OH^-$ ion from $H_2O$, and a $Cl^-$ ion does not accept a $H^+$ ion from $H_2O$.

***Salts That Produce Basic Solutions.*** Salts producing a basic pH are salts of weak acids and strong bases. These salts have a neutral cation and a basic anion. For example, NaF dissociates into $Na^+$ and $F^-$ ions. The $Na^+$ ion has no acidic or basic properties, but the fluoride ion is a conjugate base of a weak acid and therefore is a basic anion.

NaF, a salt of the NaOH, a strong base, and HF, a weak acid.

$$F^- + H_2O \rightleftharpoons HF + OH^-$$

***Salts That Produce Acidic Solutions.*** Salts producing acidic solutions are salts of strong acids and weak bases. These salts contain cations that undergo hydrolysis, and anions that do not (the neutral anions). Such a salt has a cation that is a conjugate acid of a weak base. Thus, solutions of these salts are acidic. The salt $NH_4Cl$ is a strong electrolyte and dissociates into ammonium ions and chloride ions when dissolved in water. The NH ion is a conjugate acid of the weak base, $NH_3$, and releases $H^+$ ion into solution.

$$NH_4Cl(aq) \rightarrow NH_4^+ (aq) + Cl^- (aq) \quad \text{(strong electrolyte)}$$

$$NH_4^+ \rightleftharpoons NH_3 + H^+ \qquad K_a = 5.6 \times 10^{-10}$$

The $Cl^-$ ion is neutral, having no acidic or basic properties. $NH_4Cl$ is a salt of a strong acid (HCl) and a weak base ($NH_3$).

Solutions of certain salts, such as $AlCl_3$ and $FeCl_3$, are also acidic. These contain small, highly charged cations, and anions that are conjugate bases of strong acids.

***Salts in Which Both the Anion and Cation Hydrolyze.*** Salts derived from a weak acid and a weak base contain cations and anions that will hydrolyze. Whether a solution containing such a salt is acidic, basic, or neutral depends on the relative strengths of the acidic cations and basic anions. For example, a solution of ammonium fluoride is *acidic* because $NH_4^+$ is a stronger acid than $F^-$ is a base.

$$NH_4F(aq) \rightarrow NH_4^+(aq) + F^-(aq) \quad \text{(strong electrolyte)}$$

$$NH_4^+ \rightleftharpoons NH_3 + H^+ \qquad K_a = 5.6 \times 10^{-10}$$

$$F^- + H_2O \rightleftharpoons HF + OH^- \qquad K_b = 1.4 \times 10^{-11}$$

$NH_4F$ is a salt of a weak acid and a weak base. On the other hand, another such salt ($NH_4CN$) is a *basic* salt because $CN^-$ ion is a stronger base than $NH_4^+$ is an acid.

| Examples 15.17, 15.18 |
| :---: |
| Exercises 29, 30 |

***The pH of Salt Solutions.*** The pH of a salt solution is calculated just as it is for any weak acid or base.

## Lewis Acids and Bases (Section 15.12)

***Electron-Pair Donors and Acceptors.*** A more general approach to acids and bases was proposed in 1923 by G. N. Lewis who also developed the use of electron dot structures. Lewis noticed that for a $H^+$ ion to be accepted by a base, the base must possess at least one unshared electron-pair. The electron-pair forms the bond to the proton.

A Lewis base is a molecule or ion that can donate a pair of electrons to form a bond. A Lewis acid is a molecule or ion that can accept a pair of electrons and form a bond.

$$H^+ + :NH_3 \quad \rightarrow \quad H—NH_3^+ \quad (or \ NH_4^+)$$

Electron-pair donor: $H^+$
accepts an electron-pair.

The formation of ammonium ion from $H^+$ and ammonia is a simple example. The ammonia molecule is the Lewis base because it donates a pair of electrons to $H^+$ to form a bond. By accepting the pair of electrons, the proton is a Lewis acid. Note that a new covalent bond is formed by the donation of the electron pair. Recall when both electrons of the shared pair are donated by the same atom, the bond is called a coordinate covalent bond. An acid-base reaction in the Lewis system is donation of an electron-pair to form a new covalent bond between the acid and the base.

Metal cations can be Lewis acids because they have lost their valence electrons and so have at least one vacant orbital in which to accept an electron-pair from a base. Consider the beryllium ion which can form four coordinate covalent bonds.

$$Be^{2+}(aq) + 4 \ H_2O(\ell) \rightarrow [Be(H_2O)_4]^{2+}(aq)$$

Example
15.19

Exercises
31, 32

Water molecules with unshared electron-pairs on their oxygen atoms are Lewis bases and beryllium is a Lewis acid because it accepts electron-pairs to form Be—$OH_2$ bonds. Beryllium can form four Be—$OH_2$ bonds.

## GLOSSARY LIST

| | | |
|---|---|---|
| conjugate acid-base pair | ion-product constant | |
| | pH | |
| strong acid | acid ionization constant | salt hydrolysis |
| strong base | base ionization constant | |
| weak acid | | Lewis acid |
| weak base | percent ionization | Lewis base |

## EQUATIONS

| Algebraic Equation | English Translation |
|---|---|
| $K_w = [H^+][OH^-]$ | iIon-product of water |
| $pH = -\log[H^+]$ | definition of pH |
| $pOH = -\log[OH^-]$ | definition of pOH |
| $pH + pOH = 14.00$ | |
| $\% \text{ ionization} = \dfrac{[\text{ionized acid}]}{[\text{unionized acid}]_0} \times 100\%$ | |
| $K_w = K_a K_b$ | relationship between $K_a$ of a weak acid and $K_b$ of its conjugate base |

## WORKED EXAMPLES

**EXAMPLE 15.1  Conjugate Acid-Base Pair**

Write the formula of the conjugate base of $H_2SO_4$.

**• Solution**

A conjugate base differs from its conjugate acid by the lack of a proton.

$\quad\quad H_2SO_4 \quad\quad\quad HSO_4^-$

$\quad\quad$ acid $\quad\quad\quad\quad$ conjugate base

---

**EXAMPLE 15.2  Conjugate Acid-Base Pairs**

For the reaction:

$$HSO_4^- \, (aq) + HCO_3^- \, (aq) \rightleftharpoons SO_4^{2-} \, (aq) + H_2CO_3(aq)$$

a.  Identify the acids and bases for the forward and reverse reactions.
b.  Identify the conjugate acid-base pairs.

**• Solution**

a.  In the forward reaction, $HSO_4^-$ ion is the proton donor, which makes it an acid. The $HCO_3^-$ ion is the proton acceptor, and is a base. In the reverse reaction, $H_2CO_3$ is the proton donor (acid), and $SO_4^{2-}$ is the proton acceptor (base).

b.  $HSO_4^- + HCO_3^- \rightleftharpoons SO_4^{2-} + H_2CO_3$
$\quad$ acid$_1$ $\quad$ base$_2$ $\quad\quad$ base$_1$ $\quad$ acid$_2$
$\quad$ The subscripts 1 and 2 designate the two conjugate acid-base pairs.

**Work EXERCISES & PROBLEMS: 1, 2**

---

**EXAMPLE 15.3  Using the Ion-Product Constant, $K_w$**

The $H^+$ ion concentration in a certain solution is $5.0 \times 10^{-5}$ *M*. What is the $OH^-$ ion concentration?

**• Solution**

The ion-product constant of water is applicable to all aqueous solutions.

$$K_w = [H^+][OH^-] = 1.0 \times 10^{-14}$$

When $[H^+]$ is known, we can solve for $[OH^-]$.

$$[OH^-] = \frac{1.0 \times 10^{-14}}{[H^+]} = \frac{1.0 \times 10^{-14}}{5.0 \times 10^{-5}}$$

$$[OH^-] = 2.0 \times 10^{-10} \ M$$

**Work EXERCISES & PROBLEMS: 3 – 5**

---

**EXAMPLE 15.4  pH and pOH**

The $OH^-$ ion concentration in a certain ammonia solution is $7.2 \times 10^{-4}$ *M*. What is the pOH and pH?

**• Solution**

The pOH is the negative logarithm of the $OH^-$ ion concentration.

$$pOH = - \log [OH^-]$$

$$pOH = - \log (7.2 \times 10^{-4}) = - (-3.14) = 3.14$$

The pH and pOH are related by

$$pH + pOH = 14.00$$
$$pH = 14.00 - pOH = 14.00 - 3.14 = 10.86$$

**Work EXERCISES & PROBLEMS: 6, 7**

---

**EXAMPLE 15.5  Comparing pH Values**

The pH of many cola-type soft drinks is about 3.0. How many times greater is the $H^+$ concentration in these drinks than in neutral water?

**• Solution**

First, write out the $H^+$ ion concentrations in the cola drink and in neutral water.

$$[H^+]_{cola} = 1.0 \times 10^{-3} \ M, \text{ and } [H^+]_{neut} = 1.0 \times 10^{-7} \ M.$$

Then the ratio is

$$\frac{[H^+]_{cola}}{[H^+]_{neut}} = \frac{1.0 \times 10^{-3} \ M}{1.0 \times 10^{-7} \ M} = 10^4 = 10,000$$

---

**• Comment**

Here is another approach. Since a change of 1.0 pH unit corresponds to a 10-fold change in $H^+$ concentration, then a change of 4.0 pH units corresponds to $10 \times 10 \times 10 \times 10 = 10^4$ or a 10,000-fold increase in $H^+$ concentration.

**Work EXERCISES & PROBLEMS: 8**

---

**EXAMPLE 15.6  $H^+$ Ion Concentration from pH**

What is the $H^+$ ion concentration in a solution with a pOH of 3.9?

**• Solution**

Recall that pH + pOH = 14.00 With the pOH given, the pH can be calculated, and then the $H^+$ ion concentration can be determined.

pH = 14.00 – pOH
pH = 14.00 – 3.90 = 10.10
Since pH is known, the hydrogen ion concentration is

$$[H^+] = 10^{-pH}$$
$$[H^+] = 10^{-10.10} = 7.9 \times 10^{-11} \ M$$

**Work EXERCISES & PROBLEMS: 9 – 11**

---

**EXAMPLE 15.7  The Hydrogen Ion Concentration in a Solution of a Strong Base**

What is the $H^+$ ion concentration in 0.0025 *M* NaOH?

**• Solution**

Sodium hydroxide is a base and supplies practically all the $OH^-$ ions in the solution. The $H^+$ ions come from the autoionization of water.

$$H_2O(\ell) \rightarrow H^+(aq) + OH^-aq)$$

Sodium hydroxide is a strong base and therefore, ionizes 100% in aqueous solution.

$$NaOH(aq) \rightarrow Na^+(aq) + OH^- \ (aq)$$

This means that we can calculate the hydroxide ion concentration first. Then from the ion-product constant of water ($K_w$), we can calculate the concentration of $H^+$ ions in equilibrium with the known concentration of $OH^-$ ions.

When 0.0025 mol of NaOH per liter dissociates, then 0.0025 mol $Na^+$ ions and 0.0025 mol $OH^-$ ions are formed per liter. Only the $OH^-$ affects the pH, and $[OH^-]$ = 0.0025 *M*.
Since the concentrations of $H^+$ and $OH^-$ ions are related to each other:

$$K_w = [H^+][OH^-] = 1.0 \times 10^{-14}$$

Then,

$$[H^+] = \frac{K_w}{[OH^-]} = \frac{1.0 \times 10^{-14}}{1.0 \times 10^{-3}}$$

$$[H^+] = 4.0 \times 10^{-12} \, M$$

**Work EXERCISES & PROBLEMS: 12, 13**

---

### EXAMPLE 15.8  Increasing Conjugate Base Strength

The strengths of the following acids increase in the order; $HCN < HF < HNO_3$. Arrange the conjugate bases of these acids in order of increasing base strength.

• **Solution**

Recall that as the strengths of acids increase, the strengths of their conjugate bases decrease. The acid strength increases from left to right.

$HCN < HF < HNO_3$
increasing acid strength →

Therefore, the strengths of their conjugate bases decrease from left to right.

$CN^- > F^- > NO_3^-$
decreasing base strength →

**Work EXERCISES & PROBLEMS: 14**

---

### EXAMPLE 15.9  Direction of an Acid-Base Reaction

To which side does the equilibrium lie in the following acid–base reaction?

$$HF(aq) + NO_2^- \,(aq) \rightleftharpoons HNO_2(aq) + F^- \,(aq)$$

• **Solution**

First, we need to know which is the stronger acid, the one on the left (HF) or the one on the right ($HNO_2$) of the reaction. Second, we also need to know which is the stronger base and which is the weaker base. Look this information up in Table 15.2.

$$HF(aq) + NO_2^- \,(aq) \rightleftharpoons HNO_2(aq) + F^- \,(aq)$$

| stronger | stronger | | weaker | weaker |
|----------|----------|--|--------|--------|
| acid | base | | acid | base |

Since proton transfer reactions proceed from the stronger acid and stronger base toward the weaker acid and weaker base, the reaction favors $HNO_2$ and $F^-$ ion. The equilibrium lies to the right.

**Work EXERCISES & PROBLEMS: 15**

---

### EXAMPLE 15.10  Percent Ionization of a Weak Acid

a. Calculate the concentrations of $H^+$, $F^-$, and HF in a 0.31 *M* HF (hydrofluoric acid) solution.
b. What is the percent ionization?

• **Solution**

a. First, write the ionization reaction and the acid ionization constant expression for hydrofluoric acid. The value for $K_a$ can be found in Table 15.3.

$$HF(aq) \rightleftharpoons H^+(aq) + F^-(aq)$$

$$K_a = \frac{[H^+][F^-]}{[HF]} = 7.1 \times 10^{-4}$$

The equilibrium concentrations of all species must be written in terms of a single unknown. Let $x$ equal the moles of HF ionized per liter of solution. This means that $x$ is the molarity of $H^+$ ion at equilibrium, because one $H^+$ ion is formed for each HF molecule that ionizes. Also, $[H^+] = [F^-] = x$, because one $F^-$ ion is formed for each hydrogen ion formed. It is helpful to tabulate the initial and equilibrium concentrations of all species. The initial concentrations are those that existed before any ionization of HF occurred. The change in concentration is $x$. When $x$ moles of HF have ionized per liter, $x$ moles of $H^+$ and $x$ moles of $F^-$ are formed per liter. The equilibrium concentrations are found by adding algebraically the change in concentration to the initial concentration.

| Concentration | HF $\rightleftharpoons$ | H$^+$ + | F$^-$ |
|---|---|---|---|
| Initial (*M*) | 0.31 | 0 | 0 |
| Change (*M*) | $-x$ | $+x$ | $+x$ |
| Equilibrium (*M*) | $(0.31 - x)$ | $x$ | $x$ |

The value of $x$ can be determined by substituting the equilibrium concentrations into the $K_a$ expression.

$$K_a = \frac{(x)(x)}{(0.31 - x)} = 7.1 \times 10^{-4}$$

The small magnitude of $K_a$ indicates that very little hydrofluoric acid actually ionizes. After all, it is a weak acid. Therefore, the value of $x$ is much less than 0.31, and $0.31 - x \approx 0.31$ *M*. This approximation simplifies the equation, and allows us to avoid solving a quadratic equation.

$$K_a = \frac{x^2}{0.31} = 7.1 \times 10^{-4}$$

Solving for $x$:

$$x = \sqrt{(0.31)(7.1 \times 10^{-4})} = 1.5 \times 10^{-2} \ M$$

Was our approximation valid? As a rule, if $x$ is equal to or less than 5% of the initial concentration $[HF]_0$, then neglecting $x$ in the calculation does not introduce an unacceptable amount of error into the answer.

$$[HF] = [HF]_0 - x \quad \text{and} \quad [HF] \approx [HF]_0$$

To test the validity of the approximation, divide $x$ by the initial concentration of hydrofluoric acid $[HF]_0$.

$$\frac{x}{[HF]_0} \times 100\% = \frac{1.5 \times 10^{-2}\ M}{0.31\ M} \times 100\% = 4.8\%$$

Since $x$ *is less than 5 percent* of the initial amount of HF, our approximate solution is close enough to the accurate result to be valid. Knowing $x$, we can utilize the bottom row in the table to answer part (a) of the problem.

$$[H^+] = [F^-] = 1.5 \times 10^{-2}\ M$$
$$[HF] = 0.31\ M - 0.015\ M = 0.30\ M$$

b.   The percent ionization was already calculated.

$$\% \text{ ionization} = \frac{x}{[HF]_0} \times 100\% = 4.8\%$$

• **Comment**

In this example we have followed three basic steps common to all problems of this type.

1.   Write the acid dissociation equilibrium, and express the equilibrium concentration of all species in terms of an unknown ($x$), where $x$ equals the change in the concentration of HA that is required to reach equilibrium.
2.   Substitute these concentrations into the $K_a$ expression, and look up the value of $K_a$ in a table.
3.   Solve for $x$, and calculate all equilibrium concentrations, the pH, and percent ionization.

**Work EXERCISES & PROBLEMS: 16 – 22**

---

**EXAMPLE 15.11  pH of a Weak Base Solution**

What is the pH of a 0.010 $M$ $C_5H_5N$ (pyridine) solution?

• **Solution**

Pyridine is a weak base. Write the equation for the reaction of pyridine in water and look up its $K_b$ value in Table 15.4.

$$C_5H_5N + H_2O \rightleftharpoons C_5H_5NH^+ + OH^-\quad K_b = 1.7 \times 10^{-9}$$

Let $x$ equal the moles of $C_5H_5N$ ionized per L to reach equilibrium. Note, when $x$ mol/L of $C_5H_5N$ are ionized, $x$ mol/L of $OH^-$ and $x$ mol/L of $C_5H_5N^+$ are formed. These changes are listed in the following table:

| Concentration | H₂O + C₅H₅N ⇌ | C₅H₅NH⁺ + | OH⁻ |
|---|---|---|---|
| Initial ($M$) | 0.010 | 0 | 0 |
| Change ($M$) | $-x$ | $+x$ | $+x$ |
| Equilibrium ($M$) | $(0.010 - x)$ | $x$ | $x$ |

Substitute the equilibrium concentrations into the base constant expression.

$$K_b = \frac{[C_5H_5NH^+][OH^-]}{[C_5H_5N]} = \frac{(x)(x)}{(0.010 - x)} = 1.7 \times 10^{-9}$$

Make the approximation, $[C_5H_5N] = 0.010 - x \cong 0.010$ $M$. Substituting

$$\frac{x^2}{0.010} = 1.7 \times 10^{-9}$$

$$x = \sqrt{(0.010)(1.7 \times 10^{-9})}$$

$$x = 4.1 \times 10^{-6} \ M$$

Testing the assumption that the approximation is reasonable:

$$\frac{x}{[C_5H_5N]_0} \times 100\% = \frac{4.1 \times 10^{-6} \ M}{0.010 \ M} \times 100\% = 0.041\%$$

Since $x$ is less than 5% of the initial concentration of pyridine, then the assumption is valid. At equilibrium:

$$[OH^-] = x = 4.1 \times 10^{-6} \ M$$

To find the pH, first calculate the pOH, and then use the relationship pH + pOH = 14.

pOH = $-\log(4.1 \times 10^{-6}) = 5.38$
Since, pH = 14.00 - pOH
Then, pH = 14.00 − 5.38
pH = 8.62

**Work EXERCISES & PROBLEMS: 23**

---

**EXAMPLE 15.12  Hydrolysis of Ions**

Write net ionic equations to show which of the following ions hydrolyze in aqueous solution:

a. $NO_3^-$    b. $NO_2^-$    c. $NH_4^+$

**• Solution**

Strong acids have extremely weak conjugate bases which have no affinity for $H^+$ ions. They have neutral properties. Weak acids have stronger conjugate bases with significant affinity for an $H^+$ ion. The conjugate bases of weak acids have measurable basic properties.

a.  Nitrate ion is the conjugate base of a strong acid, $HNO_3$. Therefore, $NO_3^-$ has no acidic or basic properties and has no tendency to react with water to form $HNO_3$.

$$NO_3^- + H_2O \rightarrow HNO_3 + OH^- \quad \text{(no reaction)}$$

b.  Nitrite ion is the conjugate base of a weak acid, $HNO_2$. Therefore, $NO_2^-$ is a stronger base than $NO_3^-$ and will accept a proton from water to form $HNO_2$.

$$NO_2^- + H_2O \rightleftharpoons HNO_2 + OH^-$$

c. Ammonium ion is the conjugate acid of a weak base. Therefore, it is a weak acid:

$$NH_4^+ + H_2O \rightleftharpoons NH_3 + H_3O^+$$

---

## EXAMPLE 15.13 Conversions between $K_a$ and $K_b$ for Conjugate Acid/Base Pairs

a. Calculate $K_b$ for $NO_3^-$
b. Calculate $K_a$ for $NH_4^+$

• **Solution**

a. The $K_a$ value of $HNO_2$ (the conjugate acid of $NO_3^-$) is $4.5 \times 10^{-4}$.

$$K_w = K_a K_b \qquad \text{therefore,}$$

$$K_b = \frac{K_w}{K_a} = \frac{1.0 \times 10^{-14}}{4.5 \times 10^{-4}} = 2.2 \times 10^{-11}$$

b. The $K_b$ value of $NH_3$ (the conjugate base of $NH_4^+$) is $1.8 \times 10^{-5}$.

$$K_a = \frac{K_w}{K_b} = \frac{1.0 \times 10^{-14}}{1.8 \times 10^{-5}} = 5.6 \times 10^{-10}$$

**Work EXERCISES & PROBLEMS: 24, 25**

---

## EXAMPLE 15.14 Solution Containing a Diprotic Acid

Calculate the concentrations of $H_2A$, $HA^-$, $A^{2-}$, and $H^+$ in a 1.0 $M$ $H_2A$ solution.

• **Solution**

First, write the equations for the stepwise ionization of $H_2A$, given the $K_a$ values shown.

$$H_2A \rightleftharpoons H^+ + HA^- \qquad K_{a1} = 1.3 \times 10^{-2}$$

$$HA^- \rightleftharpoons H^+ + A^{2-} \qquad K_{a2} = 6.3 \times 10^{-8}$$

Since $K_{a1} \gg K_{a2}$, we assume that essentially all the $H^+$ comes from step 1. Let $x$ be the concentration of $H_2A$ ionized to reach equilibrium. We solve for $x$ in the usual way.

| Concentration | $H_2A$ | $\rightleftharpoons$ | $H^+$ | + | $HA^-$ |
|---|---|---|---|---|---|
| Initial ($M$) | 1.0 | | 0 | | 0 |
| Change ($M$) | $-x$ | | $x$ | | $x$ |
| Equilibrium ($M$) | $(1.0 - x)$ | | $x$ | | $x$ |

$$K_{a1} = \frac{[H^+][HA^-]}{[H_2A]} = \frac{(x)(x)}{1.0 - x} = 1.3 \times 10^{-2}$$

$$K_{a1} = \frac{x^2}{1.0 - x} = 1.3 \times 10^{-2}$$

For this acid $K_{a1}$ is quite large, and the percent ionization will be greater than 5%. Rearranging the preceding equation into quadratic form:

$$ax^2 + bx + c = 0$$
$$x^2 + (1.3 \times 10^{-2})x - 1.3 \times 10^{-2} = 0$$

where,

$$a = 1, b = 1.3 \times 10^{-2}, \text{ and } c = -1.3 \times 10^{-2}.$$

Substituting into the quadratic formula, and solving for x.

$$x = \frac{-b \pm \sqrt{b^2 - 4ac}}{2a}$$

$$x = \frac{-1.3 \times 10^{-2} \pm \sqrt{(1.3 \times 10^{-2})^2 - (4)(1)(-1.3 \times 10^{-2})}}{(2)(1)}$$

$$x = 0.11 \ M$$

Therefore, at equilibrium

$$[H^+] = [HA^-] = 0.11 \ M$$

and

$$[H_2A] = 1.00 - 0.11 = 0.89 \ M$$

which corresponds to 11% ionization.

To determine the concentration of $A^{2-}$, we must consider the second stage of ionization:

$$HA^- \rightleftharpoons H^+ + A^{2-} \quad K_{a2} = 6.3 \times 10^{-8}$$

Let y be the equilibrium concentration of $A^{2-}$, and note that $x \gg y$.

| Concentration | $HA^-$ | $H^+$ | $+$ | $A^{2-}$ |
|---|---|---|---|---|
| Initial (*M*) | 0.11 | 0.11 | | 0 |
| Change (*M*) | $-y$ | $+y$ | | $+y$ |
| Equilibrium (*M*) | $(0.11 - y)$ | $(0.11 + y)$ | | $+y$ |

$$K_{a2} = \frac{[H^+][A^-]}{[HA]} = \frac{y(0.11 + y)}{(0.11 - y)} = 6.3 \times 10^{-8}$$

Since $K_{a2}$ is very small, as for a typical weak acid, we can assume that less than 5% of the $HA^-$ dissociates. Therefore,

$$0.11 \ M \pm y \approx 0.11 \ M$$

With this assumption:

$$K_{a2} = \frac{y(0.11)}{(0.11)}$$

and

$$y = K_{a2} = 6.3 \times 10^{-8} \ M$$

The concentration of $A^{2-}$ is equal to $K_{a2}$.

$$[A^{2-}] = y = 6.3 \times 10^{-8} \ M$$

Checking the validity of the assumption that,

$$0.11 \ M - (6.3 \times 10^{-8} \ M) \approx 0.11 \ M$$

The percent ionization of step 2 is:

$$\text{percent ionization} = \frac{[H^+]}{[HA^-]} \times 100\%$$

$$\frac{6.3 \times 10^{-8}}{0.11} \times 100\% = 5.7 \times 10^{-5}\%$$

• **Comment**

Because $K_{a1} \gg K_{a2}$, the species produced in step 1 will be present in much greater concentration than those from step 2. This can be seen by comparing the percent ionization of each step (11% vs. 0.000057%). Also, since $H_2A$ is a weak acid, the concentration of undissociated $H_2A$ should be greater than that of all ionic species. Therefore, we expect the following order of concentrations:

$$[H_2A] > [H^+] \approx [HA^-] > [A^{2-}]$$

**Work EXERCISES & PROBLEMS: 26, 27**

---

**EXAMPLE 15.15  Relative Acid Strength**

Which member of each of the following pairs is the stronger acid?
a.  HCl or HBr
b.  HCl or $H_2S$

• **Solution**

a.  These acids are nonmetal hydrides of elements within Group 6A of the periodic table. Variations in atomic radius are more important within a group than electronegativity variations. The stronger acid is the one with the greater radius of its nonmetal atom. HBr is the stronger acid.
b.  These acids are nonmetal hydrides of elements within the third row of the period table. The electronegativity changes more significantly going across a row than does the atomic radius. HCl with its more polar bond is the stronger acid.

---

**EXAMPLE 15.16 Relative Acid Strength**

Which member of each of the following pairs is the stronger acid?
a. $HClO_3$ or $HBrO_3$
b. $H_3PO_3$ or $H_3PO_4$

**• Solution**

a. These two oxoacids have the same structure:

Within this group, the strengths of the acids increase with increasing electronegativity of the central atom. The electronegativity of Cl is greater than that of Br. $HClO_3$ is stronger.

b. $H_3PO_3$ and $H_3PO_4$ are oxoacids of the same element. The acid strength increases as the number of O atoms increases. $H_3PO_4$ is the stronger acid because of the greater number of oxygen atoms bonded to the phosphorus atom. In $H_3PO_4$, phosphorus has a higher oxidation number than in $H_3PO_3$.

**Work EXERCISES & PROBLEMS: 28**

---

**EXAMPLE 15.17 Acid-Base Properties of Ions**

Predict whether the following aqueous solutions will be acidic, basic, or neutral.
a. KI
b. $NH_4I$
c. $CH_3COOK$

**• Solution**

a. What we must decide is whether either of the ions of the salt undergoes hydrolysis. KI is a salt of a strong acid (HI) and a strong base (KOH). Neither $K^+$ nor $I^-$ have acidic or basic properties. Thus, a solution containing KI remains neutral.

b. $NH_4I$ is the salt of a strong acid (HI) and a weak base ($NH_3$). As explained in part (a) iodine does not hydrolyze. Since ion is the conjugate acid of a weak base, it will donate protons to water.

$$NH_4^+ + H_2O \rightleftharpoons NH_3 + H_3O^+$$

Therefore, a solution containing $NH_4I$ will be acidic.

c. $CH_3COOK$ is a salt of a weak acid and a strong base. $K^+$ ion will not hydrolyze, but $CH_3COO^-$ (acetate ion) is the conjugate base of a weak acid, acetic acid. Thus, acetate ions will accept protons from water and form acetic acid and $OH^-$ ions.

$$CH_3COO^- + H_2O \rightleftharpoons CH_3COOH + OH^-$$

Solutions containing $CH_3COOK$ will be basic.

**Work EXERCISES & PROBLEMS: 29**

356 *Chemistry, Ch. 15: Acids & Bases*

## EXAMPLE 15.18  pH of a Solution Containing a Salt

Calculate the pH of 0.25 $M$ $C_6H_7O_6Na$ (sodium ascorbate) solution.

**• Solution**

First, identify the ions that undergo hydrolysis. Then, use the ionization constant to calculate the equilibrium concentrations. Sodium ascorbate is a strong electrolyte.

$$C_6H_7O_6Na(aq) \rightarrow C_6H_7O_6^-\ (aq) + Na^+(aq)$$

Ascorbate ion is a conjugate base of a weak acid (ascorbic acid, Table 15.3 text), and so it causes hydrolysis. $Na^+$ does not and is neutral.

$$C_6H_7O_6(aq) + H_2O(\ell) \rightleftharpoons C_6H_8O_6(aq) + OH^-\ (aq)$$

It is important to use the correct equilibrium constant value. Here, we need the $K_b$ for ascorbate ion. From Table 15.3 (text), we get the $K_a$ for ascorbic acid, the conjugate acid of ascorbate ion.

$$K_b = \frac{K_w}{K_a} = \frac{1.0 \times 10^{-14}}{8.0 \times 10^{-5}} = 1.3 \times 10^{-10}$$

The $OH^-$ ion concentration in a solution that is 0.25 $M$ is obtained from the equilibrium constant expression:

$$K_b = \frac{[C_6H_8O_6][OH^-]}{[C_6H_7O_6^-]}$$

Let $x$ = moles of ascorbate ion that undergo hydrolysis per L of solution.

| Concentration | $H_2O$ + $C_6H_7O_6^-$ | $\rightleftharpoons$ $C_6H_8O_6$ | + $OH^-$ |
|---|---|---|---|
| Initial ($M$) | 0.25 | 0 | 0 |
| Change ($M$) | $-x$ | $+x$ | $+x$ |
| Equilibrium ($M$) | $(0.25 - x)$ | $x$ | $x$ |

Substitute into the equilibrium constant expression.

$$K_b = 1.3 \times 10^{-10} = \frac{x^2}{(0.25 - x)}$$

Apply the approximation that $C_6H_7O_6^- = 0.25 - x \approx 0.25\ M$

$$\frac{x^2}{0.25} = 1.3 \times 10^{-10}$$

$$x = 5.7 \times 10^{-6}\ M = [OH^-]$$

$$pOH = -\log[OH^-] = 5.24 \qquad pH = 14.00 - 5.24 = 8.76$$

**Work EXERCISES & PROBLEMS: 30**

**EXAMPLE 15.19  Lewis Acids and Bases**

Identify the Lewis acids and bases in each of the following reactions.
a.  $Ag^+(aq) + Cl^-(aq) \rightarrow AgCl(s)$
b.  $Hg^{2+}(aq) + 4I^-(aq) \rightarrow HgI_4^{2-}(aq)$
c.  $BF_3(g) + NF_3(g) \rightarrow F_3N{-\!\!-}BF_3(s)$
d.  $SO_2(g) + H_2O(\ell) \rightarrow H_2SO_3(aq)$

**• Solution**

Lewis acids are species that accept electron-pairs. In each of the above reactions, the first reactant species is the Lewis acid, and the second reactant is the Lewis base (electron-pair donor). The Lewis acids in (a) and (b) are capable of accepting electron-pairs, because they are positive ions and have previously lost electrons. In (c), the boron atom has an incomplete octet. And in (d), the S atom in $SO_2$ is "oxidized," that is, its valence electrons are shifted toward the more electronegative oxygen atoms and away from the S atom. In all four reactions new coordinate covalent bonds are formed.

**Work EXERCISES & PROBLEMS: 31, 32**
**Work the rest of the EXERCISES & PROBLEMS**

## EXERCISES & PROBLEMS

1. Identify the conjugate base of $HCO_3^-$ in the following acid-base reaction:

   $$CO_3^{2-} + CH_3COOH \rightleftharpoons HCO_3^- + CH_3COO^-$$

2. Identify the Brønsted acid and base on the left side of the following equations, and identify the conjugate partner of each on the right side.

   a. $NH_4^+ (aq) + H_2O(\ell) \rightleftharpoons NH_3(aq) + H_3O^+(aq)$

   b. $HNO_2(aq) + CN^- (aq) \rightleftharpoons NO_2^- (aq) + HCN(aq)$

3. Calculate the concentration of $OH^-$ ions in an $HNO_3$ solution where $[H^+] = 0.0010\ M$.

4. The $OH^-$ ion concentration in an ammonia solution is $7.5 \times 10^{-3}\ M$. What is the $H^+$ ion concentration?

5. What are the $H^+$ ion and $OH^-$ ion concentrations in a $0.0033\ M$ NaOH solution?

6. Calculate the pH of a $0.00051\ M$ HCl solution.

7. Calculate the pH of a $0.125\ M$ NaOH solution.

8. The pH of a certain solution is 3.0. How many moles of $H_3O^+(aq)$ ions are there in 0.10 L of this solution?

9. Calculate the concentration of $H^+$ ions in an acid solution with a pH of 2.29.

10. What is the concentration of $OH^-$ ions in a NaOH solution which has a pOH of 4.90?

11. The pH of solution A is 2.0 and the pH of solution B is 4.0. How many times greater is the $H^+(aq)$ concentration in solution A than in solution B?

12. What is the pH of a $0.0025\ M$ solution of $Ca(OH)_2$?

13. What is the pH of a solution made from 3.0 g HCl dissolved in 3.0 L of solution?

14. Which of the following acids has the strongest conjugate base?
    $HNO_2$, HCN or $H_3O^+$

15. Predict the direction in which the equilibrium will lie for the following reaction.
    $$HNO_2(aq) + ClO_4^- (aq) \rightleftharpoons NO_2^- (aq) + HClO_4(aq)$$

16. What is the percent ionization of a weak acid in a $0.020\ M$ HA solution given that its $K_a$ is $2.3 \times 10^{-5}$?

17. What is the pH of a $0.050\ M$ $C_6H_5COOH$ (benzoic acid) solution?

18. HA is a weak acid. If a 0.020 $M$ HA solution is 2.5% dissociated, what is the $K_a$ value of the weak acid?

19. A 0.100 $M$ solution of chloroacetic acid, $ClCH_2COOH$, has a pH of 1.95. Calculate the $K_a$ for chloroacetic acid.

20. Which of the following is the strongest acid?
    $HF$, $CH_3COOH$, or $C_6H_5COOH$

21. Which of the following solutions has the lowest pH?
    0.10 $M$ $HNO_2$, 0.10 $M$ $CH_3COOH$, or 0.10 $M$ HCN

22. Determine accurately the percent ionization of formic acid, HCOOH, in a 0.0050 $M$ solution.

23. A 0.012 $M$ solution of an unknown base has a pH of 10.10.
    a. What are the hydroxide ion and hydronium ion concentrations in the solution?
    b. Is the base a weak base or a strong base?

24. Given that $K_a$ for formic acid (HCOOH) is $1.7 \times 10^{-4}$, find the $K_b$ value for formate ion, $HCOO^-$.

25. Obtain the $K_b$ value for $C_5H_5N$ from Table 15.4 and calculate $K_a$ for $C_5H_5NH^+$.

26. Write chemical equations for the three stages of ionization of phosphoric acid, $H_3PO_4$.

27. Calculate the pH of a 0.50 $M$ solution of phosphoric acid, $H_3PO_4$.

28. Which member of each pair is the stronger acid?
    a. $HNO_2$ or $HNO_3$
    b. $CH_4$ or $SiH_4$
    c. HOBr or HOI
    d. $CH_4$ or $NH_3$

29. Predict whether the pH is > 7, < 7, or = 7 for aqueous solutions containing the following salts.     a. $K_2CO_3$     b. $BaCl_2$     c. $Fe(NO_3)_3$

30. Calculate the pH of 0.021 $M$ NaCN solution.

31. Why are nitrogen compounds and ions such as $NH_3$, $CH_3NH_2$, and $NH_2^-$ good examples of Lewis bases?

32. Why can a $H^+$ ion, $Mg^{2+}$ ion, and $Al^{3+}$ ion act as Lewis acids?

33. What is the strongest acid and the strongest base that can exist in water?

34. Both oxide ion ($O^{2-}$) and amide ion ($NH_2^-$) are stronger bases than $OH^-$ and so react completely in water. Write equations for the reactions of these ions with water.

35. Generally, we do not need to know the $K_a$ value for a strong acid. Why?

36. Which of the following statements, if any, are true with regard to a 0.10 *M* solution of a weak acid HA?

    a. $[HA] > [H^+]$
    b. the pH = 1.0
    c. $[H^+] < [A^-]$
    d. the pH > 1.0
    e. $[OH^-] = [H^+]$

37. Why is a solution of NaF basic, whereas, a solution of NaCl is neutral?

## PRACTICE TEST QUESTIONS
See notes on taking practice test in the Preface

1. Identify the conjugate bases of the following acids:
   a. $CH_3COOH$    b. $H_2S$
   c. $HSO_3^-$    d. $HClO$

2. Identify the conjugate acid-base pairs in each of the following reactions:
   a. $NH_3 + H_2CO_3 \rightleftharpoons NH_4^+ + HCO_3^-$
   b. $CO_3^{2-} + HSO_3^- \rightleftharpoons HCO_3^- + SO_3^{2-}$
   c. $HCl + HSO_3^- \rightleftharpoons Cl^- + H_2SO_3$
   d. $HCO_3^- + HPO_4^{2-} \rightleftharpoons H_2CO_3 + PO_4^{3-}$

3. What is the $H^+(aq)$ concentration in each of the following?
   a. Neutral water
   b. 1 $M$ HCl
   c. 0.01 $M$ $HNO_3$

4. What is the $OH^-(aq)$ concentration in each of the following?
   a. Neutral water
   b. 0.01 $M$ NaOH
   c. 0.01 $M$ $Sr(OH)_2$

5. What is the $H^+(aq)$ concentration in $1.0 \times 10^{-3}$ $M$ NaOH?

6. What is the $OH^-(aq)$ concentration in 2.0 $M$ HCl?

7. What is the pH of the following solutions?
   a. 0.015 $M$ HCl
   b. 0.015 $M$ NaOH

8. What is the pOH of the following solutions?
   a. 0.30 $M$ $Ca(OH)_2$
   b. $2.0 \times 10^{-3}$ $M$ $HClO_4$

9. The pH of solution A is 2.0, and the pOH of solution B is 10.0. How many times greater is the $H^+(aq)$ concentration in solution A than in solution B?

10. The pH of a solution is 4.45. What is the $H^+$ ion concentration?

15        15

15        15

15        15

15        15

15        15

11. Classify each of the following as a weak or strong acid:
    a. $H_2SO_4$   b. $CH_3COOH$   c. $HClO$
    d. $HCl$       e. $HNO_2$      f. $HClO_4$

12. Use Table 15.2 of the textbook to arrange the following anions in order of increasing base strength.
    a. $HSO_4^-$        b. $CH_3COO^-$
    c. $Cl^-$           d. $NO_2^-$

13. Which member of each of the following pairs is the stronger acid?
    a. $H_2CO_3$ or $H_2SiO_3$
    b. $H_3AsO_4$ or $H_3AsO_3$
    c. $H_3PO_2$ or $H_3PO_3$
    d. $HClO_2$ or $HClO_3$
    e. $H_3PO_4$ or $H_3AsO_4$

14. Which member of each of the following pairs is the stronger acid?
    a. $HI$ or $HBr$
    b. $H_2O$ or $H_2S$
    c. $H_2Te$ or $HI$
    d. $H_2Se$ or $AsH_3$

15. Choose the Lewis acids and bases that could be used to form the following compounds:
    a. $HCl$
    b. $H_3O^+$
    c. $BCl_3NH_3$
    d. $Al(OH)_4^-$
    e. $MgF_2$

16. What is the pH of a 0.0055 $M$ HA (weak acid) solution that is 8.2% ionized?

17. Predict the direction in which the equilibrium will lie for the following reaction.
    $C_6H_5COO^- + HF \rightleftharpoons C_6H_5COOH + F^-$

18. Acid strength increases in the series:
    $$HCN, HF, \text{ to } HSO_4^-$$
    weakest $\longrightarrow$ strongest

    Which of the following is the strongest base?
    a. $SO_4^{2-}$   b. $F^-$   c. $CN^-$   d. $HSO_4^-$

19. What is the $H^+$ ion concentration in a 0.82 $M$ HOCl solution?
    Given $K_a = 3.2 \times 10^{-8}$

20. Calculate the percent ionization in a 1.0 $M$ solution of nitrous acid, $HNO_2$. $K_a$ for $HNO_2$ is $4.5 \times 10^{-4}$.

15

15

15

15

15

15

15

15

15

15

21. What is the pH of a 0.090 $M$ $CH_3COOH$ (acetic acid) solution? $K_a$ for $CH_3COOH$ is $1.8 \times 10^{-5}$.

22. A 1.0 $M$ HF solution is only 2.6 percent ionized. What is the $K_a$ value for HF?

23. Calculate the concentration of $H^+$, $OH^-$, $F^-$ ions and undissociated HF molecules in 0.010 $M$ HF. $K_a$ for HF is $7.1 \times 10^{-4}$.

24. What is the pH of a 2.0 $M$ $NH_3$ solution? $K_b$ for $NH_3$ is $1.8 \times 10^{-5}$.

25. The first and second ionization constants of $H_2CO_3$ are $4.2 \times 10^{-7}$ and $4.8 \times 10^{-11}$. Calculate the concentrations of $H^+$ ion, $HCO_3^-$, $CO_3^{2-}$, and undissociated $H_2CO_3$ in a 0.080 $M$ $H_2CO_3$ solution.

26. A 0.1 $M$ solution of $Na_2CO_3$ is:
   a. acidic   b. basic   c. neutral

27. Classify the solutions of the following salts as acidic, basic or neutral:
   a. $AlCl_3$     b. $Na_2SO_4$     c. $NaClO_4$
   d. KCN      e. $NH_4Cl$      f. $Na_2SO_3$

28. Calculate the pH of a 0.30 $M$ $CH_3COONa$ solution. Hint: $K_a$ for $CH_3COOH$ is $1.8 \times 10^{-5}$.

29. What is the $H^+$ ion concentration of a 0.010 $M$ NaCN solution? $K_a$ for HCN is $4.9 \times 10^{-10}$.

30. Which one of the following aqueous solutions (all 0.010 $M$) will have the highest pH value?
   a. $CH_3COOH$    b. HCl
   c. NaCl                d. KF

15

15

15

15

15

15

15

15

15

15

# Chapter 16. Acid-Base and Solubility Equilibria

**Buffer Solutions (Sections 16.2 – 16.3)**
**Titration Curves and Indicators (Sections 16.4 – 16.5)**
**Solubility and Solubility Product (Section 16.6)**
**Predicting Precipitation Reactions and Ion Separation (Sections 16.6 – 16.7)**
**Common Ion Effect and pH (Sections 16.8 – 16.9)**
**Complex Ions and Solubility (Section 16.10)**

## SUMMARY

### Buffer Solutions (Sections 16.2 – 16.3)

***The Common Ion Effect.*** In this chapter, we will examine the acid-base equilibria in buffer solutions and in titrations. Then, we will discuss two additional types of aqueous equilibria: those involving low solubility salts, and those involving the formation of metal complexes. We will start with the common ion effect.

First, consider the ionization of a weak acid such as HF for example. The species in equilibrium are HF molecules, $H^+$ ions, and $F^-$ ions:

$$HF(aq) \rightleftharpoons H^+(aq) + F^-(aq)$$

The concentration of $F^-$ ion can be increased by adding sodium fluoride, a strong electrolyte:

$$NaF(aq) \rightarrow Na^+(aq) + F^-(aq)$$

According to Le Chatelier's principle, the addition of $F^-$ ions will shift the weak acid equilibrium to the left; this consumes some of the $F^-$ ions and some $H^+(aq)$, and forms more HF. This shift lowers the percent ionization of HF. In effect, the percent ionization of a weak acid (HA) is repressed by the addition to the solution of its conjugate base, $A^-$ ion. The shift in equilibrium caused by the addition of an ion common to one of the products of a dissociation reaction is called the common ion effect.

***Buffer Solutions.*** Any solution that resists significant changes in its pH, even when a strong acid or strong base is added, is called a buffered solution or just a buffer. A buffer must contain an acid to react with any $OH^-$ ions that may be added, and a base to react with any added $H^+$ ions. A buffer solution may contain a weak acid and its salt (for example HF and NaF), or a weak base and its salt (for example $NH_3$ and $NH_4Cl$).

A solution containing HF and NaF is a buffer solution. In a typical buffer, the weak acid and its salt are in approximately equal concentrations. The weak acid HF and its conjugate base $F^-$ are called the buffer components. If you had a solution of just HF, the concentration of $F^-$ ion would not be high enough to make it a buffer solution.

$$HF(aq) \rightleftharpoons H^+(aq) + F^-(aq)$$

The additional $F^-$ ions needed in order to achieve approximately equal concentrations of the two components are supplied by the addition of NaF which is 100% dissociated in solution.

***Buffer Action.*** When a small amount of a strong acid is added to the buffer, it is neutralized by the weak base component. Also, when a small amount of a strong base is added to a buffer, it is neutralized by the weak acid component. In the HF/F⁻ ion buffer, for example, the following neutralization reactions take place:

- On addition of a strong base, the neutralization reaction is

  $$OH^- + HF \rightarrow H_2O + F^-$$

- On addition of a strong acid, the neutralization reaction is

  $$H^+ + F^- \rightarrow HF$$

In order for the buffer to be most effective, the amounts of weak acid and weak base used to prepare the buffer must be considerably greater than the amounts of strong acid or strong base that may be added later. Buffer capacity refers to the amount of acid or base that can be neutralized by a buffer solution before its pH is appreciably affected.

***The pH of a Buffer.*** To calculate the pH of a buffer requires that the concentrations of the weak acid (such as HF) and a soluble salt of the weak acid (NaF) be substituted into the ionization constant expression for the weak acid.

$$HF(aq) \rightleftharpoons H^+(aq) + F^-(aq)$$

$$K_a = \frac{[H^+][F^-]}{[HF]}$$

$$[H^+] = \frac{[HF]}{[F^-]} K_a$$

The pH is found from $pH = -\log[H^+]$.

Alternatively, the Henderson-Hasselbalch equation, which is introduced, and derived in Section 16.2 of the textbook, can be used to calculate buffer pH.

$$pH = pK_a + \log \frac{[\text{conjugate base}]}{[\text{acid}]}$$

where $pK_a = -\log K_a$.

Use of the Henderson-Hasselbalch equation is shown in Example 16.1. This equation requires the equilibrium concentration to be used, however, when dealing with buffer solutions in which the concentration of the acid and conjugate base are high (>0.1 *M*), the initial concentration is essentially the same as the equilibrium concentration.

***Preparing a Buffer.*** Each weak acid–conjugate base buffer has a characteristic pH range over which it is most effective. In terms of the Henderson-Hasselbalch equation, *the* pH of a buffer depends on the pKₐ of the weak acid, and the ratio of conjugate base concentration to weak acid concentration. A buffer has equal ability to neutralize either added acid or added base when the concentrations of its weak acid and conjugate base components are equal. When [conjugate base]/[weak acid] = 1, then,

$$\log \frac{[\text{conjugate base}]}{[\text{acid}]} = 0$$

and the Henderson-Hasselbalch equation becomes

$$pH = pK_a$$

In order to prepare a buffer of a certain pH, the weak acid must have a $pK_a$ as close as possible to the desired pH.

Next, we substitute the desired pH and the $pK_a$ of the chosen acid into the Henderson-Hasselbalch equation, in order to determine the ratio of [conjugate base] to [weak acid]. The ratio can then be converted into molar concentrations of the weak acid and its salt.

| Examples |
| :---: |
| 16.1 – 16.3 |
| Exercises |
| 1 – 5 |

***Addition of Strong Acid or Base to a Buffer.*** The whole purpose of a buffer solution is to resist the change of pH to a solution upon the addition of an acid or base. Adding 1.0 mL of 0.10 *M* NaOH to 100.0 mL of water will significantly increase the pH from 7.00 for pure water to 11.00. But, adding the same amount of base to 100 mL of a buffer consisting of 0.10 *M* CH₃COOH and 0.10 *M* CH₃COONa has very little impact, changing the pH from 4.74 for the buffer to 4.75. Example 16.3 demonstrates these calculations with the addition of an acid to a buffer.

## Titration Curves and Indicators (Sections 16.4 – 16.5)

***Titration Curves.*** Acid-base titrations were discussed in chapter 4. With the introduction to pH in the previous chapter, it is informative to follow the pH as a function of the progress of the titration. A graph of pH versus volume of titrant added is called a titration curve. Initially, the pH is that of the unknown solution. As titrant is added, the pH becomes that of a partially neutralized solution of unknown plus titrant. The pH at the equivalence point refers to the H⁺ ion concentration when just enough titrant has been added to completely neutralize the unknown. If more titrant is added after the equivalence point has been reached, the pH assumes a value consistent with the pH of excess titrant. We will review three types of titration curves.

*Strong acid-strong base titrations.* Figure 16.1 below shows a titration curve for the addition of a strong base to a strong acid. The main features of this curve can be stated briefly. (1) The pH starts out quite low because this is the pH of the original acid solution. (2) As base is slowly added, it is neutralized, and the pH is determined by the unreacted excess acid. Near the equivalence point, the pH begins to rise more rapidly. (3) At the equivalence point the pH changes sharply, increasing about 5.0 units upon the addition of only two drops of base. (4) Beyond the equivalence point the pH is determined by the amount of excess base that is added and is higher than 7.

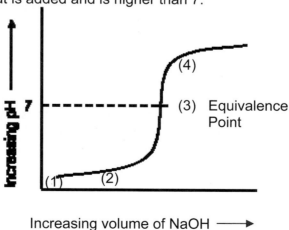

**Figure 16.1** Titration curve for the addition of NaOH to a solution of HCl.

The pH at the equivalence point of an acid-base titration is the pH of the salt solution that is formed by neutralization. NaCl, $NaNO_3$, NaBr, KCl, and KI are examples of salts that can be formed in titrations of strong acids and strong bases. These salts yield ions that do not cause hydrolysis (chapter 15). Therefore, the pH at the equivalence point in a titration of a strong acid with a strong base is 7.

*Weak acid-strong base titrations.* Figure 16.5 of the text shows a titration curve for the addition of a strong base to a weak acid. Because the acid is a weak acid, the initial pH is greater than in the titration of a strong acid. At the equivalence point, the pH is above 7.0 because of salt hydrolysis. The anion of the salt is the conjugate base of the weak acid used in the titration. In the titration of acetic acid with sodium hydroxide, the salt produced is $CH_3COONa$. The $Na^+$ ion does not hydrolyze, but $CH_3COO^-$ does:

$$CH_3COO^- + H_2O \rightleftharpoons CH_3COOH + OH^-$$

Hydrolysis of acetate ions makes the solution basic at the equivalence point.

*Strong acid-weak base titrations.* Figure 16.6 of the text shows a titration curve for the addition of a weak base to a strong acid. The first part of the curve is the same as in the strong acid versus strong base titration. However, the pH at the equivalence point is below 7.0 because the cation of the salt is the conjugate acid of the weak base used in the titration. Hydrolysis of this salt yields an acidic solution. In the titration of hydrochloric acid with ammonia, the salt produced is $NH_4Cl$. The $Cl^-$ ion does not hydrolyze, but $NH_4^+$ does:

$$NH_4^+ + H_2O \rightleftharpoons NH_3 + H_3O^+$$

**Indicators.** Indicators are used in the laboratory to reveal the equivalence point of a titration. The abrupt change in color of the indicator signals the *endpoint* of a titration which usually coincides very nearly with the equivalence point.

Indicators are usually weak organic acids that have distinctly different colors in the nonionized (molecular) and ionized forms.

$$HIn(aq) \rightleftharpoons H^+(aq) + In^-(aq) \qquad K_a = \frac{[H^+][In^-]}{[HIn]}$$

   color 1                                    color 2

To determine the pH range in which an indicator will change color we can write the $K_a$ expression in logarithmic form.

$$pH = pK_a + \log \frac{[In^-]}{[HIn]}$$

The color of an indicator depends on which form predominates. Typically, when $[In^-]/[HIn] \geq 10$ the solution will be color 2, and when $[In^-]/[HIn] \leq 0.1$, the solution will be color 1. Thus, the color change will occur between,

$$pH = pK_a + \log (10/1) = pK_a + 1.0$$

and

$$pH = pK_a + \log (1/10) = pK_a - 1.0$$

At the midpoint of the pH range over which the color changes, $[HIn] = [In^-]$, and $pH = pK_a$.

Like any weak acid, each HIn has a characteristic $pK_a$, and so each indicator changes color at a characteristic pH. Table 16.1 of the text lists a number of indicators used in acid-base titration and the pH ranges over which they change color.

The choice of indicator for a particular titration depends on the expected pH at the equivalence point. For instance, in the titration of acetic acid by sodium hydroxide, which is discussed in Example 16.5 of the text, the pH at the equivalence point is 8.72. According to Table 16.1, both cresol red and phenolphthalein change color over ranges that include pH 8.72. Therefore, either of these indicators would show the equivalence point of this titration.

| Examples |
| :---: |
| 16.5, 16.6 |
| Exercises |
| 6 – 10 |

## Solubility and Solubility Product (Section 16.6)

***The Solubility Product Constant.*** The solubility rules were discussed in chapter 4. Compounds with low solubilities in water are described as slightly soluble or insoluble. In this chapter, solubility is treated quantitatively in terms of equilibrium. In a saturated solution of a slightly soluble compound such as silver bromide, for example, the solubility equilibrium is:

$$AgBr(s) \rightleftharpoons Ag^+(aq) + Br^-(aq)$$

The equilibrium constant for the reaction in which a solid salt dissolves is called the *solubility product constant*, $K_{sp}$. For the above reaction:

$$K_{sp} = [Ag^+][Br^-]$$

The solubility product expression is always written in the form of the equilibrium constant expression for the solubility reaction. Two additional examples are $Ca(OH)_2$ and $AgCrO_4$:

$$Ca(OH)_2(s) \rightleftharpoons Ca^{2+}(aq) + 2OH^-(aq)$$

$$K_{sp} = [Ca^{2+}][OH^-]^2$$

and

$$Ag_2CrO_4(s) \rightleftharpoons 2Ag^+(aq) + CrO_4^{2-}(aq)$$

$$K_{sp} = [Ag^+]^2[CrO_4^{2-}]$$

The solubility product constant (also called a $K_{sp}$ value) can be calculated by substituting the concentrations of the ions in a saturated solution into the solubility product expression. Take AgBr again as an example. Since AgBr is a strong electrolyte, all of the solid AgBr that dissolves is dissociated into $Ag^+$ and $Br^-$ ions,

$$AgBr(s) \rightarrow Ag^+(aq) + Br^-(aq)$$

The solubility of AgBr at 25 °C is $8.8 \times 10^{-7}$ moles per liter of solution. Therefore, the ion concentrations in a saturated AgBr solution are:

$$[Ag^+] = [Br^-] = 8.8 \times 10^{-7} \ M$$

The solubility product constant is found by substituting these concentrations into the solubility product:

$$K_{sp} = [Ag^+][Br^-] = (8.8 \times 10^{-7})\,(8.8 \times 10^{-7})$$

$$K_{sp} = 7.7 \times 10^{-13}$$

This calculation points out the difference between the solubility and the $K_{sp}$ value. These two quantities are *not* the same. Here, we see that the solubility is used to determine the ion concentrations, and that substitution of these into the solubility product expression gives the solubility product constant ($K_{sp}$ value). Table 16.2 in the text and Table 16.1 in this chapter list solubility product constants for a number of slightly soluble salts, including silver bromide.

The $K_{sp}$ value can be used to calculate the solubility of a compound in moles per liter (the molar solubility), or in grams per liter. For example, the $K_{sp}$ value for $CaSO_4$ is $2.4 \times 10^{-5}$. What is the solubility of $CaSO_4$ in mol/L?

Start with the solubility equilibrium:

$$CaSO_4(s) \rightleftharpoons Ca^{2+}(aq) + SO_4^{2-}(aq) \qquad K_{sp} = [Ca^{2+}][SO_4^{2-}]$$

Let's equal the solubility in mol/L. At equilibrium, both $[Ca^{2+}]$ and $[SO_4^{2-}]$ are equal to s because when s moles of $CaSO_4$ dissolve in 1 L, s moles of $Ca^{2+}$ ions and s moles of $SO_4^{2-}$ ions are produced. Substitute into the $K_{sp}$ expression:

$$K_{sp} = [Ca^{2+}][SO_4^{2-}] = (s)(s) = s^2$$

The solubilty product is given above:  $K_{sp} = 2.4 \times 10^{-5}$

$$s^2 = 2.4 \times 10^{-5}$$

$$s = \sqrt{2.4 \times 10^{-5}} = 4.9 \times 10^{-3}\ M$$

It is important to notice that the molar solubility and $K_{sp}$ are related, but are not the same quantity.

**Solubility and $K_{sp}$.** Notice, in Table 16.1, that within a series of salts of the same type of formula (MX, MX, etc.), the solubility increases as the solubility product constant increases. Comparing the $K_{sp}$'s of the three MX salts, we can see that $CaSO_4$ has the highest $K_{sp}$ value. Therefore, its solubility equilibrium lies farthest to the right of the three salts, and it is the most soluble.

**Table 16.1** Solubility Products of Some Slightly Soluble Salts at 25 °C

| Name | Formula | $K_{sp}$ |
|------|---------|----------|
| | MX type | |
| Calcium sulfate | $CaSO_4$ | $2.4 \times 10^{-5}$ |
| Silver chloride | $AgCl$ | $1.6 \times 10^{-10}$ |
| Silver bromide | $AgBr$ | $7.7 \times 10^{-13}$ |
| | $MX_2$ and $M_2X$ type | |
| Lead chloride | $PbCl_2$ | $2.4 \times 10^{-4}$ |
| Calcium hydroxide | $Ca(OH)_2$ | $8.0 \times 10^{-6}$ |
| Lead iodide | $PbI_2$ | $1.4 \times 10^{-8}$ |
| Magnesium fluoride | $MgF_2$ | $6.4 \times 10^{-9}$ |
| Magnesium hydroxide | $Mg(OH)_2$ | $1.2 \times 10^{-11}$ |
| Silver chromate | $Ag_2CrO_4$ | $1.1 \times 10^{-12}$ |

Examples
16.7, 16.8

Exercises
11 - 14

## Predicting Precipitation Reactions and Ion Separation (Sections 16.6 – 16.7)

***Criteria for Precipitate Formation.*** When two solutions containing dissolved salts are mixed, formation of an insoluble compound is always a possibility. From a knowledge of solubility rules (Chapter 4) and solubility products (Table 16.1), you can predict whether a precipitate will form. For a dissolved salt MX(aq), the ion product (Q) is $Q = [M^+]_0[X^-]_0$, where $[\ ]_0$ stands for the initial concentrations.

Any one of three conditions for the ion product Q may exist after two solutions are mixed.

1. $Q = K_{sp}$. For a saturated solution, the value of Q is equal to $K_{sp}$. No precipitate will form in this case, as no net reaction occurs in a system at equilibrium.
2. $Q < K_{sp}$. In an unsaturated solution, the value of Q is less than $K_{sp}$. No precipitate will form in this case.
3. $Q > K_{sp}$. In a supersaturated solution, more of the salt is dissolved than the solubility allows. In this case, an unstable situation exists, and some solute will precipitate from solution until a saturated solution is attained. At this point $Q = K_{sp}$ and equilibrium is reestablished.

Suppose 500 mL of a solution containing $2.0 \times 10^{-5}$ M $Ag^+$ is mixed with 500 mL of solution containing $2.0 \times 10^{-4}$ M $Br^-$. Will a precipitate form? First, determine the new concentrations of $Ag^+$ and $Br^-$ in the mixture. The total volume is 1 L, and so accounting for dilution, the ion concentrations in the new solution are just half of what they were in the separate solutions. The initial concentrations before any precipitate forms are:

$$[Ag^+]_0 = 1.0 \times 10^{-5} M \quad [Br^-]_0 = 1.0 \times 10^{-4} M$$

---

The ion product for AgBr is

$$Q = [Ag^+]_0[Br^-]_0$$

To predict whether a precipitate of AgBr will form, calculate the ion product Q, and compare it to the $K_{sp}$ value.

$$Q = [Ag^+]_0[Br^-]_0 = (1.0 \times 10^{-5})(1.0 \times 10^{-4}) = 1.0 \times 10^{-9}$$

$$K_{sp} = 7.7 \times 10^{-13}$$

Therefore, $Q > K_{sp}$, and so the mixture corresponds to a supersaturated solution of AgBr, which means that some AgBr(s) will precipitate, that is, the equilibrium will shift to the left.

$$AgBr(s) \rightleftharpoons Ag^+(aq) + Br^-(aq)$$

until a new equilibrium is reached at which point $Q = K_{sp}$.

---

***Criteria for Separation of Ions by Precipitation.*** Ions in a solution can be separated from each other on the basis of the different solubilities of their salts. For example, when slowly adding a solution containing $OH^-$ ion to a solution that contains both $Ca^{2+}$ ions and $Mg^{2+}$ ions, we find that $Mg(OH)_2$ precipitates before $Ca(OH)_2$. $Mg(OH)_2$ with a $K_{sp}$ of $1.2 \times 10^{-11}$ precipitates first, because it is less soluble than $Ca(OH)_2$, whose $K_{sp}$ is $8.0 \times 10^{-6}$.

The concentrations of the ions in solution also help determine which ions will precipitate first. Even though $Mg(OH)_2$ is less soluble than $Ca(OH)_2$, $Ca(OH)_2$ would precipitate first when $OH^-$ ion is added, if the concentration of $Ca^{2+}$ ion was much greater than the concentration of $Mg^{2+}$ ion. It is best to determine the concentration of $OH^-$ ion needed to cause a precipitate of both compounds. Since $Ca^{2+}$ and $Mg^{2+}$ ion concentrations will be given, then the concentration of $OH^-$ that must be exceeded to initiate the precipitation of $Ca(OH)_2$ and $Mg(OH)_2$ are, respectively:

| Examples |
| 16.9 – 16.11 |
| |
| Exercises |
| 15 – 18 |

$$[OH^-] = \sqrt{\frac{K_{sp}}{[Ca^{2+}]}} \quad \text{and} \quad [OH^-] = \sqrt{\frac{K_{sp}}{[Mg^{2+}]}}$$

Examples 16.10 and 16.11 illustrate the separation of $Cl^-$ and $CrO_4^{2-}$ (chromate) ions.

## Common Ion Effect and pH (Sections 16.8 – 16.9)

***The Common Ion Effect.*** The effect of adding an ion common to one already in equilibrium in a solubility reaction is to lower the solubility of the salt. In the case of AgBr solubility:

$$AgBr(s) \rightleftharpoons Ag^+(aq) + Br^-(aq)$$

The addition of either $Ag^+$ or $Br^-$ ions will shift the equilibrium to the left, in accord with Le Chatelier's principle, thus decreasing the amount of AgBr dissolved. $Ag^+$ ions can be added by pouring in a solution of $AgNO_3$. Recall that $AgNO_3$ is very soluble, and is a strong electrolyte.

$$AgNO_3(s) \rightarrow Ag^+(aq) + NO_3^-(aq)$$

The nitrate ion will not interfere with AgBr solubility because it is not a common ion.
   Additional $Br^-$ ions could be supplied by adding NaBr, for instance. Sodium bromide is very soluble and is a strong electrolyte.

$$NaBr(s) \rightarrow Na^+(aq) + Br^-(aq)$$

Example
16.12

Exercises
19, 21

The sodium ion will not affect the solubility of AgBr because it is not a common ion.
   Here, we see that the addition of an ion that is common to one already in the solubility equilibrium shifts the equilibrium to the left, which *decreases* the solubility. This is equivalent to the case of a weak acid, where the presence of a common ion decreases the percent ionization. Several calculations involving the common ion effect appear at the end of this section.

***pH and Solubility.*** The pH can affect the solubility of a solute in two ways. One of these is through the common ion effect. Consider the solubility equilibrium of an insoluble hydroxide such as $Mg(OH)_2$ or $Ca(OH)_2$.

$$Ca(OH)_2(s) \rightleftharpoons Ca^{2+}(aq) + 2OH^-(aq)$$

Upon the addition of NaOH, for instance, the pH of the solution will increase. The equilibrium position will shift to the left because of the added $OH^-$ ion (a common ion); therefore, the solubility of $Ca(OH)_2$ will decrease proportionately.
   The other case in which pH can affect solubility is when a salt contains a basic anion such as $F^-$, $CH_3COO^-$, or $CN^-$. Any basic anion will react with $H^+$ ions present in the solution and, thereby affect the solubility of the salt. Take, for instance, the effect of adding a strong acid on the solubility of silver acetate. To explain this, we need two reaction steps. In the first one, silver acetate dissolves, and in the second, the acetate ion combines with $H^+(aq)$ ions present in the system:

$$CH_3COOAg(s) \rightleftharpoons CH_3COO^-(aq) + Ag^+(aq)$$

$$H^+(aq) + CH_3COO^-(aq) \rightarrow CH_3COOH(aq)$$

Examples
16.13–16.14

Exercises
22 - 24

In the presence of $H^+$ ions, the concentration of acetate ion is lowered by the occurrence of the second reaction. The second reaction occurs essentially completely because it forms a weak acid. According to Le Chatelier's principle, the solubility equilibrium will shift to the right, causing more silver acetate to dissolve.
   The solubilities of salts containing anions that do *not* hydrolyze, such as $Cl^-$, $Br^-$, $I^-$, and $NO_3^-$, are not affected by pH. These anions are conjugate bases of strong acids.

## Complex Ions and Solubility (Section 16.10)

*The Formation Constant.* When $Ag^+$ ion reacts with two $CN^-$ ions:

$$Ag^+(aq) + 2\,CN^-(aq) \rightleftharpoons Ag(CN)_2^-(aq)$$

it forms a complex ion $Ag(CN)_2^-$. An ion made up of the metal ion bonded to one or more molecules or ions is called a complex ion. Some other examples are $Ag(NH_3)_2^+$, $Ni(H_2O)_6^{2+}$, and $Cu(CN)_4^{2-}$.

Complex ions are extremely stable and so have an important effect on certain chemical species in solution. A measure of the tendency of a metal ion to form a certain complex ion is given by its formation constant, $K_f$.

The **formation constant** is the equilibrium constant for the reaction that forms the complex ion. For example,

$$Ag^+ + 2\,CN^- \rightleftharpoons Ag(CN)_2^-$$

$$K_f = \frac{[Ag(CN)_2^-]}{[Ag^+][CN^-]^2} = 1.0 \times 10^{21}$$

Table 16.4 in the text lists $K_f$ values for selected complex ions. The more stable the complex ion, the greater the extent of reaction, and the greater the value of $K_f$.

The formation of a complex ion has a strong effect on the solubility of a metal salt. For example, AgI, which has a very low solubility in water, will dissolve in an aqueous solution of NaCN. The stepwise process is:

$$
\begin{array}{ll}
AgI(s) \rightleftharpoons Ag^+(aq) + I^-(aq) & K_{sp} \\
Ag^+(aq) + 2CN^-(aq) \rightleftharpoons Ag(CN)_2^-(aq) & K_f \\
\hline
\text{overall}\quad AgI(s) + 2CN^-(aq) \rightleftharpoons Ag(CN)_2^-(aq) + I^-(aq) & K
\end{array}
$$

The formation of the complex ion in the second step causes a decrease in $[Ag^+]$. Therefore, the first equilibrium shifts to the right according to Le Chatelier's principle. This is why AgI is more soluble in $CN^-$ solution than in pure water.

The overall reaction is the sum of the two steps. The equilibrium constant for the overall reaction is the product of the equilibrium constants of the two steps:

$$K = K_{sp} \times K_f$$

Tables 16.1 (or 16.2 of the text) and 16.4 of the text give, respectively:

$$K_{sp} = 8.3 \times 10^{-17}$$
$$K_f = 1.0 \times 10^{21}$$

Therefore,

$$K = 8.3 \times 10^4$$

The large value of K shows that AgI is very soluble in NaCN solution.

| Examples 16.15–16.16 |
| --- |
| Exercises 25 - 28 |

## GLOSSARY LIST

| | | |
|---|---|---|
| common ion effect | ion product | complex ion |
| buffer | | formation constant |
| | molar solubility | |
| end point | solubility | qualitative analysis |
| titration curve | solubility product | |

## EQUATIONS

| Algebraic Equation | English Translation |
|---|---|
| $pH = pK_a + \log \dfrac{[\text{conjugate base}]}{[\text{acid}]}$ | Henderson-Hasselbalch Equation |

## WORKED EXAMPLES

### EXAMPLE 16.1  pH of a Buffer Solution

Calculate the pH of a buffer solution that is 0.25 *M* HF and 0.50 *M* NaF.

• **Solution**

The pH of a HF/F$^-$ buffer is given in terms of the Henderson-Hasselbalch equation.

$$HF(aq) \rightleftharpoons H^+(aq) + F^-(aq)$$

$$pH = pK_a + \log \frac{[F^-]}{[HF]}$$

where, $pK_a = -\log K_a$

$$pK_a = -\log (7.1 \times 10^{-4}) = 3.15$$

The equilibrium concentrations of F$^-$ ion and HF are determined as follows: the initial concentration of HF is 0.25 *M,* but at equilibrium, [HF] = 0.25 − *x*. We can assume that *x* is less than 5 percent of [HF]$_0$ because HF is a weak acid, and because the addition of F$^-$ ion, a common ion, represses the dissociation of HF. We assume that [HF] = [HF]$_0$ = 0.25 *M*.

The F$^-$ ion is contributed by two sources, sodium fluoride which is a strong electrolyte, and HF, a weak acid.

$$NaF(aq) \rightarrow Na^+(aq) + F^-(aq)$$

Therefore, [F$^-$] = 0.50 *M* + *x*. Since 0.50 *M* >> *x*, then [F$^-$] = 0.50 *M*.

Substituting into the Henderson-Hasselbalch equation gives:

$$pH = 3.15 + \log \frac{0.50}{0.25} = 3.15 + \log 2.0 = 3.15 + 0.30$$

$$= 3.45$$

---

**• Comment**

By comparison, the percent ionization of HF in a 0.25 *M* HF solution is 5.3%. The common ion effect lowers the percent ionization to the value found in the buffer solution. Neglecting *x* was a valid assumption.

**Work EXERCISES & PROBLEMS: 1, 2**

---

**EXAMPLE 16.2  Conjugate Base to Weak Acid Ratio in a Buffer**

What ratio of $[F^-]$ to $[HF]$ would you use to make a buffer of pH = 2.85?

**• Solution**

Start with the Henderson-Hasselbalch equation.

$$pH = pK_a + \log \frac{[\text{conjugate base}]}{[\text{acid}]}$$

Substitute in the desired pH and the $pK_a$ value of HF.

$$2.85 = 3.15 + \log \frac{[F^-]}{[HF]}$$

$$\log \frac{[F^-]}{[HF]} = -0.30$$

$$\frac{[F^-]}{[HF]} = 10^{-0.30} = 0.50$$

The ratio of $[F^-]$ to $[HF]$ should be 0.50 to 1. Any amounts of $F^-$ and HF that give a ratio of 0.5 will produce the desired pH.

**Work EXERCISES & PROBLEMS: 3, 4**

---

**EXAMPLE 16.3  Adding Strong Acid to a Buffer**

Suppose 3.0 mL of 2.0 *M* HCl is added to exactly 100 mL of the buffer described in Example 16.1, what is the new pH of the buffer after the HCl is neutralized?

**• Solution**

The HCl is neutralized in this buffer by the following reaction, which goes to completion.

$$H^+ + F^- \rightarrow HF$$

This consumes some $F^-$ ions and forms more HF. The number of moles of $H^+$ added as HCl is

$$M \cdot V = 2.0 \text{ mol/L} \times 0.0030 \text{ L} = 0.0060 \text{ mol } H^+$$

The number of moles of HF originally present in 100 mL of buffer was

$M \cdot V = 0.25 \text{ mol/L} \times 0.100 \text{ L} = 0.025 \text{ mol HF}$

The number of moles of $F^-$ originally present in 100 mL of buffer was

$M \cdot V = 0.50 \text{ mol/L} \times 0.100 \text{ L} = 0.050 \text{ mol } F^-$

After the added $H^+$ is neutralized by $H^+ + F^- \rightarrow HF$, 0.0060 $H^+$ reacts with 0.0060 mol of $F^-$ ion to form 0.0060 mol HF.

The number of moles of HF is 0.025 + 0.006 = 0.031 mol.
The number of moles of $F^-$ is 0.050 – 0.006 = 0.044 mol.

This is illustrated in the following ICF table (a means of working a limiting reactant problems introduced in chapter 3).

|   | $H^+$ | + | $F^-$ | $\rightarrow$ | HF |
|---|---|---|---|---|---|
| I | 0.0060 mol | | 0.050  mol | | 0.025  mol |
| C | - 0.0060 mol | | - 0.0060 mol | | + 0.0060 mol |
| F | All is neutralized | | 0.044  mol | | 0.031  mol |

The new pH can be found as usual using the Henderson-Hasselbalch equation, since the solution is still a buffer.

$$pH = pK_a + \log \frac{[F^-]}{[HF]}$$

$$= 3.15 + \log \frac{(0.044 \text{ mol}/0.103 \text{ L})}{(0.031 \text{ mol}/0.103 \text{ L})} = 3.15 + \log 1.42 = 3.15 + 0.15 = 3.30$$

• **Comment**

The pH has dropped only 0.15 unit from 3.45 (Example 16.1) due to the addition of 3.0 mL of 2.0 *M* HCl; thus, the use of the term buffer solution.

**Work EXERCISES & PROBLEMS: 5**

---

**EXAMPLE 16.4  Net Ionic Equations**

Write the net ionic equations for the neutralization reactions that occur during the following titrations. Predict whether the pH at the equivalence point will be above, below, or equal to 7.0.
a.   Titration of HI with $NH_3$.
b.   Titration of HI with NaOH.
c.   Titration of NaOH with HF.

**• Solution**

a.  This is a titration of a strong acid with a weak base. HI is 100 percent ionized, and $NH_3$ is a weak base. The net ionic equation for neutralization is:

$$H^+ + NH_3 \rightarrow NH_4^+$$

The result of the neutralization reaction is formation of the conjugate acid $NH_4^+$ of the weak base $NH_3$. $NH_4^+$ is an acid. The pH at the equivalence point will be below 7.0.

b.  This is a titration of a strong acid and a strong base. The net ionic equation is

$$H^+ + OH^- \rightarrow H_2O$$

The resulting solution is neutral at the equivalence point.  pH = 7.0

c.  This titration involves a weak acid and a strong base.

$$HF + OH^- \rightarrow H_2O + F^-$$

The reaction produces the conjugate base ($F^-$ ion) of the weak acid.  The solution will be basic at the equivalence point, pH > 7.0.

---

**EXAMPLE 16.5 Calculation of pH of a Titration**

Calculate the pH when 10.0 mL of 0.15 *M* $HNO_3$ is added to 50.0 mL of 0.10 *M* NaOH.

**•  Solution**

The net ionic reaction for the reaction between $HNO_3$ and NaOH is: $H^+ + OH^- \rightarrow H_2O$

This is a limiting reactant problem and can be solved using the following table, plugging moles into the table. "I" = initial moles, "C" = change, "F" = final moles:

| | $H^+$ | + | $OH^-$ | $\rightarrow$ | $H_2O$ |
|---|---|---|---|---|---|
| I | 0.0015 mol | | 0.0050 mol | | |
| C | - 0.0015 mol | | - 0.0015 mol | | |
| F | 0 | | 0.0035 mol | | |

The pH can be determined from the concentration of $OH^-$.

$[OH^-]$ = 0.0035 mol÷(0.0100 L + 0.0500 L) = 0.058 mol/L

pOH = -log(0.058 *M*) = 1.23

pH = 14.00 – 1.23 = 12.77

**EXAMPLE 16.6  Choosing an Indicator**

Choose an indicator for the titration of 50 mL of a 0.10 $M$ HI solution with 0.10 $M$ $NH_3$.

$$HI(aq) + NH_3(aq) \rightarrow NH_4I(aq)$$

• **Solution**

First, we need to know the pH at the equivalence point. This is the titration of a strong acid with a weak base. The net ionic equation is:

$$H^+(aq) \quad + \quad NH_3(aq) \quad \rightarrow \quad NH_4^+ (aq)$$
$$5.0 \times 10^{-3} \text{ mol} + 5.0 \times 10^{-3} \text{ mol} \qquad 5.0 \times 10^{-3} \text{ mol}$$

The product, $NH_4^+$ ion, is a weak acid and so we expect a slightly acidic solution at the equivalence point.

The concentration of $NH_4^+$ ion formed at the equivalence point is $5.0 \times 10^{-3}$ mol/0.100 L = $5.0 \times 10^{-2}$ $M$. The $[H^+]$ at the equivalence point is just the pH of a 0.050 $M$ $NH_4^+$ solution. Consider the ionization of the weak acid, $NH_4^+$:

$$NH_4^+ (aq) \rightleftharpoons NH_3(aq) + H^+(aq) \qquad K_a = 5.6 \times 10^{-10}$$

| Concentration | $NH_4^+$ | $\rightleftharpoons$ | $H^+$ | + | $NH_3$ |
|---|---|---|---|---|---|
| Initial ($M$) | 0.050 | | 0 | | 0 |
| Change ($M$) | $-x$ | | $+x$ | | $+x$ |
| Equilibrium ($M$) | $(0.050 - x)$ | | $x$ | | $x$ |

$$K_a = \frac{[NH_3][H^+]}{[NH_4^+]} = \frac{x^2}{(0.050 - x)} = \frac{x^2}{0.050} = 5.6 \times 10^{-10}$$

Solving for $x$:

$$x = [NH_3] = [H^+] = 5.3 \times 10^{-6} \, M$$

and the pH is 5.28 at the equivalence point. According to Table 16.1 of the text, chlorophenol blue and methyl red are indicators that will change color in the vicinity of pH 5.28.

**Work EXERCISES & PROBLEMS: 6 – 10**

### EXAMPLE 16.7  Calculating a $K_{sp}$ Value

The solubility of magnesium fluoride in water is $7.3 \times 10^{-3}$ g per 100 mL of solution. What is the solubility product constant for $MgF_2$?

**• Solution**

Write the solubility equilibrium and the $K_{sp}$ expression:

$$MgF_2(s) \rightleftharpoons Mg^{2+}(aq) + 2F^-(aq)$$

$$K_{sp} = [Mg^{2+}] [F^-]^2$$

Calculate $s$, the molar solubility of $MgF_2$, from the given solubility. The molar mass of $MgF_2$ is 62.31 g/mol.

$$s = \frac{7.3 \times 10^{-3}\,g}{100\ mL} \times \frac{1\ mL}{10^{-3}\,L} \times \frac{1\ mol}{62.31\ g} = 1.17 \times 10^{-3}\ mol/L$$

The $K_{sp}$ value can be calculated by substituting the ion concentrations in a saturated solution into the solubility product expression. When $s$ mol of $MgF_2$ dissolves in 1 L of solution, the $Mg^{2+}$ concentration is equal to $s$, and the $F^-$ ion concentration is $2s$.

$$[Mg^{2+}] = s = 1.17 \times 10^{-3}\ M$$

$$[F^-] = 2s = 2(1.17 \times 10^{-3}\ M) = 2.34 \times 10^{-3}\ M$$

Now substitute the ion concentrations into the $K_{sp}$ expression.

$$K_{sp} = (1.17 \times 10^{-3}) (2.34 \times 10^{-3})^2$$
$$K_{sp} = 6.4 \times 10^{-9}$$

**Work EXERCISES & PROBLEMS: 11 – 13**

---

### EXAMPLE 16.8  Calculating a Molar Solubility

The $K_{sp}$ value for $Ag_2CrO_4$ is $1.1 \times 10^{-12}$. Calculate the molar solubility of silver chromate.

**• Solution**

Write the solubility equilibrium,

$$Ag_2CrO_4(s) \rightleftharpoons 2Ag^+(aq) + CrO_4^{2-} (aq)$$

and the $K_{sp}$ expression.

$$K_{sp} = [Ag^+]^2[CrO_4^{2-}]$$

Let $s$ = molar solubility of $Ag_2CrO_4$. Whenever $s$ moles of $Ag_2CrO_4$ dissolve, $2s$ moles of $Ag^+$ and $s$ moles of $CrO_4^{2-}$ are produced.

$$[Ag^+] = 2s \qquad [CrO_4^{2-}] = s$$

Summarize the changes in concentrations as follows.

| Concentration | $Ag_2CrO_4(s) \rightleftharpoons 2Ag^+(aq) + CrO_4^{2-}(aq)$ | |
|---|---|---|
| Initial ($M$) | 0 | 0 |
| Change ($M$) | 2s | s |
| Equilibrium ($M$) | 2s | s |

Substitute these values into the $K_{sp}$ expression.

$$K_{sp} = [2s]^2[s] = 1.1 \times 10^{-12}$$

and solve for $s$.

$$4s^2(s) = 4s^3 = 1.1 \times 10^{-12}$$

$$s = \sqrt[3]{\frac{1.1 \times 10^{-12}}{4}} = \sqrt[3]{2.75 \times 10^{-13}}$$

$$s = 6.5 \times 10^{-5} M$$

**Work EXERCISES & PROBLEMS: 14**

---

**EXAMPLE 16.9  Predicting Formation of a Precipitate**

a.  Predict whether or not a precipitate of $PbI_2$ will form when 200 mL of 0.015 $M$ $Pb(NO_3)_2$ and 300 mL of 0.050 $M$ NaI are mixed together. Given: $K_{sp}(PbI_2) = 1.4 \times 10^{-8}$.

b.  If the answer is yes, what concentrations of $Pb^{2+}(aq)$ and $I^-(aq)$ will exist when equilibrium is reestablished?

**• Solution**

a.  Recall that $Pb(NO_3)_2$ and NaI are both strong electrolytes. When the solutions are mixed, will the following reaction occur?

$$Pb^{2+}(aq) + 2NO_3^-(aq) + 2Na^+(aq) + 2I^-(aq) \rightarrow PbI_2(s) + 2Na^+(aq) + 2NO_3^-(aq)$$

To work the problem, it is better to consider the reaction in reverse, i.e., the one that matches the solubility product constant.

$$PbI_2(s) \rightleftharpoons Pb^{2+}(aq) + 2Cl^-(aq)$$

A precipitate will form (shift left) only if the ion product Q exceeds the $K_{sp}$ value; $[Pb^{2+}]_0[I^-]_0^2 > K_{sp}$. When the two solutions are mixed, 500 mL of new solution is formed. Immediately after mixing, the initial ion concentrations will be:

$$[Pb^{2+}]_0 = 0.015 \, M \times \frac{200 \text{ mL}}{500 \text{ mL}} = 6.0 \times 10^{-3} \, M$$

$$[I^-]_0 = 0.050 \, M \times \frac{300 \text{ mL}}{500 \text{ mL}} = 3.0 \times 10^{-2} \, M$$

Substitution gives the value of the ion product.

$$Q = [Pb^{2+}]_0[I^-]_0^2 = (6.0 \times 10^{-3})(3.0 \times 10^{-2})^2 = 5.4 \times 10^{-6}$$

Since $Q > K_{sp}$, then a precipitate of $PbI_2$ will form.

b.  The concentrations of $Pb^{2+}$ and $I^-$ remaining in solution depend on the number of moles initially and the number of moles that precipitate. The initial number of moles of $Pb^{2+}$ ion was $0.015\ M \times 0.200\ L = 0.0030$ mol, and that of $I^-$ ion, $0.050\ M \times 0.300\ L = 0.015$ mol. In order to determine how many moles of $Pb^{2+}$ and $I^-$ precipitated from solution, consider the net reaction:

$$Pb^{2+}(aq) + 2I^-(aq) \rightarrow PbI_2(s)$$

Look to see if there is a limiting reactant. Then, determine the amount of the excess reactant that remains after complete reaction of the limiting reactant. $Pb^{2+}$ ion being in the smaller amount is the limiting reactant, and will react essentially completely. The 0.0030 mol of $Pb^{2+}$ will react with 0.0060 mol of $I^-$, leaving 0.009 mol of excess $I^-$ (0.015 – 0.0060) dissolved in solution.

The $I^-$ concentration is $[I^-] = 0.009$ mol/0.50 L = 0.018 M.
The $Pb^{2+}$ concentration is controlled by the $PbI_2$ solubility equilibrium. In other words, some $PbI_2$ dissolves by the reverse reaction.

$$PbI_2(s) \rightleftharpoons Pb^{2+} + 2I^-$$

$$K_{sp} = [Pb^{2+}][I^-]^2 = 1.4 \times 10^{-8}$$

The $Pb^{2+}$ ion concentration in equilibrium with 0.018 M $I^-$ ion is:

$$[Pb^{2+}] = \frac{K_{sp}}{[I^-]^2} = \frac{1.4 \times 10^{-8}}{(0.018)^2} = 4.3 \times 10^{-5}\ M$$

• **Comment**

The percentage of $Pb^{2+}$ ion remaining unprecipitated is:

$$\frac{4.3 \times 10^{-5}\ M\ (0.500\ L)}{0.0030\ mol} \times 100\% = 0.72\%$$

This confirms our assumption that essentially all the $Pb^{2+}$ ion precipitated.

**Work EXERCISES & PROBLEMS: 15, 16**

---

**EXAMPLE 16.10  Selective Precipitation of an Ion**

A solution contains 0.10 $M$ $Cl^-$ and 0.010 $M$ $CrO_4^{2-}$. If a $AgNO_3$ solution is added dropwise, which will precipitate first, $AgCl$ or $Ag_2CrO_4$?

**• Solution**

The solubility product is a number that the product of the ion concentrations can never exceed at equilibrium. First write the two equilibria of interest and their solubility constants.

$$AgCl(s) \rightleftharpoons Ag^+(aq) + Cl^-(aq) \qquad K_{sp} = [Ag^+][Cl^-] = 1.6 \times 10^{-10}$$

$$Ag_2CrO_4(s) \rightleftharpoons 2Ag^+ + CrO_4^{2-} \qquad K_{sp} = [Ag^+]^2[CrO_4^{2-}] = 1.1 \times 10^{-12}$$

The highest $Ag^+$ ion concentration possible in a solution of 0.10 $M$ $Cl^-$ is:

$$[Ag^+] = \frac{1.6 \times 10^{-10}}{[Cl^-]} = \frac{1.6 \times 10^{-10}}{0.10} = 1.6 \times 10^{-9} \ M$$

The highest $Ag^+$ ion concentration possible in a solution of 0.010 $M$ $CrO_4^{2-}$ is:

$$[Ag^+] = \sqrt{\frac{K_{sp}}{[CrO_4^{2-}]}} = \sqrt{\frac{1.1 \times 10^{-12}}{0.010}} = 1.0 \times 10^{-5} \ M$$

Silver chloride will precipitate before silver chromate because of the lower $Ag^+$ ion concentration needed to produce a saturated AgCl solution.

---

## EXAMPLE 16.11 Completeness of Precipitation

In the above example AgCl precipitates prior to $Ag_2CrO_4$ as silver ions are added dropwise to the solution. What percentage of the $Cl^-$ ion in solution will have precipitated when $Ag_2CrO_4$ just begins to precipitate?

**• Solution**

As we found above, $Ag_2CrO_4$ begins to precipitate when $[Ag^+] = 1.0 \times 10^{-5}$ $M$.

The $Cl^-$ ion concentration in equilibrium with $1.0 \times 10^{-5}$ $M$ $Ag^+$ ion is:

$$[Cl^-] = \frac{1.6 \times 10^{-10}}{[Ag^+]} = \frac{1.6 \times 10^{-10}}{1.0 \times 10^{-5}} = 1.6 \times 10^{-5} \ M$$

The percent of $Cl^-$ remaining in solution *unprecipitated* is:

$$\%Cl^- = \frac{1.6 \times 10^{-5} \ M}{0.10 \ M} \times 100\%$$

$$\%Cl^- = 0.016\%$$

Therefore, the percentage of $Cl^-$ ion precipitated is $100\% - 0.016\% = 99.98\% \approx 100\%$.

**Work EXERCISES & PROBLEMS: 17, 18**

**EXAMPLE 16.12  Common Ion Effect**

What is the solubility of $PbCl_2$ in 0.50 *M* NaCl solution?

**• Solution**

Write the solubility equilibrium reaction for $PbCl_2$, and write the $K_{sp}$ expression. The $K_{sp}$ value is in Table 16.1.

$$PbCl_2(s) \rightleftharpoons Pb^{2+}(aq) + 2Cl^-(aq)$$

$$K_{sp} = [Pb^{2+}][Cl^-]^2 = 2.4 \times 10^{-4}$$

All of the $[Pb^{2+}]$ ion comes from the dissolution reaction. Let *s* = molar solubility of $PbCl_2$. The only source of $Pb^{2+}$ is $PbCl_2$. Therefore,

$$[Pb^{2+}] = s$$

Chlorine ions are contributed by $PbCl_2$ and by NaCl (a strong electrolyte):

$$NaCl(aq) \rightarrow Na^+(aq) + Cl^-(aq)$$

Therefore, $[Cl^-] = 0.50$ *M* + 2*s*

A table here will help summarize the changes in concentrations.

| Concentration | $PbCl_2(s) \rightleftharpoons$ | $Pb^{2+}(aq)$ + | $2Cl^-(aq)$ |
|---|---|---|---|
| Initial (*M*) | | 0 | 0.50 |
| Change (*M*) | | *s* | 2*s* |
| Equilibrium (*M*) | | *s* | (0.50 + 2*s*) |

Substituting into $K_{sp}$, we get:

$$[Pb^{2+}][Cl^-]^2 = K_{sp}$$

$$(s)(0.50 + 2s)^2 = 2.4 \times 10^{-4}$$

If we assume 0.50 >> 2*s*, then solving the problem is greatly simplified:

$$(s)(0.50)^2 = 2.4 \times 10^{-4}$$

$$s = 9.6 \times 10^{-4}\ M$$

**• Comment**

Note that the assumption 0.50 >> 2*s* was valid; 0.50 >> $1.9 \times 10^{-3}$.

**Work EXERCISES & PROBLEMS: 19 – 21**

**EXAMPLE 16.13 Solubility of a Metal Hydroxide**

What is the solubility of $Pb(OH)_2$ in a buffer solution of pH 8.0 and in another of pH 9.0?
Given: $K_{sp} = 1.2 \times 10^{-15}$

**• Solution**

Write the solubility equilibrium and the $K_{sp}$ expression for $Pb(OH)_2$.

$$Pb(OH)_2(s) \rightleftharpoons Pb^{2+}(aq) + 2OH^-(aq)$$

$$K_{sp} = [Pb^{2+}][OH^-]^2 = 1.2 \times 10^{-15}$$

Let $s$ equal the molar solubility of $Pb(OH)_2$. The concentration of $Pb^{2+}$ will be equal to $s$. The concentration of $OH^-$ ion will not be $2[Pb^{2+}]$ because in a buffer solution, the $H^+$ ion and $OH^-$ ion concentrations are maintained constant by the buffer. At pH 8.0, the pOH is 6.0, and $[OH^-] = 1.0 \times 10^{-6}$ $M$. In the buffer solution, the $[OH^-]$ ion concentration is *maintained constant* at $1.0 \times 10^{-6}$ $M$. The $OH^-$ from dissolution of $Pb(OH)_2$ is neutralized by the buffer. Therefore, we can substitute as follows to obtain the solubility at pH 8.

$$[Pb^{2+}] = s$$

$$[Pb^{2+}] = \frac{K_{sp}}{[OH^-]^2}$$

$$s = \frac{1.2 \times 10^{-15}}{(1.0 \times 10^{-6})^2} = 1.2 \times 10^{-3} \ M$$

And at pH = 9, where $[OH^-] = 1.0 \times 10^{-5}$ $M$,

$$s = \frac{1.2 \times 10^{-15}}{(1.0 \times 10^{-5})^2} = 1.2 \times 10^{-5} \ M$$

**• Comment**

The lower solubility of $Pb(OH)_2$ at pH 9.0, than at pH 8.0, is due to the common ion effect.

---

**EXAMPLE 16.14 Effect of pH on Solubility**

Which of the following salts will be more soluble at acidic pH than in pure water?
a. AgBr    b. $Ba(OH)_2$    c. $MgCO_3$

**• Solution**

First, we need to identify the ions present in solutions of these compounds. Then we identify those with acid-base properties.

a.   $AgBr(s) \rightleftharpoons Ag^+ + Br^-$.  Neither $Ag^+$ nor $Br^-$ have acid-base properties and so the solubility of AgBr is not affected by pH.

b.  $Ba(OH)_2 \rightleftharpoons Ba^{2+} + 2\,OH^-$

Comparing equilibria in water and in acidic solution, as $[H^+]$ increases, $[OH^-]$ decreases, and the solubility equilibrium will shift to the right, causing the solubility of $Ba(OH)_2$ to increase.

c.  Carbonate ion is a base, and in acid solution it combines with $H^+(aq)$.

$$MgCO_3 \rightleftharpoons Mg^{2+} + CO_3^{2-}$$

$$H^+ + CO_3^{2-} \rightarrow HCO_3^-$$

With the increase in $[H^+]$ in acid solution, $[CO_3^{2-}]$ decreases, and the solubility equilibrium shifts to the right, causing more $MgCO_3$ to dissolve.

**Work EXERCISES & PROBLEMS: 22 −24**

---

**EXAMPLE 16.15  Concentration of Uncomplexed Metal Ion**

Calculate the concentration of free $Ag^+$ ions in a solution formed by adding 0.20 mol of $AgNO_3$ to 1 L of 1.0 $M$ NaCN.

• **Solution**

In this solution, $Ag^+$ ions will complex with $CN^-$ ions, and the concentration of $Ag^+$ will be determined by the equation

$$Ag^+ + 2CN^- \rightleftharpoons Ag(CN)_2^- \qquad K_f = 1.0 \times 10^{21}$$

Since $K_f$ is so large, we expect the $Ag^+$ to react essentially quantitatively to form 0.20 mol of $Ag(CN)_2^-$. We can find the concentration of free $Ag^+$ at equilibrium using the equilibrium constant expression:

$$K_f = \frac{[Ag(CN)_2^-]}{[Ag^+][CN^-]^2}$$

Rearranging gives:

$$[Ag^+] = \frac{[Ag(CN)_2^-]}{K_f[CN^-]^2}$$

Substitute all known equilibrium concentrations into the above equation.
At equilibrium;  $[Ag(CN)_2^-] = 0.20\ M$, and

$$[CN^-] = [CN^-]_0 - 2[Ag(CN)_2^-]$$

$$[CN^-] = 1.0\ M - 0.40\ M = 0.6\ M$$

Therefore,

$$[Ag^+] = \frac{(0.20)}{(1.0 \times 10^{21})(0.6)^2}$$

$$[Ag^+] = 5 \times 10^{-22} \ M$$

• **Comment**

This concentration corresponds to only three $Ag^+$ ions per 100 mL!

**Work EXERCISES & PROBLEMS: 25, 26**

---

**EXAMPLE 16.16  Effect of Complex Formation on Solubility**

Calculate the molar solubility of silver bromide in 6.0 $M$ $NH_3$.

• **Solution**

The solubility of AgBr is determined by two equilibria. The solubility and complex ion equilibria are

$$AgBr(s) \rightleftharpoons Ag^+ + Br^- \qquad\qquad K_{sp} = 7.7 \times 10^{-13}$$
$$Ag^+ + 2NH_3 \rightleftharpoons Ag(NH_3)_2^+ \qquad\qquad K_f = 1.5 \times 10^7$$

---

Overall:  $AgBr(s) + 2NH_3 \rightleftharpoons Ag(NH_3)_2^+ + Br^- \qquad K = K_{sp} \times K_f$

The equilibrium constant of this net reaction controls the solubility of AgBr.

$$K = K_{sp} \times K_f = \frac{[Ag(NH_3)_2^+][Br^-]}{[NH_3]^2} = 1.2 \times 10^{-5}$$

Let $s$ equal the solubility of AgBr in 6.0 $M$ $NH_3$.

| Concentration | $AgBr(s) + 2NH_3 \rightleftharpoons$ | $Ag(NH_3)_2^+$ | + $Br^-$ |
|---|---|---|---|
| Initial ($M$) | 6.0 | 0 | 0 |
| Change ($M$) | $-2s$ | $+s$ | $+s$ |
| Equilibrium ($M$) | $(6.0 - 2s)$ | $s$ | $s$ |

Substitute into the K expression for the overall reaction:

$$K = \frac{(s)(s)}{(6.0 - 2s)^2} = 1.2 \times 10^{-5}$$

The left-hand side is a perfect square. Taking the square root of both sides gives:

$$\sqrt{\frac{s^2}{(6.0 - 2s)^2}} = \sqrt{1.2 \times 10^{-5}}$$

$$\frac{s}{6.0 - 2s} = 3.5 \times 10^{-3}$$

$$s = 2.1 \times 10^{-2} \ M$$

**• Comment**

The solubility of AgBr in pure water is $8.8 \times 10^{-7}$ M. The enhanced solubility in this example is due to the formation of the complex ion $Ag(NH_3)_2^+$.

**Work EXERCISES & PROBLEMS 27, 28**
**Work the rest of the EXERCISES & PROBLEMS**

## EXERCISES & PROBLEMS

1.  Calculate the pH of the following solutions:
    a. 0.10 $M$ $CH_3COOH$
    b. 0.020 $M$ $CH_3COONa$ and 0.10 $M$ $CH_3COOH$

2.  a. What is the pH of a buffer solution prepared by dissolving 0.15 mol of benzoic acid ($C_6H_5COOH$) and 0.45 mol of sodium benzoate ($C_6H_5COONa$) in enough water to make 400 mL of solution?
    b. Does the volume of solution affect your answer?

3.  a. What is the change in pH when 40.0 mg of NaOH are added to 100 mL of a buffer solution consisting of 0.165 $M$ $NH_3$ and 0.120 $M$ $NH_4Cl$?
    b. Write the equation for the buffer equilibrium.
    c. Write the equation for the neutralization of NaOH by the buffer.

4.  What mole ratio of benzoate ion to benzoic acid would you need to prepare a buffer solution with a pH of 5.00?

5.  What is the optimum pH of a $H_3PO_4/H_2PO_4^-$ buffer?

6.  Approximately what range of pH should be expected at the equivalence points in the titration of weak acids with strong bases?

7.  The pH at the equivalence point in a titration was found to be approximately 5. Give the category of titration in terms of the acid and base strengths.

8.  25.0 mL of 0.222 $M$ HBr was titrated with 0.111 $M$ NaOH.
    a. Write the overall balanced equation for the reaction.
    b. After adding 30.0 mL of base solution, what is the pH?

9.  Calculate the pH in the following titration after the addition of 12.0 mL of 0.100 $M$ KOH to 20.0 mL of 0.200 $M$ $CH_3COOH$.

10. Determine the pH at the equivalence point in the titration of 0.100 $M$ NaOH with 0.100 $M$ $HNO_3$.

11. Write balanced equations and solubility product expressions for the solubility equilibria of the following compounds.
    a. $BaCO_3$    b. $CaF_2$    c. $Al(OH)_3$    d. $Ag_3PO_4$

12. The solubility of $PbBr_2$ is 0.392 g per 100 mL at 20 °C. What is the $K_{sp}$ value for $PbBr_2$?

13. Solid barium fluoride was dissolved in pure water until a saturated solution was formed. If the fluoride concentration was 0.0150 $M$, determine the $K_{sp}$ of $BaF_2$.

14. a. What is the solubility of $CaF_2$ in moles per liter given that $K_{sp} = 4.0 \times 10^{-11}$.
    b. Determine the $F^-$ ion concentration in a saturated solution of $CaF_2$.

15. If 1.0 mL of $1.0 \times 10^{-3}$ $M$ $Ba(NO_3)_2$ is added to 99.0 mL of $1 \times 10^{-4}$ $M$ $Na_2CO_3$, will $BaCO_3$ precipitate from this solution? $K_{sp}(BaCO_3) = 8.1 \times 10^{-9}$.

16. Will a precipitate form when 250 mL of $2.8 \times 10^{-3}$ $M$ $MgCl_2$ solution is added to 250 mL of $5.2 \times 10^{-3}$ $M$ NaF? Identify the precipitate if any.

17. Solid $Ba(NO_3)_2$ is slowly dissolved in a solution of $2.5 \times 10^{-4}$ $M$ $Na_2CO_3$. At what $Ba^{2+}$ concentration will a precipitate of $BaCO_3$ just begin to form?

18. A solution contains $1.0 \times 10^{-4}$ $M$ $Cu^+$ and $1.0 \times 10^{-3}$ $M$ $Pb^{2+}$.
    a. If NaI solution is added dropwise to this solution, which compound will precipitate first, CuI or $PbI_2$?
    b. Give the concentrations of $I^-$ necessary to begin precipitation of each salt.
    c. What percentage of the initial $Cu^+$ ion concentration will remain unprecipitated when $PbI_2$ just begins to precipitate? $K_{sp}$ (CuI) $= 5.1 \times 10^{-12}$, $K_{sp}$ ($PbI_2$) $= 1.4 \times 10^{-8}$

19. a. What is the molar solubility of $Ag_3PO_4$ in 0.20 $M$ $Na_3PO_4$? $K_{sp} = 1.8 \times 10^{-18}$
    b. Write the solubility equilibrium reaction for $Ag_3PO_4$.

20. What is the molar solubility of $Ag_2SO_4$ in 0.20 $M$ $AgNO_3$? $K_{sp}$ ($Ag_2SO_4$) $= 1.4 \times 10^{-5}$

21. Calculate the molar solubility AgCl in a solution containing 0.010 $M$ $CaCl_2$. $K_{sp}$ (AgCl) $= 1.6 \times 10^{-10}$

22. Which of the following salts will be more soluble in an acidic solution than in a basic solution?

    a. AgI    b. $BaCO_3$    c. $Ca(CH_3COO)_2$    d. $Zn(OH)_2$    e. $CaCl_2$

23. What is the solubility of $Pb(OH)_2$ in a solution with a pH of 10.00?
    Given: $K_{sp} = 1.2 \times 10^{-15}$

24. Calculate the minimum pH that will just prevent precipitation of $Mg(OH)_2$ from an aqueous solution containing 0.075 $M$ $Mg^{2+}$. $K_{sp}$ ($Mg(OH)_2$) $= 1.2 \times 10^{-11}$

25. Write equations for the formation reaction of each of the following complex ions.
    a. $Cu(NH_3)_4^{2+}$
    b. $Cu(CN)_4^{2-}$
    c. $Ag(CN)_2^-$

26. Calculate the concentration of $Cu^{2+}$ ions in a solution formed by adding 0.0500 mol of $CuSO_4$ to 0.500 L of 0.500 $M$ NaCN. $K_f(Cu(CN)_4^{2-} = 1.0 \times 10^{25}$.

27. Determine the equilibrium constant for the following reaction. $K_{sp}(AgBr) = 7.7 \times 10^{-13}$, $K_f Ag(CN)_2^- = 1.0 \times 10^{21}$.
    $$AgBr(s) + 2CN^- \rightleftharpoons Ag(CN)_2^- + Br^-$$

28. Calculate the molar solubility of silver bromide in 1.0 $M$ NH$_3$. $K_{sp}(AgBr) = 7.7 \times 10^{-13}$, $K_f (Ag(NH_3)_2^+) = 1.5 \times 10^7$.

29. Why is a solution of HBr and NaBr, with approximately equal concentrations of each, not a buffer?

30. What is the effect on the pH of the following solutions when:
    a. solid sodium acetate ($CH_3COONa$) is added to a dilute solution of acetic acid?
    b. solid NaCl is added to a dilute solution of NaOH?
    c. solid KOH is added to a dilute solution of acetic acid?

31. Describe how to choose an acid-base indicator for the titration of 50 mL of a 0.10 $M$ HCN solution with 0.10 $M$ NaOH.

32. Explain how changing the pH affects the solubility of $Zn(OH)_2$.

33. Why does an indicator change from its acid color to its base color over a range of pH values?

## PRACTICE TEST QUESTIONS
See notes on taking practice test in the Preface

1. A buffer solution is prepared by mixing 500 mL of 0.60 $M$ $CH_3COOH$ with 500 mL of 1.00 $M$ $CH_3COONa$ solution. What is the pH of this solution?

2. A buffer solution is prepared by mixing 300 mL of 0.10 $M$ $HNO_2$ with 200 mL of 0.40 $M$ $NaNO_2$.
   a. Calculate the pH of the resulting solution.
   b. What is the new pH after 2.0 mL of 2.0 $M$ HCl are added to this buffer?

3. What is the optimum pH of an HCN / $CN^-$ buffer?

4. Which one of the following equimolar mixtures is suitable for making a buffer solution with an optimum pH of about 9.2?
   a. $NaC_2H_3O_2$ and $HC_2H_3O_2$
   b. $NH_4Cl$ and $NH_3$
   c. HF and NaF
   d. $HNO_2$ and $NaNO_2$
   e. NaCl and HCl

5. A 20.0 mL portion of a solution of 0.0200 $M$ $HNO_3$ is titrated with 0.0100 $M$ KOH.
   a. What is the pH at the equivalence point?
   b. How many mL of KOH are required to reach the equivalence point?
   c. What will the pH be after only 10.0 mL of KOH are added?
   d. What will the pH be after 45.0 mL of KOH are added?

6. Consider the titration of 50.0 mL of 0.10 $M$ $CH_3COOH$ with 0.10 $M$ NaOH. What is the pH at the equivalence point?

7. Calculate the pH of a solution prepared by mixing 25.0 mL of 0.10 $M$ HCl and 25.0 mL of 0.25 $M$ $CH_3COONa$?

8. Which of the following is the *least* soluble in water?
   a. $SrF_2$      $K_{sp} = 2.8 \times 10^{-9}$
   b. $Zn(OH)_2$    $K_{sp} = 1.8 \times 10^{-14}$
   c. $PbI_2$      $K_{sp} = 1.4 \times 10^{-8}$
   d. $BaF_2$      $K_{sp} = 1.7 \times 10^{-6}$

9. At a certain temperature, the solubility of barium chromate ($BaCrO_4$) in water is $1.8 \times 10^{-5}$ mol/L. What is the $K_{sp}$ value at this temperature?

10. If the solubility of $Fe(OH)_2$ in water is $7.7 \times 10^{-6}$ mol/L at a certain temperature, what is its $K_{sp}$ value at that temperature?

394 Chemistry, Ch. 16: Acid-Base & Solubility Equilibria

16    16

16    16

16    16

16    16

16    16

11. What is the molar solubility of silver phosphate ($Ag_3PO_4$) in water? $K_{sp} = 1.8 \times 10^{-18}$

12. AgCl is most soluble in:
 a. 0.1 $M$ $AgNO_3$
 b. 0.2 $M$ NaCl
 c. 2.0 $M$ HCl
 d. pure water

13. Will a precipitate of $BaSO_4$ form when 400 mL of 0.020 $M$ $Na_2SO_4$ is added to 700 mL of 0.001 $M$ $BaCl_2$? $K_{sp} (BaSO_4) = 1.1 \times 10^{-10}$

14. Some municipal water supplies contain $Ca^{2+}$ at a concentration of $3 \times 10^{-3}$ $M$. Will a precipitate of $CaSO_4$ form when 0.10 mol of $Na_2SO_4$ is dissolved in 0.50 L of this water? $K_{sp}$ $CaSO_4 = 2.4 \times 10^{-5}$

15. Will a precipitate of $MgF_2$ form when 600 mL of solution that is $2.0 \times 10^{-4}$ $M$ in $MgCl_2$ is added to 300 mL of $1.1 \times 10^{-2}$ $M$ NaF solution? $K_{sp} (MgF_2) = 6.4 \times 10^{-9}$

16. Which of the following salts should be more soluble in acid solution than in pure water?
 a. $CaCO_3$    b. $MgCl_2$    c. $NaNO_3$
 d. LiBr        e. KI

17. A solution of NaOH is added dropwise to a solution that is 0.100 $M$ $Ca^{2+}$ and 0.010 $M$ $Mg^{2+}$. $K_{sp} (Ca(OH)_2) = 8.0 \times 10^{-6}$, $K_{sp} (Mg(OH)_2) = 1.2 \times 10^{-11}$.
 a. Which cation precipitates first?
 b. What concentration of $OH^-$ is necessary to begin precipitation?
 c. What is the concentration of the "first cation to precipitate" when the second cation begins to precipitate?
 d. What percentage of the initial $Mg^{2+}$ concentration remains in solution when the $Ca^{2+}$ begins to precipitate?

18. Which of the following is the most stable complex ion?
 a. $Ag(S_2O_3)_2^{3-}$    $K_f = 1.0 \times 10^{13}$
 b. $Ag(NH_3)_2^+$    $K_f = 1.5 \times 10^7$
 c. $Ag(CN)_2^-$    $K_f = 1.0 \times 10^{21}$

19. Calculate the concentration of free copper ion in a solution made from 10.0 mL of $1.0 \times 10^{-2}$ $M$ $Cu(NO_3)_2$ and 10.0 mL of 1.0 $M$ $NH_3$. $K_f (Cu(NH_3)_4^{2+}) = 5.0 \times 10^{13}$

16

16

16

16

16

16

16

16

16

20. What is the molar solubility of AgCl in 3.0 $M$ NH$_3$? $K_{sp}$ (AgCl) = 1.6 ×10$^{-10}$, $K_f$ (Ag(NH$_3$)$_2^+$) = 1.5 × 10$^7$.

21. What is the molar solubility of AgBr in 0.100 $M$ Na$_2$S$_2$O$_3$?  Given:

$$Ag^+ + 2S_2O_3^{2-} \rightleftharpoons Ag(S_2O_3)_2^{3-}$$
$$K_f = 1.0 \times 10^{13}$$
and $K_{sp}$ (AgBr) = 7.7 × 10$^{-13}$

22. a. Will a precipitate form when 400 mL of 0.20 $M$ K$_2$CrO$_4$ is added to 400 mL of 0.10 $M$ AgNO$_3$? $K_{sp}$ (Ag$_2$CrO$_4$) = 1.1 × 10$^{-12}$

b. If yes, what is the silver ion concentration left in solution?

23. What is the minimum concentration of aqueous NH$_3$ required to prevent AgCl(s) from precipitating from 1.0 L of solution prepared from 0.20 mol of AgNO$_3$ and 0.010 mol of NaCl? $K_{sp}$ (AgCl) = 1.6 × 10$^{-10}$, $K_f$ (Ag(NH$_3$)$_2^+$) = 1.5 × 10$^7$.

16

16

16

16

# Chapter 17. Chemistry in the Atmosphere

**Earth's Atmosphere and Its Regions (Sections 17.1 – 17.2)**
**Depletion of the Ozone Layer (Section 17.3)**
**The Greenhouse Effect and Acid Rain (Sections 17.5 – 17.6)**
**Photochemical Smog (Section 17.7)**
**Indoor Air Pollution (Section 17.8)**

## SUMMARY

### Earth's Atmosphere and Its Regions (Sections 17.1 – 17.2)

*Earth's Atmosphere.* Earth's atmosphere is unique in the solar system due to its high oxygen ($O_2$) content. Molecular oxygen makes up 21 percent of the atmosphere by volume. Table 17.1 in the text shows that $N_2$, $O_2$, and Ar make up over 99.9% of the atmosphere by volume. The other noble gas elements (besides argon) are only present in trace amounts. Associated with the high oxygen content is the formation of ozone ($O_3$), a molecule that absorbs ultraviolet light and filters it from sunlight. This radiation is known to be harmful to many of the life forms that inhabit Earth's surface.

Nitrogen is the major constituent of the atmosphere. Nitrogen molecules are held together by a triple bond and therefore are extremely stable. Biological and industrial nitrogen fixation converts the elemental form of nitrogen into nitrogen compounds. Lightning plays an important role by initiating the production of about 30 million tons of nitric acid each year. There are three steps in the mechanism of nitric acid production. Lightning provides the energy to break up the $N_2$ molecule and initiates the first step in the reaction sequence.

$$N_2(g) + O_2(g) \rightarrow 2\,NO(g)$$
$$2\,NO(g) + O_2(g) \rightarrow 2\,NO_2(g)$$
$$2\,NO_2(g) + H_2O(\ell) \rightarrow HNO_3(g) + HNO_2(g)$$

The nitrogen in nitric acid is an essential plant nutrient.

Carbon dioxide, the fourth most abundant component, makes up only 0.033% of the air. Carbon dioxide is extremely important to Earth's heat balance even though its relative concentration is low. In this chapter, we will review the effects of air pollutants on the ozone layer, on the Earth's temperature, and on other phenomena related to the atmosphere.

*Regions of the Atmosphere.* For convenience, the atmosphere is divided into four regions with respect to altitude. The lowest region is the troposphere. This region extends up to about 10 km. The temperature decreases with altitude in the troposphere from about 25 °C at sea level to –55 °C at 10 km. The troposphere contains about 80% of the mass of the atmosphere and virtually all the precipitation, clouds, and water vapor. All the weather occurs in the troposphere. Remember that the atmosphere gradually thins out as the altitude increases. The atmospheric pressure at 10 km is about 230 mmHg.

Above 10 km, we enter the stratosphere which extends upward to 50 km. In the stratosphere, the temperature increases with altitude from about –55 °C at 10 km to –10 °C at 50 km. The atmospheric pressure at 50 km is only 0.5 mmHg. For the most part, there is very little mixing of the contents of the troposphere and the stratosphere.

Above the stratosphere, we enter the mesosphere. As in the troposphere, the temperature decreases with altitude, falling to $-90$ °C at 80 km. From 80 km to 100 km, we find the ionosphere (also called the thermosphere). In this region, the temperature increases slightly with increasing altitude. The region gets its name from the ions that are produced there by high energy solar radiation. The temperatures associated with the different regions of the atmosphere are shown in Figure 17.3 of the text.

***Auroral Displays.*** Chemical processes in the ionosphere produce the phenomena called "northern lights," or aurora borealis, in the Northern hemisphere and aurora australis in the Southern hemisphere. In Chapter 7 you learned that excited atoms emit light when electrons jump from a higher to a lower energy state. Excited O atoms and $N_2^+$ ions are produced in the ionosphere when protons and electrons from solar flares are ejected from the Sun and eventually collide with molecules in the ionosphere.

$$N_2 + \text{energy} \rightarrow N_2^+{}^* + e^-$$
$$O_2 + \text{energy} \rightarrow 2\,O^*$$

The green and red colors seen in the auroral displays are emitted by O* atoms, and the blue and violet colors are emitted by $N_2^+{}^*$ ions. The symbols contain an asterisk to indicate excess energy.

During the emission process excited atoms (denoted by an asterisk) drop to the ground state and emit photons ($h\nu$).

$$O^* \rightarrow O + h\nu$$
$$N_2^+{}^* \rightarrow N_2 + h\nu$$

Exercises
17-1 – 17-3

The auroral displays occur near the Earth's poles because solar protons are attracted there by Earth's magnetic field.

## Depletion of the Ozone Layer (Section 17.3)

***The Ozone Layer.*** Ozone ($O_3$) is a gas that is present in the atmosphere in a layer or shell that extends around the entire Earth called the "ozone layer." This layer is centered in the stratosphere about 25 km above the Earth's surface. Ozone molecules have the ability to absorb some of the ultraviolet (UV) light present in sunlight. It is in this way that the ozone layer protects life on Earth from some of the harmful effects of sunlight. Ultraviolet light causes skin cancer and eye damage in humans and mutations in plants.

The term "ozone layer" is misleading because it may imply that the layer is pure ozone. Actually, the concentration of ozone is extremely low. Ozone is spread thinly at a concentration of about 10 ppm at elevations between 15 and 35 km. If all the ozone in the atmosphere was collected into a layer of pure ozone at standard temperature and pressure, that layer would only be 3 mm thick! The low concentration and unstable nature of ozone combine to make the ozone layer a fragile part of the atmosphere.

***Formation and Destruction of Ozone.*** Ozone is formed in the ozone layer by a two step mechanism initiated by the action of UV light on oxygen molecules. In the first step, UV-light dissociates molecular oxygen into oxygen atoms in a process called photodissociation. The energy of the UV-photon is enough to break the O—O bond. In the second step, the oxygen atom combines with an oxygen molecule, forming ozone.

$$O_2(g) + h\nu \rightarrow 2\ O(g)$$
$$O(g) + O_2(g) \rightarrow O_3(g)$$
$$O(g) + O_2(g) \rightarrow O_3(g)$$

Net:        $3\ O_2(g) \rightarrow 2\ O_3(g)$

Ozone, being an unstable molecule, tends to undergo reactions that convert it to the more stable $O_2$ molecule. Ozone absorbs UV light with wavelengths in the range 200 to 300 nm. Ozone, too, undergoes photodissociation when it absorbs ultraviolet light:

$$O_3(g) + h\nu \rightarrow O_2(g) + O(g)$$

Atomic oxygen can also destroy ozone.

$$O(g) + O_3(g) \rightarrow 2\ O_2(g)$$

**Catalytic Depletion of Ozone.** The concentration of ozone in the stratosphere depends on a balance between the rate of its production from molecular oxygen and its rate of destruction.

$$3\ O_2(g) \rightarrow 2\ O_3(g) \quad \text{production of ozone}$$
$$2\ O_3(g) \rightarrow 3\ O_2(g) \quad \text{destruction of ozone}$$

Stratospheric ozone concentrations are said to be in a steady state. This refers to the situation where the rate of ozone production is approximately equal to its rate of destruction. Any process that speeds up the destruction of ozone will cause its stratospheric concentration to decrease.

The depletion of stratospheric ozone has been associated with man-made substances called chlorofluorocarbons, or CFCs for short. The CFCs were developed in the 1930s by chemists who were seeking a new refrigerant to replace toxic and corrosive ammonia and sulfur dioxide, then in commercial use as refrigerants. Fluorocarbon-12 ($CF_2Cl_2$) was their answer, and it quickly found application in refrigerators, freezers, and air conditioners. In the 1950s, fluorocarbon-11 ($CFCl_3$) became a basic propellant for the new aerosol can industry, fostering the development of the spray can. Soon, it was also developed as a blowing agent for Styrofoam in insulation, cushions, plastic furniture, and sealants. While there are many fluorocarbon compounds, fluorocarbon-11 and fluorocarbon-12 find the widest applications. The advantage of CFCs was that they were nontoxic, noncorrosive, and nonreactive.

There are no known chemical reactions in the troposphere that destroy CFCs. Because of their inertness, CFCs accumulate in the troposphere and gradually diffuse up to the stratosphere. Since ozone filters out UV light, there is more UV light in the ozone layer than below it. In the ozone layer CFCs are bombarded by high-energy UV radiation and break apart. This photodissociation of fluorocarbon molecules breaks a C—Cl bond, and releases a chlorine atom. In the case of chlorofluorocarbon-12 the reaction is:

$$CCl_2F_2 + h\nu \rightarrow CClF_2 + Cl$$

The Cl atom attacks an ozone molecule, converting it to the more stable oxygen molecule, and forming an intermediate molecule, chlorine monoxide ClO. Compounding the problem is that once a chlorine atom has destroyed an ozone molecule, then it is regenerated when the intermediate ClO reacts with an O atom. Recall that O atoms are prevalent in the stratosphere.

$$Cl(g) + O_3(g) \rightarrow ClO(g) + O_2(g)$$
$$ClO(g) + O(g) \rightarrow O_2(g) + Cl(g)$$

net    $O(g) + O_3(g) \rightarrow 2\,O_2(g)$

Since Cl atoms accelerate the destruction of ozone and are regenerated, they are true catalysts. An increased rate of destruction of ozone has the expected effect of lowering the atmospheric ozone concentration.

The net result of ozone depletion is that UV light, which is usually absorbed by ozone, is no longer sufficiently filtered from sunlight. More UV light can reach the Earth's surface. The National Research Council predicts 10,000 additional cases of skin cancer in the U.S. for each 1% depletion in the ozone layer. This problem will be around for a long time because currently the troposphere is a giant reservoir of CFCs. Estimates are that there are already enough CFCs in the troposphere to cause depletion of ozone for the next 100 years!

Currently intense efforts are underway to develop CFC substitutes. The idea is to maintain the noncorrosive properties but make a molecule that is a little more reactive. HCFC-123 ($C_2HCl_2F_3$), where the H represents hydrogen, is a promising possibility. Introduction of a hydrogen atom into CFC molecules makes them more reactive. Chemists are synthesizing a number of HCFC molecules to try as the first generation of substitutes. The goal is to find molecules that are more easily oxidized in the troposphere, before they reach the stratosphere.

***Polar Ozone Holes.***   Since 1985 a vast hole in the ozone layer appears during the spring months over Antarctica. It fills in during the summer and fall, but opens again each spring. The size of this hole has increased every spring since then. Laboratory studies of ice particles, that mimic polar stratospheric clouds (PSCs) that form over Antarctica during the winter, again place the blame on chlorine species from CFCs for this dramatic loss in ozone. The concentration of the reaction intermediate chlorine monoxide in the ozone hole is large enough to account for the ozone loss.

Chlorine can be stored in chlorine nitrate, a so-called reservoir molecule. Chlorine nitrate is an unstable molecule that is formed when chlorine monoxide reacts with nitrogen dioxide.

$$ClO(g) + NO_2(g) \rightarrow ClONO_2(g)$$

To explain the observed ozone depletion during the Antarctic spring, the CFC theories where adjusted to include the reaction of chlorine nitrate with HCl on the surfaces of PSCs. These clouds contain HCl. Acting as a heterogeneous catalysis, these PSCs provide a surface for reactions converting HCl and chlorine nitrate to molecular chlorine.

$$HCl + ClONO_2 \rightarrow Cl_2 + HNO_3$$

The first sunlight of spring can trigger the release of chlorine atoms by photodissociation of chlorine molecules.

$$Cl_2 + h\nu \rightarrow Cl + Cl$$

To save the ozone layer a complete worldwide ban on CFC manufacturing may be necessary. To this end 24 nations met in Montreal in 1988. The Montreal Protocol is designed to achieve a 35% reduction in worldwide CFC production.

Exercises
17-4 – 17-10

### The Greenhouse Effect and Acid Rain (Sections 17.5 – 17.6)

*The Greenhouse Effect.*  Each year sees another 9 gigatons (Gtons) of carbon dioxide added to the Earth's atmosphere. This is only about 1/1,000,000 of the total mass of the atmosphere. Though relatively small in amount, carbon dioxide has an important effect on the Earth's average temperature. Carbon dioxide, water, and a few other gases, such as methane, are called "greenhouse gases."  This means that they act like the glass in a greenhouse. They let sunlight through, but do not let heat out. The main components of the air, $O_2$ and $N_2$, are not greenhouse gases.

"Greenhouse warming" occurs in the following way. Visible light from the sun passes directly through the main components of the atmosphere: $N_2$, $O_2$, $CO_2$, and $H_2O$ vapor. This radiant energy is absorbed by Earth's oceans and land areas and warms the surface. All warm objects radiate infrared radiation, called IR. The oceans and land areas emit IR, which passes directly through $N_2$ and $O_2$ into outer space. The loss of this energy would mean that the Earth would be a cooler place. However, $CO_2(g)$ and $H_2O(g)$ absorb IR, and in effect slow down the loss of heat from the surface. Therefore, Earth's surface is many degrees warmer than it would be without $CO_2(g)$ and $H_2O(g)$ in the air. This is the so-called greenhouse effect. Some carbon dioxide has always been present in the atmosphere, and some warming is considered desirable. Since atmospheric carbon dioxide is increasing at the rate of 1 ppm per year, this is expected to lead to an "enhanced greenhouse effect."

Greenhouse gases reduce the heat loss to outer space by absorbing IR radiation. Infrared radiation increases the vibrational energy level of molecules that absorb it. Then these vibrationally excited molecules reemit IR as they drop to the ground state. Statistically, one half of the IR radiation is aimed downward toward Earth's surface, and one half is aimed upward toward outer space. The IR radiation aimed back toward the surface eventually causes the temperature of the air to rise.

Scientists have been trying for many years to estimate rates of global warming and its effects. The carbon dioxide content of the atmosphere is expected to double by the year 2050. Some of the predicted effects are:

1.  A temperature increase of 3 °C to 5 °C by that year.
2.  Melting of glaciers and icecaps that will cause a 2-ft rise in sea level with its accompanying flooding of sea coasts and major cities.
3.  Widespread changes in climate.

Suggestions to lower carbon dioxide emissions center around less dependence on fossil fuels for energy. Carbon dioxide emissions could be reduced by increasing efficiency of automobiles, home heating, and electric power production. Additional reduction could result from replacing existing fossil fuel electric power plants with ones that utilize solar energy and nuclear energy.

Methane, nitrous oxide, and CFCs are also strong absorbers of IR. These gases are present in only trace amounts, but enhance the greenhouse effect significantly because they absorb IR at wavelengths that water and $CO_2$ cannot absorb. In a sense, they close an open window that would allow some IR to escape the greenhouse.

*Acid Rain.*  The term "acid rain" was coined in 1872 by an English chemist who used it to describe the increasingly acid precipitation that fell on the industrial city of Manchester. In the century since then, acid rain has grown to be an environmental problem of global proportions. Acid rain is rainwater with a pH of less than 5.5. Precipitation in the northeastern part of the U.S. has an average pH of 4.3 and in some specific storms the pH has been as low as 2.8.

The text points out that "normal" rain is slightly acid because it contains carbonic acid. As rain passes through air containing carbon dioxide, the $CO_2$ dissolves and forms carbonic acid. Carbon dioxide is an acid anhydride.

$$CO_2(g) + H_2O(\ell) \rightarrow H_2CO_3(aq)$$

Because carbonic acid is a weak acid, the pH of rain normally will not go below 5.5. The presence of several stronger acids accounts for the lower pH values of acid rain. Acid rain usually contains sulfuric acid ($H_2SO_4$) and nitric acid ($HNO_3$). We will discuss only sulfuric acid.

These acids do not start out in the atmosphere as oxoacids. The acidic oxides, $SO_2$ and $SO_3$, are the precursors of the oxoacids. Sulfur dioxide is produced by the combustion of fossil fuels such as coal and from the processing of sulfide ores at smelters.

$$2\,ZnS(s) + 3\,O_2(g) \rightarrow 2\,ZnO(s) + 2\,SO_2(g) \qquad \text{smelting an ore}$$

Coal and petroleum contain between 1% and 5% S. Combustion of sulfur in fossil fuels yields $SO_2$.

$$S(s) + O_2(g) \rightarrow SO_2(g) \qquad \text{combustion of sulfur}$$

Once in the atmosphere, some of the sulfur dioxide is oxidized to sulfur trioxide by reactions that are currently being intensely studied. The effects of light and ozone on this oxidation are mentioned in the text. The net oxidation reaction is:

$$2\,SO_2(g) + O_2(g) \rightarrow 2\,SO_3(g)$$

In the final step, sulfur trioxide combines with water to form sulfuric acid. Sulfur trioxide is converted to sulfuric acid by rainwater.

$$SO_3(g) + H_2O(\ell) \rightarrow H_2SO_4(aq)$$

Streams and lakes show the most dramatic effects of acid rain. It is known that natural waters with a low pH can kill fish eggs, salamander eggs, and frog eggs. The extent of change in acidity of a lake or stream when under the stress of acid rain is determined mainly by the buffering capacity of the surrounding soil. Watershed soils containing limestone are alkaline and can resist rapid changes in pH, making them less susceptible to harm.

Trees are also susceptible to acid rain. Acid rainfall causes damage to leaves and the growing tissues of trees. The needles of firs, spruces, and pines turn yellow and fall off. Among the possible causes are acid rain and ozone from polluted air. These pollutants damage the cell membranes of needles, allowing nutrients to escape.

Acid precipitation can acidify the soil, interfering with nutrient availability. To grow normally, trees require adequate supplies of 16 elements. Several of these, notably Ca, Mg, and K are taken up by tree roots as cations from aqueous solution in the soil. When sulfuric acid is deposited in soil by precipitation, nutrient cations can be leached from the root zone. This can greatly affect the health of forests.

Sulfur dioxide emissions can be reduced by removing $SO_2$ after combustion, but before it leaves the stack and is released into the atmosphere. Powdered limestone ($CaCO_3$) is injected into the hot gases leaving the combustion zone. Heat causes the carbonate to decompose into quicklime (CaO). The quicklime then reacts with $SO_2$ to form calcium sulfite ($CaSO_3$).

$$CaCO_3(s) \rightarrow CaO(s) + CO_2(g)$$
$$CaO(s) + SO_2(g) \rightarrow CaSO_3(s)$$

About half of the $SO_2$ is removed by contact with the dry CaO. The remaining $SO_2$ must be removed by spraying the hot gases with a suspension of quicklime. This process, called "scrubbing," creates huge amounts of calcium sulfite to dispose of. Retrofitting of scrubbers onto established power plants is very expensive and significantly raises the cost of electricity.

Exercises<br>17-11 – 17-14

## Photochemical Smog (Section 17.7)

Photochemical smog is formed by the reactions of automobile exhaust in the presence of sunlight. Reactions initiated by photons are called photochemical reactions. Smog begins with certain primary pollutants. These substances may or may not be objectionable by themselves. These are transformed by sunlight or by ordinary chemical reactions into secondary pollutants. Secondary pollutants are involved directly in the buildup of smog. Nitric oxide is a good example of a primary pollutant. It is formed at the high temperatures inside an internal combustion engine when nitrogen and oxygen from air react. Nitric oxide does not build up in the air because it is rapidly converted to nitrogen dioxide.

$$N_2(g) + O_2(g) \rightarrow 2\ NO(g)$$
$$2\ NO(g) + O_2(g) \rightarrow 2\ NO_2(g)$$

Nitrogen dioxide is involved in a chain of reactions that produces ozone. First, $NO_2$ is photochemically decomposed by sunlight.

$$NO_2(g) + h\nu \rightarrow NO(g) + O(g)$$

Oxygen atoms initiate a number of reactions in polluted air. An important one is the formation of ozone.

$$O(g) + O_2(g) \rightarrow O_3(g)$$

Exposure to 0.1 to 1.0 ppm of ozone produces headaches, burning eyes, and irritation to the respiratory passages. Another secondary pollutant is peroxyacetyl nitrate, better known as PAN. PAN literally brings tears to your eyes. It is an example of a lachrymator, a compound that causes burning of the eyes and tears. Onions, as you know, contain a lachrymator.

Unburned hydrocarbons in automobile exhaust also produce secondary air pollutants. The oxidation of unburned hydrocarbons produces various alcohols and organic acids. These can condense to produce an aerosol. Aerosols are liquid droplets dispersed in air. They are objectionable because they reduce visibility and make the air look hazy.

Efforts to control smog are usually focused on reducing the source of primary pollutants. The catalytic converters on automobiles are designed to remove NO, CO, and unburned hydrocarbons from automobile exhaust. The catalyst contains platinum and palladium metals. Nitrogen monoxide, a primary pollutant, is converted back to nitrogen and oxygen by a reaction whose rate is increased by the catalyst in the converter.

$$2\ NO(g) \rightarrow N_2(g) + O_2(g)$$

Carbon monoxide and unburned hydrocarbons are oxidized to carbon dioxide by the catalytic converter, as shown by the following equations ($C_5H_{12}$ represents a typical hydrocarbon).

Exercises<br>17-15 – 17-17

$$2 \, CO(g) + O_2(g) \rightarrow 2 \, CO_2(g)$$
$$C_5H_{12}(g) + 8 \, O_2(g) \rightarrow 5 \, CO_2(g) + 6 \, H_2O(\ell)$$

## Indoor Air Pollution (Section 17.8)

***Radon Gas.*** All isotopes of radon are radioactive. Concern about radon in homes began in 1984 when a worker at a nuclear power plant in Pennsylvania found that he was setting off the plant's radiation monitor alarms in the morning upon *arriving* at the plant. The source of the radioactivity was traced to the worker's home, which measured an extremely high level of radioactivity in the basement air. The air in his home contained an isotope of radon. Radon-222 atoms undergo radioactive decay by emitting alpha particles. Radon is always associated with uranium deposits. Some uranium occurs naturally in most soils and rocks in widely varying amounts. Radon-222 has a half-life of 3.8 days. This gives it time to migrate up out of the ground and to enter buildings through cracks in foundations. It can also dissolve in ground water and enter a home via well water.

The element radon has been known since its discovery in 1900. As a member of the noble gas group of elements, radon is a colorless, odorless, and tasteless gas with very little tendency to combine with other elements. Therefore, it tends to stay in the air. Radon atoms in the air can be inhaled into the lungs. If radon atoms decay in the lungs, we get a dose of radiation. In addition, its decay products, particularly polonium-218 and polonium-214, are deposited in the lungs. These isotopes are radioactive and contribute more to the radiation dose than that received from radon itself. These alpha-emitting isotopes are of the greatest health concern because they are solids and become trapped in respiratory passages. Here their radioactivity can damage nearby cells, which over a long period of time leads to lung cancer. In order to better assess the danger to people, health workers are trying to sort out the importance of thethree main causes of lung cancer: cigarette smoking, secondary smoke, and radon gas.

> Exercises
> 17-18 – 17-20

## GLOSSARY LIST

| | | |
|---|---|---|
| nitrogen fixation | thermosphere | greenhouse effect |
| troposphere | ionosphere | photochemical smog |
| stratosphere | mesosphere | |

## EXERCISES & PROBLEMS

1. What is nitrogen fixation? Write an equation for an example.

2. Name the regions of the atmosphere in order of increasing altitude. What region has dramatic weather changes?

3. a. What chemical species emits green and red light in auroral displays?
   b. What chemical species emits blue and violet light in auroral displays?

4. a. What region of the atmosphere contains the ozone layer?
   b. Ozone is distributed over approximately what range of altitudes?

5. What is the average concentration of ozone in the ozone layer?

6. What is the overall reaction for the formation of ozone in the stratosphere?

7. Give three reactions that are examples of photodissociation.

8. What does the abbreviation CFC stand for? Write the formulas of two CFCs.

9. What is a "molecular reservoir" of atmospheric chlorine?

10. What is the Montreal Protocol?

11. Give three sources of $CO_2$ in the atmosphere.

12. Name three "greenhouse" gases besides carbon dioxide and water.

13. Below what pH is rain considered to be "acid rain"?

14. $SO_2$ and $SO_3$ are acid anhydrides. Write reactions to show their conversion into acids.

15. Name three primary pollutants.

16. Name three pollutants that are removed by catalytic converters.

17. Is ozone in a city a primary or secondary pollutant?

18. What is the formula for radon gas? To what group of elements does it belong?

19. Radon decays into two extremely dangerous radioisotopes. What are they?

20. What is the principle health effect of radon gas?

21. Ozone occurs naturally in the atmosphere; so is it correct to call it a pollutant?

22. Is it more feasible to try to reduce the effects of acid rain by reducing $SO_2$ emissions from power plants or by neutralizing acidic lakes, streams, and soils?

## PRACTICE TEST QUESTIONS
See notes on taking practice test in the Preface

1. What noticeable change marks the transition from the troposphere to the stratosphere?

2. a. Write out the steps in the mechanism by which ozone is formed in the stratosphere.
   b. What other naturally occurring molecule absorbs UV light in the stratosphere?

3. a. Write out the steps in the mechanism of ozone destruction by chlorine atoms.
   b. Identify the catalyst and the intermediate.

4. What does the abbreviation HCFC stand for?
   Write the formula of HCFC-123.

5. What is the greenhouse effect?

6. Name two ways in which deforestation contributes to the greenhouse effect that involve $CO_2$ levels.

7. Give two sources of methane in the atmosphere.

8. By what means does Earth lose heat?

9. Write chemical equations that show what happens when acid rain reacts with iron and with limestone.

10. a. List three primary pollutants removed from automobile exhaust by catalytic converters.
    b. Explain briefly the role of each in the formation of smog.

11. The strength of the C—Cl bond in $CF_2Cl_2$ is 318 kJ/mol. What is the wavelength of a photon that has enough energy to break a C—Cl bond? Is this in the UV or visible region of the spectrum?

12. List three effects of acid rain.

17                    17

# Chapter 18.  Entropy, Free Energy & Equilibrium

**Entropy and the Second Law of Thermodynamics (Sections 18.2 – 18.4)**
**Gibbs Free Energy (Section 18.5)**
**Free Energy and Equilibrium (Section 18.6)**


## SUMMARY

### Entropy and the Second Law of Thermodynamics (Sections 18.2 – 18.4)

*Spontaneous Processes.*  An important part of experimental chemistry deals with spontaneous reactions, that is, reactions that take place without a continual supply of energy from outside the system. One goal of thermodynamics is to be able to predict whether a reaction will take place when a given set of reactants is brought together. In this chapter, we discuss those properties of a system that can be used as criteria for predicting spontaneous reactions. The release of heat by a reaction was once thought to be an indication that the reaction was spontaneous. The sign of the enthalpy change, $\Delta H$, by itself is not an adequate guide to spontaneity because, while some spontaneous reactions are known to be exothermic ($\Delta H < 0$), many endothermic reactions ($\Delta H > 0$) are known to be spontaneous as well.

It is also important to remember that the term *spontaneous* doesn't necessarily mean a fast reaction rate. The rate of a reaction is controlled by several factors as we discussed in chapter 13: reactant concentrations, catalysts, and temperature. Thermodynamics only tells us whether or not the reaction will occur spontaneously; it tells us nothing about how fast.

*Entropy.*  In addition to the heat absorbed or released in a chemical process, another factor, called entropy, must be considered. Entropy is a measure of the disorder or randomness of a system. The entropy (S) is a state function that increases in value as the disorder or randomness of the system increases. Entropy has the units J/K·mol. Intuitively, we consider a system to be "ordered" if it is arranged according to some plan or pattern. The system is "disordered" when its parts are haphazard and their arrangement is random.

Order and disorder in chemical systems are discernible at the molecular level. Crystalline solids are highly ordered, with molecules or ions occupying fixed lattice sites, and with the unit cell repeated identically over and over again. Liquids are less ordered than solids because the solid lattice has broken down, and molecules or ions have kinetic energy and therefore, are in random motion. The molecular motion in liquids increases the disorder compared to that of solids.

Gases are more random than liquids. On vaporization, the molar volume increases almost 1000-fold, and it is much more difficult to locate the position of any one molecule. The entropy of a substance increases as its molecules are distributed over an increasingly greater volume. For a given substance, the molecular disorder increases from the solid to the gaseous state.

$$S_{solid} < S_{liquid} \ll S_{gas}$$

***Entropy Changes.*** We can estimate whether the change in entropy in a process is positive or negative. For chemical reactions in which solids or liquids are converted to gases, the entropy change, $\Delta S$, is positive. It is negative for the condensation of a gas or the freezing of a liquid. When a crystal dissolves in water, the disorder increases due to the increase in freedom of motion of molecules in the solution, compared to the highly ordered crystal. Heating also increases the entropy of a system. The higher the temperature, the more energetic the molecular motion and its accompanying disorder. Figure 18.3 in the text shows several processes that lead to an increase in entropy.

---

**Hints for estimating the sign of $\Delta S$:**

1. If a reaction produces an increase in the number of moles of gaseous compounds, $\Delta S$ is positive.
2. If the total number of moles of gaseous compounds is decreased, $\Delta S$ is negative.
3. If there is no net change in the total number of gas molecules, then $\Delta S$ is either a small positive or a small negative number.

---

***The Second Law.*** The second law of thermodynamics is concerned with predicting the direction of spontaneous change. It states that the entropy of the universe increases in a spontaneous change. The term universe used here refers to a system and all of its surroundings. For an isolated system:

$$\Delta S_{univ} = \Delta S_{sys} + \Delta S_{surr} \geq 0$$

where $\Delta S_{sys}$ stands for the entropy change of the system, and $\Delta S_{surr}$ is the entropy change of the surroundings. For any spontaneous change:

$$\Delta S_{sys} + \Delta S_{surr} > 0$$

and for a reaction at equilibrium (no net change):

$$\Delta S_{sys} + \Delta S_{surr} = 0$$

***Entropy Changes in Chemical Reactions.*** Not only can the sign $\Delta S$ for a chemical reaction be estimated, but the actual value of $\Delta S$ can also be calculated. The absolute value of the entropy S of one mole of an element or compound can be determined by very careful experimentation. Appendix 3 of the textbook lists experimental values of the absolute entropy of a number of substances in their standard states at 1 atm and 25 °C. Recall that the degree superscript " ° " refers to the standard state of the substance. These values are called absolute, or sometimes, standard entropies.

One can confirm that the entropy of a gas is greater than that of a liquid by comparing the absolute entropy of $H_2O$ in the liquid and gas phases.

$S°$ of $H_2O(\ell)$ = 69.94 J/K•mol
$S°$ of $H_2O(g)$ = 188.72 J/K•mol

For a reaction:

$$aA + bB \rightarrow cC + dD$$

the standard entropy change of the reaction is given by:

$$\Delta S^{\circ}_{rxn} = \sum nS^{\circ} \text{ (products)} - \sum mS^{\circ} \text{ (reactants)}$$

Examples
18.1 – 18.3

Exercises
18-1 – 18-3

where m and n are stoichiometric coefficients. When applied to the above reaction, we get:

$$\Delta S^{\circ}_{rxn} = [cS^{\circ}(C) + dS^{\circ}(D)] - [aS^{\circ}(A) + bS^{\circ}(B)]$$

## Gibbs Free Energy (Section 18.5)

*Gibbs Free Energy.* The Gibbs free energy, expressed in terms of enthalpy and entropy, refers only to the system, yet can be used to predict spontaneity. The Gibbs free energy change ($\Delta G$) for a reaction carried out at constant temperature and pressure is given by:

$$\Delta G = \Delta H - T\Delta S$$

where both $\Delta H$ and $\Delta S$ refer to the system.

The Gibbs free energy change is equal to the maximum possible work (w) that can be obtained from a process. Any process that occurs spontaneously can be utilized to perform useful work. The Gibbs free energy of the system will decrease ($\Delta G < 0$) in a spontaneous process, and will increase ($\Delta G > 0$) in a nonspontaneous process. The free energy criteria (at constant temperature and pressure) are summarized as follows:

If $\Delta G < 0$, the forward reaction is spontaneous.
If $\Delta G = 0$, the reaction is at equilibrium.
If $\Delta G > 0$, the forward reaction is nonspontaneous. The reverse reaction will have a negative $\Delta G$ and will be spontaneous.

The standard free energy change of reaction, $\Delta G^{\circ}_{rxn}$, is the free energy change for a reaction when it occurs under standard state conditions, when reactants in their standard states are converted to products in their standard states.

*Calculation of $\Delta G^{\circ}_{rxn}$.* The free energy change for a reaction can be calculated in two ways. When both $\Delta H$ and $\Delta S$ are known, then $\Delta G = \Delta H - T\Delta S$ will give the free energy change at the temperature T. $\Delta G^{\circ}_{rxn}$ can also be calculated from standard free energies of formation in a manner analogous to the calculation of $\Delta H^{\circ}$ (Chapter 6) by using enthalpies of formation of the reactants and products. The standard free energies of formation ($\Delta G^{\circ}_f$) of selected compounds are tabulated in Appendix 3 of the text. Just as for the standard enthalpies of formation, the free energies of formation of elements in their standard states are equal to zero.

For a general reaction:

$$aA + bB \rightarrow cC + dD$$

The standard free energy change is given by:

$$\Delta G^{\circ}_{rxn} = [c\,\Delta G^{\circ}_f(C) + d\,\Delta G^{\circ}_f(D)] - [a\,\Delta G^{\circ}_f(A) + b\,\Delta G^{\circ}_f(B)]$$

In general:

$$\Delta G^{\circ}_{rxn} = \sum n\,\Delta G^{\circ}_f \text{ (products)} - \sum m\,\Delta G^{\circ}_f \text{ (reactants)}$$

where n and m are stoichiometric coefficients. Example 18.4 illustrates this type of calculation.

***Temperature and the Free Energy Change.*** From the equation, $\Delta G = \Delta H - T\Delta S$, we can see that temperature, too, will influence the spontaneity of reaction. If both $\Delta H$ and $\Delta S$ are positive then, as long as $\Delta H > T\Delta S$, at low temperature, $\Delta G$ is positive and the process will be nonspontaneous. However, as temperature increases, the $T\Delta S$ term increases and eventually $\Delta H = T\Delta S$. At this point, $\Delta G$ is zero. With further T increase, $T\Delta S > \Delta H$, making $\Delta G < 0$, and the reaction becomes spontaneous. Table 18.1 summarizes the four possible situations affecting the $\Delta G$ of a reaction.

**Table 18.1**  Enthalpy and Entropy Factors that Affect the sign of $\Delta G^{\circ}_{rxn}$

| $\Delta H$ | $\Delta S$ | $\Delta G$ |
|---|---|---|
| + | + | $\Delta G$ is positive at low temperatures and negative at high temperatures. |
| + | − | $\Delta G$ is positive at all temperatures (reverse reaction is spontaneous). |
| − | + | $\Delta G$ is negative at all temperatures (forward reaction is spontaneous). |
| − | − | $\Delta G$ is negative at low temperatures and positive at high temperatures. |

In the above situation, where both $\Delta H^{\circ}$ and $\Delta S^{\circ}$ are positive, the temperature at which $\Delta H^{\circ} = T\Delta S^{\circ}$, and also at which $\Delta G^{\circ} = 0$, can be calculated from the equation, $T = \Delta H^{\circ}/\Delta S^{\circ}$. Above this temperature, the reaction favors the products at equilibrium. See Worked Example 18.5.

***Phase Transitions.*** For a phase transition, $\Delta G = 0$ when the two phases coexist in equilibrium. For example, at the boiling point ($T_{bp}$) the liquid and vapor phases are in equilibrium, and

$$\Delta H_{vap} - T_{bp} \Delta S_{vap} = 0$$

rearranging gives  $\Delta S_{vap} = \dfrac{\Delta H_{vap}}{T_{bp}}$

This equation allows the calculation of the entropy of vaporization from knowledge of the heat of vaporization and the boiling point.

Examples
18.4 – 18.6

Exercises
18-4 – 18-8

## Free Energy and Equilibrium (Section 18.6)

***$\Delta G$ and $\Delta G^{\circ}$.*** Recall that $\Delta G^{\circ}$ refers to the standard free energy change. All the values we have calculated so far relate to processes in which the reactants are present in their standard states and are converted to products in their standard states. However, in many cases neither the reactants nor the products are present at standard concentration (1 *M*) and standard pressure (1 atm). Under nonstandard state conditions, we use the symbol $\Delta G$.

The relationship between $\Delta G$ and $\Delta G^{\circ}$ is:

$$\Delta G = \Delta G^{\circ} + RT \ln Q$$

where R is the gas constant (8.314 J/K•mole), T is the absolute temperature, and Q is the reaction quotient. For a certain reaction at a given temperature the value of $\Delta G°$ is constant, but the value of Q depends on the composition of the reacting mixture; therefore, $\Delta G$ will depend on Q. To calculate $\Delta G$, first, find $\Delta G°$, then calculate Q from the given concentrations of reactants and products, and substitute into the preceding equation.

Under special conditions, this equation reduces to an extremely important relationship. At equilibrium, $\Delta G = 0$ and therefore, Q = K. The equation then becomes:

$$0 = \Delta G° + RT \ln K$$

or

$$\Delta G° = - RT \ln K$$

| Conversion between | Tool |
|---|---|
| Free energy $\longleftrightarrow$ equilibrium constant | $\Delta G° = - RT \ln K$ |

This equation relates the equilibrium constant of a reaction to its standard free energy change. Thus, if $\Delta G°$ can be calculated, K can be determined, and vice versa. In the equation $K_p$ is used for gases and $K_c$ for reactions in solution.

Three possible relationships exist between $\Delta G°$ and K, because $\Delta G°$ can be negative, positive, or zero.

1.  When $\Delta G°$ is *negative*, ln K is positive and K > 1. The products are favored over reactants at equilibrium. The extent of reaction is large.
2.  When $\Delta G°$ is *positive*, ln K is negative and K < 1. The reactants are favored over products at equilibrium. The extent of reaction is small.
3.  When $\Delta G° = 0$, ln K is zero and K = 1. The reactants and products are equally favored at equilibrium.

When calculating $\Delta G$ and $\Delta G°$ using the two equations above, we need to express both free energy changes in units of kJ/mol. The "per mole" refers to "a mole of reaction" which is the reaction of exactly the molar amounts written in the balanced equation.

| Examples |
| 18.7, 18.8 |
| Exercises |
| 18-9 – 18-13 |

## GLOSSARY LIST

| | |
|---|---|
| entropy | Gibbs free energy |
| second law of thermodynamics | standard free energy of reaction |
| standard entropy of reaction | standard free energy of formation |
| free energy | third law of thermodynamics |

## EQUATIONS

| Algebraic Equation | English Translation |
|---|---|
| $\Delta S_{univ} = \Delta S_{sys} + \Delta S_{surr} > 0$ | 2$^{nd}$ law of thermodynamics (spontaneous process) |
| $\Delta S_{univ}^{eq} = \Delta S_{sys} + \Delta S_{surr} = 0$ | 2$^{nd}$ law of thermodynamics (equilibrium process) |
| $\Delta S_{rxn}^{\circ} = \sum nS^{\circ} \text{ (products)} - \sum mS^{\circ} \text{ (reactants)}$ | standard entropy change of a reaction. |
| $\Delta G = \Delta H - T\Delta S$ | free energy change at constant temperature |
| $\Delta G_{rxn}^{\circ} = \sum n\,\Delta G_{f}^{\circ} \text{ (products)} - \sum m\,\Delta G_{f}^{\circ} \text{ (reactants)}$ | standard free energy change of a reaction. |
| $\Delta G = \Delta G^{\circ} + RT \ln Q$ | relationship between free energy change and standard free energy change and reaction quotient |
| $\Delta G^{\circ} = -RT \ln K$ | relationship between standard free energy change and the equilibrium constant |

## WORKED EXAMPLES

**EXAMPLE 18.1  Changes in Entropy**

Predict the sign of $\Delta S_{sys}$ for each of the following reactions using just the qualitative ideas discussed in this chapter.

a.  $H_2O_2(\ell) \rightarrow H_2O(\ell) + \frac{1}{2} O_2(g)$
b.  $H^{+}(aq) + OH^{-}(aq) \rightarrow H_2O(\ell)$
c.  $CaO(s) + CO_2(g) \rightarrow CaCO_3(s)$

**• Solution**

a.  The number of moles of gaseous compounds in the products is greater than in the reactant. Entropy increases during this reaction. The sign of $\Delta S$ is +.
b.  Two reactants combine into one product in this reaction. Order is increased and so entropy decreases. The sign of $\Delta S$ is −.
c.  The number of gas molecules is decreased. The sign of $\Delta S$ is −.

**Work EXERCISES & PROBLEMS: 1, 2**

## EXAMPLE 18.2  The Second Law

The solubility of silver chloride is so low that it precipitates spontaneously from many solutions. The entropy change of the system is negative for this process.

$$Ag^+(aq) + Cl^-(aq) \rightarrow AgCl(s) \quad \Delta H° = -65 \text{ kJ/mol}$$

Since $\Delta S$ decreases in this spontaneous reaction, shouldn't the reaction be nonspontaneous?

### • Solution

For a spontaneous reaction, the second law states that $\Delta S_{univ} > 0$:

$$\Delta S_{sys} + \Delta S_{surr} > 0$$

In order to predict whether a reaction is spontaneous, both $\Delta S_{sys}$ and $\Delta S_{surr}$ must be considered, not just $\Delta S_{sys}$. Since the entropy change of the system is negative, the only way for the inequality to be true is if $\Delta S_{surr}$ is positive and greater in amount than $\Delta S_{sys}$. In this case, this is a reasonable assumption because the reaction is exothermic. This means that heat is released to the surroundings, which causes increased thermal motion and disorder of molecules in the surroundings. Therefore, the sum of $\Delta S_{sys}$ and $\Delta S_{surr}$ is greater than zero, even though $\Delta S_{sys}$ is negative.

## EXAMPLE 18.3  Calculation of the Entropy Change for a Reaction

Use absolute entropies to calculate the standard entropy of reaction $\Delta S°_{rxn}$ :

$$H_2(g) + \tfrac{1}{2}O_2(g) \rightarrow H_2O(\ell)$$

### • Solution

The standard entropy change is given by:

$$\Delta S°_{rxn} = S°(H_2O) - [S°(H_2) + \tfrac{1}{2}S°(O_2)]$$

Look up the standard entropy values in Appendix 3 of the text.

$$\Delta S°_{rxn} = 1 \text{ mol}\left(\frac{69.9 \text{ J}}{K \cdot mol}\right) - \left[1 \text{ mol}\left(\frac{130.1 \text{ J}}{K \cdot mol}\right) + \tfrac{1}{2}\text{mol}\left(\frac{205.0 \text{ J}}{K \cdot mol}\right)\right]$$

$$\Delta S°_{rxn} = 69.9 \text{ J/K} - 233.5 \text{ J/K} = -163.6 \text{ J/K} \cdot mol$$

### • Comment

This reaction is spontaneous. The value of $\Delta S°_{rxn}$ applies only to the system, not $\Delta S_{univ}$.

**Work EXERCISES & PROBLEMS: 3**

---

**EXAMPLE 18.4  Calculation of the Free Energy Change for a Reaction**

Calculate $\Delta G^{\circ}_{rxn}$ at 25 °C for the following reaction using Appendix 3 and given:

$$\Delta G^{\circ}_{f}(Fe_2O_3) = -741.0 \text{ kJ/mol}$$

$$2Al(s) + Fe_2O_3(s) \rightarrow Al_2O_3(s) + 2Fe(s)$$

• **Solution**

$$\Delta G^{\circ}_{rxn} = [\Delta G^{\circ}_{f}(Al_2O_3) + 2\,\Delta G^{\circ}_{f}(Fe)] - [2\,\Delta G^{\circ}_{f}(Al) + \Delta G^{\circ}_{f}(Fe_2O_3)]$$

$$= [(-1576.41 \text{ kJ/mol}) + 0] - [0 + (-741.0 \text{ kJ/mol})]$$

$$\Delta G^{\circ}_{rxn} = -1576.41 \text{ kJ} + 741.0 \text{ kJ} = -835.4 \text{ kJ/mol}$$

**Work EXERCISES & PROBLEMS: 4**

---

**EXAMPLE 18.5  Effect of Temperature on $\Delta G$**

Hydrated lime $Ca(OH)_2$ can be reformed into quicklime $CaO$ by heating.

$$Ca(OH)_2(s) \rightarrow CaO(s) + H_2O(g)$$

At what temperatures is this reaction spontaneous under standard conditions (that is, where $H_2O$ is formed at 1 atm pressure)?  Given the following data on $Ca(OH)_2$ that is not in the Appendix:

$$\Delta H^{\circ}_{f}[Ca(OH)_2] = -986.2 \text{ kJ/mol}$$

$$S^{\circ}[Ca(OH)_2] = 83.4 \text{ J/K·mol}$$

• **Solution**

This reaction is nonspontaneous at room temperature. The temperature above which the reaction becomes spontaneous under standard conditions corresponds to $\Delta G^{\circ} = 0$, and is given by:

$$T = \frac{\Delta H^{\circ}}{\Delta S^{\circ}}$$

$\Delta H^{\circ}$ and $\Delta S^{\circ}$ must be calculated separately. From Appendix 3 and the given data:

$$\Delta H^{\circ} = [\Delta H^{\circ}_{f}(CaO) + \Delta H^{\circ}_{f}(H_2O)] - [\Delta H^{\circ}_{f}(Ca(OH)_2)]$$

$$\Delta H^{\circ} = (-635.55 \text{ kJ/mol}) + (-241.83 \text{ kJ/mol}) - (-986.2 \text{ kJ/mol})$$

$= 108.82$ kJ/mol (or $1.088 \times 10^5$ J/mol)

$\Delta S° = S°(CaO) + S°(H_2O) - S°(Ca(OH)_2)$

$\Delta S° = (39.8$ J/K·mol$) + (188.7$ J/K·mol$) - (83.4$ J/K·mol$)$

$= + 145.1$ J/K · mol

The temperature at which $\Delta G°$ is equal to zero is:

$$T = \frac{\Delta H°}{\Delta S°} = \frac{1.088 \times 10^5 \text{ J}}{145.1 \text{ J/K}} = 750 \text{ K}$$

At temperatures above 750 K the reaction is spontaneous.

**• Comment**

Recall that this is an approximate value because of the assumption that neither $\Delta H°$ nor $\Delta S°$ change appreciably from their values calculated at 25 °C.

**Work EXERCISES & PROBLEMS: 5, 6**

---

**EXAMPLE 18.6  Entropy of Fusion**

The heat of fusion of water ($\Delta H_{fus}$) at 0 °C is 6.02 kJ/mol. What is $\Delta S_{fus}$ for 1 mole of $H_2O$ at the melting point?

**• Solution**

$$\Delta S_{fus} = \frac{\Delta H_{fus}}{T_{mp}} = \frac{6.02 \times 10^3 \text{ J/mol}}{273 \text{ K}} = +22.1 \text{ J/K•mol}$$

**• Comment**

The increase in entropy upon melting of the solid corresponds to the higher degree of molecular disorder in the liquid state as compared to the solid state.

**Work EXERCISES & PROBLEMS: 7, 8**

---

**EXAMPLE 18.7  Calculating the Equilibrium Constant**

The standard free energy change for the reaction,

$$\tfrac{1}{2} N_2(g) + \tfrac{3}{2} H_2(g) \rightleftharpoons NH_3(g)$$

is $\Delta G°_{rxn} = 26.9$ kJ/mol at 700 K. Calculate the equilibrium constant at this temperature.

**• Solution**

The equilibrium constant is related to the standard free energy change by the equation:

$$\Delta G^{\circ}_{rxn} = -RT \ln K_p$$

Since the gas constant R has units involving joules and the free energy change has units involving kilojoules, we must be careful to use consistent units. In terms of joules, we get

$$26.9 \times 10^3 \text{ J/mol} = -(8.314 \text{ J/mol} \cdot \text{K})(700 \text{ K}) \ln K_p$$

$$-4.62 = \ln K_p$$

Taking the antilog of both sides:

$$K_p = e^{-4.62} = 9.8 \times 10^{-3}$$

**Work EXERCISES & PROBLEMS 9 – 11**

---

### EXAMPLE 18.8  $\Delta G$ at Nonstandand State Conditions

Using data given in the preceding example, calculate $\Delta G$ at 700 K if the reaction mixture consists of 30.0 atm of $H_2$, 20.0 atm of $N_2$, and 0.500 atm of $NH_3$.

• **Solution**

Under nonstandard conditions, $\Delta G$ is related to the reaction quotient Q by the equation:

$$\Delta G = \Delta G^{\circ} + RT \ln Q_p$$

where,

$$Q_p = \frac{P_{NH_3}}{P_{N_2}^{1/2} P_{H_2}^{3/2}} = \frac{(0.500)}{(20.0)^{1/2}(30.0)^{3/2}} = 6.80 \times 10^{-4}$$

From Example 18.7, $\Delta G^{\circ} = 26.9$ kJ/mol. Substitution yields:

$$\Delta G = 26.9 \text{ kJ/mol} + (8.314 \text{ J/K} \cdot \text{mol})(700 \text{ K}) \ln (6.80 \times 10^{-4})$$

$$= 26.9 \text{ kJ/mol} - (42,400 \text{ J/mol} \times \frac{1 \text{ kJ}}{10^3 \text{ J}}) = 26.9 \text{ kJ/mol} - 42.4 \text{ kJ/mol}$$

$$= -15.5 \text{ kJ/mol}$$

• **Comment**

By making the partial pressures of $N_2$ and $H_2$ high and that of $NH_3$ low, the reaction is spontaneous in the forward reaction. This condition corresponds to $Q_p < K_p$, and so the reaction proceeds in the forward direction until $Q_p = K_p$.

**Work EXERCISES & PROBLEMS: 12**
**Work the rest of the EXERCISES & PROBLEMS**

## EXERCISES & PROBLEMS

1. For each pair of substances, choose the one having the higher standard entropy value at 25 °C.
   a. $CS_2(s)$ or $CS_2(\ell)$   b. $SO_2(g)$ or $CO_2(g)$   c. $BaSO_4(s)$ or $BaSO_4(aq)$

2. Predict, using the intuitive ideas about entropy, whether $\Delta S_{rxn}$ will be positive, negative, or essentially zero for each of the following:
   a. $CuSO_4(s) \rightarrow Cu^{2+}(aq) + SO_4^{2-}(aq)$
   b. $SO_2(g) + \frac{1}{2}O_2(g) \rightarrow SO_3(g)$
   c. $Ca(OH)_2(s) + CO_2(g) \rightarrow CaCO_3(s) + H_2O(g)$
   d. $Ag^+(aq) + 2CN^-(aq) \rightarrow Ag(CN)_2^-(aq)$

3. Using tabulated values, calculate $\Delta S°$ for the following reaction:

   $N_2(g) + 3 H_2(g) \rightarrow 2 NH_3(g)$

4. Calculate $\Delta G_{rxn}^°$ for the following reaction:

   $3 NO_2(g) + H_2O(\ell) \rightarrow 2 HNO_3(\ell) + NO(g)$

   Given the following free energies of formation:

   | | $\Delta G_f^°$ (kJ/mol) |
   |---|---|
   | $H_2O(\ell)$ | −237.2 |
   | $HNO_3(\ell)$ | −79.9 |
   | $NO(g)$ | 86.7 |
   | $NO_2(g)$ | 51.8 |

5. Calculate $\Delta G_{rxn}^°$ for the following reaction at 298 K:

   $O_3(g) \rightarrow O_2(g) + O(g)$

   Given: $\Delta H° = 106.5$ kJ and $\Delta S° = 127.3$ J/K.

6. The following reaction is nonspontaneous at 25 °C.

   $Cu_2O(s) \rightarrow 2Cu(s) + \frac{1}{2}O_2(g)$   $\Delta H° = 166.69$ kJ/mol

   If $\Delta S° = 68.3$ J/K, above what temperature will the reaction become spontaneous?

7. The enthalpy of vaporization of mercury is 58.5 kJ/mol and the normal boiling point is 630 K. What is the entropy of vaporization of mercury?

8. What is the sign of $\Delta G$ for the melting of ice at 5 °C?

9.  Hydrogen peroxide ($H_2O_2$) decomposes according to the equation:

    $$H_2O_2(\ell) \rightarrow H_2O(\ell) + \frac{1}{2}O_2(g)$$

      $\Delta H° = -98.2$ kJ/mol
      $\Delta S° = +70.1$ J/K•mol

    a. Is this reaction spontaneous at 25 °C?
    b. From the following data, calculate the value of $K_p$ for this reaction at 25 °C.

10. The autoionization of water at 25 °C has the equilibrium constant,

    $$2H_2O(\ell) \rightleftharpoons H_3O^+(aq) + OH^-(aq) \quad K_c = 1.0 \times 10^{-14}$$

    Calculate the value of $\Delta G°$ for this reaction.

11. The equilibrium constant for the reaction:

    $$AgBr(s) \rightleftharpoons Ag^+(aq) + Br^-(aq)$$

    is the solubility product constant, $K_{sp} = 7.7 \times 10^{-13}$ at 25 °C. Calculate $\Delta G$ for the reaction when $[Ag^+] = 1.0 \times 10^{-2}$ $M$ and $[Br^-] = 1.0 \times 10^{-3}$ $M$. Is the reaction spontaneous or nonspontaneous at these concentrations?

12. Explain the difference between $\Delta G$ and $\Delta G°$.

13. Calculate $\Delta G$ for the following reaction at 25 °C if the pressure of $CO_2$ is 0.001 atm

    $$CaCO_3(s) \rightarrow CaO(s) + CO_2(g)$$

    Given: $\Delta H° = 177.8$ kJ/mol and $\Delta S° = 160.5$ J/K•mol.

14. When the environment is contaminated by a toxic chemical spill or an oil spill, the substance tends to disperse. How is this consistent with the second law of thermodynamics? In the same regard, which requires less work:  cleaning the environment after a spill, or keeping the substance contained before a spill?

## PRACTICE TEST QUESTIONS
See notes on taking practice test in the Preface

1. Which of the following processes are spontaneous?
   a. melting of ice at $-10\ °C$ and 1 atm pressure
   b. evaporation of water at $30\ °C$ when the relative humidity is less than 100 percent
   c. Water + NaCl(s) $\rightarrow$ salt solution

2. Choose the substance that has the larger standard entropy at $25\ °C$ from each pair:
   a. $H_2O(\ell)$ or $H_2O(g)$
   b. $SiO_2(s)$ or $CO_2(g)$
   c. $Ag^+(g)$ or $Ag^+(aq)$
   d. $F_2(g)$ or $Cl_2(g)$     e. $2Cl(g)$ or $Cl_2(g)$

3. Predict, using the intuitive ideas about entropy, whether $\Delta S_{sys}$ will be positive, negative, or essentially zero for each of the following:
   a. $Ca(OH)_2(s) + CO_2(g) \rightarrow CaCO_3(s) + H_2O(g)$
   b. $CuSO_4(s) \rightarrow Cu^{2+}(aq) + SO_4^{2-}(aq)$
   c. $2\ HCl(g) + Br_2(\ell) \rightarrow 2\ HBr(g) + Cl_2(g)$
   d. $3H_2(g) + N_2(g) \rightarrow 2NH_3(g)$
   e. $Cu^{2+}(aq) + 4\ NH_3(aq) \rightarrow Cu(NH_3)_4^{2+}(aq)$

4. At the boiling point, $35\ °C$, the heat of vaporization of $MoF_6$ is 25 kJ/mol. Calculate $\Delta S$ for the vaporization of $MoF_6$.

5. Calculate $\Delta G^{\circ}_{rxn}$ for the following reaction at 298 K:

   $$2\ H_2(g) + CO(g) \rightleftharpoons CH_3OH(g)$$

   given that $\Delta H^{\circ} = -90.7$ kJ/mol and $\Delta S^{\circ} = -221.5$ J/K•mol for this process.

6. For the reaction at 298 K,

   $$Mg(s) + \tfrac{1}{2}O_2(g) \rightarrow MgO(s)$$

   $\Delta H^{\circ} = -602$ kJ/mol and $\Delta G^{\circ} = -569$ kJ/mol. Calculate $\Delta S^{\circ}$.

7. Using Appendix 3 of the text, calculate $\Delta G^{\circ}$ values for the following reactions:
   a. $3\ CaO(s) + 2\ Al(s) \rightarrow 3\ Ca(s) + Al_2O_3(s)$
   b. $ZnO(s) \rightarrow Zn(s) + \tfrac{1}{2}O_2(g)$

8. Which of the following three reactions will have the largest equilibrium constant?
   a. $N_2 + O_2 \rightleftharpoons 2\ NO$
   b. $N_2 + 2O_2 \rightleftharpoons N_2O_4$
   c. $N_2 + \tfrac{1}{2}O_2 \rightleftharpoons N_2O$

   Given: $\Delta G^{\circ}_f(NO) = +86.7$ kJ/mol
   $\Delta G^{\circ}_f(N_2O_4) = +98.3$ kJ/mol
   $\Delta G^{\circ}_f(N_2O) = +103.6$ kJ/mol

9. Given the equilibrium constant at $400\ °C$ for the reaction:

   $$H_2(g) + I_2(g) \rightleftharpoons 2\ HI(g) \quad K_p = 64$$

   calculate the value of $\Delta G^{\circ}_{rxn}$ at this temperature.

10. Calculate $\Delta G_{rxn}^{\circ}$ and $K_p$ at 25 °C for the following reaction:

$$NO(g) + \frac{1}{2} O_2(g) \rightarrow NO_2(g)$$

---

11. The synthesis of $O_2(g)$ can be carried out by the decomposition of $KClO_3$:

$$KClO_3(s) \rightarrow KCl(s) + \frac{1}{2} O_2(g)$$

for which $\Delta H° = -44.7$ kJ/mol and $\Delta S° = +247.2$ J/K•mol. Is this reaction spontaneous at 25 °C under standard conditions?

---

12. For the reaction:   $N_2 + O_2 \rightarrow 2\ NO$
the following are given:
$\Delta H° = 180.7$ kJ/mol  and
$\Delta S° = 24.7$ J/K•mol.
a. Is this reaction spontaneous at 25 °C?
b. Above what temperature will this reaction become spontaneous under standard conditions?

---

13. For the reaction,

$$2\ SO_2(g) + O_2(g) \rightleftharpoons 2\ SO_3(g),$$

$K_p = 7.4 \times 10^4$ at 700 K. If, in a reaction vessel at 700 K, we have the following partial pressures, what is $\Delta G$?

$P_{SO_2} = 1.2$ atm    $P_{O_2} = 0.5$ atm
$P_{SO_3} = 50$ atm
Predict the direction of reaction.

# Chapter 19. Electrochemistry

**Balancing Redox Equations (Section 19.1)**
**Galvanic Cells and Reduction Potentials (Sections 19.2 – 19.3)**
**Spontaneity of Redox Reactions (Section 19.4)**
**Effect of Concentration on Cell EMF (Section 19.5)**
**Electrolysis and Its Quantitative Aspects (Section 19.8)**


## SUMMARY

### Balancing Redox Equations (Section 19.1)

*Half-Reactions.* Oxidation-reduction reactions were discussed in chapter 4. It will be useful to review that material as you start this chapter. Recall that oxidation is a loss of electrons by a chemical species and reduction is a gain of electrons (Leo says ger). Redox reactions are electron transfer reactions involving transfer of electrons from a reducing agent to an oxidizing agent.

For example, when copper metal is immersed in a solution containing $Ag^+$ ions a reaction occurs in which electrons are transferred spontaneously from Cu atoms to the $Ag^+$ ions, forming Ag atoms that plate out on the copper surface. The newly formed $Cu^{2+}$ ions go into solution.

$$Cu(s) + 2\ Ag^+(aq) \rightarrow Cu^{2+}(aq) + 2\ Ag(s)$$

Cu is oxidized to $Cu^{2+}$ ion, and $Ag^+$ ion is reduced to Ag.

All redox reactions can be divided into half-reactions. The half-reactions are:

oxidation $\quad Cu \rightarrow Cu^{2+} + 2e^-$
reduction $\quad 2e^- + 2\ Ag^+ \rightarrow 2\ Ag$

Copper metal is the reducing agent because it supplies electrons to the silver ions. The silver ions are the oxidizing agent because they accept electrons from copper. Two $Ag^+$ ions are required to accept the two electrons from one Cu atom. Before we introduce galvanic cells, we will discuss how to balance redox equations.

*Ion-Electron Method.* In this method, the overall reaction is divided into two half-reactions, one involving oxidation and the other reduction. Each half-reaction is balanced according to mass and charge, and then the two half-reactions are added together to give the overall balanced equation. We will balance the equation below to illustrate the steps involved. This equation shows just the essential changes and is sometimes called a skeletal equation. To separate the equation into half-reactions, first identify which element is oxidized and which is reduced. Do this by writing oxidation numbers above each element.

$$\overset{+1\ -2}{H_2S} + \overset{+5\ -2}{NO_3^-} \rightarrow \overset{0}{S} + \overset{+2\ -2}{NO}$$

Note that S atoms and N atoms undergo oxidation number changes: S $(-2 \rightarrow 0)$ and N $(+5 \rightarrow +2)$. Sulfur is oxidized and nitrogen is reduced. Now we write the half-reactions.

oxidation      $H_2S \rightarrow S$
reduction      $NO_3^- \rightarrow NO$

Each half-reaction must be balanced by mass. Add $H_2O$ to balance the O atoms and in acidic solution, always add $H^+$ to balance H atoms. Here, we add $H^+$ to the oxidation half-reaction.

oxidation      $H_2S \rightarrow S + 2\,H^+$

The oxidation half-reaction is now balanced by mass. The reduction half-reaction needs 2 $H_2O$ on the right side to balance the O atoms and then, $4H^+$ on the left-hand side to balance the H atoms.

reduction      $4\,H^+ + NO_3^- \rightarrow NO + 2\,H_2O$

This half-reaction is now balanced according to mass.
   Next, the ionic charges must be balanced by adding electrons. In the oxidation half-reaction there is a net charge of +2 on the right-hand side and zero on the left-hand side. To balance the charge, 2 electrons are added to the right.

oxidation      $H_2S \rightarrow S + 2\,H^+ + 2e^-$

Notice that both sides of the equation have the same charge, in this case, 0.
   The reduction half-reaction can be charge balanced by adding three electrons to the left-hand side.

reduction      $3e^- + 4\,H^+ + NO_3^- \rightarrow NO + 2\,H_2O$

Note that electrons are added to the reactant side of a reduction half-reaction and to the product side of an oxidation half-reaction.
   The balanced redox equation is obtained by adding the two half-reactions. But, first the number of electrons shown in the two half-reactions must be the same, since all the electrons lost during oxidation must be gained during reduction. Multiplying the oxidation half-reaction by 3 and the reduction half-reaction by 2 will make the number of electrons in the two half-reactions equal. That is, 6 electrons are given up during oxidation and 6 are gained during reduction.

$3 \times (H_2S \rightarrow S + 2\,H^+ + 2e^-)$          gives  $3\,H_2S \rightarrow 3\,S + 6\,H^+ + 6e^-$

$2 \times (3e^- + 4\,H^+ + NO_3^- \rightarrow NO + 2\,H_2O)$  gives  $6e^- + 8\,H^+ + 2\,NO_3^- \rightarrow 2\,NO + 4\,H_2O$

The sum of the half-reactions is the overall balanced redox equation.

$$3\,H_2S \rightarrow 3\,S + 6\,H^+ + 6e^-$$
$$6e^- + 8\,H^+ + 2\,NO_3^- \rightarrow 2\,NO + 4\,H_2O$$
$$\overline{\phantom{3\,H_2S + 8\,H^+ + 2\,NO_3^- \rightarrow 3\,S + 6\,H^+ + 2\,NO + 4\,H_2O}}$$
$$3\,H_2S + 8\,H^+ + 2\,NO_3^- \rightarrow 3\,S + 6\,H^+ + 2\,NO + 4\,H_2O$$

The 6 $H^+$ on the right side will cancel 6 of the 8 $H^+$ on the left side yielding:

$$3\,H_2S + 2\,H^+ + 2\,NO_3^- \rightarrow 3\,S + 2\,NO + 4\,H_2O$$

**The following steps are useful in balancing redox equations by the ion-electron method.**

1.  Write the skeletal equation containing the oxidizing and reducing agents and the products in ionic form.
2.  Separate the equation into two half-reactions.
3.  Balance the atoms other than O and H in each half-reaction separately.
4a. For reactions in acidic medium, add $H_2O$ to balance O atoms and $H^+$ to balance H atoms.
4b. For reactions in basic medium, first, balance the atoms as you would for an acidic solution. Then, for each $H^+$ ion, add an $OH^-$ ion to both sides of the half-reaction. Whenever $H^+$ and $OH^-$ appear on the same side, combine them to make $H_2O$.
5.  Add electrons to one side of each half-reaction to equalize the charges. Electrons are added to the reactant side of a reduction half-reaction, and to the product side of an oxidation half-reaction. The number of electrons added to one side of a half-reaction should make the total charge of that side equal to the charge on the other side. This procedure is called a charge balance.
6.  Add the two half-reactions together. Before this can be done, the number of electrons shown in both half-reactions must be the same. The number of electrons in the two half-reactions can be equalized by multiplying one or both reactions by appropriate coefficients. Now, add the two half-reactions.
7.  Check the final equation by inspection. Recall that a properly balanced equation consists of a set of the smallest possible whole numbers.

> Examples
> 19.1, 19.2
>
> Exercises
> 1, 2

## Galvanic Cells and Reduction Potentials (Sections 19.2 – 19.3)

***Galvanic Cells.*** Electrochemistry is the area of chemistry that deals with the interconversion of electrical energy and chemical energy. Electrochemical processes are redox reactions in which chemical energy is converted into electricity or in which electricity is used to cause a chemical reaction to occur.

A device which utilizes a spontaneous redox reaction to supply a constant flow of electrons is called a galvanic cell, or a voltaic cell. The design of a galvanic cell is such that the reactants are prevented from direct contact with each other. The oxidation half-reaction occurs at an electrode called the anode, and the reduction half-reaction occurs at an electrode called the cathode.

> Memory device: <u>a</u>node and <u>o</u>xidation both start with a vowel; <u>c</u>athode and <u>r</u>eduction both start with a consonant.

Figure 19.1 shows a copper-silver galvanic cell. The anode is a bar of copper metal that is partially immersed in a solution of $CuSO_4$. The cathode is a small bar of silver that is partially immersed in a solution of $AgNO_3$. The reducing agent, Cu metal, does not come into direct contact with the oxidizing agent, $Ag^+$ ion. Cu atoms lose two electrons and become copper ions that enter the solution. Electrons travel through the outer circuit to the Ag electrode, where $Ag^+$ ions from the solution are reduced to silver atoms at the surface of the electrode. These atoms plate out on the electrode. As the reaction proceeds, the copper electrode loses mass and the silver electrode gains mass.

While electrons travel through the outer circuit from Cu to Ag (from the anode to the cathode), negative ions move through a porous barrier or a "salt bridge" from the $AgNO_3$ solution into the $CuSO_4$ solution. This motion resupplies negative charge to the anode compartment and maintains electrical neutrality in the solutions surrounding the electrodes. An electric current will flow until either the Cu metal or the $Ag^+$ ions have completely reacted.

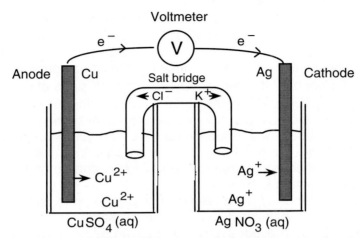

**Figure 19.1** A galvanic cell consisting of a copper electrode (anode) and a silver electrode (cathode).

The fact that electrons flow from the anode to the cathode means that a difference in electrical potential energy exists between the two electrodes. Electrons flow from where they are at higher potential to where they are at a lower potential. The difference in electrical potential between the anode and cathode is measured in volts, and is called the cell potential or cell emf which stands for <u>e</u>lectro<u>m</u>otive <u>f</u>orce. The actual difference in electrical potential depends on the nature of the species involved in the half-reactions and their concentrations. The cell potential is represented by the symbol $E_{cell}$.

***Cell Diagrams.***  Rather than always representing a cell by a sketch, we can use a cell *diagram*. For instance, the copper-silver cell discussed above and shown in Figure 19.1 is represented by:

Example
19.3

Exercise
3

$$\underset{\text{Anode}}{Cu(s) \mid CuSO_4(aq)} \mathbin{\|} \underset{\text{Cathode}}{AgNO_3(aq) \mid Ag(s)}$$

In the diagram, the anode is on the left, and the cathode is on the right. The single vertical lines indicate phase boundaries, which in this cell, are between the solid electrodes and the aqueous solutions. The sequence Cu(s) | $CuSO_4$(aq) represents oxidation, and the sequence $AgNO_3$(aq) | Ag(s) represents reduction. The double vertical lines indicate a salt bridge or porous barrier between the two solutions.

***Standard Electrode Potentials.***  The emf of a galvanic cell is dependent upon the nature of the half-reactions, the concentrations of the reactants and products, and the temperature. The standard cell emf is the voltage generated under standard state conditions. The standard cell emf has the symbol $E^{\circ}_{cell}$. The superscript "°" denotes that all reactants and products are in their standard states. The standard state of all solute ions and molecules is a concentration of 1 *M*, and for all gases it is a partial pressure of 1 atm. *Unless stated* *otherwise, the temperature is assumed to be 25 °C.* When the concentrations of $Cu^{2+}$ and $Ag^+$ ions in the cell (Figure 19.1) are both 1.0 *M*, the cell emf is 0.46 V at 25 °C.

Rather than construct a cell and measure its voltage every time a cell emf is needed, a method has been devised that allows the use of a table of half-reactions and standard electrode potentials. The standard reduction potential, $E^{\circ}_{red}$, is the voltage associated with a reduction reaction when all metals, solutes, and gases are in their standard states. This is the reduction potential when compared to a standard hydrogen electrode (the arbitrarily established "zero-point" for reduction potential). A number of standard reduction potentials are listed in Table 19.1 of the textbook, and a partial version is given in Table 19.1 in this study guide.

The following list summarizes the information contained in the table of standard reduction:

1. Standard electrode potentials are reduction potentials, and measure the ease of reduction under standard conditions. The $E^{\circ}_{red}$ values apply to the half-reactions when read in the forward direction. The more positive the value of $E^{\circ}_{red}$, the greater the tendency for the half-reaction to occur in the forward direction.
2. All half-cell reactions are reversible. Depending on the reducing strength of the other electrode that is chosen, any given electrode may act as an anode or as a cathode.
3. The species appearing on the left-hand side of the half-reactions are oxidizing agents, and that $E^{\circ}_{red}$ is related to the tendency of these oxidizing agents to accept electrons. The very strong oxidizing agents are at the top of the table. $F_2$ is the strongest oxidizing agent and $Li^+$ is the weakest.

| Table 19.1 Standard Reduction Potentials at 25°C | |
| --- | --- |
| Half-Reaction | E° (V) |
| $F_2(g) + 2e^- \rightarrow 2F^-(aq)$ | +2.87 |
| $Au^{3+}(aq) + 3e^- \rightarrow Au(s)$ | +1.50 |
| $Br_2(\ell) + 2e^- \rightarrow 2Br^-(aq)$ | +1.07 |
| $Ag^+(aq) + e^- \rightarrow Ag(s)$ | +0.80 |
| $Cu^{2+}(aq) + 2e^- \rightarrow Cu(s)$ | +0.34 |
| $Cu^{2+}(aq) + e^- \rightarrow Cu^+9aq)$ | +0.15 |
| $2H^+(aq) + 2e^- \rightarrow H_2(g)$ | 0.00 |
| $Pb^{2+}(aq) + 2e^- \rightarrow Pb(s)$ | −0.13 |
| $Sn^{2+}(aq) + 2e^- \rightarrow Sn(s)$ | −0.14 |
| $Ni^{2+}(aq) + 2e^- \rightarrow Ni(s)$ | −0.15 |
| $Fe^{2+}(aq) + 2e^- \rightarrow Fe(s)$ | −0.44 |
| $Zn^{2+}(aq) + 2e^- \rightarrow Zn(s)$ | −0.76 |
| $Al^{3+}(aq) + 3e^- \rightarrow Al(s)$ | −1.66 |
| $Mg^{2+}(aq) + 2e^- \rightarrow Mg(s)$ | −2.37 |
| $Ca^{2+}(aq) + 2e^- \rightarrow Ca(s)$ | −2.87 |
| $Li^+(aq) + e^- \rightarrow Li(s)$ | −3.05 |

Increasing strength of reducing agents

4. The species shown on the right-hand side of the half-reactions are reducing agents. For a reducing agent to supply electrons, the reaction must occur in the reverse direction. The strength of the reducing agents increases as you read down the table. Therefore, $F^-$ ion is the weakest reducing agent and Li metal is the strongest reducing agent.
5. When predicting the direction of reaction under standard state conditions, you can use the diagonal rule. This rule states that any species on the left-hand side of a given half-cell reaction (an oxidizing agent) will react spontaneously with a species that is shown on the right of any half-reaction (a reducing agent) located below it in the table.

See the "fish" drawn on Table 19.1. Follow the arrow. The 1st down stroke gives the reactants. The 2nd down stroke gives the products for a spontaneous redox reaction. In the example we see the reaction $Sn^{2+}(aq) + Fe(s) \rightarrow Sn(s) + Fe^{2+}(aq)$ is a spontaneous reaction.

6. The standard reduction potential is not affected when stoichiometric coefficients of a half-cell reaction are changed. For example,

Examples
19.4, 19.5

Exercises
4 – 6

$$Ag^+ + e^- \rightarrow Ag(s) \qquad E^{\circ}_{red} = 0.80 \text{ V}$$
$$2\,Ag^+ + 2e^- \rightarrow 2\,Ag(s) \qquad E^{\circ}_{red} = 0.80 \text{ V}$$

Electrode potentials are intensive properties, and therefore, do not depend on the size of the electrode or the manner in which the half-reaction is balanced. By definition of standard reduction potential, the concentrations must be 1.0 *M*.

Notice the 0.0 V potential assigned to hydrogen ion. A positive reduction potential means that the oxidizing agent (such as $F_2$) has a greater tendency to be reduced than the hydrogen ion, and is therefore a stronger oxidizing agent. A negative reduction potential for an oxidizing agent (such as $Mg^{2+}$) indicates that the hydrogen ion is more readily reduced than that substance.

***Calculation of the Standard Cell EMF.*** Standard reduction potentials are used to calculate the potential of a galvanic cell. The standard cell emf (or standard cell potential), $E^{\circ}_{cell}$, is the difference between the standard reduction potential of the cathode, and the standard reduction potential of the anode. The subtraction (or negative sign) at the anode is due to the fact that oxidation occurs at the anode. Oxidation is the reverse reaction. If the reaction is reversed, the sign is changed.

$$E^{\circ}_{cell} = E^{\circ}(\text{cathode}) - E^{\circ}(\text{anode})$$

The standard emf of the silver-copper cell discussed above is calculated as follows. The overall reaction is:

$$Cu + 2\,Ag^+ \rightarrow Cu^{2+} + 2\,Ag$$

The half-reactions are:

| | |
|---|---|
| oxidation | $Cu \rightarrow Cu^{2+} + 2e^-$ |
| reduction | $2\,Ag^+ + 2e^- \rightarrow 2\,Ag$ |

Therefore, the standard cell emf is given by:

$$E^{\circ}_{cell} = E^{\circ}_{cathode} - E^{\circ}_{anode} = E^{\circ}_{Cu^{2+}/Cu} - E^{\circ}_{Ag^+/Ag}$$

The symbols $Cu^{2+}/Cu$ and $Ag/Ag^+$ represent the reduction half-reactions that are related to the reduction potentials.

The reduction potentials needed here are found in Table 19.1 of the text.

$$E^{\circ}_{cathode} = E^{\circ}_{Ag^+/Ag} = 0.80 \text{ V}$$

$$E^{\circ}_{anode} = E^{\circ}_{Cu^{2+}/Cu} = 0.34 \text{ V}$$

$$E^{\circ}_{cell} = 0.80 \text{ V} - 0.34 \text{ V}$$

$$E^{\circ}_{cell} = 0.46 \text{ V}$$

Example
19.6

Exercise
7

A positive $E^{\circ}_{cell}$ indicates that the reaction is spontaneous under standard conditions.

## Spontaneity of Redox Reactions (Section 19.4)

***Criterion for Spontaneity.*** For a process carried out at constant temperature and pressure, the Gibbs free energy change is equal to the maximum amount of work ($w_{max}$) that can be done by the process.

$$\Delta G = w_{max}$$

Any reaction that occurs spontaneously can be utilized to perform useful work. Both combustion of fuels and the oxidation of food components provide energy to do work. The difference in free energy between the reactants and the products is the energy available to do work.

Electrical work is equal to the product of the cell emf ($E_{cell}$) and the total charge in coulombs carried through the circuit. The total charge in coulombs is nF, where n is the number of moles of electrons transferred from the reducing agent to the oxidizing agent, according to the balanced equation, and F is the Faraday constant. The Faraday constant equals the charge in coulombs carried by one mole of electrons which is 96,500 C/mol of electrons. In most thermodynamic calculations, F is expressed in coulomb per mole of $e^-$ units: 96,500 J/V·mol $e^-$.

The electrical work is given by:

$$w_{ele} = -nFE_{cell}$$

The maximum work from an electrochemical process is:

$$w_{max} = w_{ele} = -nFE_{cell}$$

The negative sign is in accord with the sign convention that when work is done on the surroundings, the system loses energy.

The Gibbs free energy change for a redox reaction is:

$$\Delta G = -nFE_{cell}$$

For a spontaneous redox reaction, $\Delta G$ will be negative, and E must be positive. For reactions in which reactants and products are in their standard states:

$$\Delta G° = -nFE°_{cell}$$

| Conversion between | Tool |
|---|---|
| Free energy ←→ cell potential | $\Delta G° = -nFE°_{cell}$ |

***The Equilibrium Constant.*** In chapter 18 you learned that $\Delta G°$ is related to the equilibrium constant K for the reaction.

$$\Delta G° = -RT \ln K$$

Therefore, the equilibrium constant of a redox reaction is related to the standard cell emf (E°) by the equation.

$$-nFE° = -RT \ln K$$

and $\quad E° = \dfrac{RT}{nF} \ln K$

Substituting for R and F at 25 °C, gives the term RT/F which equals 0.0257 V.

$$E° = \frac{0.0257 \text{ V}}{n} \ln K$$

| Conversion between | Tool |
|---|---|
| equilibrium constant ←→ cell potential | $E° = \dfrac{0.0257 \text{ V}}{n} \ln K$ |

Therefore, reactions with large equilibrium constants generate higher standard cell emf's. When E° > 1, K > 1. Rearranging,

$$\ln K = \frac{nE°}{0.0257 \text{ V}}$$

If any one of the three quantities $\Delta G°$, K, or E° is known, both of the others can be calculated. Table 19.2 summarizes the criteria for spontaneous redox reactions.

**Table 19.2** Criteria for Spontaneous Redox Reactions

| $\Delta G°$ | K | $E°_{cell}$ | Reaction under Standard Conditions |
|---|---|---|---|
| Negative | > 1 | Positive | Spontaneous |
| 0 | = 1 | 0 | At equilibrium |
| Positive | < 1 | Negative | Nonspontaneous (Reaction is spontaneous in the reverse direction.) |

Example 19.7

Exercises 8 – 9

## Effect of Concentration on Cell EMF (Section 19.5)

***The Nernst Equation.***  We mentioned earlier that the emf of a cell depends on the nature of the reactants and products, and on their concentrations. Since the cell emf is a measure of the spontaneity of the cell reaction, we might reasonably expect the voltage to fall as reactants are consumed and products accumulate.

The equation that relates the cell emf to the concentrations of reactants and products is named after Walter Nernst. At 298 K, for a redox reaction of the type:

aA + bB → cC + dD

The Nernst equation is:

$$E = E° - \frac{0.0257 \text{ V}}{n} \ln Q$$

where E° is the standard cell emf, E is the nonstandard cell emf, 0.0257 V is a constant at 298 K, and n is the number of moles of electrons transferred according to the balanced equation. The reaction quotient Q contains the concentrations of reactants and products.

$$Q = \frac{[C]^c [D]^d}{[A]^a [B]^b}$$

The Nernst equation predicts that E will decrease as reactant concentrations decrease, or as product concentrations increase. Both changes cause Q to increase. As Q increases, ln Q increases. Therefore, an increasingly larger number is subtracted from E°. Therefore, E is not a constant, but decreases as a reaction proceeds. E° is a constant for a reaction and is characteristic of the reaction.

At equilibrium, this equation reduces to one we have seen before. When the reaction is at equilibrium, no net reaction occurs and no net transfer of electrons occurs, and E = 0. Also, at equilibrium, Q = K, and the Nernst equation becomes:

$$0 = E° - \frac{0.0257 \text{ V}}{n} \ln K$$

$$E° = \frac{0.0257 \text{ V}}{n} \ln K$$

Therefore, the standard cell emf is related to the equilibrium constant.

When both $E_{cell}$ and $E^o_{cell}$ are known, the Nernst equation can be used to calculate an unknown concentration.

> Examples
> 19.8, 19.9
>
> Exercises
> 10 – 12

## Electrolysis and Its Quantitative Aspects (Section 19.8)

***Electrolysis of Molten Salts.*** Electrolysis is the process in which electrical energy is used to cause nonspontaneous redox reactions to occur. An electrolytic cell is the apparatus for carrying out electrolysis. A schematic diagram of an electrolytic cell is shown in Figure 19.2. A battery or other DC power supply serves as an "electron pump" that supplies electrons to the cathode, where chemical species are reduced. Electrons resulting from the oxidation of chemical species are withdrawn from the anode and return to the battery. In the electrolytic cell, the cathode is negative and the anode is positive. This is the opposite of a galvanic cell.

Molten calcium chloride can be decomposed into the elements calcium and chlorine in an electrolytic cell. Let's examine the redox processes and the movement of ions and electrons in the apparatus. First, no current can flow through the cell unless it is filled with a liquid, because ions in a solid crystal do not move from place to place. $CaCl_2$ melts at 780 °C. The electrode in the electrolysis cell that is attached to the battery's negative terminal is the cathode. Electrons flow from the battery onto the cathode. $Ca^{2+}$ ions are attracted to the cathode and are reduced to Ca metal. The anode of the electrolysis cell is attached to the positive terminal of the battery. Chloride ions drift toward the anode where they are oxidized. The electrons travel from the anode to the battery. For every electron leaving the battery, one must return. The reactions at the electrodes are:

| | |
|---|---|
| cathode/reduction | $Ca^{2+}(\ell) + 2e^- \rightarrow Ca(s)$ |
| anode/oxidation | $2 \, Cl^-(\ell) \rightarrow Cl_2(g) + 2e^-$ |

―――――――――――――――――――――――――――――――――――――――――――

| | |
|---|---|
| overall | $Ca^{2+}(\ell) + 2 \, Cl^-(\ell) \rightarrow Ca(s) + Cl_2(g)$ |

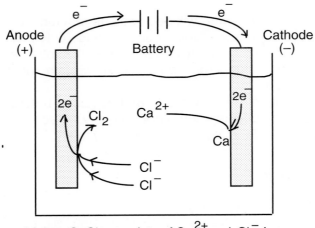

Molten $CaCl_2$ consists of $Ca^{2+}$ and $Cl^-$ ions

**Figure 19.2** Schematic diagram of an electrolytic cell. During electrolysis the cations ($M^+$) migrate to the cathode where they are reduced. Anions ($X^-$) migrate to the anode, where they are oxidized.

***Voltage Requirement.*** The minimum voltage required to bring about electrolysis can be *estimated* from the standard reduction potentials. For the above reaction:

$$E^\circ_{cell} = E^\circ_{cathode} - E^\circ_{anode} = E^\circ_{Ca^{2+}/Ca} - E^\circ_{Cl_2/Cl^-}$$

$$E^\circ_{cell} = -2.87\ V - 1.36\ V = -4.23\ V$$

Note: the above values apply to species in aqueous solution. Here we are dealing with molten $CaCl_2$. The negative cell potential means that this is a nonspontaneous reaction.

This reaction can be forced to occur by connecting the electrodes to an external source of electrical energy such as a battery or voltage supply, as in Figure 19.2. The electrolysis will occur when the external voltage is greater than the negative voltage of the nonspontaneous reaction. The external source of electrical energy must provide more than 4.23 V.

***Competing Electrode Reactions in Aqueous Solution.*** A complicating factor in the electrolysis of aqueous solutions containing a dissolved salt is that water molecules can be reduced to $H_2$ or oxidized to $O_2$ in preference to the solute species.

Reduction of water:     $2\ H_2O(\ell) + 2e^- \rightarrow H_2(g) + 2\ OH^-$      $E^\circ_{red} = -0.83\ V$

Oxidation of water:     $H_2O(\ell) \rightarrow \frac{1}{2}O_2(g) + 2\ H^+(aq) + 2e^-$      $E^\circ_{ox} = -E^\circ_{red} = -1.23\ V$

Because oxidation is the reverse of reduction, the potentials measuring the ease of oxidation are $-E^\circ_{red}$.

We must consider whether water molecules are oxidized or reduced instead of ions of the salt. In the electrolysis of aqueous KI solution, for example, $H_2(g)$ is formed at the cathode, and $I_2$ at the anode. $H_2(g)$ is coming from the reduction of water. Water has a higher (more positive) standard reduction potential than $K^+$, which indicates that it is reduced more easily than potassium ion. $K^+$ ions drift to the cathode, but no K metal is formed.

$$2\,H_2O(\ell) + 2e^- \rightarrow H_2(g) + 2\,OH^-(aq) \qquad E^\circ_{red} = -0.83\ V$$
$$K^+(aq) + e^- \rightarrow K(s) \qquad E^\circ_{red} = -2.93\ V$$

Molecular iodine is formed at the anode during electrolysis of aqueous KI. This means the anode reaction is:

$$2I^-(aq) \rightarrow I_2(s) + 2e^-$$

rather than:

$$H_2O(\ell) \rightarrow \frac{1}{2}O_2(g) + 2\,H^+(aq) + 2e^-$$

This preference reflects that $I^-$ is easier to oxidize than $H_2O$.

$$2\,I^- \rightarrow I_2(s) + 2e^- \qquad E^\circ_{ox} = -E^\circ_{red} = -0.53\ V$$
$$H_2O(\ell) \rightarrow \frac{1}{2}O_2(g) + 2\,H^+(aq) + 2e^- \qquad E^\circ_{ox} = -E^\circ_{red} = -1.23\ V$$

The higher (more positive) the value of $-E^\circ_{red}$, the more favorable is the oxidation. Iodide ions have a higher oxidation potential than $H_2O$, and so iodide ions are oxidized preferentially.

In summary, if there is more than one reducible species in solution, the species with the greater reduction potential will be preferentially reduced. If there is more than one oxidizable species in solution, the species with the higher $-E^\circ_{red}$ will be preferentially oxidized.

***Overvoltage.*** The higher standard oxidation potential of $I^-$ indicates that it should be more readily oxidized than $H_2O$. However, the existence of an overvoltage is a complicating factor. Overvoltage is the additional voltage over and above the standard potential required to cause electrolysis. Overvoltage affects $H_2O$ oxidation more than any other process by lowering the ease of oxidation by about 0.6 V. According to the oxidation potentials $(-E^\circ_{red})$, $I^-$ ions and $Br^-$ ions will be oxidized in the presence of $H_2O$, and $Cl^-$ ions would not be. Due to overvoltage effects that reduce the tendency of $H_2O$ to be oxidized, $Cl^-$ ions can also be oxidized in aqueous solution. Keep in mind that standard oxidation potentials do not always make the correct prediction of which species will be oxidized when one of the oxidizable species is $H_2O$.

***Quantitative Aspects of Electrolysis.***  Our knowledge of the quantitative relationships in electrolysis is due mostly to the work of Michael Faraday. He observed that the quantity of a substance undergoing chemical change during electrolysis is proportional to the quantity of electrical charge that passes through the cell. The quantity of electrical charge is expressed in coulombs. The electrical charge of 1 mole of electrons is called 1 Faraday which is equal to 96,500 coulombs. 1 mole $e^- = 1\ F = 96,500\ C$.

| Conversion between | Tool |
| --- | --- |
| charge $\longleftrightarrow$ moles electrons | Faraday constant |

TOOL

The number of electrons shown in the half-reaction is the link between the number of moles of substance reacted and the number of faradays of electricity required. We obtain the following conversion factors for the two half-reactions shown below:

$$K^+ + e^- \rightarrow K \qquad\qquad Ca^{2+} + 2e^- \rightarrow Ca$$

$$\frac{1\ mol\ K}{1\ mol\ e^-} \qquad\qquad \frac{1\ mol\ Ca}{2\ mol\ e^-}$$

| Conversion between | Tool |
|---|---|
| moles substance ←→ moles electrons | Coefficients of balanced ½ reaction |

The number of coulombs passing through an electrolytic cell in a given period of time is related to the electrical current in amperes (A), where 1 ampere is equal to 1 coulomb per second.

$$1\ A = 1\ C/s$$

| Conversion between | Tool |
|---|---|
| charge ←→ time | Amperes |

In an electrolysis experiment, the measured quantities are current and the time that the current flows. The number of coulombs passing through the cell is

Total charge in coulombs (C) = current (A) × time (t)

$$Charge\ in\ coulombs = \frac{C}{s} \times s$$

When an electrical current of known amperes flows for a known time, the amount of chemical change can be calculated according to the following road map.

current × time → coulombs → [moles e⁻ transferred] → moles of → grams of
                                                       product      product

The number of moles of electrons is the key term. It links the experimental variables to the theoretical yield. Reversing the road map allows one to calculate the time required to produce a specific amount of chemical change with a given electrical current.

| Examples |
|---|
| 19.10 – 19.12 |
| |
| Exercises |
| 13 – 18 |

## GLOSSARY LIST

| | | |
|---|---|---|
| electrochemistry | cell voltage | battery |
| galvanic cell | electromotive force | fuel cell |
| cathode | standard reduction potential | corrosion |
| anode | standard emf | |
| half-cell reaction | | electrolysis |
| | Faraday | electrolytic cell |
| | Nernst equation | overvoltage |

## EQUATIONS

| Algebraic Equation | English Translation |
|---|---|
| $E^\circ_{cell} = E^\circ_{cathode} - E^\circ_{anode}$ | calculating the standard emf of a galvanic cell |
| $\Delta G = -nFE_{cell}$ | relating free-energy change to the emf of the cell |
| $\Delta G^\circ = -nFE^\circ_{cell}$ | relating the standard free-energy change to the standard emf of the cell |
| $E^\circ_{cell} = \dfrac{RT}{nF}\ln K$ | relating the standard emf of the cell to the equilibrium constant |
| $E_{cell} = E^\circ_{cell} - \dfrac{RT}{nF}\ln Q$ | Nernst equation for calculating the emf of a cell under non-standard-state conditions |

## CONVERSION TOOLS

| Conversion between | Tool |
|---|---|
| free energy $\leftrightarrow$ equilibrium constant | $\Delta G^\circ = -RT\ln K$ |
| charge $\leftrightarrow$ moles electrons | Faraday constant |
| moles substance $\leftrightarrow$ moles electrons | Coefficients of balanced ½ reaction |
| charge $\leftrightarrow$ time | Amperes |

## WORKED EXAMPLES

---

**EXAMPLE 19.1   Balancing a Redox Reaction in Acidic Solution**

Balance the following redox reaction by the ion-electron (half-reaction) method.

$$Cu + HNO_3 \rightarrow Cu^{2+} + NO_2 \quad \text{(in acidic solution)}$$

**• Solution**

Steps 1 and 2. The skeletal equation is given. Separate the given equation into half-reactions. Note that Cu is oxidized and nitrogen is reduced. However, it is not necessary to know which substance is reduced and which substance is oxidized. This will be evident as you work on balancing the half reactions. The half-reactions are:

$$\begin{array}{ll} \text{oxidation} & Cu \rightarrow Cu^{2+} \\ \text{reduction} & HNO_3 \rightarrow NO_2 \end{array}$$

Step 3.  The oxidation reaction's elements are already balanced. The nitrogen atoms in the reduction reaction are balanced.

Step 4a. Balance the reduction half-reaction by adding one $H_2O$ to the right-hand side, and one $H^+$ to the left-hand side. (Step 4b is unnecessary, since the solution is acidic.)

$$\begin{array}{ll} \text{oxidation} & Cu \rightarrow Cu^{2+} \\ \text{reduction} & H^+ + HNO_3 \rightarrow NO_2 + H_2O \end{array}$$

Step 5. To balance according to charge, add electrons to balance the changes. For the oxidation reaction, add two electrons to the reactant side, to give it the same total neutral charge that exist on the reactant side. For the reduction reaction, one electron must be added to the reactant side, to give it the same neutral charge that is on the product side. *At this point, it becomes evident, which reaction is the oxidation reaction and which is the reduction reaction.* The loss of electron (copper reaction) is oxidation and the gain or electrons (nitrogen reaction) is reduction.

$$\begin{array}{ll} \text{oxidation} & Cu \rightarrow Cu^{2+} + 2e^- \\ \text{reduction} & e^- + H^+ + HNO_3 \rightarrow NO_2 + H_2O \end{array}$$

Step 6.  Next, equalize the number of electrons in the two half-reactions by multiplying the reduction half-reaction by 2.

$$2e^- + 2H^+ + 2HNO_3 \rightarrow 2NO_2 + 2H_2O$$

Now add the two half-reactions to obtain a balanced overall equation.

$$\begin{array}{ll} \text{oxidation} & Cu \rightarrow Cu^{2+} + 2e^- \\ \text{reduction} & 2e^- + 2\,H^+ + 2\,HNO_3 \rightarrow 2\,NO_2 + 2\,H_2O \end{array}$$

$$\overline{Cu + 2\,H^+ + 2\,HNO_3 \rightarrow Cu^{2+} + 2\,NO_2 + 2\,H_2O}$$

Step 7. Check that the electrons on both sides cancel and that the equation is balanced in terms of atoms (mass) and charge.

---

**EXAMPLE 19.2  Balancing a Redox Reaction in Basic Solution**

Balance the following redox reaction by the ion-electron (half-reaction) method.

$MnO_4^- + SO_2 \rightarrow Mn^{2+} + SO_4^{2-}$ (in basic solution)

**• Solution**

Steps 1 and 2. The skeletal equation is given. Separate the given equation into half-reactions. We will not label the half-reactions as oxidation and reduction yet. We will just refer to them as the sulfur and manganese half-reactions.

$SO_2 \rightarrow SO_4^{2-}$
$MnO_4^- \rightarrow Mn^{2+}$

Step 3.  The elements other than O and H are already balanced. Sulfur half-reaction has one sulfur atom on each side, manganese half-reaction has one manganese on each side.

Step 4a. Balance the reduction half-reaction by adding one $H_2O$ to the right-hand side, and one $H^+$ to the left-hand side.

$2H_2O + SO_2 \rightarrow SO_4^{2-} + 4H^+$
$`8 H^+ + MnO_4^- \rightarrow Mn^{2+} + 4H_2O$

Step 4b. Add an $OH^-$ to both sides of the reaction for every $H^+$, which is in the half-reaction. This is accomplished because the reaction is in basic solution.

$4OH^- + 2H_2O + SO_2 \rightarrow SO_4^{2-} + 4H^+ + 4OH^-$
$8OH^- + 8H^+ + MnO_4^- \rightarrow Mn^{2+} + 4H_2O + 8OH^-$

Step 4b, continued. Combine $OH^-$ and $H^+$, which appear on the same side of the equation, to make water.

$4OH^- + 2H_2O + SO_2 \rightarrow SO_4^{2-} + 4H_2O$  or  $4OH^- + SO_2 \rightarrow SO_4^{2-} + 2H_2O$
$8H_2O + MnO_4^- \rightarrow Mn^{2+} + 4H_2O + 8OH^-$  or  $4H_2O + MnO_4^- \rightarrow Mn^{2+} + 8OH^-$

Step 5. Add electrons to balance the charge of each half-reaction.
- The sulfur reaction has a total of (-4) charge on the reactant side and a (-2) charge on the product side. Add 2 electrons to the product side to balance the charge.
- The manganese reaction has a total of (-1) charge on the reactant side and a (+2 − 8 = -6) charge on the product side. Add 5 electrons to the reactant side to balance the charge.

$4OH^- + SO_2 \rightarrow SO_4^{2-} + 2H_2O + 2e^-$
$5e^- + 4H_2O + MnO_4^- \rightarrow Mn^{2+} + 8OH^-$

At this point, we can label which is oxidation (loss of electrons, the sulfur reaction) and which is reduction (gain of electrons, manganese reaction.)

oxidation     $4OH^- + SO_2 \rightarrow SO_4^{2-} + 2H_2O + 2e^-$
reduction     $5e^- + 4H_2O + MnO_4^- \rightarrow Mn^{2+} + 8OH^-$

Step 6. Next, equalize the number of electrons in the two half-reactions by multiplying the oxidation reaction by 5 and the reduction half-reaction by 2.

oxidation    $20OH^- + 5SO_2 \rightarrow 5SO_4^{2-} + 10H_2O + 10e^-$
reduction    $10e^- + 8H_2O + 2MnO_4^- \rightarrow 2Mn^{2+} + 16OH^-$

Now, add the two half-reactions (cancel substance common to both sides of the equation) to obtain a balanced overall equation.

$$
\begin{array}{ll}
& \quad\quad\quad\quad 4 \quad\quad\quad\quad\quad\quad\quad\quad\quad\quad 2 \\
\text{oxidation} & \quad\quad 20OH^- + 5SO_2 \rightarrow 5SO_4^{2-} + 10H_2O + 10e^- \\
\text{reduction} & 10e^- + 8H_2O + 2MnO_4^- \rightarrow 2Mn^{2+} + 16OH^- \\
\hline
& 4OH^- + 5SO_2 + 2MnO_4^- \rightarrow 5SO_4^{2-} + 2Mn^{2+} + 2H_2O
\end{array}
$$

Step 7. Check that the electrons on both sides cancel, and that the equation is balanced in terms of atoms (mass) and charge.

**Work EXERCISES & PROBLEMS: 1, 2**

---

**EXAMPLE 19.3  Galvanic Cell Reactions**

Consider a galvanic cell constructed from the following half-cells, linked by a $KNO_3$ salt bridge. A Cu electrode immersed in 1.0 $M$ $Cu(NO_3)_2$, and a Sn electrode in 1.0 $M$ $Sn(NO_3)_2$. As the reaction proceeds, the Cu electrode gains mass and the Sn electrode loses mass.
a.  Which electrode is the anode and which is the cathode?
b.  Write a balanced chemical equation for the overall cell reaction.
c.  Sketch the half-cells and show the direction of flow of the electrons in the external circuit.
d.  Which electrodes do the positive and the negative ions diffuse toward?  Include a salt bridge in your sketch.

**• Solution**

a.  Reduction occurs at the cathode and oxidation occurs at the anode. Since the Cu electrode gains mass, its half-reaction must involve the plating out of $Cu^{2+}$ ions as Cu atoms. The copper electrode is the cathode.

$$Cu^{2+}(aq) + 2e^- \rightarrow Cu(s) \quad\quad\quad\quad \text{reduction}$$

Since the Sn electrode loses mass, Sn metal must be dissolving as it is oxidized, and so Sn is the anode. The $Sn^{2+}$ ions go into solution.

$$Sn(s) \rightarrow Sn^{2+}(aq) + 2e^- \quad\quad\quad\quad \text{oxidation}$$

b. Adding the two half-reactions:

$$Sn(s) \rightarrow Sn^{2+}(aq) + 2e^-$$
$$Cu^{2+}(aq) + 2e^- \rightarrow Cu(s)$$

overall    $Sn + Cu^{2+}(aq) \rightarrow Sn^{2+}(aq) + Cu$

c. Electrons flow from the anode to the cathode. See Figure 19.2.
d. The positive ions (cations) diffuse toward the Cu electrode (the cathode) and the negative ions (anions) diffuse toward the Sn electrode (the anode).

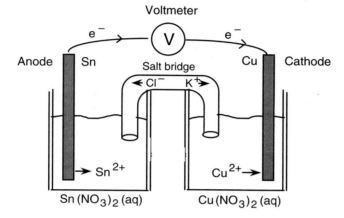

**Figure 19.2** A galvanic cell consisting of a tin electrode (anode) and a copper electrode (cathode). Notice that in Figure 19.1, copper was an anode.

**Work EXERCISES & PROBLEMS: 3**

---

**EXAMPLE 19.4  Relative Strength of Oxidizing and Reducing Agents**

Consider the following species:   Mg, $MnO_4^-$, $Cl^-$, and $Cu^{2+}$. Using Table 19.1 of the text, separate these into oxidizing and reducing agents. Then, arrange the reducing agents according to increasing strength, and the oxidizing agents according to strength.

**• Solution**

First, we will take the half-reactions from Table 19.1 that contain the above species and list them here. The oxidizing agents appear on the left-hand side of the half-reactions in Table 19.1, and the reducing agents appear on the right-hand side. Examination of the potentials, shows that $MnO_4^-$ and $Cu^{2+}$ ion are oxidizing agents, and $Cl^-$ ion and Mg are reducing agents.

| Oxidizing Agents | Reducing Agents | $E^\circ_{red}$ (V) |
|---|---|---|
| $MnO_4^- + 8H^+ + 5e^- \rightleftharpoons 2Mn^{2+}(aq) + 4H_2O$ | | +1.51 |
| $Cl_2(g) + 2e^- \rightleftharpoons 2Cl^-(aq)$ | | +1.36 |
| $Cu^{2+}(aq) + 2e^- \rightleftharpoons Cu(s)$ | | +0.34 |
| $Mg^{2+}(aq) + 2e^- \rightleftharpoons Mg(s)$ | | −2.37 |

The greater the value of the reduction potential, the greater the tendency for the half-reaction to occur as written, and therefore, the greater the strength of the oxidizing agent. Comparing the oxidizing agents, we see that permanganate ion, $MnO_4^-$ is stronger than $Cu^{2+}$ ion.

For a substance on the right-hand side of the half-reactions in Table 19.1 to act as a reducing agent, it must react in the reverse direction. The reducing agent strength increases as we read down the list. We see that Mg is a stronger reducing agent than chloride ion, $Cl^-$.

• **Comment**

All half-reactions are reversible. The product of a strong oxidizing agent such as $MnO_4^-$ is a weak reducing agent $Mn^{2+}$.

---

**EXAMPLE 19.5  Predicting Spontaneous Redox Reactions**

Predict whether the following reactions are spontaneous under standard conditions.

a.   $2 Fe^{3+}(aq) + 2 I^-(aq) \rightarrow 2 Fe^{2+}(aq) + I_2(s)$
b.   $Cu(s) + 2 H^+(aq) \rightarrow Cu^{2+}(aq) + H_2(g)$

• **Solution**

For a redox reaction to be spontaneous, it must have a positive cell emf. For the reaction in part (a) we will apply the diagonal rule. In part (b) we will calculate the standard emf of the cell.

a. According to the diagonal rule when applied to Table 19.1 in the text; any oxidizing agent will react spontaneously with a reducing agent of any half-reaction located below it in the table. In reaction (a) we ask, can $Fe^{3+}$ oxidize $I^-$ ion? In the table, $I^-$ ion appears below and to the right of $Fe^{3+}$. We conclude that $Fe^{3+}$ will oxidize $I^-$ ion.

b. To calculate the cell emf, separate the reaction into half-reactions:

reduction/cathode        $2 H^+(aq) + 2e^- \rightarrow H_2(g)$
oxidation/anode                    $Cu(s) \rightarrow Cu^{2+}(aq) + 2e^-$

---

$Cu(s) + 2 H^+(aq) \rightarrow Cu^{2+}(aq) + H_2(g)$

$E^\circ_{cell} = E^\circ_{cathode} - E^\circ_{anode} = E^\circ_{H^+/H_2} - E^\circ_{Cu^{2+}/Cu}$

$E^\circ_{cell} = 0.00\ V - (0.34\ V) = -0.34\ V$

The negative standard cell emf indicates that this reaction is not spontaneous at standard conditions. Cu will not oxidize in 1 *M* HCl.

---

**• Comment**

Since the cell emf changes sign when the reaction is reversed, the reaction
$Cu^{2+} + H_2 \rightarrow Cu + 2\,H^+$ will have a positive cell emf, and the reduction of $Cu^{2+}$ ion by $H_2$ will
be spontaneous under standard conditions.

**Work EXERCISES & PROBLEMS: 4 – 6**

---

**EXAMPLE 19.6  Standard Cell EMF**

Calculate the standard cell emf (cell potential) for a galvanic cell having the following overall
cell reaction of tin metal and lead(II) ion:

$$Sn(s) + Pb^{2+}(aq) \rightarrow Sn^{2+}(aq) + Pb(s)$$

**• Solution**

The standard emf of the cell is $E^{\circ}_{cell} = E^{\circ}_{cathode} - E^{\circ}_{anode}$

Separate the overall reaction into half-reactions, and identify the anode and cathode. We
see that $Pb^{2+}$ is reduced and tin is oxidized. Look up the necessary reduction potentials in
Table 19.1 of the text.

$$Pb^{2+}(aq) + 2e^- \rightarrow Pb(s) \qquad \text{reduction/cathode}$$
$$Sn(s) \rightarrow Sn^{2+}(aq) + 2e^- \qquad \text{oxidation/anode}$$

---

$$Sn(s) + Pb^{2+}(aq) \rightarrow Sn^{2+}(aq) + Pb(s) \quad \text{overall}$$

$$E^{\circ}_{cell} = E^{\circ}_{cathode} - E^{\circ}_{anode} = E^{\circ}_{Pb^{2+}/Pb} - E^{\circ}_{Sn^{2+}/Sn}$$

$$E^{\circ}_{cell} = -0.13\ V - (-0.14\ V) = 0.01\ V$$

**• Comment**

The positive value of the standard cell potential indicates that the reaction is spontaneous
under standard conditions.

**Work EXERCISES & PROBLEMS: 7**

---

**EXAMPLE 19.7  Free Energy Change and the Standard Cell EMF**

Calculate $\Delta G^{\circ}$ and the equilibrium constant at 25 °C for the reaction, given the standard cell
potential:

$$2\ Br^-(aq) + I_2(s) \rightarrow Br_2(\ell) + 2\ I^-(aq) \qquad\qquad E^{\circ}_{cell} = -0.54\ V$$

**• Solution**

The standard Gibbs free energy change is related to the standard cell potential by the
following equation in which F is the Faraday constant and n is the number of moles of
electrons transferred:

$$\Delta G° = -nFE°$$

The equilibrium constant at 25 °C is also related to $E°_{cell}$:

$$\ln K = \frac{nE°}{0.0257 \text{ V}}$$

Examination of the half-reactions clearly indicates that 2 mol of electrons are transferred in the balanced equation. $n = 2$.

The half-reactions are:

| | |
|---|---|
| reduction | $2e^- + I_2(s) \rightarrow 2\,I^-(aq)$ |
| oxidation | $2\,Br^-(aq) \rightarrow Br_2(\ell) + 2e^-$ |

overall $\quad 2\,Br^-(aq) + I_2(s) \rightarrow Br_2(\ell) + 2\,I^-(aq)$

Substituting into $\Delta G° = -nF\,E°_{cell}$

$$\Delta G° = -2(96{,}500 \text{ J/V·mol})(-0.54 \text{ V})$$

The units of the Faraday constant must be expressed in J/(V·mol) in order to cancel the units of volts from $E°_{cell}$. Continuing with our calculation:

$$\Delta G° = 1.04 \times 10^5 \text{ J} = 104 \text{ kJ/mol}$$

The positive value of $\Delta G°$ indicates the reaction is not spontaneous under standard conditions. Next, we calculate the equilibrium constant.

$$\ln K = \frac{nE°}{0.0257 \text{ V}} = \frac{2(-0.54 \text{ V})}{0.0257 \text{ V}} = -42.02$$

Taking the antilog of both sides we get:

$$K = e^{-42.0} = 5.6 \times 10^{-19}$$

---

### EXAMPLE 19.8  EMF for a Nonstandard Cell

Calculate the cell emf at 25 °C for a galvanic cell in which the following overall reaction occurs at the concentrations given:

$$Zn(s) + 2\,H^+(aq,\ 1.0 \times 10^{-4}\ M) \rightarrow Zn^{2+}(aq.\ 1.5\ M) + H_2(g,\ 1 \text{ atm})$$

• **Solution**

The cell emf can be calculated by use of the Nernst equation.

$$E = E° - \frac{0.0257 \text{ V}}{n} \ln \frac{[Zn^{2+}]P_{H_2}}{[H^+]^2}$$

Where,

$$E°_{cell} = E°_{cathode} - E°_{anode} = E°_{H^+/H_2} - E°_{Zn^{2+}/Zn} = 0.0 \text{ V} - (-0.76 \text{ V}) = 0.76 \text{ V}$$

n = 2 (moles of electrons transferred according to the balanced equation)
$P_{H_2} = 1$, because hydrogen is in its standard state, 1 atm
$[Zn^{2+}] = 1.5 \text{ } M$ and $[H^+] = 1.0 \times 10^{-4} \text{ } M$

Substitution into the Nernst equation yields:

$$E = 0.76 \text{ V} - \frac{0.0257 \text{ V}}{2} \ln \frac{1.5}{(1.0 \times 10^{-4})^2}$$

$$= 0.76 \text{ V} - 0.24 \text{ V} = 0.52 \text{ V}$$

• **Comment**

The low concentration of $H^+$ ($1.0 \times 10^{-4}$ $M$) compared to its standard state value (1 $M$) means that the driving force for reaction will be less than in the standard state, and as we see E < E°.

**Work EXERCISES & PROBLEMS: 10**

---

## EXAMPLE 19.9 Determining Ion Concentrations

A galvanic cell is constructed from a silver half-cell and a copper half-cell. The copper half-cell contains 0.10 $M$ $Cu(NO_3)_2$ and the concentration of silver ions in the other half-cell is unknown. If the Ag electrode is the cathode, and the cell emf is measured and found to be 0.10 V at 25 °C, what is the $Ag^+$ ion concentration?

• **Solution**

Since $E_{cell}$ is known and $E°_{cell}$ can be easily calculated, the Nernst equation can be used to calculate an unknown concentration. To write the reaction quotient, Q, we need the cell reaction. Since Ag is the cathode, the reduction half-reaction is:

reduction/cathode $\qquad Ag^+ + e^- \rightarrow Ag$

Then Cu must be oxidized:

oxidation/anode $\qquad Cu \rightarrow Cu^{2+} + 2e^-$

The overall reaction is: $Cu + 2 Ag^+ \rightarrow Cu^{2+} + 2 Ag$.  Therefore, $Q = [Cu^{2+}]/[Ag^+]^2$.

The standard cell potential when calculated as we have shown several times is 0.46 V. Also, n = 2. The Nernst equation is:

$$E = E° - \frac{0.0257 \text{ V}}{n} \ln \frac{[Cu^{2+}]}{[Ag^+]^2}$$

Substitution gives:

$$0.10 \text{ V} = 0.46 \text{ V} - \frac{0.0257 \text{ V}}{2} \ln \frac{(0.10)}{[Ag^+]^2}$$

This equation must be solved for the unknown concentration $[Ag^+]$.

$$\frac{2(0.10 \text{ V} - 0.46 \text{ V})}{-0.0257 \text{ V}} = \ln \frac{(0.10)}{[Ag^+]^2}$$

$$28 = \ln \frac{(0.10)}{[Ag^+]^2}$$

Take $e^x$ of both sides:

$$e^{28} = \frac{(0.10)}{[Ag^+]^2}$$

$$1.4 \times 10^{12} = \frac{(0.10)}{[Ag^+]^2}$$

And

$$[Ag^+] = \sqrt{\frac{0.10}{1.4 \times 10^{12}}}$$

$$[Ag^+] = 2.6 \times 10^{-7} \text{ } M$$

**Work EXERCISES & PROBLEMS: 11, 12**

---

**EXAMPLE 19.10  Electrolysis of Aqueous Solutions**

Predict the products of the electrolysis of aqueous $MgCl_2$ solution.

**• Solution**

In aqueous solution the metal ion is not the only species that can be reduced. $H_2O$ can be reduced as well. The two possible reduction reactions are:

$$Mg^{2+} + 2e^- \rightarrow Mg \qquad\qquad E^\circ_{red} = -2.37 \text{ V}$$
$$2 \text{ } H_2O + 2e^- \rightarrow H_2 + 2 \text{ } OH^- \qquad E^\circ_{red} = -0.83 \text{ V}$$

Of the two, water has a higher standard reduction potential, and therefore, has a greater tendency to be reduced than $Mg^{2+}$. The two oxidizable species are $H_2O$ and $Cl^-$ ion. The possible oxidation reactions and their $-E^\circ_{red}$ potentials are:

$$H_2O \rightarrow \tfrac{1}{2}O_2 + 2 \text{ } H^+ + 2e^- \qquad -E^\circ_{red} = -1.23 \text{ V}$$
$$2 \text{ } Cl^- \rightarrow Cl_2 + 2e^- \qquad\qquad -E^\circ_{red} = -1.36 \text{ V}$$

The oxidation potentials indicate that $H_2O$ is more readily oxidized than chloride ions. However, in this particular case, as discussed in Section 19.7 of the text, $Cl^-$ ions are actually oxidized. The large overvoltage for $O_2$ formation prevents its production when $Cl^-$ ion is there to compete. The overall reaction is:

$$2 H_2O + 2 Cl^- \rightarrow H_2 + 2 OH^- + Cl_2$$

• **Comment**

The only example of overvoltage that you will need to be concerned with is in the oxidation of $H_2O$. It lowers the $-E^{\circ}_{red}$ by about 0.6 V to −1.83 V.

**Work EXERCISES & PROBLEMS: 13**

---

### EXAMPLE 19.11 Quantitative Aspects of Electrolysis

How many moles of electrons are transferred in an electrolytic cell when a current of 12 amps flows for 16 hours?

• **Solution**

First, find the number of coulombs passing through the cell in 16 hours, and then, convert to moles of electrons. Our road map is:

current × time → coulombs → moles of electrons

Recall: 12 A = 12 C/s

The number of coulombs in 16 hours is given by multiplying the current times the time (in seconds).

$$\frac{12\ C}{1\ s} \times 16\ h \times \frac{3600\ s}{1\ h} = 691,000\ C$$

The Faraday constant is the charge of one mole of electrons. To three significant figures, $F = 9.65 \times 10^4$ C.
The number of moles of electrons is:

$$691,000\ C \times \frac{1\ mol\ e^-}{9.65 \times 10^4 C} = 7.2\ mol\ e^-$$

**Work EXERCISES & PROBLEMS: 14 – 16**

---

### EXAMPLE 19.12 Quantitative Aspects of Electrolysis

How many grams of copper metal will be deposited from a solution of $CuSO_4$ by the passage of 3.0 A of electrical current through an electrolytic cell for 2.0 hours?

• **Solution**

$Cu^{2+}$ is reduced according to the half-reaction:

$$Cu^{2+} + 2e^- \rightarrow Cu$$

This half-reaction equation tells us that 2 moles of electrons are required to produce 1 mol of Cu.

$$\frac{1\ mol\ Cu}{2\ mol\ e^-}$$

The number of coulombs passing through a cell is given by the current times the time.

$$3.0 \text{ A} \times 2.0 \text{ h} \times \frac{1 \text{ C/s}}{1 \text{ A}} \times \frac{3600 \text{ s}}{1 \text{ h}} = 2.16 \times 10^4 \text{ C}$$

From this number of coulombs we can find the number of moles of electrons passing through the cell. The road map will be:

current × time → coulombs → [moles e⁻] → mol Cu → g Cu

The number of moles of electrons is the link from the current measurements to the molar amount of Cu formed. The number of moles of electrons is:

$$2.16 \times 10^4 \text{ C} \times \frac{1 \text{ mol e}^-}{9.65 \times 10^4 \text{C}} = 0.224 \text{ mol e}^-$$

The number of moles of Cu is:

$$0.224 \text{ mol e}^- \times \frac{1 \text{ mol Cu}}{2 \text{ mol e}^-} = 0.112 \text{ mol Cu}$$

From balanced half-reaction

The number of grams of Cu is

$$0.112 \text{ mol Cu} \times \frac{63.5 \text{ g}}{1 \text{ mol Cu}} = 7.11 \text{ g Cu}$$

Of course, this calculation could be carried out using dimensional analysis.

$$3.0 \text{ A} \times \frac{1 \text{ C/s}}{1 \text{ A}} \times 2.0 \text{ h} \times \frac{3600 \text{ s}}{1 \text{ h}} \times \frac{1 \text{ mol e}^-}{9.65 \times 10^4 \text{C}} \times \frac{1 \text{ mol Cu}}{2 \text{ mol e}^-} \times \frac{63.5 \text{ g}}{1 \text{ mol Cu}} = 7.11 \text{ g Cu}$$

**Work EXERCISES & PROBLEMS: 17, 18**
**Work the rest of the EXERCISES & PROBLEMS**


## EXERCISES & PROBLEMS

1.  Balance the following redox equations.

    a. $Cr_2O_7^{2-}$ (aq) + Br⁻(aq) → $Cr^{3+}$(aq) + $Br_2$(g)   (acidic solution)
    b. $MnO_4^-$(aq) + I⁻(aq) → $MnO_2$(s) + $I_2$(aq)       (basic solution)

2.  Balance the following redox equations.

    a. $H_2C_2O_4$(aq) + $MnO_4^-$ (aq) → $Mn^{2+}$(aq) + $CO_2$(g)   (acidic solution)
    b. Zn(s) + ClO⁻(aq) → $Zn(OH)_2$(s) + Cl⁻(aq)       (basic solution)

3. Consider a voltaic cell constructed from the following half-cells, linked by a KCl salt bridge. Answer the following questions.

   - Cu electrode in 1.0 *M* $Cu(NO_3)_2$ solution
   - Pb electrode in 1.0 *M* $Pb(NO_3)_2$ solution

   a. Which electrode is the cathode?
   b. Write a balanced chemical equation for the reaction occurring when the cell is operating.
   c. Which electrode gains mass?
   d. Which direction do electrons flow in the external circuit?
   e. Which electrode will $K^+$ of the KCl salt bridge move toward?

4. Given the following standard reduction potentials in acid solution:

$$E^\circ_{red} \text{ (V)}$$

| | $E^\circ_{red}$ (V) |
|---|---|
| $Sn^{4+} + 2e^- \rightleftharpoons Sn^{2+}$ | +0.14 |
| $Al^{3+} + 3e^- \rightleftharpoons Al(s)$ | −1.66 |
| $Fe^{3+} + e^- \rightleftharpoons Fe^{2+}$ | +0.77 |

   a. Write the formula of the strongest reducing agent.
   b. Write the formula of the strongest oxidizing agent.

5. Predict whether $Br^-$ ion can reduce $I_2$ under standard state conditions.

   $$2\,Br^-(aq) + I_2(s) \rightarrow Br_2(\ell) + 2\,I^-(aq)$$

6. Predict whether the following reactions will occur spontaneously in aqueous solution at 25 °C under standard state conditions.

   a. $Ag(s) + Fe^{3+}(aq) \rightarrow Ag^+(aq) + Fe^{2+}(aq)$
   b. $Cu^{2+}(aq) + H_2(g) \rightarrow Cu(s) + 2\,H^+(aq)$

7. Calculate the standard emf ($E^\circ_{cell}$), at 25°C, of cells that have the following overall reactions:
   a. $2\,Co^{2+}(aq) + Zn^{2+}(aq) \rightarrow 2\,Co^{3+}(aq) + Zn(s)$
   b. $Cl_2(g) + 2\,Br^-(aq) \rightarrow 2\,Cl^-(aq) + Br_2(\ell)$

8. Calculate the standard Gibbs free energy changes for the two reactions in problem 7.

9. Calculate the equilibrium constant for the following redox reaction at 25 °C.
   $$2\,Fe^{2+}(aq) + Ni^{2+}(aq) \rightleftharpoons 2\,Fe^{3+}(aq) + Ni(s)$$

10. Determine the cell potential ($E_{cell}$), at 25 °C, for the following reaction where the concentrations are as shown.

    $$2\,Ag^+(1\,M) + H_2(1\,atm) \rightarrow 2\,Ag(s) + 2\,H^+ \ (pH = 7.12)$$

11. The cell emf, at 25 °C, for the following reaction is 0.98 V.

    $2\,Ag^+(aq, 1\ M) + H_2(g, 1\ atm) \rightarrow 2\,Ag(s) + 2\,H^+(aq, ?\ M)$
    Calculate the concentration of hydrogen ions.

12. Calculate $E^\circ_{cell}$, $E_{cell}$, and $\Delta G$ for the following cell reaction at 25°C.

    $3\,Zn(s) + 2\,Cr^{3+}(0.0010\ M) \rightarrow 3\,Zn^{2+}(0.010\ M) + 2\,Cr(s)$

13. a. Write the half-reaction that occurs at the anode during the electrolysis of aqueous lithium bromide.
    b. Write the half-reaction that occurs at the cathode during the electrolysis of aqueous lithium bromide?

14. How many coulombs are required to cause reduction of 0.20 mol of $Cr^{3+}$ to Cr?

15. How many moles of electrons are transferred in an electrolytic cell when a current of 5.0 amps flows for 10.0 hours?

16. How many moles of electrons are required to electroplate 29.4 g of nickel from a solution containing $Ni^{2+}$ ion?

17. How many minutes are required to electroplate 25.0 g of metallic Cu from a solution containing $Cu^{2+}$ ions using a constant current of 20.0 A?

18. A current of 0.80 A was applied to an electrolytic cell containing molten $CdCl_2$ for 2.5 hours. Decide on the appropriate half-reaction, and calculate the mass of cadmium metal deposited.

19. Consider the solution in the anode half-cell of Figure 19.1.
    a. How is the concentration of $Cu^{2+}$ ions related to the concentration of $SO_4^{2-}$ ions before any reaction occurs?
    b. How are the concentrations of these two ions related after reaction occurs?
    c. How does the solution stay electrically neutral?

20. Electrolysis experiments provide one of the most accurate ways to determine Avogadro's number. Electrolysis of a $NiSO_4$ solution by a current of 1.02 A for 252 s yielded a deposit of 0.0766 g Ni. Use this information to calculate a value for Avogadro's number. Do not use the Faraday constant, but you will need the electron charge: $1.602 \times 10^{-19}$ C.

21. Why are different products obtained when molten $AlBr_3$ and aqueous $AlBr_3$ are electrolyzed using inert electrodes?

## PRACTICE TEST QUESTIONS
See notes on taking practice test in the Preface

1. Draw the cell diagrams for each of the redox reactions given below. You may use platinum as an inert electrode.
   a. $2 Al(s) + 3 H_2SO_4(aq) \rightarrow Al_2(SO_4)_3(aq) + 3 H_2(g)$
   b. $Fe_2(SO_4)_3 + 3 Pb(s) \rightarrow 3 PbSO_4(s) + 2 Fe(s)$
   c. $CuSO_4(aq) + H_2(g) \rightarrow Cu(s) + H_2SO_4(aq)$
   d. $2 NaBr(aq) + I_2(g) \rightarrow Br_2(\ell) + 2 NaI(aq)$
   e. $2 FeCl_2(aq) + SnCl_2(aq) \rightarrow 2 FeCl_3(aq) + Sn(s)$
   f. $2 FeCl_2(aq) + SnCl_4(aq) \rightarrow 2 FeCl_3(aq) + SnCl_2(aq)$

2. Consider a galvanic cell constructed from the following half-cells that are linked by a porous membrane. (1) an Au electrode dipped into 1.0 *M* $Au(NO_3)_3$ and (2) an Fe electrode dipped into 1.0 *M* $FeSO_4$. Answer the following questions:
   a. Which electrode is the cathode?
   b. Write a balanced equation for the reaction occurring while the cell is discharging.
   c. What emf should the cell generate?
   d. In what direction will electrons flow in the outer circuit?
   e. Toward which electrode will positive ions migrate?

3. Arrange the following species in order of increasing strength as oxidizing agents:
   $Ce^{4+}$, $O_2$, $H_2O_2$, $SO_4^{2-}$

4. Arrange the following species in order of increasing strength as reducing agents:
   $Zn$, $Ni$, $H_2$, $F^-$

5. a. Will $O_2(g)$ oxidize $I^-$ ions in acid solution under standard conditions?
   b. Will $O_2(g)$ oxidize $Br^-$?
   c. Will $O_2(g)$ oxidize $Cl_2$?

6. Calculate $\Delta G°$ and the equilibrium constant (K) for the following reaction at 25 °C.

   $$Fe^{3+}(aq) + Ag(s) \rightarrow Fe^{2+}(aq) + Ag^+(aq)$$

7. At what ratio of $[Fe^{2+}]/[Fe^{3+}]$ would the reaction given in problem 6 become spontaneous if $[Ag^+] = 1.0$ *M*?

8. Calculate the cell emf for the following reaction:

   $$2 Ag^+(0.10 \ M) + H_2(1 \ atm) \rightarrow$$
   $$2 Ag(s) + 2 H^+(pH = 8)$$

19

19

19

19

19

19

19

19

19

9. What is the minimum voltage required to bring about electrolysis of a solution of $Cu^{2+}$ and $Br^-$ at standard concentrations?

10. What change in the emf of a hydrogen-copper cell will occur when $NaOH(aq)$ is added to the solution in the hydrogen half-cell?

11. Consider an uranium-bromine cell in which U is oxidized and $Br_2$ is reduced. The half-reactions are:

$U^{3+}(aq) + 3e^- \rightarrow U(s)$    $E°_{U^{3+}/U} = ?$

$Br_2(\ell) + 2e^- \rightarrow 2Br^-(aq)$   $E°_{Br_2/Br^-} = 1.07$ V

If the standard cell emf is 2.91 V, what is the standard reduction potential for uranium?

12. When the concentration of $Zn^{2+}$ ion is 0.15 $M$, the measured voltage of the Zn-Cu cell is 0.40 V. What is the $Cu^{2+}$ ion concentration?

13. A hydrogen electrode with 1.0 atm $H_2$ is immersed in an acetic acid solution. This electrode is connected to another, consisting of an iron nail dipped into 0.10 $M$ $FeCl_2$. If $E_{cell}$ is found to be 0.64 V, what is the pH of the acetic acid solution?

14. How many moles of electrons are transferred in an electrolytic cell when a current of 2.0 amps flows for 6 hours?

15. How many moles of electrons are required to electroplate 6.0 g of chromium from a solution containing $Cr^{3+}$?

16. a. How many grams of nickel can be electroplated by passing a constant current of 5.2 A through a solution of $NiSO_4$ for 60.0 min?
   b. How many grams of cobalt can be electroplated by passing a constant current of 5.2 A through a solution of $CoCl_3$ for 60.0 min?

17. How long will it take to produce 54 kg of Al metal by the reduction of $Al^{3+}$ in an electrolytic cell using a current of 500 amps?

18. Balance the following redox reactions, in acidic solution, by the ion-electron method.
   a. $H_2O_2 + I^- \rightarrow I_2 + H_2O$
   b. $Cr_2O_7^{2-} + H_3AsO_3 \rightarrow Cr^{3+} + H_3AsO_4$
   c. $Ag + NO_3^- \rightarrow NO_2 + Ag^+$

19. Balance the following redox reaction, in basic solution, by the ion-electron method.

$$Cl_2 \rightarrow ClO_4^- + Cl^-$$

20. Given the reduction potential of the standard hydrogen electrode, and the half-reaction equation:

$$2H_2O + 2e^- \rightarrow H_2(g) + 2OH^-(aq)$$
$$E° = -0.83 \text{ V}$$

calculate the value of the ionization constant for water at 25 °C.

$$H_2O \rightleftharpoons H^+(aq) + OH^-(aq)$$

21. A cell was constructed using a standard hydrogen electrode as one half-cell, and another hydrogen half-cell with a weak acid solution and 1.00 atm $H_2$ gas. If the cell voltage is 0.20 V, what is the pH of the weak acid solution?

22. A 40 watt light bulb is powered by a lead storage battery that has 20.0 g of Pb available for the anode. For how many hours will the bulb provide illumination? Assume the voltage is constant at 1.5 V. Recall: 1 watt = 1 J/s.

19          19

19          19

# Chapter 20. Metallurgy and the Chemistry of Metals

**Metallurgical Processes (Sections 20.1 – 20.2)**
**The Band Theory of Conductivity (Section 20.3)**
**The Alkali Metals (Section 20.5)**
**The Alkaline Earth Metals (Section 20.6)**
**Aluminum (Section 20.7)**

## SUMMARY

### Metallurgical Processes (Sections 20.1 – 20.2)

*Preliminary Treatment.* Metallurgy is the science and technology of extracting metals from their ores and preparing these metals for use. Most metals are found in nature in the combined state as minerals. A mineral is a naturally occurring substance with a characteristic range of composition. An ore is a mineral deposit of high enough concentration to make an economical recovery of a desired metal. Ores generally must be cleaned before the metal is extracted. The waste materials, called gangue, are removed by flotation, ferromagnetic separation, or distillation of mercury amalgams.

*Reduction of Metals.* The process of producing a free metal is always one of reduction. Reduction of metal oxides is usually less complex than reduction of sulfides. Roasting is a preliminary operation used to convert a metal sulfide or carbonate into an oxide. Two types of reductions are ordinarily employed:

1. Chemical reduction involves the use of a reducing agent to prepare the elemental form of a metal from a compound. Calcium, magnesium, aluminum, hydrogen, and carbon are often used in chemical reductions.
2. Electrolytic reduction is necessary to produce the very electropositive metals such as sodium, magnesium and aluminum.

*Iron.* Iron is prepared by a chemical reduction process. The raw materials for making iron are (1) iron ore, either hematite $Fe_2O_3$ or magnetite $Fe_3O_4$, (2) coke, and (3) limestone. These three materials are mixed and fed into a huge blast furnace (Figure 20.3 text). A strong blast of air preheated to 1500 °C is blown in at the bottom, where oxygen in the air reacts with the coke. This reaction supplies heat for the furnace, and carbon monoxide which is the reducing agent in the reaction of the iron oxides.

$$2 \ C(s) + O_2(g) \rightarrow 2 \ CO(g)$$
$$Fe_2O_3(s) + 3 \ CO(g) \rightarrow 2 \ Fe(\ell) + 3 \ CO_2(g)$$

A temperature gradient is set up in the furnace. The temperature at the bottom is 1500 °C, and it decreases with height; falling to about 250 °C at the top. Much of the carbon dioxide produced is reduced by reaction with excess coke as follows:

$$CO_2(g) + C(s) \rightarrow 2 \ CO(g)$$

thereby forming more of the principal reducing agent. The gas that escapes at the top of the furnace is mostly nitrogen and carbon monoxide. The molten iron runs to the bottom where it is withdrawn periodically.

Limestone serves as a flux in the removal of impurities such as silica ($SiO_2$) and alumina ($Al_2O_3$). The limestone decomposes at temperatures above 900 °C, forming calcium oxide and carbon dioxide.

$$CaCO_3(s) \rightarrow CaO(s) + CO_2(g)$$

Calcium oxide unites with silica and alumina to form a glassy, molten substance called slag which is composed mainly of $CaSiO_3$ and some calcium aluminate. Slag is less dense than iron, and so it collects as a pool on top of the metal. It is drawn off, leaving the molten iron. Iron prepared in this manner contains many impurities and is called pig iron, or cast iron. Cast iron is made into steel by further treatment in a basic oxygen furnace.

***Purification of Metals.***   Once a metal is prepared it may need to be further purified. Distillation, electrolysis, and zone refining are three common procedures. Metals with relatively low boiling points, such as mercury (357 °C), cadmium (767 °C), and zinc (907 °C), are referred to as volatile. They can be purified by fractional distillation. On heating, these metals vaporize, leaving behind any nonvolatile impurities. Condensation of the vapor yields the purified metal.

Metals can also be purified by electrolysis. The metals that plate out on a cathode can be controlled by the voltage at which the electrolysis is carried out. In the purification of copper, $Cu^{2+}$ is reduced much more readily than iron or zinc ions, which are common impurities. At just the right voltage Cu is plated out and the metal impurities remain in solution.

Zone refining takes advantage of the fact that when liquids begin to freeze, the impurities tend to remain in the liquid phase. See Figure 20.8 of the textbook. Extremely pure metals can be obtained by repeating this process a number of times.

| Examples 20.1, 20.2 |
|---|
| Exercises 1 – 5 |

# The Band Theory of Conductivity (Section 20.3)

***The Conduction Band.***   The band theory is the result of the application of molecular orbital theory to metals. In metals, the atoms lie in a three-dimensional array and take part in bonding that spreads over the entire crystal. Take, for example, an alkali metal atom that carries a single valence electron in an s orbital. Recall that the atomic orbitals of two atoms will overlap when the atoms are close together. This results in the formation of two molecular orbitals:  a bonding orbital and an antibonding orbital. The total number of molecular orbitals produced always equals the number of atomic orbitals that overlap to produce them. The overlap of four atomic orbitals from four atoms, for instance, will form four molecular orbitals:  two bonding and two antibonding orbitals. A crystal made up of $N$ atoms with overlapping atomic orbitals will have a total of $N$ molecular orbitals. These bonding and antibonding orbitals are so closely spaced in terms of energy that they are called a "band."  The formation of a band is shown in Figure 20.1.

The band containing the valence shell electrons (3s for Na) is called the valence band. For alkali metals, there are $N$ valence electrons and $N$ orbitals in the band. These $N$ orbitals have a capacity for $2N$ electrons. Any band that is either vacant or partially filled is called a conduction band.

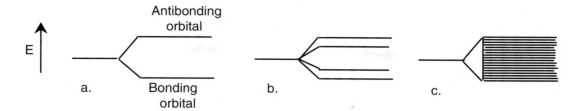

**Figure 20.1** The formation of an energy band by the successive overlap of atomic orbitals. a. When the orbitals of two atoms overlap, one bonding orbital and one antibonding orbital of significantly different energies are formed. b. When the orbitals of four atoms overlap, four molecular orbitals are formed. c. When the orbitals of *N* atoms overlap where *N* is a large number, *N* orbitals are formed. With so many orbitals, the orbital energies differ only infinitesimally from each other and form a virtually continuous band.

***Electrical Conduction.*** The band theory explains conduction in the alkali metals in the following way. When a voltage is applied across a piece of sodium metal, conduction occurs. The current is the result of electrons in the 3s band being free to jump from atom to atom. The free movement of electrons is possible for two reasons. In alkali metals the conduction band and the valence band are the same. The orbitals within the band are so similar in energy that an electron does not need to gain appreciable energy to reach the conduction band. Also, as discussed above, the conduction band is only half filled to capacity which means that an electron is free to move through the entire metal.

In an insulator, such as glass or plastic, the valence band is filled. Thus, the next vacant higher-energy band becomes the conduction band. An energy gap exists between the valence band and the conduction band, as shown in Figure 20.10 of the text. This large separation prevents electrons in insulators from entering the conduction band.

A semiconductor, such as Si and Ge, has a filled valence band and an empty conduction band, but in contrast to an insulator, a relatively small gap exists between these bands. A relatively small amount of thermal energy will promote an electron into the conduction band. Thus as temperature increases, the conductivity of semiconductors increases correspondingly.

> Example
> 20.3
>
> Exercises
> 6 – 7

## The Alkali Metals (Section 20.5)

***Properties of the Alkali Metals.*** As a group, the alkali metals are extremely reactive and never occur naturally in the elemental form. They are the least electronegative group of elements and exist as +1 cations combined with halides, sulfate, carbonate, and silicate ions. A number of properties of these elements are listed in Table 20.4 of the text. Compared with other metals, they are very soft and have low melting points and low densities. Lithium, sodium and potassium will float on water. The low density is a result of their large atomic volume and the fact that they all possess a body-centered crystal structure which has a low packing efficiency.

***Sources and Preparations.*** The preparation of Group 1A metals requires large amounts of energy because their positive ions are difficult to reduce. Lithium, being the most active metal, is prepared by electrolysis of molten LiCl.

Sodium and potassium occur in a wide variety of minerals, and are the sixth and seventh most abundant elements in earth's crust. Sodium compounds are so abundant and widespread that it is difficult to find matter free of this element. Sodium chloride makes up about two-thirds of the solid matter dissolved in sea water. The minerals carnallite ($KMgCl_3 \cdot 6H_2O$), and sylvite (KCl) are found in ancient lake and seabeds and serve as commercial sources of potassium and its compounds.

Both metals, sodium and potassium, were first prepared in 1807 by the English chemist Humphry Davy. He used the electrolysis of the corresponding moist hydroxides. As their discoverer, Davy was allowed to name the elements giving them the previous names. However, you will notice that neither has an element symbol consistent with its name. At the time, both potash and soda were recognized as carbonates of these unisolated metal elements. Potassium obtained its name because it was the metallic element in potash, potassium carbonate. The metallic element of soda (sodium carbonate), Davy called sodium.

***Sodium and Potassium.***  Metallic sodium is prepared commercially by the electrolysis of molten sodium chloride. Most of the sodium made in the United States is produced in the Downs cell (Figure 19.17 text), by the electrolysis of a mixture of sodium chloride and calcium chloride. This electrolyte mixture melts at 505 °C, whereas pure NaCl melts at 801 °C. The lower temperature reduces the cost of production.

A Downs cell consists of a carbon anode and an iron cathode. The chloride ions are oxidized at the anode, and the sodium ions are reduced at the cathode. The half-reactions are:

anode $\qquad$ $2\ Cl^- \rightarrow Cl_2(g) + 2e^-$
cathode $\qquad$ $2\ Na^+ + 2e^- \rightarrow 2\ Na(\ell)$

The liquid sodium is drawn off and kept from contact with chlorine and oxygen. Sodium can also be obtained by the electrolysis of molten sodium hydroxide.

Potassium is made by reaction of potassium chloride with sodium vapor in the absence of air at 900 °C.

$$KCl(\ell) + Na(g) \rightarrow K(g) + NaCl(\ell)$$

This reaction should occur to only a small extent. However, removal of potassium vapor as it is formed drives the reaction to completion.

***Reactions of Alkali Metals.***  The alkali metals are extremely reactive. The reactivity increases with atomic mass. Sodium forms two oxides on reaction with oxygen:  sodium oxide ($Na_2O$) and sodium peroxide ($Na_2O_2$).

$$2\ Na(s) + \tfrac{1}{2}\ O_2(g) \rightarrow Na_2O(s)$$
$$2\ Na(s) + O_2(g) \rightarrow Na_2O_2(s)$$

The reaction of potassium with oxygen forms the peroxide ($K_2O_2$) but, in addition, when K burns in air it forms potassium superoxide.

$$K(s) + O_2(g) \rightarrow KO_2(s)$$

Potassium superoxide is a source of $O_2$ in breathing equipment. $KO_2$ reacts with water:

$$2\ KO_2(s) + 2\ H_2O(\ell) \rightarrow 2\ KOH(aq) + O_2(g) + H_2O_2(aq)$$

Exercises
8 – 9

Sodium chloride, sodium carbonate, sodium bicarbonate, sodium hydroxide, sodium nitrate, potassium hydroxide, and potassium nitrate are important compounds of these elements.

## The Alkaline Earth Metals (Section 20.6)

***Properties of the Alkaline Earth Metals.*** The alkaline earth metals are somewhat less reactive than the alkali metals. Except for beryllium, the alkaline earth metals have similar chemical properties. The oxidation number of the Group 2A metals in almost all compounds is +2. The electronegativity of these elements is low enough that they form predominately ionic compounds with nonmetals. Table 20.5 in the text lists some common properties of these metals. As with the other groups of elements in the periodic table, metallic character increases as you read down the group. The properties of magnesium and calcium are reviewed below. The formulas of the important ores of magnesium and calcium are given in Table 20.1.

**Table 20.1** Ores Containing Magnesium and Calcium

| Element | Rank (Earth's crust) | Ore | Formula |
|---|---|---|---|
| Magnesium | 6th | brucite | $Mg(OH)_2$ |
|  |  | dolomite | $MgCO_3 \cdot CaCO_3$ |
|  |  | epsomite | $MgSO_4 \cdot 7H_2O$ |
| Calcium | 5th | limestone | $CaCO_3$ |
|  |  | gypsum | $CaSO_4 \cdot 2H_2O$ |
|  |  | fluorite | $CaF_2$ |

***Preparation of Magnesium and Calcium.*** Seawater is an important source of magnesium. It contains 0.13 percent $Mg^{2+}$ by weight. $Mg(OH)_2$ being an insoluble compound is precipitated from seawater by adding $Ca(OH)_2$.

$$Mg^{2+}(aq) + Ca^{2+}(aq) + 2\ OH^-(aq) \rightarrow Mg(OH)_2(s) + Ca^{2+}(aq)$$

Magnesium hydroxide is then converted to magnesium chloride by neutralization with hydrochloric acid. Metallic magnesium is obtained by electrolysis of molten magnesium chloride.

$$Mg^{2+}(aq) + 2\ Cl^-(aq) \xrightarrow{\text{electrolysis}} Mg(\ell) + Cl_2(g)$$

Calcium is obtained from limestone. Heating limestone causes it to decompose into calcium oxide and carbon dioxide. Calcium metal is prepared by a thermal reduction process, rather than electrolysis. Calcium oxide is reduced with aluminum at high temperature (1200 °C).

$$CaCO_3(s) \rightarrow CaO(s) + CO_2(g)$$
$$6\ CaO(s) + 2\ Al(\ell) \rightarrow 3\ Ca(\ell) + Ca_3Al_2O_6(s)$$

***Properties of Magnesium and Calcium.*** Magnesium is less reactive than calcium. Magnesium does not react with cold water, but it reacts slowly with steam. Calcium, on the other hand, reacts with cold water.

$$Mg(s) + H_2O(g) \rightarrow MgO(s) + H_2(g)$$
$$Ca(s) + 2\ H_2O(\ell) \rightarrow Ca(OH)_2(aq) + H_2(g)$$

Both MgO and CaO react with water to give hydroxides.

$$MgO(s) + H_2O(\ell) \rightarrow Mg(OH)_2(s) \qquad \text{(reacts with steam)}$$
$$CaO(s) + H_2O(\ell) \rightarrow Ca(OH)_2(s) \qquad \text{(reacts with cold water)}$$

When magnesium burns in air, considerable nitride is formed along with the oxide.

$$2\ Mg(s) + O_2(g) \rightarrow 2\ MgO(s)$$
$$3\ Mg(s) + N_2(g) \rightarrow Mg_3N_2(s)$$

***Uses of the Alkaline Earth Metals.*** Beryllium has excellent alloying qualities and is used to make alloys that are corrosion resistant. Beryllium is used as a "window" in X-ray tubes.

The major uses of magnesium are also in alloys. Because of its low atomic mass, it is a good lightweight structural metal. It is also used in batteries, and for cathodic protection of buried metal pipelines and storage tanks.

Metallic calcium finds use mainly in alloys. Calcium salts are used as dehydrating agents; anhydrous calcium chloride, for example, has a strong affinity for water. Quicklime, CaO, is used in steel production, and in the removal of $SO_2$ from coal-fired power plants. Slaked lime, $Ca(OH)_2$, is used in water treatment.

Strontium nitrate and carbonate are used in fireworks and highway flares to provide their brilliant red color. Barium metal is the most active of the alkaline earth metals so it has very few uses.

| Examples 20.4, 20.5 |
| Exercises 10 – 11 |

## Aluminum (Section 20.7)

***Source and Preparation.*** Aluminum is the third most abundant element in the earth's crust, making up 7.5 percent by mass. It is too active chemically to occur free in nature, but is found in compounds in over 200 different minerals. The most important ore of Al is bauxite, which contains hydrated aluminum oxide ($Al_2O_3 \cdot 2\ H_2O$), along with silica and hydrated iron oxide ($Fe_2O_3 \cdot 2\ H_2O$). Another important mineral is cryolite ($Na_3AlF_6$), which is used in the metallurgy of aluminum. The Hall process for the production of aluminum metal is described in Figure 20.18 of the text.

Aluminum is obtained from bauxite ore ($Al_2O_3 \cdot 2\ H_2O$). The first step in its preparation is to separate pure aluminum oxide from the silica and iron impurities. First, bauxite is pulverized and digested with sodium hydroxide solution. This converts the silica into soluble silicates, and aluminum oxide is converted to the aluminate ion $AlO_2^-$ which remains in solution. However this digestion treatment has no effect on the iron which remains as insoluble $Fe_2O_3(s)$. It is removed by filtration. Aluminum hydroxide is then precipitated by acidification (with carbonic acid) to about pH 6. The precipitate is heated strongly to produce pure $Al_2O_3$. The chemical changes involving aluminum are:

$$Al_2O_3(s) + 2\ OH^-(aq) \rightarrow 2\ AlO_2^-\ (aq) + H_2O(\ell)$$
$$AlO_2^-\ (aq) + H^+(aq) + H_2O(\ell) \rightarrow Al(OH)_3(s)$$
$$2\ Al(OH)_3(s) \rightarrow Al_2O_3(s) + 3\ H_2O(g)$$

The aluminum ions in $Al_2O_3$ can be reduced to metallic aluminum efficiently only by electrolysis. The melting point of $Al_2O_3$ is 2050 °C, which makes electrolysis of pure molten $Al_2O_3$ extremely expensive owing to the need to maintain the high temperature. In the Hall process, $Al_2O_3$ is dissolved in molten cryolite ($Na_3AlF_6$) which melts at 1000 °C. The use of cryolite makes it possible to lower the temperature of electrolysis by 1050 °C! The mixture is electrolyzed to produce aluminum and oxygen.

| cathode | $2\ Al^{3+} + 6e^- \rightarrow 2\ Al(\ell)$ |
|---|---|
| anode | $3\ O^{2-} \rightarrow \frac{3}{2}\ O_2 + 6e^-$ |

| overall | $2\ Al^{3+} + 3\ O^{2-} \rightarrow 2\ Al(\ell) + \frac{3}{2}\ O_2$ |
|---|---|

**Properties.** Pure aluminum is a silvery-white metal with low density ($2.7\ g/cm^3$) and high tensile strength. It is malleable and can be rolled into thin foils. Its electrical conductivity is about 65 percent that of copper.

Aluminum is an amphoteric element, reacting with both acids and bases.

$$2\ Al(s) + 6\ HCl(aq) \rightarrow AlCl_3(aq) + 3\ H_2(g)$$
$$2\ Al(s) + 2\ NaOH(aq) + 2\ H_2O(\ell) \rightarrow 2\ NaAlO_2(aq) + 3\ H_2(g)$$

Aluminum has a strong affinity for oxygen, thus, it is usually covered by an oxide film:

$$4\ Al(s) + 3\ O_2(g) \rightarrow 2\ Al_2O_3(s)$$

This layer of $Al_2O_3$ forms a compact, adherent, protective, surface coating on the metal and is responsible for preventing further oxidation and corrosion. Because of this surface oxide layer, Al is practically insoluble in weak or dilute acids.

The large enthalpy of formation of aluminum oxide ($\Delta H_f^{\circ} = -1670\ kJ/mol$) makes the metal an excellent reducing agent. Thus, a variety of metals can be produced in a series of similar reactions involving Al powder with the corresponding metal oxides. So much heat is liberated in these reactions, called aluminothermic reactions, that the metal is usually obtained in the molten state. The thermite reaction, used in welding steel and iron, is one example.

$$2\ Al(s) + Fe_2O_3(s) \rightarrow Al_2O_3(\ell) + 2\ Fe(\ell)$$

Aluminum hydroxide is an amphoteric hydroxide, dissolving in both acid and base.

$$Al(OH)_3(s) + 3\ H^+(aq) \rightarrow Al^{3+}(aq) + 3\ H_2O(\ell)$$
$$Al(OH)_3(s) + OH^-(aq) \rightarrow Al(OH)_4^-(aq)$$

**Uses.** Aluminum metal is used in high-voltage transmission lines because it is cheaper and lighter than copper. Its chief use is in aircraft construction. Aluminum is employed also as a solid propellant in rockets. This is another example of the great affinity that aluminum has for oxygen. Ammonium perchlorate, $NH_4ClO_4$, is the oxidizer.

The formation of aluminum hydroxide precipitate is used in water treatment plants. The process requires large amounts of aluminum sulfate.

$$Al_2(SO_4)_3(aq) + 3\ Ca(OH)_2(aq) \rightarrow 2\ Al(OH)_3(s) + 3\ CaSO_4(aq)$$

Aluminum hydroxide is a gelatinous substance. As it settles in treatment pools, it coprecipitates suspended matter such as bacteria and colloidal sized particles. This process clarifies drinking water.

Alums are compounds that have the general formula:

$$M^+M^{3+}(SO_4)_2 \cdot 12\ H_2O$$

Examples
20.6, 20.7

Exercises
12 – 14

where $M^+ = K^+$, $Na^+$, $NH_4^+$, and $M^{3+} = Al^{3+}$, $Cr^{3+}$, $Fe^{3+}$. They are used in the dying industry. The formation of the gelatinous aluminum hydroxide "fixes" the dye to the cloth.

## GLOSSARY LIST

| | | |
|---|---|---|
| metallurgy | pyrometallurgy | n-type semiconductor |
| alloy | amalgam | p-type semiconductor |
| mineral | band theory | |
| ore | conductor | acceptor impurity |
| ferromagnetic | insulator | donor impurity |
| | semiconductor | |

## WORKED EXAMPLES

### EXAMPLE 20.1  Chemical Reduction

Chemical reduction is the use of a reducing agent to prepare the elemental form of a metal from a compound. Look up the reduction potentials of calcium, magnesium, and hydrogen, listed in Table 19.1 in the text and suggest five metals that calcium, and magnesium can reduce, but that hydrogen cannot reduce.

• **Solution**

Ca and Mg are strong reducing agents. Their reduction potentials are quite low which means that they have a strong tendency to transfer electrons to other metal cations.

$$Ca^{2+}(aq) + 2e^- \rightarrow Ca(s) \qquad E^\circ_{red} = -2.87$$

$$Mg^{2+}(aq) + 2e^- \rightarrow Mg(s) \qquad E^\circ_{red} = -2.37 \text{ V}$$

$$2\,H^+ + 2e^- \rightarrow H_2(g) \qquad E^\circ_{red} = 0 \text{ V}$$

$H_2$ is not as strong a reducing agent as the others.

Ca(s) and Mg(s) can be used to reduce ions of metals having higher reduction potentials (the diagonal rule) such as $Zn^{2+}$, $Cr^{3+}$, $Co^{2+}$, $Ni^{2+}$, and $Sn^{2+}$, but not $Li^+$ and $K^+$. $H_2$ cannot reduce any of these ions, but is used to reduce more noble metals such as $Cu^{2+}$.

### EXAMPLE 20.2  Steel Production

How are the impurities present in iron ore, such as silica and sulfur, removed during steel production?

• **Solution**

Acidic impurities such as silica and sulfides combine with CaO, a base, to form a molten slag.

$$SiO_2 + CaO \rightarrow CaSiO_3$$
$$FeS + CaO \rightarrow CaS + FeO$$

The slag has a lower density than molten iron. Therefore, it collects as a pool on top of the metal. This slag is drawn off and used as a component in making cement.

**Work EXAMPLES & PROBLEMS: 1 – 5**

### EXAMPLE 20.3  Band Theory

Show for sodium that the valence band is half-filled.

**• Solution**

Sodium atoms have filled 1s, 2s, and 2p orbitals; therefore, the corresponding bands in the solid are filled. The valence band for Na is the 3s. In the metal, the 3s orbitals of $N$ sodium atoms overlap to form a total of $N$ molecular orbitals that make up the valence band. The capacity of this band is $2N$ electrons. Since each Na atom has one 3s electron, the $N$ Na atoms have $N$ electrons in the valence band. This means that the valence band is half-filled. The 3s band of Na is the conduction band.

**• Comment**

The highest occupied band is called the valence band.  This unfilled valence band is called the conduction band of sodium metal.

**Work EXERCISES & PROBLEMS: 6 - 9**

### EXAMPLE 20.4  Properties of Alkaline Earth Metals

Identify the element among the alkaline earth metals (excluding radium) that will have the following properties:
a.  Most reactive oxide towards water
b.  Lowest electronegativity
c.  Smallest atomic radius
d.  Lowest first ionization energy

**• Solution**

Applying peroidic trends in these properties.
a.  Ba because reactivity increases as you read down in the group.
b.  Ba because electronegativity decreases as you read down in the group.
c.  Be because atomic radius increases as you read down in the group.
d.  Ba because ionization energy decreases as you read down in the group.

### EXAMPLE 20.5  Electrolytic Reduction of Calcium

Starting with $Ca^{2+}$ ions in limestone, write equations to show how you would obtain pure calcium, using electrolysis, rather than thermal reduction.

**• Solution**

Electrolysis of molten metal chlorides is often used to prepare pure metals. First, limestone, chalk, or sea shells, all of which contain $CaCO_3$, must be decomposed at 900 °C.

$$CaCO_3(s) \rightarrow CaO(s) + CO_2(g)$$

The CaO(s) is slaked to yield calcium hydroxide.

$$CaO(s) + H_2O(\ell) \rightarrow Ca(OH)_2(s)$$

This base is then neutralized with hydrochloric acid to give the desired salt $CaCl_2$.

$$Ca(OH)_2(aq) + 2\ HCl(aq) \rightarrow\ CaCl_2(aq) + 2\ H_2O(\ell)$$

After drying, the $CaCl_2$ can be melted and then electrolyzed.

electrolysis
$$CaCl_2(\ell)\ \rightarrow\ Ca(\ell) + Cl_2(g)$$

**Work EXERCISES & PROBLEMS: 10, 11**

## EXAMPLE 20.6  Amphoterism

Explain the meaning of the term amphoterism. Write equations for the amphoteric behavior of aluminum.

### • Solution

An amphoteric substance can react either as an acid or as a base. Aluminum reacts with hydrochloric acid and with strong bases as follows:

$$2\ Al(s) + 6\ HCl(aq) \rightarrow AlCl_3(aq) + 3\ H_2(g)$$
$$2\ Al(s) + 2\ NaOH(aq) + 2\ H_2O(\ell) \rightarrow 2\ NaAlO_2(aq) + 3\ H_2(g)$$

## EXAMPLE 20.7  Oxidation of Aluminum

Aluminum is a good reducing agent, and its reduction potential is quite negative, $(E^{\circ}_{red} = -1.66\ V)$ which means that aluminum should react with water and liberate hydrogen. But, we know that airplanes do not dissolve in thunderstorms. Explain.

### • Solution

Aluminum readily forms the oxide $Al_2O_3$ when exposed to air. The oxide forms a tenacious surface film that protects the aluminum metal from further corrosion due to water. Therefore, it is the presence of this oxide layer that makes aluminum practically insoluble even in dilute acids.

**Work EXERCISES & PROBLEMS 12 – 15**

## EXERCISES & PROBLEMS

1. Distinguish between an ore and a mineral.

2. Why must electrolytic reduction be used in the production of some metals such as lithium and sodium?

3. What is zone refining used for?

4. How is limestone used in iron production?

5. Write an equation for the chemical reduction of $Cr_2O_3$ with Al metal yielding chromium.

6. Define the terms: valence band and conduction band.

7. Define the terms: conductor, insulator, and semiconductor.

8.  List four physical properties of the alkali metals.

9.  List three chemical properties of the alkali metals.

10. Give the sources of magnesium and calcium.

11. Write chemical formulas for limestone, quicklime, and slaked lime.

12. What is the role of cryolite, $Na_3AlF_6$, in aluminum production?

13. Write chemical equations that show the amphoteric nature of $Al(OH)_3$.

14. The enthalpy of formation of $Al_2O_3$ is large and exothermic. Why doesn't aluminum metal just completely oxidize away?

15. What is an alum?  Which alum is used in baking powder?  Hint:  Read a label. What is its function in baking powder?

## PRACTICE TEST QUESTIONS
See notes on taking practice test in the Preface

1.  Distinguish between:
    a.  Leaching and flotation
    b.  Roasting and reduction

2.  Can carbon be used to reduce $Al_2O_3$ to $Al(s)$?  Hint:  Is the following reaction spontaneous?

    $2 Al_2O_3(s) + 2 C(s) \rightarrow 4 Al(s) + 3 CO_2(g)$

3.  When a solution of NaOH is added, dropwise to a test tube containing an aqueous solution of $Al^{3+}$, a white gelatinous precipitate is formed. Upon continued addition of NaOH, the precipitate disappears. Write the chemical equations to explain this observation.

4.  In the production of sodium, the metal is prevented from coming into contact with $Cl_2$ and $O_2$. Write chemical equations for the reactions of these elements with sodium.

5.  What is the agent that actually reduces Fe in the blast furnace?  Write a chemical equation to show the reaction.

6.  Distinguish between an alloy and an amalgam.

20

20

20

20

20

20

7. According to the band theory, why is copper a conductor?

8. Considering that electrons in metals have random thermal motion like molecules of gas, explain why electrical conductivity of conductors decreases with increasing temperature.

9. Why does the electrical conductivity of semiconductors increase with increasing temperature?

10. What are the sources of aluminum and potassium?

11. What is quicklime? How is it made?

12. Aluminum metal was first prepared in 1825 by the action of potassium metal on aluminum chloride. Write a balance equation for this reaction.

13. What is the chemical form of aluminum in bauxite?

14. How is aluminum separated from $Fe_2O_3$?

15. What is the role of aluminum sulfate in water purification?

16. What is the pH of a 0.10 *M* aluminum chloride solution, if it is hydrolyzed to the extent of 10 percent at room temperature?
$[Al(H_2O)_6]^{3+} \rightarrow H^+ + [Al(H_2O)_5OH]^{2+}$

20          20

20          20

20          20

20          20

20          20

# Chapter 21. Non-Metallic Elements and their Compounds

**Hydrogen (Sections 21.1 – 21.2)**
**Carbon (Section 21.3)**
**Nitrogen and Phosphorus (Section 21.4)**
**Oxygen and Sulfur (Section 21.5)**
**The Halogens (Section 21.6)**


## SUMMARY

### Hydrogen (Sections 21.1 – 21.2)

***General Properties of Nonmetals.*** The nonmetal elements, with the exception of hydrogen, are located in the upper right hand corner of the periodic table. The properties of nonmetals are more varied than those of metals. Several nonmetal elements are gases, and nonmetals exhibit both positive and negative oxidation numbers. In this chapter, we will discuss the chemistry of a number of important nonmetallic elements: hydrogen, carbon, nitrogen, phosphorus, oxygen, sulfur, fluorine, chlorine, bromine, and iodine.

***Hydrogen.*** Hydrogen is the most abundant element in the universe, and the tenth most abundant element in earth's crust, where it is found in combination with other elements. At ordinary temperatures, elemental hydrogen is a diatomic molecule. Commercially, the most important large-scale preparation is the reaction between propane and steam in the presence of a catalyst at 900 °C:

$$C_3H_8(g) + 3\ H_2O(g) \rightarrow 3\ CO(g) + 7\ H_2(g)$$

In another process steam is passed over a bed of red-hot coke (the water gas reaction):

$$C(s) + H_2O(g) \xrightarrow{1000\ °C} CO(g) + H_2(g)$$

Small quantities of hydrogen are prepared in the laboratory by the reaction of zinc with hydrochloric acid.

$$Zn(s) + 2\ HCl(aq) \rightarrow ZnCl_2(aq) + H_2(g)$$

***Properties of Hydrogen.*** Hydrogen can be oxidized to $H^+$ ions and can be reduced to $H^-$ ions (hydride ions) as well. $H^+$ ions have the same charge as the metal alkali metal ions, and the $H^-$ ions are similar to the halide ions. It also forms diatomic molecules as do the halogens. As a result of these properties, there is no totally satisfactory position for hydrogen in the periodic table.

Hydrogen forms binary compounds called hydrides with a large number of elements. It reacts directly with most active metals to form ionic hydrides in which hydrogen is an $H^-$ ion.

$$2\ Na(s) + H_2(g) \rightarrow 2\ NaH(s)$$
$$2\ Al(s) + 3\ H_2(g) \rightarrow 2\ AlH_3(s)$$

The ionic character of these metal hydrides increases as we move down a group of the periodic table. Metal hydrides react with water to form hydrogen gas and the metal hydroxide.

$$NaH(s) + H_2O(\ell) \rightarrow NaOH(aq) + H_2(g)$$

Hydrogen in compounds with nonmetals such as $CH_4$, $NH_3$, and $H_2S$ has an oxidation state of +1. These are covalent hydrides.

***Uses of Hydrogen.*** Hydrogen is used in the synthesis of ammonia, an important fertilizer. The Haber process for synthesizing ammonia was discussed in chapter 14. Ammonia which results from the reaction of hydrogen with nitrogen is an example of a covalent hydride.

$$N_2(g) + 3\ H_2(g) \rightarrow 2\ NH_3(g)$$

Hydrogen is used in the production of margarine and cooking oil to convert polyunsaturated fats, which contain $C = C$ double bonds into partially hydrogenated products.

Hydrogen may play an important role in our energy future in terms of the "hydrogen economy." As fossil fuel supplies dwindle, hydrogen gas could take on more of a role as a fuel. Nuclear energy could be used to generate electricity, which in turn could be used to electrolyze water to obtain hydrogen.

$$\text{electrical energy} + H_2O(\ell) \rightarrow H_2(g) + \tfrac{1}{2} O_2(g)$$

The combustion of hydrogen as a fuel would release energy.

$$H_2(g) + \tfrac{1}{2} O_2(g) \rightarrow H_2O(\ell) \qquad \Delta H° = -285.8 \text{ kJ/mol}$$

> Examples
> 21.1, 21.2
>
> Exercises
> 1 – 4

## Carbon (Section 21.3)

***Properties of Carbon.*** Carbon makes up only 0.09% of earth's crust by mass. Carbon occurs in nature as the free element in the allotropic forms, graphite and diamond. Carbon occurs as $CO_2$ in the atmosphere and as carbonate ions in limestone and chalk. Fossil fuels contain a large percentage of carbon. Carbon is the essential element in living organisms.

Pure carbon in the form of diamond is the hardest substance known. The microscopic structure of diamond is such, that each carbon atom is covalently linked to four other carbon atoms (see Figure 11.28 in the text). Therefore, a diamond is a single giant molecule. The melting point of a diamond is extremely high, 3550 °C. Graphite is the "lead" in lead pencils.

On heating carbon and silicon at 1500 °C a compound called silicon carbide or carborundum (SiC) is formed. Silicon carbide is almost as hard as diamond and has the diamond structure. Its melting point is about 2700 °C. The hardness of diamond and carborundum are put to important uses. Synthetic diamonds are not of gem quality, but are used as abrasives in cutting concrete and many other hard substances. Carborundum is used for cutting, grinding, and polishing metals and glasses.

Carbon forms two important classes of compounds called carbides and cyanides. $CaC_2$ and $Be_2C$ contain carbon in the form of $C_2^{2-}$ and $C^{4-}$ ions, respectively. The carbide ion contains a triple bond, $C \equiv C^{2-}$. These ions are strong Brønsted bases and react with water as follows:

$$C_2^{2-}\ (aq) + 2\ H_2O(\ell) \rightarrow 2\ OH^-(aq) + C_2H_2(g)$$
$$C^{4-}\ (aq) + 4\ H_2O(\ell) \rightarrow 4\ OH^-(aq) + CH_4(g)$$

Cyanides contain the cyanide anion, $C \equiv N^-$. Hydrogen cyanide, sodium cyanide, and potassium cyanide are well-known cyanide compounds.

***Oxides of Carbon.*** Carbon forms two important oxides, $CO$ and $CO_2$. Carbon monoxide is a colorless, odorless, and poisonous gas. It is formed by incomplete combustion of carbon compounds. Carbon monoxide burns readily in air.

Carbon dioxide is a colorless and odorless gas. Its concentration in air is 350 ppm. At this concentration, it is not toxic. In enclosed low-lying places such as abandoned mines, $CO_2$ concentrations can build up to suffocating levels. Carbon dioxide is formed by the complete combustion of carbon-containing compounds and fuels. If oxygen is restricted from reaching the combustion zone, carbon monoxide is formed instead. Since $CO_2$ cannot be oxidized further, it finds use in fire extinguishers. Dry ice (solid carbon dioxide) is used as a refrigerant.

Exercises
5 – 7

## Nitrogen and Phosphorus (Section 21.4)

***Nitrogen ($N_2$).*** The elemental form of nitrogen is extremely stable owing to its triple bond, $N \equiv N$. Molecular nitrogen is obtained from air, of which nitrogen makes up 78 percent by volume. When air under a pressure of 30 atm is cooled to $-190$ °C, it liquefies. Fractional distillation of liquid air yields pure $N_2$. Nitrogen is used in the laboratory to create atmospheres devoid of oxygen. This is desirable when working with chemicals that are attacked and degraded by oxygen, or that burn spontaneously in it. Liquid $N_2$ is used as a coolant, providing a temperature of $-196$ °C.

***Nitrides ($N^{3-}$).*** Most nitrogen compounds are covalent, however, nitrogen forms ionic compounds with certain metals on heating. $Li_3N$, $Mg_3N_2$, and $Ca_3N_2$ are examples. The nitride ion is a very strong Brønsted base and readily accepts protons from water, forming ammonia:

$$Mg_3N_2 + 6\ H_2O \rightarrow 3\ Mg(OH)_2 + 2\ NH_3$$

***Ammonia ($NH_3$).*** The quantity of ammonia produced each year is second in tonnage only to sulfuric acid production. Ammonia is synthesized by the Haber process, which uses nitrogen gas and hydrogen gas (from hydrocarbon cracking).

$$3\ H_2 + N_2 \rightarrow 2\ NH_3$$

The main use of ammonia is in the production of ammonium and nitrate fertilizers. Ammonia is an important ingredient in such diverse substances as explosives and nylon.

Liquid ammonia (b.p. $-33.4$ °C) is a solvent resembling water. It undergoes autoionization

$$2\ NH_3(\ell) \rightleftharpoons NH_4^+ + NH_2^-$$

The value of the equilibrium constant at $-50$ °C is $K = [NH_4^+][NH_2^-] = 1 \times 10^{-33}$. Alkali metals dissolve in liquid ammonia, producing beautiful blue solutions containing a positive ion and an electron.

$$Li(s) \rightarrow Li^+ + e^-$$

Both the cations and the electrons are highly solvated by ammonia. The solvated electrons are responsible for the characteristic color of these solutions.

***Nitrogen Oxides ($N_2O$, NO, & $NO_2$).*** "Fixed nitrogen" refers to nitrogen present in chemical compounds. Nitrogen is fixed, that is incorporated into compounds, by the action of lightning in the atmosphere, by high temperature, by the Haber process, and by certain types of soil bacteria. An appreciable amount of nitric oxide (NO) is formed in electrical storms and in internal combustion engines.

$$N_2 + O_2 \rightarrow 2\ NO$$

Nitric oxide reacts very rapidly with oxygen to yield nitrogen dioxide.

$$2\ NO + O_2 \rightarrow 2\ NO_2$$

Nitrogen dioxide is a key molecule in the formation of photochemical smog. $NO_2$ is prepared in the laboratory by the action of concentrated nitric acid on copper.

$$Cu(s) + 4\ HNO_3(aq) \rightarrow Cu(NO_3)_2(aq) + 2\ H_2O(\ell) + 2\ NO_2(g)$$

Nitric acid is a major inorganic acid. Over 8.0 million tons were produced in the United States in 1985. Nitric acid is synthesized by the Ostwald process, which uses ammonia and oxygen as starting materials. Reaction occurs at 1000 °C in the presence of a platinum-rhodium catalyst.

$$4\ NH_3(g) + 5\ O_2(g) \rightarrow 4\ NO(g) + 6\ H_2O(g)$$

Then oxygen is introduced.

$$2\ NO(g) + O_2(s) \rightarrow 2\ NO_2(s)$$

The resulting $NO_2$ readily dissolves in water, yielding a mixture of nitrous acid and nitric acid.

$$2\ NO_2(g) + H_2O(\ell) \rightarrow HNO_2(aq) + HNO_3(aq)$$

Then, nitrous acid is eliminated by heating.

$$3\ HNO_2(aq) \rightarrow HNO_3(aq) + H_2O(\ell) + 2\ NO(g)$$

Nitric acid is a strong acid, and an oxidizing acid. Aqua regia, the only acid that can dissolve gold, is 1 part nitric acid and 3 parts hydrochloric acid. Nitric acid is used to manufacture fertilizers and explosives such as nitroglycerin and TNT.

Nitrous oxide ($N_2O$) is a colorless gas with a pleasing odor and a sweet taste. It is used as an anesthetic in dental and minor surgery.

***Phosphorus.*** There are several allotropic forms of phosphorus. The most important ones are white and red phosphorus. Their structures are shown in Figure 21.9 in the text. White phosphorus consists of discrete $P_4$ molecules. The white solid melts at 44 °C and is insoluble in water. It is usually stored under water to prevent contact with air. White phosphorus is extremely reactive and bursts into flame when exposed to air.

$$P_4(s) + 5\ O_2(g) \rightarrow P_4O_{10}(s)$$

When white phosphorus is heated to 300 °C in the absence of air it is slowly converted to red phosphorus. Red phosphorus has a polymeric structure. It is much more stable than white phosphorus. It melts at 600 °C and is stable toward oxygen at room temperature. Red phosphorus must be heated to about 400 °C before it will ignite in air.

***Oxides of Phosphorus.*** Tetraphosphorus hexaoxide ($P_4O_6$) and tetraphosphorus decaoxide ($P_4O_{10}$) are two important oxides of phosphorus, (Figure 21.10 text). Both oxides are acidic.

$$P_4O_6(s) + 6\ H_2O(\ell) \rightarrow 4\ H_3PO_3(aq)$$
$$P_4O_{10}(s) + 6\ H_2O(\ell) \rightarrow 4\ H_3PO_4(aq)$$

***Phosphoric Acid.*** Industrially, phosphoric acid is prepared by the reaction of sulfuric acid with calcium phosphate:

$$Ca_3(PO_4)_2(s) + 3\ H_2SO_4(aq) \rightarrow 2\ H_3PO_4(aq) + 3\ CaSO_4(aq)$$

Phosphoric acid is not a particularly strong acid, as indicated by its first ionization constant.

$$H_3PO_4 \rightleftharpoons H^+ + H_2PO_4^- \qquad K_{a1} = 7.5 \times 10^{-3}$$

Large quantities of phosphoric acid are produced commercially every year. Its main uses are in the production of phosphate fertilizers, cattle feed additives, and water softeners (builders) in household detergents.

> Examples
> 21.3, 21.4
>
> Exercises
> 8 – 10

## Oxygen and Sulfur (Section 21.5)

***Allotropic Forms of Oxygen.*** Oxygen is the most abundant element in earth's crust, making up about 46% by mass. As the free element, oxygen exists in two allotropic forms, diatomic oxygen ($O_2$) and ozone ($O_3$). Diatomic oxygen makes up 21 percent of earth's atmosphere by volume. Industrially, oxygen is obtained by liquefaction of air followed by fractional distillation. In the laboratory, oxygen is obtained by heating potassium chlorate in the presence of $MnO_2$ catalyst.

$$KClO_3(s) \rightarrow KCl(s) + \tfrac{3}{2} O_2(g)$$

Ozone is a less stable form of oxygen. It is produced by the action of an electrical discharge or ultraviolet (UV) light on diatomic oxygen.

$$O_2(g) + UV\ light \rightarrow O(g) + O(g)$$
$$O(g) + O_2(g) \rightarrow O_3(g)$$

Ozone's presence in the atmosphere prevents the sun's UV light from reaching Earth's surface. Ozone absorbs the light in the following reaction:

$$O_3(g) + UV\ light \rightarrow O_2(g) + O(g)$$

Both ozone and diatomic oxygen are strong oxidizing agents.

$$O_2(g) + 4\ H^+ + 4e^- \rightarrow 2\ H_2O(\ell) \qquad E^\circ_{red} = 1.23\ V$$
$$O_3(g) + 2\ H^+ + 2e^- \rightarrow O_2(g) + H_2O(\ell) \qquad E^\circ_{red} = 2.02\ V$$

Ozone is used mainly to purify drinking water, to deodorize air and sewage gases, and to bleach waxes, oils, and textiles. About 65 percent of all the diatomic oxygen produced in the United States is used in the basic oxygen process for making steel. Pure oxygen gas is blown over the surface of molten iron to lower the carbon content by burning it to $CO_2$. At the same time, silicon and phosphorus impurities are oxidized and later removed as slag.

**Hydrogen Peroxide.** Besides forming water, oxygen forms another important binary compound with hydrogen: hydrogen peroxide ($H_2O_2$). Hydrogen peroxide is never used as a pure substance because of its tendency to explode.

Metal peroxides contain the peroxide ion, $O_2^{2-}$. These ions are strong Brønsted bases. All the metal peroxides hydrolyze in acidic solutions to form hydrogen peroxide. Hydrogen peroxide can be prepared in the laboratory from barium peroxide octahydrate and sulfuric acid.

$$BaO_2 \cdot 8\ H_2O(s) + H_2SO_4(aq) \rightarrow BaSO_4(aq) + H_2O_2(aq) + 8\ H_2O(\ell)$$

In aqueous solutions, $H_2O_2$ can act both as an oxidizing agent and a reducing agent. For example, $H_2O_2$ oxidizes iron(II) ion to iron(III) ion.

$$H_2O_2(aq) + 2\ Fe^{2+}(aq) + 2\ H^+(aq) \rightarrow 2\ Fe^{3+}(aq) + 2\ H_2O(\ell)$$

It reduces silver oxide to metallic silver:

$$H_2O_2(aq) + Ag_2O(s) \rightarrow 2\ Ag(s) + H_2O(\ell) + O_2(g)$$

Hydrogen peroxide readily decomposes on heating and in the presence of various catalysts.

$$2\ H_2O_2(aq) \rightarrow 2\ H_2O(\ell) + O_2(g)$$

This is a disproportionation reaction in which one O atom in $H_2O_2$ is reduced and the other one is oxidized.

**Sulfur.** Although it is not very abundant, sulfur is readily available because of its occurrence in large deposits of the free element. Sulfur is extracted from these underground deposits by the Frasch process. There are several allotropic forms of sulfur. The most important are rhombic sulfur and monoclinic sulfur. Rhombic sulfur is the more stable form at room temperature. In this form, sulfur atoms are linked to each other as puckered, eight-membered rings, $S_8$. Above 96 °C, monoclinic sulfur is more stable. This form melts at 119 °C. Above 150 °C, the pale yellow liquid becomes more viscous as the rings break open and the newly formed molecular chains entangle. Sulfur has a wide range of oxidation numbers in its compounds, as shown in Table 21.3 of the text. The major end uses of sulfur are in fertilizers (60%) and in chemicals (20%).

**Hydrogen Sulfide.** $H_2S$ (oxidation number of S = –2) is a colorless, highly toxic gas, with an offensive odor. When dissolved in water, the resulting solution is called hydrosulfuric acid. $H_2S(aq)$ is a weak acid. In the laboratory, $H_2S(aq)$ is used to separate groups of metal ions for their qualitative analysis. It is often prepared in the laboratory by the hydrolysis of thioacetamide.

$$CH_3CSNH_2(aq) + 2\ H_2O + H^+(aq) \rightarrow CH_3COOH(aq) + NH_4^+(aq) + H_2S(aq)$$

**Oxides of Sulfur.** The major oxides of sulfur are sulfur dioxide and sulfur trioxide, $SO_2$ and $SO_3$. Sulfur dioxide is a colorless gas with a pungent odor and is quite toxic. It dissolves in water forming sulfurous acid.

$$SO_2(g) + H_2O(\ell) \rightarrow H_2SO_3(aq)$$

Sulfur dioxide is slowly oxidized in air to sulfur trioxide.

$$2\ SO_2(g) + O_2(g) \rightarrow 2\ SO_3(g)$$

Sulfur trioxide dissolves in water to form sulfuric acid.

$$SO_3(g) + H_2O(\ell) \rightarrow H_2SO_4(aq)$$

Sulfur dioxide is a major air pollutant. It is formed during combustion of sulfur-containing oil and coal. Smelters and coal-fired electric power plants are the primary sources of sulfur oxides.

**Sulfuric Acid.** More sulfuric acid is produced each year than any other industrial chemical. It is formed from $SO_3$ and $H_2O$ as shown above. First, sulfur is burned in air to make $SO_2$. Second, a vanadium oxide catalyst is used to convert $SO_2$ to $SO_3$ in the presence of oxygen. Third, $SO_3$ and water react as above.

Hot, concentrated sulfuric acid is a strong oxidizing agent. In such a solution, the oxidizing agent is actually the sulfate ion. Copper is oxidized by sulfuric acid by the following reaction:

Examples
21.5, 21.6

Exercises
11 – 16

$$Cu(s) + 2\ H_2SO_4(aq) \rightarrow CuSO_4(aq) + SO_2(g) + 2\ H_2O(\ell)$$

## The Halogens (Section 21.6)

**Occurrence and Preparation.** The elements in Group 7 are called the halogens. In the elemental state, they form diatomic molecules, $X_2$. Because of their high reactivity, the halogens are always found combined with other elements. Chlorine, bromine, and iodine occur as halides in seawater, and fluorine occurs in the minerals fluorspar ($CaF_2$) and cryolite ($Na_3AlF_6$).

Fluorine is prepared by the electrolysis of liquid HF at 70 °C.

electrolysis
$$2\,HF(\ell) \;\rightarrow\; H_2(g) + F_2(g)$$

Chlorine is prepared industrially by the electrolysis of a concentrated aqueous NaCl solution (brine). This is called the chlor-alkali process (NaOH is the alkali).

electrolysis
$$2\,NaCl(aq) + 2\,H_2O(\ell) \;\rightarrow\; 2\,NaOH(aq) + H_2(g) + Cl_2(g)$$

Because chlorine is a stronger oxidizing agent than bromine or iodine, free $Br_2$ and $I_2$ can be generated by bubbling $Cl_2$ gas through solutions containing bromide and iodide ions.

$$Cl_2(g) + Br^-(aq) \rightarrow 2\,Cl^-(aq) + Br_2(\ell)$$
$$Cl_2(g) + 2\,I^-(aq) \rightarrow 2\,Cl^-(aq) + I_2(s)$$

Chlorine, bromine, and iodine can be prepared in the laboratory by heating the corresponding alkali halides in concentrated sulfuric acid in the presence of $MnO_2$.

$$2\,NaCl(aq) + 2\,H_2SO_4(aq) + MnO_2(s) \rightarrow Cl_2(g) + Na_2SO_4(aq) + MnSO_4(aq) + 2\,H_2O(\ell)$$

***Oxoacids of the Halogens.*** Most of the halides (halogens with an oxidation number of $-1$, as in chloride) can be classified into two categories: ionic and covalent. The fluorides and chlorides of many metallic elements, especially those belonging to the Group 1A and 2A metals, are ionic compounds. The halides of nonmetals are covalent compounds.

Because fluorine is the most electronegative element and because of its unusually weak F—F bond, fluorine has a chemistry somewhat different from the rest of the halogens. Five differences are pointed out in Section 21.6 of the text.

- Fluorine is the most reactive of all of the halogens.
- Hydrogen fluoride exhibits hydrogen bonding. None of the other hydrogen halides do.
- Hydrofluoric acid is a weak acid. All other HX acids are strong.
- Fluorine reacts with cold NaOH solution.
- Silver fluoride is soluble. All other silver halides are insoluble.

The halogens form a series of oxoacids with the following general formulas:

| HOX | $HXO_2$ | $HXO_3$ | $HXO_4$ |
|---|---|---|---|
| hypohalous acid | halous acid | halic acid | perhalic acid |

Table 21.5 in the text lists the oxoacids of the halogens.

Hypochlorous, hypobromous, and hypoiodous acids can be prepared by the reaction of the elemental halogen with water.

$$X_2 + H_2O \;\rightleftharpoons\; \underset{\substack{\text{hydrohalic} \\ \text{acid}}}{HX} \;+\; \underset{\substack{\text{hypohalous} \\ \text{acid}}}{HOX}$$

The smaller the radius of the halogen atom, the farther the equilibrium lies to the right. Mixtures of these halogen acids react with bases to produce hypohalite salts (NaOX)

$$HX + HOX + 2\ NaOH \rightarrow NaX + NaOX + 2\ H_2O$$

When the elemental form of a halogen, $X_2$, is dissolved in a base the net reaction is:

$$X_2 + 2\ NaOH \rightarrow NaX + NaOX + H_2O$$

This is another example of a type of redox reaction called a disproportionation reaction. In this type of reaction, the same element (the halogen) is both oxidized and reduced. One halogen atom in X—X is the reducing agent and the other is the oxidizing agent.

Interhalogens are compounds formed when molecules of two different halogens react. The interhalogen molecules consist of an atom of the larger halogen bound to an odd number of atoms of the smaller halogen. They have the following formulas: $XX'$, $XX'_3$, $XX'_5$, and $XX'_7$, where X and X' are different halogens and X is the larger atom of the two.

***Uses of the Halogens.*** The halogens and their compounds find many uses in industry, the home, and in the field of health. Fluorine is used in fluoride toothpaste, in uranium isotope separation as $UF_6$, and in the manufacture of Teflon for electrical insulators, plastics, and cooking utensils.

Chlorine is widely used as a bleaching agent and as a disinfectant for water purification. Household bleach is 5.25 percent NaOCl by mass. The major end uses of chlorine are in manufacturing polymers such as polyvinyl chloride, in chemical solvents, in the pulp and paper industry, and in water treatment.

Bromine is used to prepare dibromoethane ($BrCH_2CH_2Br$) which is a gasoline additive, and silver bromide (AgBr), used in photographic film. Iodine finds uses as a medicinal antiseptic. NaI is added to table salt to prevent goiter, and silver iodide is used in cloud seeding.

| Examples 21.7, 21.8 |
| Exercises 17 – 21 |

## GLOSSARY LIST

| | | |
|---|---|---|
| hydrogenation | carbide | chlor-alkali process |
| catenation | cyanide | |

# WORKED EXAMPLES

---

### EXAMPLE 21.1  Hydrogen Compounds

Give an example of an ionic hydride and an example of a covalent hydride. What is the oxidation number of hydrogen in each?

#### • Solution

In ionic compounds, hydrogen has an oxidation number of $-1$. The hydrides of the alkali metals and some of the alkaline earth metals are ionic hydrides. $CaH_2$ is an example of an ionic hydride. Hydrogen has an electronegativity of 2.1 and that of calcium is 1.0. Thus, hydrogen has an oxidation number of $-1$ in $CaH_2$.

$H_2S$ is a covalent hydride. The electronegativity of sulfur is 2.5. The lower polarity of the H—S bond is indicated by a $\Delta$ value of only 0.4. The oxidation number of hydrogen is $+1$ in $H_2S$.

---

### EXAMPLE 21.2  Comparison of Ionic and Covalent Hydrides

Compare the reactions of water with covalent hydrides and ionic hydrides.

#### • Solution

Covalent hydrides either do not react with water at all, which is the case with $CH_4$, or are *acids* in water. One of them, ammonia, is a weak base.

$$H_2S(aq) + H_2O(\ell) \rightleftharpoons HS^-(aq) + H_3O^+(aq)$$

Ionic hydrides are strong Brønsted bases; they readily accept a proton from water.

$$H^-(aq) + H_2O(\ell) \rightarrow H_2(g) + OH^-(aq)$$

Lithium hydride is an ionic hydride.

$$LiH(s) + H_2O(\ell) \rightarrow LiOH(aq) + H_2(g)$$

**Work EXERCISES & PROBLEMS: 1 – 4**

**EXAMPLE 21.3  Oxidation Numbers of Nitrogen**

List the known oxidation numbers of nitrogen and write the formula of a compound or ion that exemplifies each oxidation number.

• **Solution**

| Oxidation Number | Compound | Formula |
|---|---|---|
| + 5 (maximum) | nitrate ion | $NO_3^-$ |
| +4 | nitrogen dioxide | $NO_2$ |
| +3 | nitrite ion | $NO_2^-$ |
| +2 | nitric oxide | $NO$ |
| +1 | nitrous oxide | $N_2O$ |
| 0 | nitrogen gas (not a compound) | $N_2$ |
| −1 | chloramine | $NH_2Cl$ |
| −2 | hydrazine | $N_2H_4$ |
| −3 | ammonia and nitride | $NH_3$ and $N^{3-}$ |

**EXAMPLE 21.4  Allotropic Forms**

Starting with white phosphorus, write equations for the steps needed to prepare phosphoric acid.

• **Solution**

First, burn the white phosphorus in excess oxygen to make tetraphosphorus decaoxide.

$$P_4(s) +\ 5\ O_2(g) \rightarrow P_4O_{10}(s)$$

The oxide is an acidic oxide and is converted to phosphoric acid in water.

$$P_4O_{10}(s) + 6\ H_2O(\ell) \rightarrow 4\ H_3PO_4(aq)$$

**Work EXERCISES & PROBLEMS: 5 – 10**

**EXAMPLE 21.5  Oxygen Compounds**

Distinguish between the three types of oxides:  the normal oxide, the peroxide, and the superoxide.

• **Solution**

Normal oxides contain the $O^{2-}$ ion in which oxygen has an oxidation state of −2. Peroxides contain an oxygen-oxygen bond. Metal peroxides are ionic and contain the $O_2^{2-}$ ion. Hydrogen peroxide is a covalent peroxide  (H—O—O—H) with oxygen in the −1 oxidation state. The superoxide ion $O_2^-$ has oxygen in the −1/2 oxidation state.

**EXAMPLE 21.6  Oxidation Numbers of Sulfur**

List the known oxidation numbers of sulfur and write the formula of a compound or ion that exemplifies each oxidation number.

**• Solution**

| Oxidation Number | Compound | Formula |
|---|---|---|
| +6 | sulfate ion | $H_2SO_4$ |
| +4 | sulfite ion | $CaSO_3$ |
| +2 | sulfur dichloride | $SCl_2$ |
| +1 | disulfur dichloride | $S_2Cl_2$ |
| 0 | sulfur (the element) | $S_8$ |
| −2 | sulfide ion | $Na_2S$ |

**Work EXERCISES & PROBLEMS: 11 – 16**

**EXAMPLE 21.7  Preparation of $F_2$**

The chemistry of fluorine differs in many ways from that of the rest of the halogens. For instance, it cannot be prepared by the electrolysis of an aqueous solution of its sodium salt, as $Cl_2$ is prepared. Write chemical equations to show the steps in the industrial process of making $F_2$. Give the source of the fluorine, and mention why it cannot be prepared in aqueous solution.

**• Solution**

The industrial process starts with the mineral fluorspar, $CaF_2$. This is treated with sulfuric acid to make hydrogen fluoride.

$$CaF_2(s) + H_2SO_4(\ell) \rightarrow 2\ HF(\ell) + CaSO_4(s)$$

The hydrogen fluoride is mixed with KF and electrolyzed:

$$2\ HF(\ell) \rightarrow H_2(g) + F_2(g)$$

The $F_2$ molecule has such a strong attraction for electrons that it reacts violently with water, hot glass, and most metals. If $F_2$ were formed in an aqueous solution, it would immediately oxidize water to oxygen gas.

**EXAMPLE 21.8  Reaction of Chlorine with Sodium Hydroxide**

When chlorine is dissolved in base, a solution of aqueous sodium hypochlorite is formed. A 5.25% solution is sold as liquid laundry bleach. Write the chemical equation for the reaction. Assign oxidation numbers and identify the oxidizing agent and the reducing agent.

**• Solution**

$$Cl_2(g) + 2\,NaOH(aq) \rightarrow NaCl(aq) + NaOCl(aq) + H_2O(\ell)$$

If we assign oxidation numbers to all the chlorine atoms, we can see that chlorine is both oxidized and reduced. None of the other elements in the equation changes oxidation state. Thus, chlorine is the oxidizing agent and the reducing agent.

This is an example of a disproportionation reaction.

## EXERCISES & PROBLEMS

1. Write an equation to show how aluminum can be used to prepare hydrogen in the laboratory.

2. What is the oxidation number of hydrogen in $AlH_3$, in $H_2S$?

3. Give an example of an ionic hydride.

4. Describe the hydride ion as a Brønsted acid or a base. Write an equation for the reaction of the hydride ion with water.

5. Diamond is the highest melting substance. Why is the melting point of diamond so high?

6. Write formulas for the cyanide ion and two kinds of carbide ions.

7. Which oxide of carbon is flammable?

8. What are the oxidation numbers of nitrogen in nitrate ion and nitrite ion?

9. Write three equations that show the steps in the production of nitric acid from ammonia.

10. Compare the stability of white and red phosphorus.

11. Which is a stronger oxidizing agent, diatomic oxygen or ozone?

12. What is the principle use of $O_2$ in the United States?

13. What is the oxidation number of O in $H_2O_2$, $H_2O$ and $O_2$?

14. Hydrogen peroxide can decompose by a disproportionation reaction. Write the equation for this reaction.

15. Write formulas of oxide, peroxide, and superoxide compounds that include a metal.

16. Starting with elemental sulfur, $S(s)$, show the steps needed to prepare sulfuric acid.

17. Which halogens are prepared by electrochemical oxidation?

18. Write a chemical equation for the preparation of elemental bromine from seawater.

19. Which of the hydrogen halides are strong acids in water?

20. Which halogen is the strongest oxidizing agent among all the elements?

21. Show the similarity of chlorine and bromine by writing equations for their reactions:
    a. with $I^-$ ion.
    b. with $NaOH(aq)$.

22. Explain why hydrogen does not have a unique position in the periodic table.

23. Explain how $Br_2$ can act as both an oxidizing agent and a reducing agent.

## PRACTICE TEST QUESTIONS
See notes on taking practice test in the Preface

1. Write chemical equations to show the two commercial methods of preparation of $H_2(g)$.

2. What are two important uses of hydrogen?

3. Write a chemical equation to show the synthesis of boric acid from kernite.

4. Coke is used to reduce $Fe_2O_3$ in iron production. What is it? How can it be converted to graphite?

5. List the major uses of:
   a. $N_2$    b. $NH_3$    c. $HNO_3$

6. What is *aqua regia?*

7. What are the acid anhydrides of $HNO_3$, $HNO_2$, $H_3PO_4$, and $H_3PO_3$?

8. Write a balanced overall equation that shows the synthesis of nitric acid from ammonia.

9. Why is calcium phosphate not used directly as a fertilizer?

10. What chemical compounds contain phosphate in commercial fertilizers?

11. Give the formulas and names of the allotropic forms of oxygen and sulfur.

12. List several uses of $O_2$.

13. Distinguish between the peroxide ion and the superoxide ion.

14. Write chemical equations to show the preparations of the elemental forms of the halogens.

15. Write the chemical formulas of the halic acids.

16. Describe the colors and the standard physical states at 25 °C of the diatomic halogens.

17. Write equations for the preparation of three interhalogens: $ClF_3$, $BrF_5$, and $IF_7$.

# Chapter 22. Transition Metal Chemistry and Coordination Compounds

Properties of the Transition Metals (Section 22.1)
Coordination Compounds (Section 22.3)
The Structure of Coordination Compounds (Section 22.4)
Bonding in Coordination Compounds (Section 22.5)

## SUMMARY

### Properties of the Transition Metals (Section 22.1)

***Electron Configurations of Atoms and Ions.*** In this chapter, we will focus on the first-row transition elements from scandium to copper. Recall that the transition metals have atoms or ions with incompletely filled d subshells.

The electron configurations of the first-row transition elements were discussed in Section 7.9 of the text. For the representative elements, potassium and calcium, the 4s orbital is lower in energy than the 3d orbital. The 3d orbitals begin to fill only after the 4s orbital is complete. For elements after Ca $[Ar]4s^2$, electrons are added one at a time to the 3d subshell beginning with scandium, the first transition element.

The electron configurations are given in Table 22.1 below. The configurations of chromium $[Cr]4s^13d^5$ and copper $[Cu]4s^13d^{10}$ are exceptions to the order of filling. The basis for these exceptions is the stability associated with a half-filled or completely filled 3d subshell.

**Table 22.1** Electron Configurations of Atoms and Ions

|     | Atom | +2 ion | +3 ion |
| --- | --- | --- | --- |
| Sc | $4s^23d^1$ | $3d^1$ | [Ne] |
| Ti | $4s^23d^2$ | $3d^2$ | $3d^1$ |
| V | $4s^23d^3$ | $3d^3$ | $3d^2$ |
| Cr | $4s^13d^5$ | $3d^4$ | $3d^3$ |
| Mn | $4s^23d^5$ | $3d^5$ | $3d^4$ |
| Fe | $4s^23d^6$ | $3d^6$ | $3d^5$ |
| Co | $4s^23d^7$ | $3d^7$ | $3d^6$ |
| Ni | $4s^23d^8$ | $3d^8$ | ____ |
| Cu | $4s^13d^{10}$ | $3d^9$ | ____ |

To understand the electron configurations of the ions shown in Table 22.1, it helps to recall that electrons are removed from the 4s orbital before they are taken out of the 3d. The energies of the 3d and 4s orbitals are not as close together in the ions of the transition metals as they are in the neutral atoms. In fact, for the ions of transition elements the 3d orbitals are lower in energy than the 4s orbitals. Therefore, the electrons most easily lost are those in the outermost principal energy level, the ns. Additional electrons may then be lost from the (n − 1)d orbital.

**Properties of Transition Elements.** For the representative elements, properties such as the atomic radius, ionization energy, and electronegativity, vary markedly from element to element as the atomic number increases across any period. In contrast, the chemical and physical properties of the transition metal elements vary only slightly as we read across a period. Table 22.1 in the text gives a number of properties of the transition elements for comparison. The characteristics of transition metals are summarized below.

1. *General Physical Properties.* Transition metals have relatively high densities, high melting and boiling points, and high heats of fusion and vaporization.
2. *Atomic Radius.* As discussed in chapter 7, the atomic radii of representative elements decrease markedly as we read across a period of elements. In contrast, the atomic radii of transition metals decrease only slightly as we read across a series of these elements.

　　For representative elements, the decrease in radii is consistent with an increase in effective nuclear charge as atomic number increases.

　　All the first-row transition elements have 4s orbitals as the outermost occupied orbitals. The atomic radius will depend on the strength of the nuclear charge felt by the 4s electrons. The greater the charge, the smaller the atom. The effective nuclear charge experienced by the outermost electrons depends on the shielding provided by inner electrons.

　　For transition metals, the nuclear charge increases from scandium to copper, and electrons are being added to an inner 3d subshell. These 3d electrons shield the 4s electrons from the increasing nuclear charge for the most part. Consequently, the 4s electrons of the first-row transition elements, feel only a slightly increasing effective nuclear charge as the atomic number increases, and the atomic radii makes only a gradual decrease.

3. *Ionization Energy.* The first ionization energies of the first transition metal series are remarkably similar, increasing very gradually from left to right. There is a slight increase over the first five elements then the ionization energy barely changes from iron to copper. See Table 22.1 in the text.
4. *Variable Oxidation States.* The common oxidation states for the transition elements from Sc to Cu are +2 and +3 (Figure 22.2 text). Some of the elements exhibit the +4 state, and Mn even shows +5, +6, and +7. Transition metals usually exhibit their highest oxidation states in compounds with oxygen, fluorine, or chlorine. $KMnO_4$ and $K_2Cr_2O_7$ are examples. The variability of oxidation states for transition metal ions results from the fact that the 4s and 3d subshells are similar in energy. Therefore, an atom can form ions of roughly the same stability by losing different numbers of electrons. The +3 oxidation states are more stable at the beginning of the series, but toward the end, the +2 oxidation states are more stable.

> Examples
> 22.1, 22.2
>
> Exercises
> 1 – 3

## Coordination Compounds (Section 22.3)

**Terminology.** A complex ion consists of a central metal cation to which several anions or molecules are bonded. The complex ion may be positively or negatively charged. The metal cation is called the central atom, and the attached anions and molecules are called ligands. The free ligands each have at least one unshared pair of electrons which can be donated to the electron-deficient metal ions. The donor atom is the atom in the ligand that is directly bonded to the metal. Some typical ligands are $Cl^-$, $CN^-$, $NH_3$, $H_2O$, and $H_2NCH_2CH_2NH_2$.

　　A neutral species containing a complex ion is called a coordination compound. These compounds usually have complicated formulas. Two examples are:

$[Ag(NH_3)_2]Cl$ and $K_3[Fe(CN)_6]$

The complex ion is shown enclosed in brackets. In the silver compound, $Cl^-$ is a free (meaning uncomplexed) chloride ion, and in the iron compound each $K^+$ is a free potassium ion. $K^+$ and $Cl^-$ ions in the above formulas are examples of counter ions. They serve to balance or neutralize the charge of the complex ion. The coordination number is defined as the number of donor atoms surrounding the central metal atom. As a result, the coordination number of $Pt^{2+}$ in $[Pt(NH_3)_4]^{2+}$ is 4, and that of $Co^{2+}$ in $[Co(NH_3)_6]^{2+}$ is 6.

Some ligands contain more than one donor atom. Ligands that coordinate through two bonds are called bidentate ligands. Those with more than two donor atoms are referred to as polydentate ligands. Ethylenediamine (en) and oxalate ion (ox) are bidentate ligands:

$$\cdot\cdot \qquad\qquad \cdot\cdot$$
$$H_2N-CH_2-CH_2-NH_2$$

Ethylenediamine (en)

$$:O:\quad:O:$$
$$\|\quad\|$$
$$C-C$$
$$/\qquad\backslash$$
$$^-:O:\qquad:O:^-$$

Oxalate ion (ox), $C_2O_4{}^{2-}$

Bidentate and polydentate ligands are often called chelating ligands (pronounced key-lateing). The name derives from the Greek word *chele* meaning "claw." Chelate complexes are extra stable because two bonds must be broken to separate a metal from a ligand. $EDTA^{4-}$ is an excellent chelating ligand because it has six donor atoms. Its structure is shown in Figure 22.8 in the text.

***Oxidation Number.*** Each complex ion carries a net charge that is the sum of the charges on the central atom (or ion) and on each of the ligands. In $Fe(CN)_6^{3-}$, for example, each cyanide ion ($CN^-$) contributes a $-1$ charge, and the iron ion a $+3$ charge. The charge of the complex ion then is; $(+3) + 6(-1) = -3$. The oxidation number of iron is $+3$ which is the same as its ionic charge. See Example 22.3 for a calculation.

| Example 22.3 |
| :---: |
| Exercises 4 – 7 |

***Naming Coordination Compounds.*** Thousands of coordination compounds are known. The rules that have been developed for naming them are summarized below:

1. The cation is named before the anion.
2. Within the complex ion, the ligands are named first, followed by the metal ion. Ligands are listed in alphabetical order.
3. The names of anionic ligands end with the letter o, whereas neutral ligands names are usually the same as the names of the molecules. The exceptions are $H_2O$ (aquo), CO (carbonyl), and $NH_3$ (ammine). Table 22.4 in the textbook lists the names of some common ligands.
4. When several ligands of a particular kind are present, use the Greek prefixes di-, tri-, tetra-, penta-, and hexa. Thus, the ligands in $[Co(NH_3)_4Cl_2]^+$ are "tetraamminedichloro." If the ligand itself contains a Greek prefix, use the prefixes; bis (2), tris (3), and tetrakis (4) to indicate the number of ligands present. The ligand ethylenediamine already contains the term di; therefore, bis(ethylenediamine) is used to indicate two ethylenediamine ligands.

5. The oxidation number of the metal is written in Roman numerals following the name of the metal.
6. If the complex is an anion, attach the ending -ate to the name of the metal.

   $Fe(CN)_6^{3-}$ is named hexacyanoferrate (III) ion.

Examples
22.4, 22.5

Exercises
8 – 10

## The Structure of Coordination Compounds (Section 22.4)

***Common Structures.*** The geometry of a coordination complex is defined by the arrangement of the donor atoms of the ligands around the central metal atom. Complexes with coordination numbers of two, four, and six are the most common. Complexes with a general formula $ML_2^{n+}$, where M is the metal atom or ion and L is a donor atom, are linear with ligands on opposite sides of the metal atom. Other geometries are as follows:

| General formula | Coodination number | Structure |
| --- | --- | --- |
| $ML_2^{n+}$ | 2 | Linear |
| $ML_4^{n+}$ | 4 | Tetrahedral or square planar |
| $ML_6^{n+}$ | 6 | Octahedral |

***Geometric Isomers.*** When two or more compounds have the same composition but a different arrangement of atoms, the compounds are called isomers. Isomerism is a characteristic feature of coordination compounds. Stereoisomers are compounds that have the same types and numbers of atoms bonded together in the same sequence, but with different spatial arrangements. There are two types of stereoisomers: geometric isomers and optical isomers.

Geometric, or *cis-trans*, isomers are distinguished by the position of like ligands or groups. The isomer with the like-groups in adjacent positions is called the *cis* isomer, and the one with like-groups across from each other is called the *trans* isomer. *Cis* and *trans* isomers of coordination compounds generally have different physical and chemical properties.

Two forms of the complex $Pt(NH_3)_2Cl_2$ have been prepared. Both have square planar structures. The *trans* form has no dipole moment, whereas the *cis* form has an appreciable dipole moment.

The simplest type of geometric isomerism of octahedral complexes occurs in cases where four of the six ligands of the complex are the same. Then the *cis* and *trans* forms correspond to those shown on the next page. The two X ligands are closest to each other in the *cis* isomer. In the *trans* isomer they are across the complex ion from each other.

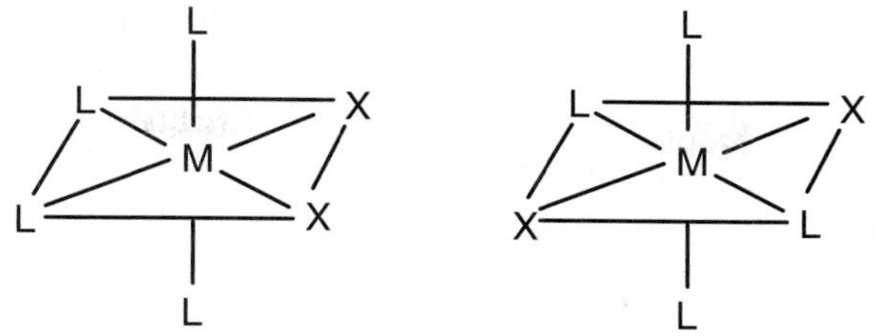

cis          trans

**Optical Isomers.** Optical isomers are nonsuperimposable mirror images of one another. They are also called enantiomers. Enantiomers have the same relationship to one another as do your right and left hands. If you place your left hand parallel to your right hand facing each other, you get the same effect that you would get if you placed one hand in front of a mirror. Your left hand is the mirror image of your right hand. However, your left hand is not superimposable upon your right hand. Your hands are enantiomers, that is, nonsuperimposable mirror images of one another.

The enantiomers of *cis*-dichlorobis(ethylenediamine)cobalt(III) ion are shown in Figure 22.13 in the text. The mirror image of the *trans* isomer can be superimposed on the original after rotating it by 90°. Therefore, there is only one form of the *trans* isomer. The mirror image of the *cis* isomer cannot be superimposed on the original no matter how it is rotated. The *cis* isomer exists as enantiomers.

Not all objects are enantiomers. An ordinary chair, for example, looks the same in the mirror as when viewed directly. A chair is superimposable on its mirror image. An object which is not identical with its mirror image is said to be chiral, from the Greek word for hand (see Figure 22.12 in the text). Example 23. 7 illustrates chiral complexes.

Enantiomers are called optical isomers because they differ with respect to the direction in which they rotate the plane of polarization of plane-polarized light, when light is passed through the substance. The isomer that rotates the plane of polarized light to the right is said to be dextrorotatory. Its mirror image will rotate the plane to the left, levorotatory. A racemic mixture is an equimolar mixture of the two optical isomers. Such a mixture produces no net rotation of plane polarized light.

| Examples 22.6, 22.7 |
| Exercises 11, 12 |

## Bonding in Coordination Compounds (Section 22.5)

**Crystal Field Theory.** In the case of coordination compounds a satisfactory theory must account for properties such as color, magnetism, and stereochemistry. Although the crystal field theory is not the only theory of coordination compounds, it explains satisfactorily their colors and magnetic properties.

The crystal field model considers the ligand to metal bonding in a complex ion to be primarily electrostatic, rather than covalent. In an isolated transition metal ion all five 3d orbitals have the same energy. The effect of the ligands is to change the relative energies of these orbitals through electrostatic interactions. There are two types of electrostatic interactions. One is the force that holds the ligands to the metal ion. This is the attraction between the positive charge of the metal ion and the lone electron pairs of the ligands. Second, there is the repulsion between the lone pairs on the ligand donor atoms and the electrons in the 3d orbitals of the metal. It is this latter interaction that gives rise to crystal field splitting, and its effect on the color and magnetic properties of the complex ion.

Consider the $Fe(CN)_6^{3-}$ complex ion, for example. An isolated $Fe^{3+}$ ion has five 3d electrons, all with the same energy.

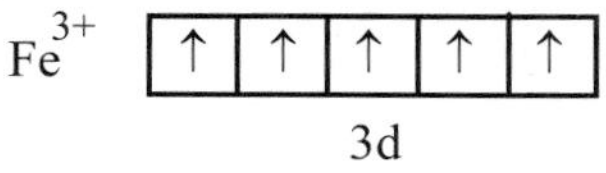

$Fe^{3+}$ | ↑ | ↑ | ↑ | ↑ | ↑ |
3d

When the six cyanide ligands become positioned in an octahedral arrangement, all the 3d orbitals are raised in energy due to the repulsion just mentioned. The repulsion energy is not the same for all 3d orbitals. Rather, those orbitals that are directed straight toward a ligand, experience greater repulsion than those directed between ligands. Thus, in an octahedral complex, the five 3d orbitals are split into two groups in terms of energy. One group, comprising the $d_{xy}$, $d_{yz}$ and $d_{xz}$ orbitals, is not raised in energy as much as the other group consisting of the $d_{z^2}$ and $d_{x^2-y^2}$ orbitals. The situation is shown here schematically in Figure 22.1. The energy difference between these two sets of d orbitals is called the crystal field splitting energy and is given the symbol $\Delta$.

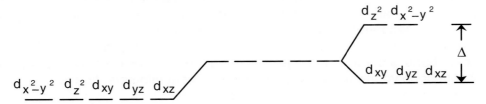

**Figure 22.1**  Crystal field splitting. On the left are the energy levels of the five 3d orbitals in a free $Fe^{3+}$ ion. On the right are the same d orbitals in the same ion, when it is a part of the octahedral complex ion $Fe(CN)_6^{3-}$.

***Color.***  The color of a coordination compound results from electron transitions from the lower-energy set of orbitals to the higher-energy orbitals. Absorption of light occurs when the energy of an incoming photon is equal to the difference in energy $\Delta$, and an electron transition occurs. The requirement for light absorption is:

$$\Delta = E_{photon}$$

Recall from chapter 7 that the energy of a photon is $E_{photon} = h\nu$, where h is Planck's constant, and $\nu$ is the frequency of the radiation. Therefore, $\Delta = h\nu$.

A substance has a color because it absorbs light at certain wavelengths in the visible part of the electromagnetic spectrum (from 400 to 700 nm) and it reflects the other wavelengths. Light that is a combination of all visible wavelengths appears white. When white light impinges on a coordination compound, and light of a certain wavelength is absorbed, the reflected light is missing that component and no longer appears white to the eye.

If the complex absorbs all colors except orange, the complex appears orange. The complex will also appear orange if it absorbs only light of the color blue. In a complementary manner, if the complex ion absorbed only orange, it would appear blue. Blue and orange are referred to as complementary colors.

The following "color wheel" allows you to estimate the color of reflected light when light of a given color (or wavelength) is absorbed. The colors that appear across from each other are complementary colors. For example, the hexacyanoferrate(II) ion absorbs light in the visible region of the spectrum at about 410 nm. The absorbed light is violet. The color of the complex is the color directly across the wheel from violet. The complex will appear yellow, the complementary color of violet.

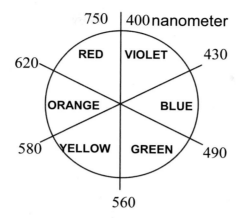

The extent of crystal field splitting, that is, the magnitude of $\Delta$, determines the wavelength of light absorbed, and consequently, the color of the complex. Each ligand has a different strength of interaction with the d orbitals of the metal ion. Thus, each splits the orbital energies by a different amount. The ligands can be arranged according to increasing values of $\Delta$. This establishes a spectrochemical series.

$$I^- < Br^- < Cl^- < OH^- < F^- < H_2O < NH_3 < en < CN^- < CO$$

CO and $CN^-$ are called strong-field ligands because they produce a relatively large splitting, whereas, $I^-$ and $Br^-$ are weak-field ligands because they split d orbital energies to a lesser extent.

***Magnetic Properties.*** In chapter 7 you learned that paramagnetic substances have at least one unpaired electron and diamagnetic substances have all their electrons paired. The $Fe^{3+}$ ion has a $3d^5$ configuration. Two of its octahedral complexes, $[FeF_6]^{3-}$ and $[Fe(CN)_6]^{3-}$ are both paramagnetic, and yet they have different magnetic properties. $[FeF_6]^{3-}$ has five unpaired electrons, but $[Fe(CN)_6]^{3-}$ has only one. The former is called a high-spin complex, and the latter a low-spin complex.

The reason for this difference lies in the spectrochemical series and the value of $\Delta$. Remember, that according to Hund's rule, electrons in orbitals with similar energies prefer to be unpaired. Energy is required to make electrons pair up. Therefore, if $\Delta$ is small, as in the case of the weak-field ligand, $F^-$, the electrons enter all five orbitals one at a time without pairing, in accord with Hund's rule. For small $\Delta$ values, pairing of electrons would not occur until the sixth electron is added.

However, when $\Delta$ is large enough, Hund's rule no longer applies to the entire set of d orbitals. The lowest energy configuration corresponds to placing all of the electrons in the three lowest-energy 3d orbitals first, before any enter the two higher-energy orbitals. This is the case for $[Fe(CN)_6]^{3-}$, because $CN^-$ is a strong-field ligand and produces large $\Delta$ values. Figure 22.2 shows the energy level diagrams for hexafluoroferrate (III) ion and hexacyanoferrate (III) ion.

So far, we have focused on octahedral complexes. Crystal field splitting occurs for tetrahedral and square planar complexes as well. See Figures 22.23 and 22.24 in the text for the splitting diagrams of these complexes.

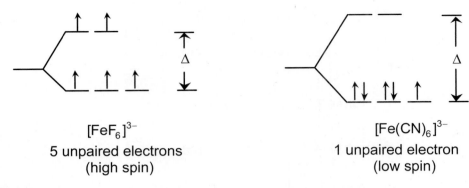

$[FeF_6]^{3-}$
5 unpaired electrons
(high spin)

$[Fe(CN)_6]^{3-}$
1 unpaired electron
(low spin)

Examples
22.8, 22.9

Exercises
13 – 17

**Figure 22.2** Effect of crystal field splitting on pairing of d electrons in two Fe(III) complex ions. When Δ is small, Hund's rule applies to all five orbitals. When Δ is large, the three lower-energy orbitals fill before the two higher orbitals.

## GLOSSARY LIST

| | | |
|---|---|---|
| ligand | bidentate ligands | polarimeter |
| donor atom | polydentate ligands | optical isomers |
| inert complex | chelating ligands | chiral |
| labile complex | | enantiomers |
| | isomer | racemic |
| coordination compound | geometric isomers | |
| coordination number | stereoisomers | crystal field theory |
| | | spectrochemical  series |

## EQUATIONS

| Algebraic Equation | English Translation |
|---|---|
| $\Delta = h\nu$ | Calculating crystal field splitting |

## WORKED EXAMPLES

### EXAMPLE 22.1 Transition Metal Elements

What distinguishes a transition element from a representative element?

• **Solution**

The representative elements are those in which all the inner subshells are filled and the last electron enters an outer s or p subshell. The transition elements are those in which an inner d subshell is partially filled. For example, the outer electron configuration of Sc is $4s^2 3d^1$ and for Co it is $4s^2 3d^7$.

### EXAMPLE 22.2 Properties of Transition Elements

Why do the transition metals have higher densities than the Group 1A and 2A metals?

• **Solution**

Densities of solids are related to atomic and ionic radii. As the atoms get smaller, the solids tend to get more dense. The effective nuclear charge experienced by the outermost electrons in transition metal atoms is greater than that experienced in Group 1A and 2A metal atoms. The atomic radii of the transition elements, in general, are smaller than those of the Group 1A and 2A elements of the same period.

**Work EXERCISES & PROBLEMS: 1 – 3**

### EXAMPLE 22.3 Terminology of Coordination Compounds

A certain coordination compound has the formula $[Co(NH_3)_4Cl_2]Cl$.
a. Which atom is the central atom?
b. Name the ligands and point out the donor atoms.
c. What is the charge on the complex ion?
d. What is the oxidation number (O.N.) of the central atom?

• **Solution**

a. Inside the brackets, the metal atom is written first, and this is the central atom to which all ligands are bonded. Cobalt is the central atom.
b. The ligands are written next. These are ammonia and chloride ions. The donor atoms are N and Cl, respectively.
c. Since only one chloride ion is needed to balance the charge of the complex ion, the complex ion must have a charge of +1.
d. The charge of the central ion plus the sum of charges of ligands = the charge of the complex ion. Ammonia is a neutral ligand and will not affect the complex ion charge. Since two chloride ions are –2, in order for the complex ion to have a +1 charge, then Co must be a +3 ion.

$$\text{O.N.(Co)} + 2 \text{ O.N.(Cl)} = \text{complex ion charge}$$
$$\text{O.N.(Co)} + 2(-1) = +1$$
$$\text{O.N.(Co)} = +3$$

**Work EXERCISES & PROBLEMS: 4 – 7**

### EXAMPLE 22.4 Naming Coordination Compounds

Give a systematic name for each of the following compounds:
a. $[Ag(NH_3)_2]Br$
b. $Ni(CO)_4$
c. $K_2[Cd(CN)_4]$

• **Solution**

a. The two ammonia ligands are represented by the term "diammine." The oxidation state of the Ag ion is +1. The bromide ion is not complexed with silver. It is a counter ion. The systematic name is diamminesilver(I) bromide.
b. The term tetracarbonyl signifies that there are four carbon monoxide ligands. The oxidation state of the nickel atom is zero. Systematic name: tetracarbonylnickel(0).
c. In this case the complex ion is an anion, as a result, cadmium will be referred to as cadmate. Since K is +1 and each cyanide ligand is –1, Cd must have a +2 charge. The name is potassium tetracyanocadmate(II).

**Work EXERCISES & PROBLEMS: 8, 9**

### EXAMPLE 22.5 Writing Formulas from the Names

Write formulas for the following coordination compounds:
a. diaquodicyanocopper(II)
b. potassium hexachloropalladate(IV)
c. dioxalatocuprate(II) ion

• **Solution**

a. *Diaquo* refers to two water molecules, and dicyano to two cyanide ions. Since the copper ion has a +2 charge, this coordination compound is neutral: $Cu(H_2O)_2(CN)_2$
b. The ending *-ate* indicates that the complex ion must be an anion: $K_2[Pd(Cl)_6]$.
c. The complex ion is an anion: $[Cu(C_2O_4)_2]^{2-}$

**Work EXERCISES & PROBLEMS: 10**

**EXAMPLE 22.6 Geometric Isomers**

Sketch the geometric isomers of $[Cr(en)_2Br_2]^+$.

**• Solution**

First, identify the ligands. Recall that "en" stands for ethylenediamine, a neutral bidentate ligand. Bromide ion is the other ligand. Then, the central ion is $Cr^{3+}$ and its coordination number is 6 because it bonds to six donor atoms. The complex will be octahedral. N — N represents ethylenediamine.

      For an octahedral complex, ligands at adjacent corners are referred to as *cis* to each other. The ligands at opposite corners are *trans* to each other. Two geometric isomers are possible. One has two $Br^-$ ions in *cis* positions, and in the other isomer the $Br^-$ ligands are *trans*.

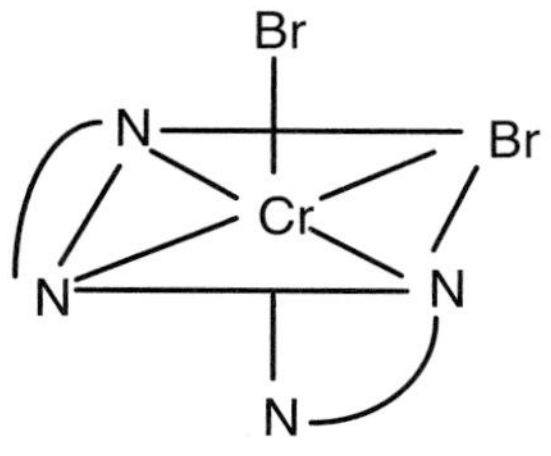

cis 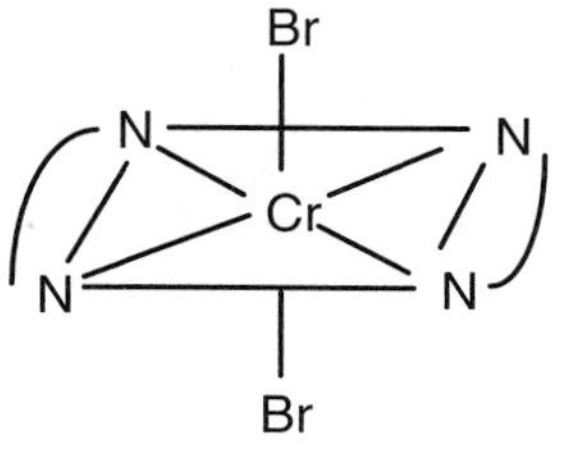                   trans

**EXAMPLE 22.7 Optical Isomers**

For the complex ion in the preceding example, indicate if either of the isomers exhibit optical isomerism.

**• Solution**

Sketch the mirror images of the *cis* and *trans* isomers, and look for nonsuperimposable mirror images, the enantiomers.

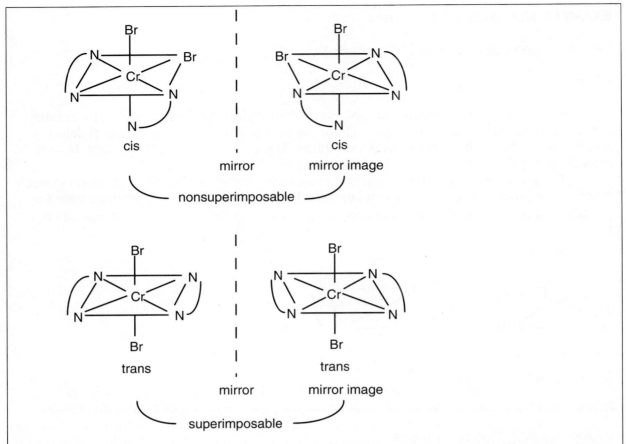

cis     mirror     mirror image     cis

nonsuperimposable

trans     mirror     mirror image     trans

superimposable

The mirror image of the *trans* isomer can be superimposed on the original. Therefore, there is only one form of the *trans* isomer. The mirror image of the *cis* isomer cannot be superimposed on the original no matter how it is rotated. The *cis* isomer exists in two forms (enantiomers). The *cis* isomer is chiral. The *trans* isomer is not.

**Work EXERCISES & PROBLEMS: 11, 12**

---

**EXAMPLE 22.8  Crystal Field Theory**

$\left[Ti(H_2O)_6\right]^{3+}$ ion absorbs light at a wavelength of 498 nm.
a.  Calculate the crystal field splitting energy $\Delta$.
b.  Determine the color of the complex.

**• Solution**

a.  Absorption of light occurs when the energy of an incoming photon is equal to the difference in energy $\Delta$ between the lower energy 3d orbitals, and the higher energy 3d orbitals of the titanium ion. The requirement for light absorption is:

$$\Delta = E_{photon}$$

$$\Delta = h\nu = \frac{hc}{\lambda}$$

where h is Planck's constant, c is the velocity of light, and $\lambda$ is the wavelength of light, 498 nm.

$$\Delta = \frac{(6.63 \times 10^{-34} \text{ J s})(3.00 \times 10^8 \text{ m/s})}{498 \text{ nm} \times \left( \dfrac{10^{-9} \text{ m}}{1 \text{ nm}} \right)}$$

$$\Delta = 3.99 \times 10^{-19} \text{ J}$$

b.  The complex absorbs light with a wavelength of 498 nm. On the color wheel this is green light. A solution that absorbs green light will reflect all the colors except green. According to the color wheel the solution will be red.

## EXAMPLE 22.9  Crystal Field Theory

$[CoF_6]^{4-}$ is a high-spin paramagnetic complex, and $[Co(en)_3]^{2+}$ is a low-spin paramagnetic complex. Draw the crystal field splitting diagrams for these two complex ions and show the proper relationship of the two $\Delta$ values.

• Solution

The crystal field splitting diagram for a $Co^{2+}$ ion ($3d^7$) must show 7 electrons in the 3d orbitals. In a high-spin complex, $\Delta$ is small enough so that electrons can enter 3d orbitals one at a time, according to Hund's rule. Pairing of electrons will not occur until after the fifth electron has been added. In the low-spin complex, $\Delta$ is much larger and the electrons enter the lower-energy orbitals first. Only after 6 electrons are positioned in the lower-energy set of orbitals, can electrons then be placed in the upper two orbitals.

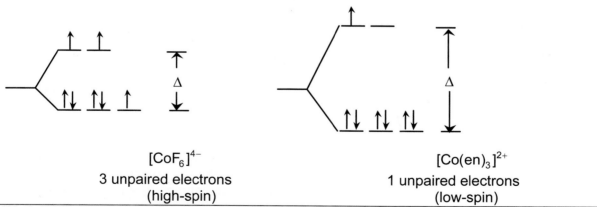

$[CoF_6]^{4-}$
3 unpaired electrons
(high-spin)

$[Co(en)_3]^{2+}$
1 unpaired electrons
(low-spin)

**Work EXERCISES & PROBLEMS: 13 – 17**

## EXERCISES & PROBLEMS

1. Using your periodic table, write the electron configurations of the following:
   a. a Mn atom   b. a $Mn^{2+}$ ion   c. a $Mn^{3+}$ ion

2. Which atom in each pair has the larger atomic radius?  a. V or Fe   b. V or Nb

3. Which has the higher ionization energy, Ti or Fe?

4. What is the charge on the complex ion in $K_2[PtCl_4]$?

5. What is the coordination number of Hg in $[Hg(en)_2]^{2-}$?

6. What is the oxidation number of Fe in $[Fe(CN)_6]^{4-}$?

7. What is the coordination number and oxidation number of the central atom in $K[Cr(NH_3)_3Cl_3]$?

8. Give the systematic name for $[CoCl_3Br_3]^{4-}$.

9. Give the systematic name for $[Ni(en)_2(NH_3)_2]^{2+}$.

10. Write formulas for the following coordination compounds:
    a. diammineoxolatocopper(II)
    b. potassium amminetrichloroplatinate(II)

11. Distinguish between geometric isomers and optical isomers.

12. Descirbe the term chiral.

13. If a complex ion absorbs blue light, what color is the complex ion?

14. Which ligand is the stronger-field ligand, ethylenediamine or bromide ion?

15. Distinguish between a high-spin complex and a low-spin complex.

16. In the complex ion, $ML_6^{n+}$, $M^{n+}$ has six d electrons and L is a weak-field ligand. How many unpaired electrons are there in this complex?

17. The color of oxyhemoglobin is due to complex ions involving Fe(II) ions. Oxyhemoglobin (with $O_2$ bound to iron) is a low-spin, Fe(II) complex. Five of the donor atoms are N atoms and do not change. The sixth donor atom is O in the $O_2$ molecule. Deoxyhemoglobin (without the $O_2$) has a water ligand in place of $O_2$ and is a high-spin complex. Both complexes are octahedral $Fe^{2+}$ complexes.
    a. How many unpaired electrons occupy the 3d orbitals of the iron(II) ion in each case?
    b. Oxyhemoglobin is red and deoxyhemoglobin is blue. Explain in terms of the crystal field theory why the two forms of hemoglobin have different colors.

## PRACTICE TEST QUESTIONS
See notes on taking practice test in the Preface

1. Using a periodic table, write the electron configuration of an Fe atom; an $Fe^{2+}$ ion; an $Fe^{3+}$ ion.

2. Write the formula of the coordination compound containing a $Pt^{4+}$ central atom, three chlorides, three ammonia ligands, and one uncomplexed nitrate ion.

3. Find the oxidation number of the central atom in $Na_2[Co(H_2O)_2I_4]$.

4. For the coordination compound $[Co(NH_3)_6]Cl_3$, give the oxidation state and coordination number of the central atom.

5. For the coordination compound $[Ni(en)_2(NH_3)_2]Cl_2$, give the oxidation state and coordination number of the central atom.

6. Name the following coordination compounds and complex ions.
    a. $[Pt(NH_3)_4Cl_2]Cl_2$
    b. $[Ni(NH_3)_6]^{2+}$
    c. $[Cr(OH)_4]^-$
    d. $[Co(CN)_6]^{3-}$
    e. $K_2[Cu(CN)_4]$

7. Write the formula for each of the following coordination compounds and ions.
    a. tetrahydroxoaluminate(III) ion
    b. tetraiodomercurate(II) ion
    c. potassium dichlorodioxalatocobaltate(III)
    d. tris(ethylenediamine)nickel(II) sulfate

8. Sketch two geometrical isomers for $[Ni(NH_3)_2(CN)_4]^{2-}$.

9. Sketch the structures of the square planar complex *trans*-$[Pt(Cl)_2(NH_3)_2]$, and its mirror image. Are the two structures enantiomers?

10. Why are chiral substances said to be optically active?

11. Illustrate each of the following by giving formulas, geometry, or energy diagrams of:
  a. octahedral complex of $NiCl_6^{4-}$
  b. *cis* and *trans* isomers of $Ni(CO)_2(CN)_2$
  c. $[Fe(H_2O)_6]^{3+}$ is a high-spin complex.
  d. a bidentate ligand
  e. a coordination number of 2

12. If green light is absorbed by a complex in solution, what is the color of the solution?

13. Sketch the crystal field splitting diagram for the low-spin $[Fe(CN)_6]^{4-}$ ion.

14. Sketch the crystal field splitting diagram for the tetrahedral complex $[FeCl_4]^-$.

15. Distinguish between inert and labile complexes.

16. Complexes of zinc are never paramagnetic. Explain.

17. In the complex ion, $[ML_6]^{n+}$, $M^{n+}$ has four d electrons, and L is a weak-field ligand. How many unpaired electrons are there in this complex?

18. $[Ni(H_2O)_6]^{2+}$ is green and $[Ni(en)_3]^{2+}$ is violet. Which has the larger value of $\Delta$?

# Chapter 23. Nuclear Chemistry

**The Nature of Nuclear Reactions (Section 23.1)**
**The Stability of Nuclei (Section 23.2)**
**Natural Radioactivity (Section 23.3)**
**Nuclear Transmutation (Section 23.4)**
**Nuclear Fission and Fusion (Sections 23.5 – 23.6)**
**Biological Effects of Radiation (Section 23.8)**

## SUMMARY

### The Nature of Nuclear Reactions (Section 23.1)

***Nuclear Reactions and Nuclear Equations.*** This chapter emphasizes changes that occur within the nucleus of an atom. To begin, we will discuss two types of nuclear reactions: radioactive decay (Section 23.2) and nuclear transmutation (Section 23.3). Radioactive decay, or radioactivity, is described as the spontaneous emission of particles and/or radiation by unstable atomic nuclei. These processes often result in the formation of a new element. The kinds of particles emitted from various nuclei are shown in Table 23.1.

**Table 23.1** Particles from Radioactive Decay

| Particle | Mass (amu) | Charge | Symbol |
|----------|-----------|--------|--------|
| alpha    | 4.0       | +2     | $_2^4 \alpha$ or $_2^4 \text{He}$ |
| beta     | 0.0005    | −1     | $_{-1}^0 \beta$ or $_{-1}^0 \text{e}$ |
| positron | 0.0005    | +1     | $_{+1}^0 \beta$ or $_{+1}^0 \text{e}$ |
| gamma    | 0         | 0      | $_0^0 \gamma$ |

A nucleus can also undergo change by nuclear transmutation. In this process, one nucleus reacts with another nucleus, an elementary particle, or a photon (gamma particle) to produce one or more new nuclei.

Radioactive decay and nuclear transmutation processes are described by nuclear equations. These equations use isotopic and elementary particle symbols to represent the reactants and products of nuclear reactions. For example, in the first nuclear transmutation ever observed (in 1919) alpha particles were used to bombard nitrogen-14 nuclei. The observed products were oxygen-17 nuclei and protons. The nuclear equation is:

$$_7^{14}\text{N} + {}_2^4\text{He} \rightarrow {}_8^{17}\text{O} + {}_1^1\text{H}$$

The balancing rules for nuclear equations are given below and are applied to the above equation:

1. The sum of the mass numbers of the reactants must equal the sum of the mass numbers of the products (conservation of mass number). 14 + 4 = 18 = 17 + 1

2.  The sum of the nuclear charges of the reactants must equal the sum of the nuclear charges of the products (conservation of atomic number).
$$7 + 2 = 9 = 8 + 1$$

These rules will be illustrated further in the example problems.

Example
23.1

Exercises
1 – 3

***Comparison of Chemical and Nuclear Reactions.*** Table 23.1 of the text lists a number of comparisons between chemical and nuclear processes. Keep in mind that in chemical reactions the number of atoms of each element is conserved. Only changes in chemical bonding occur. However, in nuclear reactions the compositions of the atomic nuclei are altered, and so elements are converted from one to another. Since a nucleus is so extremely small, and in some cases contains large numbers of positively charged protons, the energy changes associated with nuclear changes are much greater than energy changes in chemical reactions.

## The Stability of Nuclei (Section 23.2)

***Nuclear Stability.*** Little is known about the forces that hold a nucleus together. However, some interesting facts emerge if we examine the numbers of protons and neutrons found in those nuclei that are stable. Nuclei can be classified according to whether they contain even or odd numbers of protons and neutrons. The number of stable isotopes of each of the four types of nuclei classified in this way are shown in Table 23.2.

**Table 23.2** The Number of Stable Isotopes

| Number of protons | even | even | odd | odd |
|---|---|---|---|---|
| Number of neutrons | even | odd | even | odd |
| Number of stable isotopes | 157 | 52 | 50 | 8 |

The following rules are useful in predicting nuclear stability:

1.  Nuclei with even numbers of both protons and neutrons are generally more stable than those with odd numbers of these particles. Nuclei that contain certain specific numbers of protons and neutrons ensure an extra degree of stability. These so-called magic numbers for protons and for neutrons are 2, 8, 20, 28, 50, 82, and 126.
2.  Nuclei with even numbers of protons or neutrons are generally more stable than those with odd numbers of these particles.
3.  All isotopes of elements after bismuth ($Z = 83$) are radioactive.

***The Belt of Stability.*** The principal factor for determining whether a nucleus is stable is the neutron to proton ratio. Figure 23.1 shows a plot of the number of neutrons versus the number of protons in various isotopes. The stable nuclei are located in an area of the graph known as the belt of stability. In this figure, we see that at low atomic numbers, stable nuclei possess a neutron to proton ratio of about 1.0. Above $Z = 20$ the number of neutrons always exceeds the number of protons in stable isotopes. The n : p ratio increases to about 1.5 at the upper end of the belt of stability.

If you were given the symbol of a radioisotope, without any experience, it would be impossible to tell its mode of decay. But with knowledge of the belt of stability, you can make accurate predictions of the expected mode of decay.

Isotopes with too many neutrons lie above the belt of stability. The nuclei of these isotopes decay in such a way as to lower their n : p ratio. For example, one neutron may decay into a proton and a beta particle.

$$\ _{0}^{1}\text{n} \rightarrow\ _{1}^{1}\text{p} +\ _{-1}^{0}\beta$$

The proton remains in the nucleus, and the beta particle is emitted from the atom. The loss of a neutron and the gain of a proton produces a new isotope with two important properties. It has a lower n : p ratio, and thus is more likely to be stable. Also, the daughter product has an atomic number that is one greater than the decaying isotope, due to the additional proton. Consider the decay of carbon-14 for example. (Carbon-14 is continually produced in the upper atmosphere by the interaction of cosmic rays with nitrogen.) Carbon-14 has a higher n : p ratio than either of carbon's stable isotopes (C-12 and C-13), and decays by beta decay.

$$\ _{6}^{14}\text{C} \rightarrow\ _{7}^{14}\text{N} +\ _{-1}^{0}\beta$$

Note that the product isotope, $_{7}^{14}\text{N}$, is one atomic number greater than carbon. Also, it is stable. Its n : p ratio is 1.0.

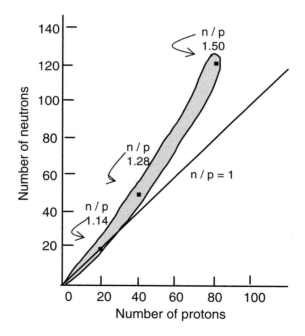

**Figure 23.1** A graph showing all the stable isotopes when plotted as neutron number versus proton number. The shaded area represents the belt of stability. In order for isotopes with Z = 20 to be stable, the ratio of n : p must be about 1.1. For isotopes with Z = 40 to be stable, the ratio of n : p must be about 1.3 : 1.0. For isotopes with Z = 80 to be stable, the ratio of n : p must be about 1.5 : 1.0.

Isotopes with too many protons have low n : p ratios and lie below the belt of stability. These isotopes tend to decay by positron emission because this process produces a new isotope with a higher n : p ratio. During positron emission a proton emits a positron, $_{+1}^{0}\beta$, and becomes a neutron.

$$_{1}^{1}p \rightarrow {}_{0}^{1}n + {}_{+1}^{0}\beta$$

The neutron remains in the nucleus, and the positron is ejected from the atom. Thus, newly produced nucleus will contain one less proton and one more neutron than the parent nucleus. The n : p ratio increases due to positron decay.

Electron capture accomplishes the same end, that is, a higher n : p ratio. Some nuclei decay by capturing an orbital electron of the atom.

$$_{1}^{1}p + {}_{-1}^{0}e \rightarrow {}_{0}^{1}n$$

Lanthanum-138, a naturally occurring isotope with an abundance of 0.089 percent, decays by electron capture.

> Examples
> 23.2, 23.3
>
> Exercises
> 4 – 6

$$_{57}^{138}La + {}_{-1}^{0}e \rightarrow {}_{56}^{138}Ba + h\nu$$

Electron capture is accompanied by x-ray emission, represented in the above reaction as $h\nu$ (the equation for the energy of a photon.)

***Nuclear Binding Energy.***  One of the important consequences of Einstein's theory of relativity was the discovery of the interconvertability of mass and energy. The total energy content (E) of a system of mass m is given by Einstein's theory:

$$E = mc^2$$

where c is the velocity of light ($3.0 \times 10^8$ m/s). Therefore, the mass of a nucleus is a direct measure of its energy content. It was discovered in the 1930s that the measured mass of a nucleus is always smaller than the sum of the separate masses of its constituent nucleons (nuclear particles.) This difference in mass is called the mass defect. When the mass defect is expressed as energy by applying Einstein's equation, it is called the binding energy of the nucleus. The binding energy is the energy required to break up a nucleus into its component, protons and neutrons. The binding energy provides a quantitative measure of nuclear stability. The greater its binding energy, the more stable a nucleus is toward decomposition.

In terms of the $_{8}^{17}O$ nucleus for instance, the binding energy (BE) is the energy required for the process:

$$_{8}^{17}O + BE \rightarrow 8{}_{1}^{1}p + 9{}_{0}^{1}n$$

The mass defect, $\Delta m$, is equal to the total mass of the products minus the total mass of the reactants.

$$\Delta m = [8(\text{proton mass}) + 9(\text{neutron mass})] - ({}_{8}^{17}O \text{ nuclear mass})$$

Here, we can substitute atomic masses, which include the electrons for the nuclear masses.

$$\Delta m = [8(^1_1\text{H atomic mass}) + 9(\text{neutron mass})] - (^{17}_8\text{O atomic mass})$$

This works because the mass of 8 electrons in the 8 hydrogen atoms is canceled by the mass of the 8 electrons in the oxygen atom.

$$8(^1_1\text{H atomic mass}) = 8(\text{proton mass}) + 8(\text{electron mass})$$

$$(^{17}_8\text{O atom mass}) = (^{17}_8\text{O nuclear mass}) + 8(\text{electron mass})$$

Continue by using atomic masses given in the text (Section 23.2) and in Table 23.3, below.

$$\Delta m = [8(^1_1\text{H atomic mass}) + 9(\text{neutron mass})] - (^{17}_8\text{O atomic mass})$$

$$\Delta m = [8(1.007825 \text{ amu}) + 9(1.008665 \text{ amu})] - (16.999131 \text{ amu}) = 0.141454 \text{ amu}$$

The eight protons and nine neutrons have more mass than the oxygen-17 nucleus. The binding energy is:

$$\Delta E = (\Delta m)c^2$$

In comparing the stabilities of any two different nuclei, we must take into account the different numbers of nucleons per nucleus. A satisfactory comparison of nuclear stabilities can be made by using the binding energy per nucleon, that is, the binding energy of each nucleus divided by the total number of protons plus neutrons in the nucleus (nucleons). This is one of the most important properties of a nucleus. When the BE per nucleon is plotted as a function of the atomic mass, we get the curve of binding energy as shown in Figure 23.2 of the text. Note that at first, it rises rapidly with increasing atomic mass, reaching a maximum at mass 56. Above mass 56, the binding energy drops slowly as atomic mass increases. Table 23.3, below, compares the total binding energy and the binding energy per nucleon for several isotopes.

**Table 23.3** Binding Energies of Selected Isotopes

| | Mass (amu) | Binding Energy (J) | |
|---|---|---|---|
| | | Total | Per Nucleon |
| $^2_1\text{H}$ | 2.01410 | $3.57 \times 10^{-13}$ | $1.78 \times 10^{-13}$ |
| $^3_2\text{He}$ | 3.01603 | $1.24 \times 10^{-12}$ | $4.13 \times 10^{-13}$ |
| $^4_2\text{He}$ | 4.00260 | $4.52 \times 10^{-12}$ | $1.13 \times 10^{-12}$ |
| $^{16}_8\text{O}$ | 15.99491 | $2.04 \times 10^{-11}$ | $1.28 \times 10^{-12}$ |
| $^{17}_8\text{O}$ | 16.999131 | $2.10 \times 10^{-11}$ | $1.24 \times 10^{-12}$ |
| $^{56}_{26}\text{Fe}$ | 55.934939 | $7.90 \times 10^{-11}$ | $1.41 \times 10^{-12}$ |
| $^{238}_{92}\text{U}$ | 238.0508 | $2.89 \times 10^{-10}$ | $1.22 \times 10^{-12}$ |

Example 23.4

Exercises 7 – 10

## Natural Radioactivity (Section 23.3)

***Natural Radioactivity.***  A number of isotopes exist in nature that have an n : p ratio that places them outside the belt of stability. These isotopes, called radioisotopes, occur naturally on earth and give rise to natural radioactivity. Uranium, thorium, radon, potassium-40, carbon-13, and tritium (H-3) are naturally occurring radioisotopes. For example, the radioactive decay of uranium-238, which is fairly abundant in Earth's crust, begins a sequence of decay reactions that ultimately changes U-238 to a stable isotope of lead. The uranium series is shown in Figure 23.2.

All of the radioisotopes in this series decay by alpha or beta emission. In the first step, U-238 decays by alpha emission.

$$^{238}_{92}U \rightarrow {}^{234}_{90}Th + {}^{4}_{2}\alpha$$

The decay product, thorium-234, is also radioactive and decays by beta emission.

$$^{234}_{90}Th \rightarrow {}^{234}_{91}Pa + {}^{0}_{-1}\beta$$

The product is also radioactive and the series continues through a number of transitions to end at Pb-206.

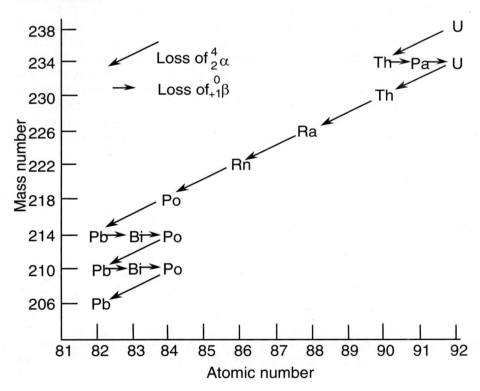

**Figure 23.2.** The uranium decay series involves 14 steps as uranium decays eventually into lead.

***Rates of Decay.*** Radioactive decay rates obey first-order kinetics.

decay rate = number of atoms disintegrating per unit time = $\lambda N$

where $\lambda$ is the first-order rate constant, analogous to k in chemical kinetics (Chapter 3) and N is the number of atoms of the particular radioisotope present in the sample being studied. Recall that the half-life is related to the rate constant by:

$$t_{\frac{1}{2}} = \frac{0.693}{\lambda}$$

The integrated first-order equation is

$$\ln \frac{N}{N_0} = -\lambda t$$

where N is the number of atoms of the radioisotope present in the sample after time $t$ has elapsed, and $N_0$ is the number of atoms of the radioisotope present initially. If $N_0$, N, and $t$ are known, we can calculate the rate constant, $\lambda$.

The purpose of radioactive dating is to determine the age of geological and archaeological samples and specimens. The age ($t$ in the calculation) of certain rocks, for instance, can be estimated from analysis of the number of atoms of a particular radioisotope present now (N), as compared to the number present when the rock was formed originally ($N_0$).

Rearranging the integrated first-order equation, the age $t$ is given by:

$$t = -\frac{1}{\lambda} \ln \frac{N}{N_0}$$

| |
|---|
| Example 23.5 |
| Exercises 11 – 17 |

The value of the initial number of atoms $N_0$ is the sum, N + D, where D is the number of daughter nuclei resulting from the decay of atoms of the radioisotope. The original number of atoms of a radioisotope present in a rock sample is equal to the number N remaining at time $t$, plus the number of daughter atoms (D).

## Nuclear Transmutation (Section 23.4)

***Artificial Radioactivity.*** Nuclear transmutation is the process of converting one element into another. Artificial radioactivity results when an unstable nucleus is produced by transmutation. Irene Curie and Frederic Joliot discovered this phenomenon in 1933 while bombarding light elements with $\alpha$ particles from radioactive sources. For example, when aluminum atoms are bombarded with alpha particles, the product is phosphorus-30, which is radioactive, equation (1). Phosphorus-30 decays by positron emission, equation (2), and has a half-life of 2.5 min. This isotope does not occur naturally in phosphorus compounds.

$$^{27}_{13}\text{Al} + ^{4}_{2}\text{He} \rightarrow ^{30}_{15}\text{P} + ^{1}_{0}\text{n} \qquad (1)$$
$$^{30}_{15}\text{P} \rightarrow ^{30}_{14}\text{Si} + ^{0}_{+1}\beta \qquad (2)$$

Neutrons readily produce "artificial" radioactivity because they are easily captured by stable nuclei, with the result that new nuclei with higher n : p ratios are formed. This leads to products that decay by beta decay. Neutron capture by chlorine-37 yields chlorine-38.

$$_0^1n + \ _{17}^{37}Cl \rightarrow \ _{17}^{38}Cl$$

$$_{17}^{38}Cl \rightarrow \ _{18}^{38}Ar + \ _{-1}^{0}\beta$$

Chlorine-38 has a short half-life, and is not found naturally on earth. Note that neutron capture followed by beta decay, yields a new element (Ar) with an atomic number one greater than the original element. This procedure of neutron capture followed by beta decay of the product nucleus, has been used to synthesize elements that were "missing" from the periodic table, such as Tc and Pm, element numbers 43 and 61, respectively. All isotopes of these elements are radioactive.

Using neutrons as bombarding particles is convenient because neutrons have no charge and therefore are not repelled by target nuclei. When projectiles are positively charged, such as, is the case for protons and alpha particles, they require considerable kinetic energy in order to overcome the electrostatic repulsion they feel upon approach and collision with the target nucleus.

Nuclear transmutation processes can be abbreviated according to the following format.

Target nucleus (bombarding particle, ejected particle) Product nucleus

The alpha bombardment of aluminum, see equation (1) above, can be written:

$$_{13}^{27}Al \ (\alpha,n) \ _{15}^{30}P$$

| |
|---|
| Example 23.6 |
| Exercises 18 – 20 |

The symbols n, p, d, $\alpha$, e (or $\beta$), and $\gamma$ are used in this notation to represent neutron, proton, deuteron ($_1^2H$), alpha particle, electron, and gamma ray.

## Nuclear Fission and Fusion (Sections 23.5 – 23.6)

*Nuclear Fission.* The curve of binding energy per nucleon versus mass number (Figure 23.2 in text) shows that the most stable nuclei are those with masses close to iron-56, which is the most stable nucleus. During fission, a heavy nucleus of mass greater than about 230 amu splits into two lighter nuclei whose masses are usually between 80 and 160 amu. Since the two smaller nuclei are more stable than the larger nucleus, energy is released in the process.

Although many nuclei heavier than uranium can undergo fission, the most important ones for practical purposes are U-233, U-235, and Pu-239. These isotopes undergo fission upon capture of a neutron. It is important to realize that naturally occurring uranium consists of two isotopes, U-235 and U-238, but that only U-235 is fissionable with thermal neutrons. The term thermal neutron refers to those existing at temperatures around 25 °C. Since nuclei do not repel neutrons, as they do alpha particles, for instance, neutron-induced fission will occur at ordinary temperatures.

The reaction is quite complex because the same two products are not formed by all fissioning nuclei. The two reactions on the next page show just two out of many possibilities.

$$\begin{array}{ll}
(1) & \rightarrow \ ^{141}_{56}Ba + {}^{92}_{36}Kr + 3{}^{1}_{0}n \\
^{235}_{92}U + {}^{1}_{0}n & \\
(2) & \rightarrow \ ^{137}_{52}Te + {}^{96}_{40}Zr + 2{}^{1}_{0}n
\end{array}$$

The actual distribution of product yields is shown in Figure 23.7 of the text.

A significant feature of fission is that on the average 2 to 3 neutrons are released per fission event. Since neutrons are required to initiate fission, and because neutrons are also products of fission, a nuclear chain reaction is possible.

The energy released during nuclear fission depends somewhat on just what products are formed. The energy released from the fission of one mole of U-235 atoms can be calculated from the equation, $\Delta E = \Delta mc^2$. The calculation shows that about $2.0 \times 10^{10}$ kJ are released per mole of uranium. This amount of energy is *70 million times* the amount released in the exothermic chemical reaction in which 1 mol of $H_2$ reacts with 1/2 mol of $O_2$ to form 1 mol of water!

*Nuclear Fusion.* Radioactivity and nuclear fission are processes in which matter "comes apart." The energy releasing processes that occur on the Sun are ones in which matter is fused, or "comes together." Nuclear fusion is the combining of small nuclei, such as hydrogen nuclei, to form larger, more stable nuclei. These larger nuclei will have a higher average binding energy per nucleon, so fusion reactions will be exothermic. Because all nuclei are positively charged, they must collide with enormous force in order to combine (fuse). This means that the atoms that will undergo fusion must be heated to millions of degrees. Fusion reactions are called thermonuclear reactions because they occur only at very high temperatures, such as those in the Sun.

The reaction that accounts for the tremendous release of energy by the sun is believed to be the stepwise fusion of four hydrogen nuclei to produce one helium nucleus. The net process is:

$$4\ {}^{1}_{1}H \rightarrow {}^{4}_{2}He + 2\ {}^{0}_{1}e \quad \Delta E = -4.3 \times 10^{-12}\ J$$

One gram of hydrogen, upon fusion, releases the energy equivalent to the combustion of 20 tons of coal. The fusion of four moles of H atoms by the preceding equation releases $2.6 \times 10^9$ kJ of energy.

Our Sun is made up of mostly hydrogen (90 percent) and helium (9 percent). In its interior, the temperatures are estimated to reach 15 million degrees Celsius. At these temperatures, hydrogen will fuse to form helium; but helium, with its greater nuclear charge, will not fuse to form the heavier elements. The heavier elements up to iron, are formed by fusion reactions that occur in exploding stars called nova and supernova, where the temperatures can reach 2 billion degrees Celsius.

| Examples 23.7, 23.8 |
| Exercises 21 – 23 |

## Biological Effects of Radiation (Section 23.8)

*Interaction of Radiation with Matter.* In passing through matter; alpha, beta, and gamma rays lose energy, chiefly by interaction with electrons. Alpha and beta particles, on colliding with electrons, forcefully eject these electrons from atoms and molecules, and thereby produce ions. These particles lose only a small fraction of their energy in a single collision with an electron. Because alpha and beta particles are extremely energetic, thousands of collisions are required to bring them to rest. These particles produce "tracks" of ionization. Alpha, beta, and gamma rays are known as ionizing radiation.

Most devices for detecting radioactivity depend on the formation of ions. The best known instrument for detecting radiation is the Geiger counter, in which ions produced by a particle trigger a pulse of electricity that is counted electronically. Darkening of photographic plates, discharging of electroscopes, and damage to biological tissue all involve ionization.

*Units of Radiation Dose.* Two units used to measure radiation dose are the rad and the rem. The rad (radiation absorbed dose) is defined as 0.01 joule of energy absorbed per kilogram of any absorbing material. The millirad is one-thousandth of a rad. Because beams of different radiations cause very different biological damage, even when the body absorbs the same amount of energy from each type, it is necessary to define a unit specifically for biological tissue. The unit of biologically effective dose is the rem (<u>r</u>adiation <u>e</u>quivalent in <u>m</u>an), which is the absorbed dose in rads multiplied by the <u>r</u>elative <u>b</u>iological <u>e</u>ffectiveness factor, *RBE*. The millirem is one thousandth of a rem. For beta and gamma rays, *RBE* = 1.0; for fast neutrons and alpha particles, *RBE* = 10. Thus, a dose of one rad of alpha radiation is equivalent to 10 rem.

dose (in rem) = *RBE* × dose (in rad)

## GLOSSARY LIST

| | | |
|---|---|---|
| transmutation | transuranium elements | thermonuclear reaction |
| positron | nuclear chain reaction | breeder reactor |
| nucleon | nuclear fission | |
| nuclear binding energy | nuclear fusion | plasma |
| mass defect | critical mass | radical |
| radioactive decay series | moderators | tracer |

## EQUATIONS

| Algebraic Equation | English Translation |
|---|---|
| $\Delta E = \Delta mc^2$ | Nuclear binding energy is mass defect times the speed of light squared. |

## WORKED EXAMPLES

---

### EXAMPLE 23.1  Nuclear Equations

Complete the following nuclear equations. Label the nuclear reaction as radioactive decay or nuclear transmutation.

a. $^{14}_{7}N + ^{1}_{0}n \rightarrow ^{14}_{6}C + $ ____
b. $^{226}Ra \rightarrow ^{4}_{2}\alpha + $ ____

### • Solution

a.  According to rule 1, the sum of the mass numbers of the reactants must equal the sum of the mass numbers of the products. 14 + 1 = 15 = 14 + A. Therefore, the unknown product will have a mass number of 1. According to rule 2, the sums of the nuclear charges on both sides of the equation must be the same. 7 + 0 = 7 = 6 + Z. Therefore, the nuclear charge of the unknown product must be 1, making it a proton.

$$^{14}_{7}N + ^{1}_{0}n \rightarrow ^{14}_{6}C + ^{1}_{1}p$$

This reaction is a nuclear transmutation.

b.  Note that the atomic number of radium is missing. The periodic table indicates that Ra is element number 88. Balancing the mass numbers first, we find that the unknown product must have a mass number of 222. Balancing the nuclear charges next, we find that the atomic number of the unknown must be 86. Element number 86 is radon.

$$^{226}_{88}Ra \rightarrow ^{4}_{2}\alpha + ^{222}_{86}Rn$$

This reaction is a radioactive decay.

**Work EXERCISES & PROBLEMS: 1 – 3**

---

### EXAMPLE 23.2  Nuclear Stability

Using the stability rules, rank the following isotopes in order of increasing nuclear stability.

$$^{39}_{20}Ca \qquad ^{40}_{20}Ca \qquad ^{30}_{15}P$$

### • Solution

Phosphorus-30 should be the least stable because it has odd numbers of both protons and neutrons. Calcium-39 has an even number of protons (20) and an odd number of neutrons. With an even number of protons and a "magic number" at that, it should be more stable than P-30. Of the three isotopes, calcium-40 should be the most stable. It has an even number of protons and of neutrons. Both numbers are "magic numbers."

$$^{30}_{15}P < ^{39}_{20}Ca < ^{40}_{20}Ca$$

**Work EXERCISES & PROBLEMS: 4**

---

### EXAMPLE 23.3  Types of Radioactive Decay

The only stable isotope of sodium is sodium-23. What type of radioactivity would you expect from sodium-25?

### • Solution

Sodium-23 has 11 protons and 12 neutrons and is in the belt of stability. Sodium-25 must have two more neutrons, and so has a higher n : p ratio than the stable isotope. Sodium-25 will decay by $_{-1}^{0}\beta$ emission.

---

**Work EXERCISES & PROBLEMS: 5, 6**

---

### EXAMPLE 23.4  Nuclear Binding Energy

Calculate the nuclear binding energy of the light isotope of helium, helium-3. The atomic mass of $_{2}^{3}\text{He}$ is 3.01603 amu.

### • Solution

The binding energy is the energy required for the process

$$_{2}^{3}\text{He} \rightarrow 2\,_{1}^{1}\text{p} + \,_{0}^{1}\text{n}$$

where, $\Delta m$ = [2(proton mass) + (neutron mass)] $- \,_{2}^{3}\text{He}$ (nuclear mass)

The mass difference is calculated using atomic masses:

$$\Delta m = [2(_{1}^{1}\text{H atomic mass}) + (\text{neutron mass})] - (_{2}^{3}\text{He atomic mass})$$

$$\Delta m = [2(1.007825 \text{ amu}) + 1.008665 \text{ amu}] - 3.01603 \text{ amu}$$
$$\Delta m = 8.29 \times 10^{-3} \text{ amu}$$

> Masses are given in Table 23.3.

Using Einstein's equation:

$$\Delta E = (\Delta m)c^2 = (8.29 \times 10^{-3} \text{ amu}) \times (3.00 \times 10^8 \text{ m/s})^2$$

$$= 7.46 \times 10^{14} \text{ amu m}^2/\text{s}^2$$

$$= 7.46 \times 10^{14} \text{ amu m}^2/\text{s}^2 \times \frac{1.00 \text{ kg}}{6.022 \times 10^{26} \text{ amu}} \times \frac{1 \text{ J}}{1 \text{ kg m}^2/\text{s}^2}$$

The binding energy is:

$$\Delta E = 1.24 \times 10^{-12} \text{ J/atom}$$

Each $_{2}^{3}\text{He}$ atom contains three nucleons. The binding energy per nucleon is:

$$\text{BE per nucleon} = \frac{1.24 \times 10^{-12} \text{ J/atom}}{3 \text{ nucleons/atom}}$$

The binding energy per nucleon is:

$$\Delta E = 4.13 \times 10^{-13} \text{ J/nucleon}$$

### • Comment

By combining the above conversion factors into one constant, the number of steps in future calculations can be lessened.

$$? \text{ J/amu } = (3.00 \times 10^8 \text{ m/s})^2 \times \frac{1.00 \text{ kg}}{6.022 \times 10^{26} \text{ amu}} \times \frac{1 \text{ J}}{1 \text{ kg m}^2/\text{s}^2}$$

$$= 1.49 \times 10^{-10} \text{ J/amu}$$

This constant is a useful factor relating energy to mass in amu. Applying this to the mass defect (0.14145 amu) calculated earlier in the discussion for $^{17}_{8}\text{O}$, the binding energy is $2.10 \times 10^{-11}$ J/atom. See Table 23.3.

**Work EXERCISES & PROBLEMS: 7 – 10**

---

### EXAMPLE 23.5  Radioactive Dating

The rubidium-87/strontium-87 method of dating rocks was used to analyze lunar samples. The half-life of Rb-87 is $4.9 \times 10^{10}$ yr.

$$^{87}_{37}\text{Rb} \rightarrow ^{87}_{38}\text{Sr} + ^{0}_{-1}\beta$$

Estimate the age of moon rocks in which the mole ratio of Rb-87 to Sr-87 is 40.

### • Solution

The age of the rock can be calculated from the first-order equation relating time to the number of nuclei remaining:

$$t = -\frac{1}{\lambda} \ln \frac{N}{N_0}$$

Since the rate constant k is not given, we must use the equation relating k and $t_{1/2}$:

$$\lambda = \frac{0.693}{t_{1/2}} = 1.41 \times 10^{-11}/\text{yr}$$

This can be substituted for $\lambda$, yielding:

$$t = -\frac{1}{1.41 \times 10^{-11} / \text{yr}} \ln \frac{N^{Rb}}{N_0^{Rb}}$$

Since Rb decays to Sr, the initial number of Rb atoms is equal to the sum of the Rb atoms and Sr atoms present.

$$N_0^{Rb} = N^{Rb} + N^{Sr}$$

Given that: $N^{Rb}/N^{Sr} = 40$, then:

$$N^{Sr} = \tfrac{1}{40} N^{Rb}$$

Therefore, after substitution for $N^{Sr}$, we get:

$$N_0^{Rb} = N^{Rb} + \tfrac{1}{40} N^{Rb} = 1.025\ N^{Rb}$$

Substituting into the rate equation, we get:

$$t = -\frac{1}{1.41 \times 10^{-11}/\text{yr}} \ln \frac{N^{Rb}}{1.025\ N_0^{Rb}} = -\frac{1}{1.41 \times 10^{-11}/\text{yr}} \ln \frac{1}{1.025}$$

$$t = 1.7 \times 10^9 \text{ yr}$$

**Work EXERCISES & PROBLEMS: 11 – 17**

---

### EXAMPLE 23.6  Synthesizing a Transuranium Element

The first transuranium element to be synthesized by scientists was neptunium, atomic number 93. Devise a means to produce neptunium starting with U-238 and neutrons.

#### • Solution

Neutron capture by a nucleus followed by beta decay produces a new nucleus of one atomic number higher than the original. Neptunium is one atomic number beyond uranium. So start with uranium and bombard it with neutrons.

$$^{238}_{92}\text{U} + ^{1}_{0}\text{n} \rightarrow ^{239}_{92}\text{U}$$
$$^{239}_{92}\text{U} \rightarrow ^{239}_{93}\text{Np} + ^{0}_{-1}\beta$$

#### • Comment

Uranium's other naturally occuring isotope, U-235, will not produce neptunium in the same way because it undergoes neutron-induced fission.

**Work EXERCISES & PROBLEMS: 18–20**

---

### EXAMPLE 23.7  Relationship of Mass and Energy

Calculate the mass of hydrogen that must undergo nuclear fusion each day, in order to provide just the fraction of the daily energy output of the Sun that reaches the earth, which is $1.5 \times 10^{22}$ J.

#### • Solution

Using the following equation:

$$4\ ^{1}_{1}\text{H} \rightarrow ^{4}_{2}\text{He} + 2\ ^{0}_{+1}\text{e} \qquad \Delta E = -4.3 \times 10^{-12} \text{ J}$$

we see that fusion of 4 H atoms yields $4.3 \times 10^{-12}$ J. We can set up the calculation:

$$g\ H\ atoms = \frac{1.5 \times 10^{22}\,J}{day}$$

$$g\ H\ atoms = \frac{1.5 \times 10^{22}\,J}{day} \times \frac{4\,H\,atoms}{4.3 \times 10^{-12}\,J} \times \frac{1.66 \times 10^{-24}\,g}{H\,atom}$$

$$g\ H\ atoms = 2.3 \times 10^{10}\,g$$

Expressing the answer in tons, we get:

$$2.3 \times 10^{10}\,g = 25{,}500\ tons\ of\ hydrogen$$

---

**EXAMPLE 23.8  Comparing Nuclear Fission and Fusion**

Compare fission and fusion with respect to the temperatures required and the nature of the by-products of these processes.

**• Solution**

Neutron-induced fission occurs at ordinary temperatures, whereas nuclear fusion requires temperatures in the millions of degrees. Fission of heavy elements yields hundreds of isotopes of elements with intermediate atomic numbers. These isotopes, by and large, have an excess of neutrons and are beta emitters. On the other hand, fusion of "light" nuclei yields stable isotopes of low-to-medium atomic mass.

**Work EXERCISES & PROBLEMS 21 – 24**

## EXERCISES & PROBLEMS

1. Write symbols for alpha particles, beta particles, and gamma rays.

2. How many protons (p), neutrons (n), and electrons (e) does an atom of the radioisotope phosphorus-32 contain?

3. Complete the following nuclear reactions:
   a. $^{14}_{7}N + ^{1}_{0}n \rightarrow ^{14}_{6}C +$ _____
   b. $^{210}_{84}Po \rightarrow ^{4}_{2}\alpha +$ \_\_\_\_\_
   c. $^{4}_{2}He + ^{9}_{4}Be \rightarrow ^{12}_{6}C +$ \_\_\_\_
   d. $^{15}_{7}N \rightarrow$ \_\_\_\_ $+ ^{0}_{-1}\beta$

4. For each pair of nuclei, predict which one is the more stable.
   a. $^{3}_{2}He$ or $^{4}_{2}He$
   b. $^{26}_{13}Al$ or $^{27}_{13}Al$

5. When a nucleus decays by positron emission, the atomic number of the decay product will be (higher or lower) than the original radioisotope?

6.  Fluorine has only one stable isotope, fluorine-19.
    a. The nucleus of fluorine-18 lies below the belt of stability. Write an equation for the decay of fluorine-18.
    b. The nucleus of fluorine-21 lies above the belt of stability. Write an equation for the decay of fluorine-21.

7.  How many atomic mass units are in one kilogram?

8.  Which has more mass, the nucleus of aluminum-27 or 13 protons and 14 neutrons?

9.  What is the binding energy of $^{27}_{13}Al$? Calculate its binding energy per nucleon. The atomic mass of aluminum-27 is 26.981541 amu.

10. How much energy is released when one Po-214 atom decays by alpha emission?

$$^{214}_{84}Po \rightarrow {}^{4}_{2}\alpha + {}^{210}_{82}Pb$$

Given the atomic masses:   Pb-214 = 213.99519 amu, Po-210 = 209.98286 amu, and He-4 = 4.00260 amu.

11. The uranium decay series starts with uranium-238 and ends with lead-206. Each step in the series involves the loss of either an alpha or a beta particle. In the entire series, how many alpha particles and how many beta particles are emitted?

12. The rates of decay of all radioisotopes follow the same type of rate equation. What type of equation is it?

13. Cobalt-60 has a half-life of 5.26 years. a. Calculate the rate constant for this isotope. b. How much cobalt-60 will remain from a 20.0 mg sample after 8.75 years?

14. A 2.52 mg sample of pure uranium-238 has a decay rate (activity) of 31.2 dps due to U–238. a. Compute the rate constant for U-238.  b. Compute the half-life of U–238.

15. The C-14 activity of some ancient corn was found to be 10 disintegrations per minute per gram of C. If present day plant life gives 15.3 dpm/g C, how old is the corn? The half-life of C-14 is 5730 yr.

16. Estimate the age of rocks in which the mole ratio of U-238 to Pb-206 is 0.75. The half-life of U-238 is $4.5 \times 10^9$ yr.

17. Analysis of a sample of uranite ore yields 3.2 g of U-238 and 1.5 g Pb-206. Assuming there was no Pb-206 present initially, how old is the rock? The half-life of U-238 is $4.5 \times 10^9$ yr.

18. Predict the product of the following bombardment reactions.

    a. $^{27}_{13}Al + {}^{1}_{0}n \rightarrow X + {}^{1}_{1}H$
    b. $^{11}_{5}B + {}^{4}_{2}He \rightarrow X + {}^{1}_{1}H$

19. Predict the product of the following bombardment reactions.

    a. $^{11}_{5}\text{B}$ (p, $\alpha$)X

    b. $^{10}_{5}\text{B}$ (n,p)X

20. Devise a scheme by which bombardment of molybdenum-98 with neutrons and subsequent $\beta$ decay of the product could be used to prepare an isotope of the missing element technetium, $^{99}_{43}\text{Tc}$.

21. Which process, nuclear fusion or nuclear fission, requires temperatures of millions of degrees?

22. Complete the following fission reaction of plutonium.

    $$^{239}_{94}\text{Pu} + {}^{1}_{0}\text{n} \rightarrow {}^{144}_{58}\text{Ce} + \text{X} + 3\,{}^{1}_{0}\text{n}$$

23. Complete the following nuclear fission reaction.

    $$^{2}_{1}\text{H} + {}^{3}_{1}\text{H} \rightarrow \text{X} + {}^{1}_{0}\text{n}$$

24. How does a rem differ from a rad?

## PRACTICE TEST QUESTIONS
See notes on taking practice test in the Preface

1. What similarities and differences exist between beta particles and positrons?

2. Complete and balance the following nuclear equations.

   a. $^{239}_{94}Pu \rightarrow {}^{4}_{2}He + \underline{\quad}$

   b. $\underline{\quad} + {}^{6}_{3}Li \rightarrow 2\,{}^{4}_{2}He$

   c. $^{90}_{38}Sr \rightarrow \underline{\quad} + {}^{0}_{-1}\beta$

   d. $^{10}_{5}B + {}^{4}_{2}He \rightarrow \underline{\quad} + {}^{1}_{0}n$

   e. $^{56}_{26}Fe + {}^{1}_{0}n \rightarrow \underline{\quad}$

3. Rank the following nuclides in order of increasing nuclear stability:

   $^{40}_{20}Ca \qquad {}^{39}_{20}Ca \qquad {}^{11}_{5}B$

4. With reference to the belt of stability, state the modes of decay you would expect for the following:

   a. $^{13}_{7}N$   b. $^{26}_{13}Al$   c. $^{28}_{13}Al$

5. a. Calculate the binding energy and the binding energy per nucleon of $^{27}_{13}Al$. The atomic mass of Al-27 is 26.98154 amu.
   b. Compare this result to the binding energy of $^{28}_{14}Si$, which has an even number of protons and neutrons. The atomic mass of Si-28 is 27.976928 amu.

6. Cobalt-60 is used in radiation therapy. It has a half-life of 5.26 years.
   a. Calculate the rate constant for radioactive decay.
   b. What fraction of a certain sample will remain after 12 years?

7. Radioactive decay follows first-order kinetics. If 20 percent of a certain radioisotope decays in 4.0 years, what is the half-life of this isotope?

8. Estimate the age of a bottled wine that has a tritium ($^{3}_{1}H$) content that is 3/4 that of environmental water obtained from the area where the grapes were grown. $t_{1/2}$ = 12.3 yr

9. Analysis of a sample of uranite ore shows that the ratio of U-238 atoms to Pb-206 atoms is 3.8. Assuming there was no Pb-206 present initially, how old is the rock?

   $t_{1/2}$ (U-238) = $4.5 \times 10^9$ yr

10. Consider the following fusion reactions:

    a. $^{1}_{1}H + {}^{2}_{1}H \rightarrow {}^{3}_{2}He$

    b. $^{3}_{2}He + {}^{3}_{2}He \rightarrow {}^{4}_{2}He + 2\,{}^{1}_{1}H$

    c. $^{2}_{1}H + {}^{3}_{1}H \rightarrow {}^{4}_{2}He + {}^{1}_{0}n$

    Given the atomic masses (in amu):

    $^{3}_{2}He$ = 3.016029; $^{4}_{2}He$ = 4.002603;

    $^{3}_{1}H$ = 3.017005; $^{2}_{1}H$ = 2.01410. Which of the above has the largest change in energy as indicated by its change in mass $\Delta m$?

11. One atom of element 109 was prepared by bombardment of a target of bismuth-209 with accelerated nuclei of iron-58. Write a balanced nuclear equation to show the formation of the isotope of element 109 with a mass number of 266.

12. Write nuclear equations that show how Pu-239 is formed in a "breeder" reactor.

13. $^{137}_{55}$Cs is a fission product of $^{235}_{92}$U. If it is formed along with three neutrons, what is the other isotope formed?

14. What is the mode of decay expected for "light" nuclei which are unstable because of low n : p ratio?

15. One natural radioactive series begins with $^{238}_{92}$U and ends with $^{206}_{82}$Pb. All steps in the series are either alpha or beta decay. How many $\alpha$ particles and $\beta$ particles are emitted?

16. The $\beta$ particles emitted by carbon-14 atoms have a maximum energy of $2.5 \times 10^{-14}$ J per particle. What is the dose in rads when $8.0 \times 10^{10}$ carbon-14 atoms decay, and all the energy from their decay is absorbed by 2.0 kg of matter?

17. A sample of biological tissue absorbs a dose of 1.0 rad of alpha radiation. How many rems is this?

18. Radium-226 is an $\alpha$ emitter with a half-life of 1600 years. If 2.00 g of radium were allowed to undergo decay for 10.0 years, and all of the $\alpha$ particles were collected over that time as helium gas, what would be the mass and the volume of the He at STP?

23          23

23          23

23          23

23          23

# Chapter 24. Organic Chemistry

**Aliphatic Hydrocarbons (Sections 24.1 – 24.2)**
**Aromatic Hydrocarbons (Section 24.3)**
**Functional Groups (Section 24.4)**

## SUMMARY

### Aliphatic Hydrocarbons (Sections 24.1 – 24.2)

***Organic Chemistry.*** The heart of organic chemistry is the carbon atom. Carbon is a key ingredient of about seven million chemical compounds, primarily because of its ability to form long chains of self-linked atoms. The existence of structural and geometric isomers contributes strongly to the number of organic or carbon-containing compounds.

The hydrocarbons are an important class of organic compounds that consist only of the elements, carbon and hydrogen. Hydrocarbons are divided into two classes: aliphatic and aromatic. Aromatic hydrocarbons contain one or more benzene rings. Aliphatic hydrocarbons do not contain benzene rings. Four groups of aliphatic hydrocarbons are known. These are alkanes, alkenes, alkynes, and cycloalkanes.

***Alkanes.*** Alkanes contain only single bonds between carbon atoms.The general formula for an alkane is $C_nH_{2n+2}$, where n is the number of carbon atoms in the molecule, n = 1, 2, 3.... When n = 1, we have the simplest member of the alkane family methane ($CH_4$). The alkanes make up a homologous series; a series of compounds differing in the number of carbon atoms. As n increases one at a time, we can generate the formulas of the entire series of alkanes. The names and formulas of the first ten straight-chain alkanes are given in Table 24.1. The first part of each name represents the number of C atoms in the molecule. The ending '-ane' is common to all alkanes. If you learn these names, they will prove to be very useful in naming other organic compounds.

**Table 24.1** Names of the First Ten Alkanes

| Formula | Name |
|---------|------|
| $CH_4$ | methane |
| $C_2H_6$ | ethane |
| $C_3H_8$ | propane |
| $C_4H_{10}$ | butane |
| $C_5H_{12}$ | pentane |
| $C_6H_{14}$ | hexane |
| $C_7H_{16}$ | heptane |
| $C_8H_{18}$ | octane |
| $C_9H_{20}$ | nonane |
| $C_{10}H_{22}$ | decane |

***Alkenes.*** Alkenes contain $C = C$ double bonds, and members of this homologous series have the general formula $C_nH_{2n}$. Alkenes are named with the same root word as alkanes to indicate the number of carbon atoms, but all names end in "ene."

| $C_2H_4$ | $CH_2\!=\!CH_2$ | ethene (ethylene) |
| $C_3H_6$ | $CH_3CH\!=\!CH_2$ | propene |
| $C_4H_8$ | $CH_3CH_2CH\!=\!CH_2$ | l-butene |

**Alkynes.** Alkynes contain $C\!\equiv\!C$ triple bonds, and have the general formula $C_nH_{2n-2}$. The names of alkynes end with "yne."

| $C_2H_2$ | $CH\!\equiv\!CH$ | ethyne (acetylene) |
| $C_3H_4$ | $CH_3C\!\equiv\!CH$ | propyne |
| $C_4H_6$ | $CH_3CH_2C\!\equiv\!CH$ | l-butyne |

**Cycloalkanes.** There is also a type of alkane that has atoms bonded into ring configurations. These are the cycloalkanes. They have the same general formula as the alkenes, $C_nH_{2n}$, but do not have double bonds. Neither are these aromatics. Two cycloalkanes are shown below.

```
   H2C—CH2              H2C—CH2
    \   /                |    |
     CH2                H2C—CH2
  cyclopropane          cyclobutane
```

**Structural Isomers.** Until now, we have considered only aliphatic hydrocarbons that have straight chains of carbon atoms. Branching of hydrocarbon chains is very common. Butane ($C_4H_{10}$) can be straight chained, and branched chained.

```
CH3—CH2—CH2—CH3   and   CH3—CH—CH3
                                |
                               CH3
     n-butane, C4H10      2-methylpropane, C4H10
```

Note that both molecules have the same molecular formula, but, they have different arrangements of atoms (different structures). These are actually two distinguishable compounds with their different structures producing slightly different chemical and physical properties. Molecules that have the same molecular formula, but a different structure are called structural isomers. Straight-chain hydrocarbons are called *normal* and use the symbol *n*. The straight-chain form of $C_4H_{10}$ has the name *n*-butane. Naming branched chain hydrocarbons is covered in the next section. In the alkane series, as the number of C atoms increases, the number of possible structural isomers increases dramatically. For example, $C_4H_{10}$ has 2 isomers, $C_6H_{14}$ has 5, and $C_{10}H_{22}$ has 75 isomers.

---

**Nomenclature.** The rules for naming hydrocarbons according to the IUPAC system are briefly summarized as follows:

1. The systematic name of an alkane is based on the number of carbon atoms in the longest carbon chain. For alkanes, the longest carbon chain is given a name corresponding to the alkane with the same number of C atoms. The parent name of a compound with 5 carbon atoms in the longest carbon chain is pentane.
2. Groups attached to the main chain are called substituent groups. Substituent groups that contain only hydrogen and carbon atoms are called alkyl groups. When an H atom is removed from an alkane the fragment is called an alkyl group. This group can be

attached to the longest chain. Alkyl groups are named by dropping the ending *-ane* and adding *yl* to the alkane name. Therefore, $CH_3-$ is a methyl group, $C_2H_5-$ is an ethyl group, and $C_3H_7-$ is a propyl group. Table 24.2 in the text lists the names of six common alkyl groups.

3. The locations of alkyl groups that are attached to the main chain must also be included in the name. Number the carbon atoms in the longest chain. The numbering should start at the end of the chain such that the side groups will have the smallest numbers.

4. Prefixes such as *di*, *tri*, and *tetra* are used when more than one substituent of the same kind is present. Combine the names of substituent groups, their locations, and the parent chain into the hydrocarbon name. Arrange the substituent groups alphabetically, followed by the parent name of the main chain.

5. Alkanes can have many different types of substituents besides alkyl groups. Halogenated hydrocarbons contain fluorine, chlorine, bromine and iodine atoms which are named fluoro, chloro, bromo, and iodo, respectively. $NH_2-$, $NO_2-$, and $CH_2 = CH-$ are called amino, nitro, and vinyl, respectively.

6. When naming alkenes we indicate the positions of the carbon-carbon double bonds, and in alkynes the position of the triple bond is numbered. For alkenes and alkynes, the parent name is derived from the longest chain that contains the double or triple bond. Then, parent name will end in *-ene* for alkenes and *-yne* for alkynes. Number the C atoms in the main chain by starting at the end nearer to the double or triple bond, and use the lower number of the carbon atom in the $C = C$ bond.

Examples
24.1 – 24.2

Exercises
1 - 5

**Addition Reactions.** Alkenes, alkynes, and aromatics are called unsaturated hydrocarbons. This means that they can acquire more hydrogen atoms in an addition reaction called hydrogenation.

$$CH_2 = CH_2 + H_2 \rightarrow CH_3 - CH_3$$
$$CH \equiv CH + 2H_2 \rightarrow CH_3 - CH_3$$

Alkanes are called saturated hydrocarbons because they cannot acquire additional hydrogen atoms. The carbon atoms in a saturated hydrocarbon are already bonded to the maximum number of H atoms.

$$CH_3 - CH_3 + H_2 \rightarrow \text{no reaction}$$

In an addition reaction, a small molecule such as $H_2$ is added to an unsaturated hydrocarbon. The addition reaction occurs at the $C = C$ double bond. One atom of the small molecule links to one of the carbon atoms of the double bond, while the other atom attaches to the other carbon atom. Other examples are:

$$CH_2 = CH_2 + Cl_2 \rightarrow CH_2Cl - CH_2Cl$$
$$CH_2 = CH_2 + H_2O \rightarrow CH_3 - CH_2OH$$
$$CH_2 = CH_2 + HCl \rightarrow CH_3 - CH_2Cl$$

**Markovnikov's Rule.** The two carbons of a double bond suffer different fates during addition reactions with unsymmetrical reagents such as HCl, HBr, and H—OH (water). For example, two different compounds might possibly be formed by reaction of 1-butene with HCl.

$$CH_3CH_2CH=CH_2 + HCl \rightarrow CH_3CH_2CHCl{-}CH_3 \text{ or } CH_3CH_2CH_2{-}CH_2Cl$$

observed product       not found

The rule that predicts which of the two products is formed is called Markovnikov's rule. This rule states:  In the addition of unsymmetrical reagents to alkenes, the positive group of the reagent adds to the carbon that already has the most hydrogen atoms. In HCl, HBr, and $H_2O$ the hydrogen atoms have a partially positive charge because they are bonded to more electronegative atoms. Thus, in the above reaction with $CH_3CH_2CH=CH_2$, the H atom from HCl bonds to the terminal C atom. The Cl atom bonds to the CH group.

> Example
> 24.4
>
> Exercises
> 7 – 9

***Geometric Isomers of Alkenes.***  C=C double bonds are completely rigid. Consider the structures of *cis-* and *trans-2*-butene.

```
CH3     H              CH3     CH3
  \    /                 \    /
   C=C                    C=C
  /    \                 /    \
 H      CH3             H       H
 trans-2-butene        cis-2-butene
```

Neither structure can rotate around the C=C bond to become the other structure. These two molecules have different physical properties and can be separated from one another. Therefore, these two structures represent two different compounds and are isomers. Many pairs of such isomers have been isolated, and they have no tendency to undergo interconversion.

The rigidity of the double bond gives rise to *cis-* and *trans-* geometric isomerism. (This is the same geometric isomerism discussed in chapter 22, in reference to coordination compounds. Recall that,  the *cis* configuration has like groups on the same side of the double bond.

```
   CH3     CH3
     \    /
- - - C=C - - -     cis
     /    \
    H      H
```

The *trans* configuration has like groups on the opposite sides of the double bond. *Trans* means across.

```
   CH3     H
     \    /
- - - C=C - - -     trans
     /    \
    H      CH3
```

> Exercise
> 10

## Aromatic Hydrocarbons (Section 24.3)

**Benzene.** Benzene is the parent compound of a class of hydrocarbons called aromatic hydrocarbons. The hydrogen to carbon ratio in the molecular formula of benzene $C_6H_6$ suggests that it is highly unsaturated. In 1865 Kekule suggested that benzene had a cyclic structure with three double bonds.

Benzene can be represented as two resonance structures.

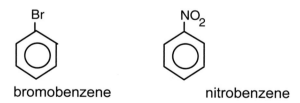

Alternatively, benzene can be represented in terms of delocalized molecular orbitals.

Regardless of which symbol is used, the hydrogen atoms are usually not explicitly written. Rather, we must remember that there is one attached to each carbon.

Aromatic hydrocarbons contain benzene rings and do not have a simple general formula. Benzene is the simplest compound in this group.

**Nomenclature of Aromatic Hydrocarbons.** The naming of monosubstituted benzenes is straightforward.

bromobenzene                    nitrobenzene

If more than one substituent is present, we must indicate the position of the two groups relative to each other. The prefixes *ortho- (o), meta- (m),* and *para- (p)* are used to indicate their relative positions.

o-dichlorobenzene          *m*-dichlorobenzene   *p*-dichlorobenzene

**Reactions of Aromatic Compounds.** Halogens react with benzene by a substitution reaction rather than an addition reaction. In this reaction, an atom or group of atoms replaces an H atom in the benzene ring. Chlorine atoms can be substituted for hydrogen atoms by reacting benzene and chlorine in the presence of an $FeCl_3$ catalyst.

| Exercises |
| 11 - 12 |

## Functional Groups (Section 24.4)

**Alcohols.** Alkanes are quite inert toward most substances, the main exception being oxygen. The alkane portion of organic compounds is combustible. Certain groups of atoms within carbon compounds always react in the same way regardless of the chain length of the alkane portion. The group of atoms that is largely responsible for the chemical behavior of a molecule is called a functional group. For instance, alcohols contain an alkane group and a hydroxyl group, OH. Methanol ($CH_3OH$), ethanol ($C_2H_5OH$), and n-propanol ($C_3H_7OH$) are the first three alcohols of a homologous series. In each of the three, chemical reactions occur at the OH group, rather than at the less reactive alkane group. The hydroxyl group is the functional group in each, and the three alcohols react toward other reagents in the same way.

Reactions of alcohols discussed in the textbook are oxidation by oxidizing agents, displacement of hydrogen by alkali metals, and esterification. An oxidizing agent removes two H atoms, one from the hydroxyl group and one from the adjacent carbon atom, forming an aldehyde (shown below). Permanganate ion is an oxidizing agent.

$$CH_3CH_2CH_2-OH \xrightarrow{MnO_4^-} CH_3CH_2CHO$$

Quite often, in organic reactions, we do not write completely balanced equations. Here, the emphasis is on the chemical change of the alcohol functional group.

Alkali metals displace $H_2(g)$ from alcohols. Potassium is a reducing agent.

$$2\ CH_3CH_2OH + 2\ K \rightarrow 2\ CH_3CH_2OK + H_2$$

Esterification will be discussed under esters.

**Ethers.** The functional group in ethers is C—O—C or, more generally, R—O—R, where R stands for any alkyl group such as those listed in Table 24.2 of the textbook. Ethers and alcohols are structural isomers. As illustrated in the following diagram, both dimethyl ether and ethanol have the same chemical formula.

$$CH_3OCH_3 \quad \leftarrow [C_2H_6O] \rightarrow \quad CH_3CH_2OH$$

dimethyl ether                      ethanol
$T_b$ −25 °C                        $T_b$ 78 °C

The boiling points given, show that ethers are much more volatile than the corresponding alcohols. This is because of the absence of hydrogen bonding in ethers. Alcohols contain the polar hydroxyl group, $O^{\delta-} - H^{\delta+}$, which can participate in hydrogen bonding to neighboring alcohol molecules. For many years diethyl ether, $CH_3CH_2OCH_2CH_3$, was used as an anesthetic.

***Aldehydes and Ketones.*** The funtional group in these compounds is the carbonyl group, $C = O$. In aldehydes the carbonyl group is at the end of the alkane chain. In ketones it is not a terminal group.

$$
\begin{array}{cc}
O & O \\
\| & \| \\
CH_3CH_2CH_2CH \quad \text{or } CH_3CH_2CH_2CHO & CH_2CH_2CCH_3 \text{ or } CH_3CH_2COCH_3 \\
\text{butanal (an aldehyde)} & \text{2-butanone (a ketone)}
\end{array}
$$

Aldehydes are easily oxidized to carboxylic acids.

$$
\begin{array}{cc}
O & O \\
\| \quad MnO_4^- & \| \\
CH_3CH_2CH & \rightarrow \quad CH_3CH_2C-OH
\end{array}
$$

Ketones, generally, are less reactive than aldehydes.

$$
\begin{array}{c}
O \\
\| \quad MnO_4^- \\
CH_3CH_2C-CH_3 \rightarrow \text{ no reaction}
\end{array}
$$

***Carboxylic Acids.*** The carboxyl group is the functional group in organic acids.

$$
\begin{array}{c}
O \\
\| \\
-C-OH \quad \text{abbreviated as } -COOH
\end{array}
$$

This group is weakly acidic. The ionizable hydrogen atom accounts for the acidic properties of carboxylic acids.

$$CH_3COOH(aq) \rightleftharpoons CH_3COO^-(aq) + H^+(aq)$$

A homologous series of organic acids can be generated starting with formic acid.

| | |
|---|---|
| HCOOH | methanoic acid (formic acid) |
| $CH_3COOH$ | ethanoic acid (acetic acid) |
| $CH_3CH_2COOH$ | propanoic acid |
| $CH_3CH_2CH_2COOH$ | butanoic acid |

Fatty acids are carboxylic acids that contain more than four carbon atoms. Two fatty acids are:

$$CH_3(CH_2)_{14}COOH \qquad \text{palmitic acid}$$
$$CH_3(CH_2)_{16}COOH \qquad \text{stearic acid}$$

**Esters.**  The ester functional group is:

$$\underset{\text{—C—OR}}{\overset{\displaystyle O}{\overset{\|}{\phantom{x}}}} \quad \text{or} \quad \text{—COOR}$$

where R stands for any alkyl group. This functional group resembles the carboxylic acid group except that the R group replaces the H atom. Carboxylic acids react with alcohols to form the compounds called esters. The reaction that changes an acid to an ester is called an esterification reaction.

$$\underset{\text{acetic acid}}{CH_3\overset{\displaystyle O}{\overset{\|}{C}}\!-\!OH} + \underset{\text{methanol}}{H\!-\!OCH_3} \rightarrow \underset{\text{methyl acetate}}{CH_3\overset{\displaystyle O}{\overset{\|}{C}}\!-\!OCH_3} + H_2O$$

Esterification is an example of a condensation reaction. A "condensation" is a reaction in which parts of two molecules become joined by formation of a new covalent bond to make a new, larger molecule. Esterification reactions are reversible, and at equilibrium a mixture is formed that contains all four substances.

Saponification is the alkaline hydrolysis of an ester. In this reaction, the base reacts directly with an ester to split it into a salt of an organic acid and an alcohol.

$$C_2H_5COOCH_3 + NaOH \rightarrow C_2H_5COONa + CH_3OH$$

**Amines.**  Organic bases contain the —$NH_2$ functional group, which is called an amino group. Molecules containing an amino group are called amines. The amines may be considered as derivatives of ammonia ($NH_3$). Amines have the general formula, $R_3N$, where R may be H, an alkyl group, or an aromatic group. Molecules of the type $RNH_2$ are called primary amines. Those with $R_2NH$ are secondary amines, and $R_3N$ is a tertiary amine.

| Examples 24.4, 24.5 |
| Exercises 13 – 19 |

$$\underset{\substack{\text{methylamine}\\\text{a primary amine}}}{\overset{\displaystyle H}{\overset{\|}{CH_3\!-\!N\!-\!H}}} \qquad \underset{\substack{\text{dimethylamine}\\\text{a secondary amine}}}{\overset{\displaystyle CH_3}{\overset{\|}{CH_3\!-\!N\!-\!H}}} \qquad \underset{\substack{\text{trimethylamine}\\\text{a tertiary amine}}}{\overset{\displaystyle CH_3}{\overset{\|}{CH_3\!-\!N\!-\!CH_3}}}$$

## GLOSSARY LIST

| | | |
|---|---|---|
| organic chemistry | cycloalkane | aldehyde |
| hydrocarbon | unsaturated hydrocarbon | ketone |
| aliphatic hydrocarbon | addition reaction | carboxylic acid |
| aromatic hydrocarbon | substitution reaction | ester |
| alkane | condensation reaction | |
| saturated hydrocarbon | | saponification |
| structural isomer | functional group | amine |
| alkene, alkyne | alcohol | substituent group |

## WORKED EXAMPLES

---

### EXAMPLE 24.1  Naming Hydrocarbons

Name the molecule that has the following structure:

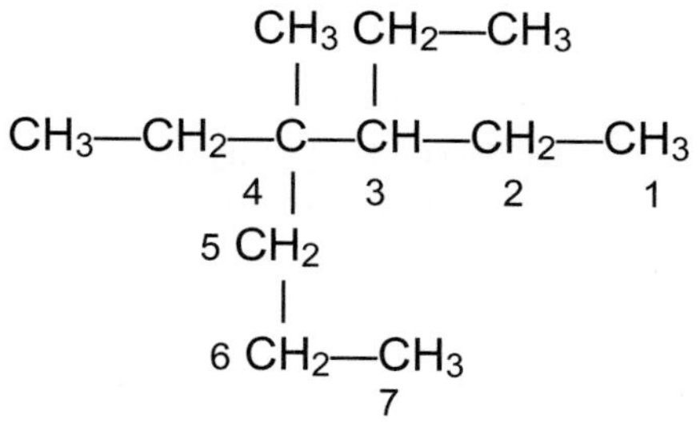

$$
\begin{array}{c}
\text{CH}_3\ \text{CH}_2\text{—CH}_3 \\
|\quad | \\
\text{CH}_3\text{—CH}_2\text{—C—CH—CH}_2\text{—CH}_3 \\
4\ |\quad 3\quad\ 2\qquad 1 \\
5\ \text{CH}_2 \\
| \\
6\ \text{CH}_2\text{—CH}_3 \\
7
\end{array}
$$

#### • Solution

First identify the longest continuous carbon chain. Two equivalent chains containing seven carbon atoms are noticeable. Number the carbon atoms as shown. There are two ethyl groups, one at carbon 3 and one at carbon 4, and one methyl group also bonded to carbon 4. The parent name is heptane. Placing the names of the substituent groups (alphabetically, ethyl is before methyl) and their position numbers in front of the parent alkane name we get: 3,4-diethyl-4-methylheptane.

#### • Comment

If the chain had been numbered in reverse order the name would be 4,5-diethyl-5-methylheptane. Since 3,4 is smaller than 4,5 the first name is the correct one.

---

### EXAMPLE 24.2  Naming Hydrocarbons

Name the following hydrocarbon:

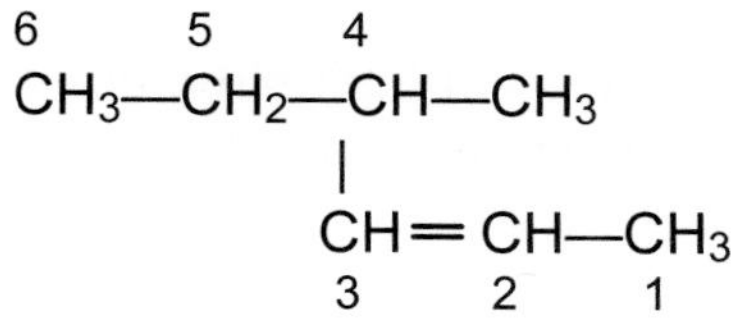

$$
\begin{array}{c}
6\qquad 5\qquad 4 \\
\text{CH}_3\text{—CH}_2\text{—CH—CH}_3 \\
| \\
\text{CH}=\text{CH—CH}_3 \\
3\qquad 2\qquad 1
\end{array}
$$

#### • Solution

Locate the longest chain that contains the double bond. This chain contains six C atoms so the root name will be hexene. Number the chain so that the double bond has the smallest number, as shown. Note the methyl group on carbon 4. The name is 4-methyl-2-hexene.

#### • Comment

The 2 indicates the position of the $C=C$ double bond, which connects carbons numbered 2 and 3.

**Work EXERCISES & PROBLEMS: 1 – 6**

## EXAMPLE 24.3  Addition Reactions

Give the structure of the product of the following reaction.

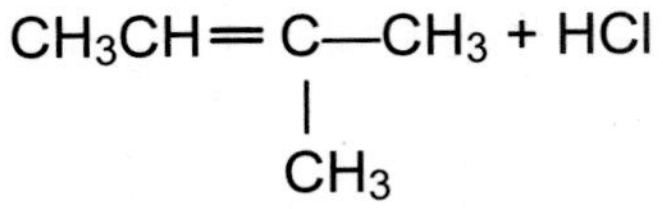

$$CH_3CH=C-CH_3 + HCl$$
$$\underset{CH_3}{|}$$

• **Solution**

This solution involves the addition of an unsymmetrical reagent to a double bond. According to Markovnikov's rule the positive portion of the reagent (in this case an H atom) adds to the carbon atom in the double bond that already has the most hydrogen atoms. The Cl atom adds to the other C atom. The product will be

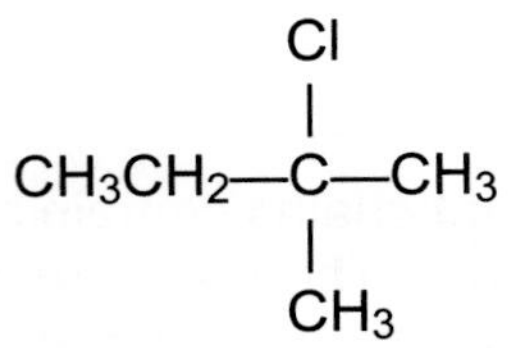

$$\underset{\underset{CH_3}{|}}{\overset{\overset{Cl}{|}}{CH_3CH_2-C-CH_3}}$$

**Work EXERCISES & PROBLEMS: 7 – 9**

## EXAMPLE 24.4  Identifying Functional Groups

Indicate the functional groups in the following formulas:
a.  $C_5H_{11}OH$
b.  $CH_3CHO$
c.  $C_3H_7OCH_3$
d.  $CH_3COC_2H_5$
e.  $CH_3COOCH_3$

• **Solution**

Learning to recognize functional groups requires memorization of their structural formulas. Table 24.4 (textbook) shows a number of the important functional groups.
a.  $C_5H_{11}OH$ contains a hydroxyl group. It is an alcohol.

b.  $CH_3CHO$ is a way to represent $CH_3\overset{\overset{O}{\|}}{C}-H$ on a single line of type.
    $C=O$ is a carbonyl group. $CH_3CHO$ is an aldehyde.
c.  $C_3H_7OCH_3$ contains a C—O—C group and is an ether.
d.  $CH_3COC_2H_5$ is a way to represent a ketone on a single line. $C=O$ is a carbonyl group.
e.  $CH_3COOCH_3$ is a condensed structural formula for:

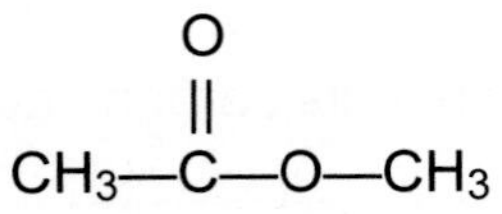

$$CH_3-\overset{\overset{O}{\|}}{C}-O-CH_3$$

$CH_3COOCH_3$ is an ester.

**Work EXERCISES & PROBLEMS: 13 – 15**

---

**EXAMPLE 24.5  Predicting Reaction Products**

Predict the product or products of the following reactions:

a. $CH_3CH(OH)CH_3 + Cr_2O_7^{2-} \rightarrow$

b. $CH_3CHO + MnO_4^- \rightarrow$

c. $HCOOH + CH_3OH \rightarrow$

d. $CH_3CH_2COOCH_3 + NaOH \rightarrow$

**• Solution**

a. This is the reaction of a secondary alcohol with an oxidizing agent. The product is a ketone:

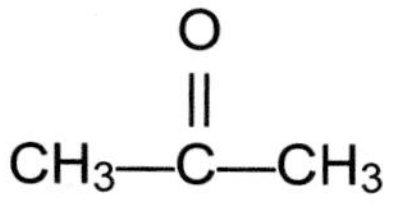

$$
\begin{array}{c}
\quad\quad O \\
\quad\quad \| \\
CH_3\!-\!C\!-\!CH_3
\end{array}
$$

b. This is the reaction of an aldehyde with an oxidizing agent. The product is a carboxylic acid: $CH_3COOH$.

c. The reaction of an alcohol with an acid yields an ester and water.

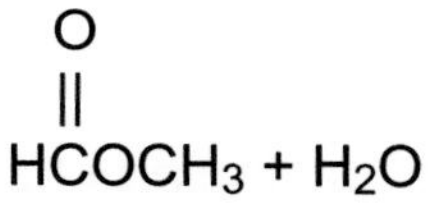

$$
\begin{array}{c}
O \\
\| \\
HCOCH_3 + H_2O
\end{array}
$$

d. This is a saponificaton of an ester to form the sodium salt of a carboxylic acid and an alcohol.

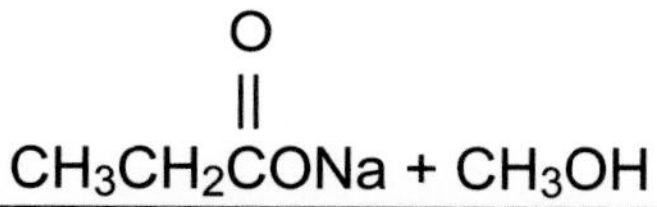

$$
\begin{array}{c}
\quad\quad O \\
\quad\quad \| \\
CH_3CH_2CONa + CH_3OH
\end{array}
$$

**Work EXERCISES & PROBLEMS: 16 – 19**

## EXERCISES & PROBLEMS

1. Write the general formulas of alkanes, alkenes, and alkynes.

2. Give the IUPAC name for the compound with the following structure.

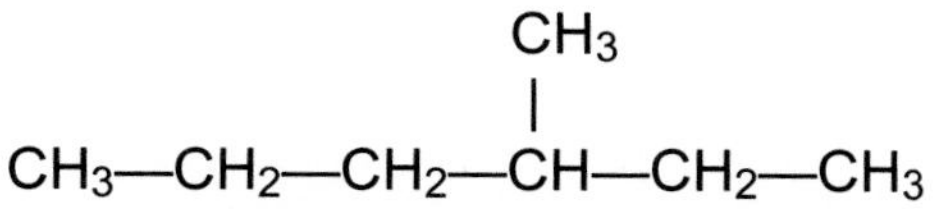

$$
\begin{array}{c}
\quad\quad\quad\quad\quad CH_3 \\
\quad\quad\quad\quad\quad | \\
CH_3\!-\!CH_2\!-\!CH_2\!-\!CH\!-\!CH_2\!-\!CH_3
\end{array}
$$

3. Give the IUPAC name for the following compound.

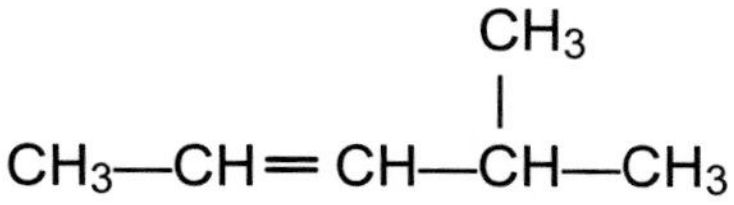

$$
\begin{array}{c}
\quad\quad\quad\quad CH_3 \\
\quad\quad\quad\quad | \\
CH_3\!-\!CH\!=\!CH\!-\!CH\!-\!CH_3
\end{array}
$$

4. Name the following compounds.
   a. $CH_3CHFCH_3$    b. $(CH_3)_3CCH_2CHClCH_3$

5. Write the structure for 3,3-dimethyl pentane.

6. Name the following:  a. $CH_3(CH_2)_3CH=CHCH_3$    b. $CH_3(CH_2)_3C\equiv CH$

7. Which of the classes of hydrocarbons in number 1 are unsaturated hydrocarbons?

8. What is the product of the following addition reaction:  $Cl_2 + CH_3CH=CH_2$?

9. What is the product of the addition reaction: $HBr + CH_3CH=CH_2$?

10. Draw all the isomers of $C_4H_8$.

11. Name the following compounds.

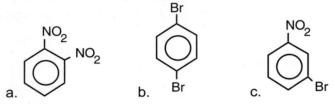

a.    b.    c.

12. How does benzene differ in reactivity from unsaturated aliphatic hydrocarbons?

13. Draw structures for the following functional groups: aldehyde, amine, carboxylic acid, ester.

14. What elements are present in an amine group?

15. Identify the functional group or type of molecule in the following.
    a. $CH_3OH$
    b. $CH_3CHO$
    c. $CH_3COCH_3$
    d. $CH_3COOCH_3$
    e. $CH_3CH_2CH_3$

16. Give two chemical properties of alcohols.

17. Oxidation of what type of compound yields a ketone?

18. Which of the following functional groups exhibit hydrogen bonding?  carboxylic acids, alcohols, ethers.

19. What are the reactants in a saponificaton reaction that yields $C_6H_{13}COONa$ and $C_2H_5OH$?

**PRACTICE TEST QUESTIONS**

See notes on taking practice test in the Preface

1. Give the systematic name for the following structural formulas:

   a. $CH_3(CH_2)_7CH_3$

   b.
   $$CH_3-CH_2-\underset{\underset{CH_3}{|}}{CH}-CH=CH_2$$

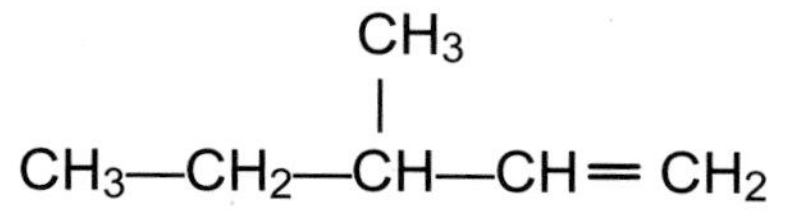

2. Give the systematic name for the following structural formulas:

   a. $CH_3-CH_2-\underset{\underset{CH_2-CH_3}{|}}{CH}-CH_2-OH$

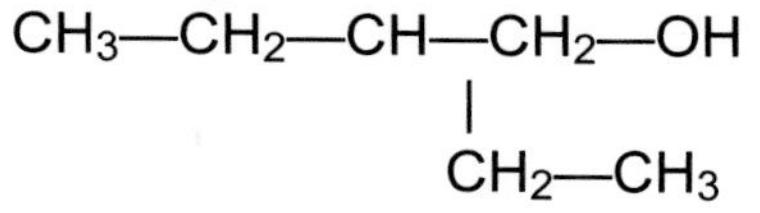

   b. $CH_3-CH_2-C\equiv C-CH_2-\underset{\underset{C_3H_7}{|}}{CH}-CH_3$

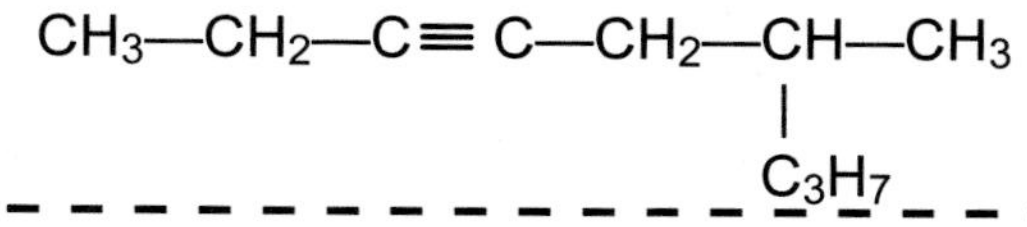

3. Give the systematic name for the following:

   $$CH_3-CH_2-CH-\underset{\underset{\underset{CH_3}{|}}{\underset{|}{CH_2-CH_2}}}{CH}-\underset{\underset{CH_2-CH_3}{|}}{CH}-CH_3$$

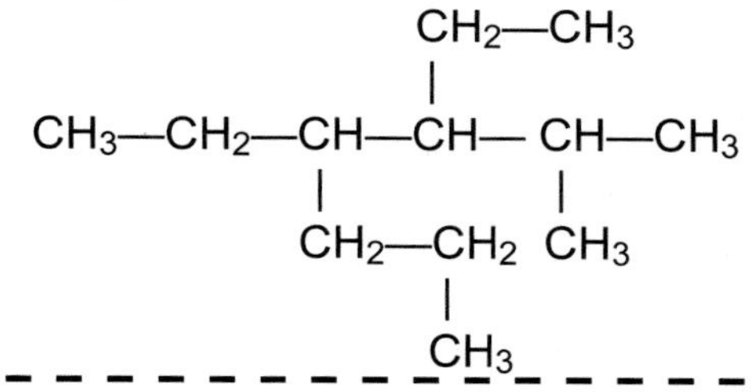

4. Give the systematic name for the following:

   $$CH_3-CH_2-\underset{\underset{CH_2-CH_2-CH_3}{|}}{CH}-\underset{\underset{CH-CH_2-CH_3}{|}}{\overset{CH_3}{|}}$$

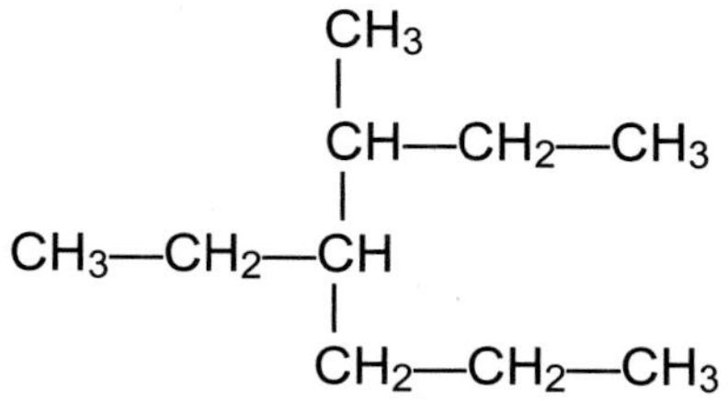

5. Give the systematic name for the following:

   $$\underset{CH_3}{\overset{H}{\diagdown}}C=C\underset{H}{\overset{CH_2CH_3}{\diagup}}$$

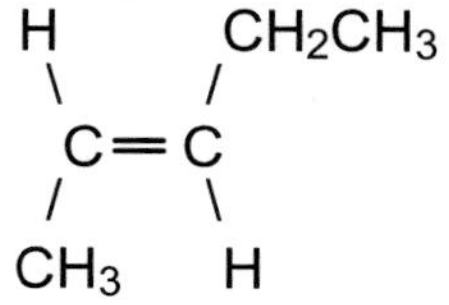

6. Give the systematic name for the following:

   a. $CH_3-\underset{\underset{OH}{|}}{CH}-CH_2-CH_3$

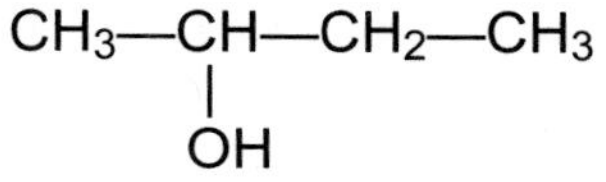

   b. $CH_3-OH$

7. Draw structural formulas for:

   a. 3-ethyl-4-methyl-4-isopropylheptane
   b. 3-bromo-2,5-dimethyl-*trans*-3-hexene
   c. 3-methyl-3-hexanol
   d. trichloroacetic acid

8. Draw structural formulas of all isomers of $C_5H_{10}$. Include both structural and geometric isomers.

9. List the functional groups by name that are shown in the following molecule:

   $$CH_3-CH=CH-\underset{\underset{OH}{|}}{CH}-CH_2-NH_2$$

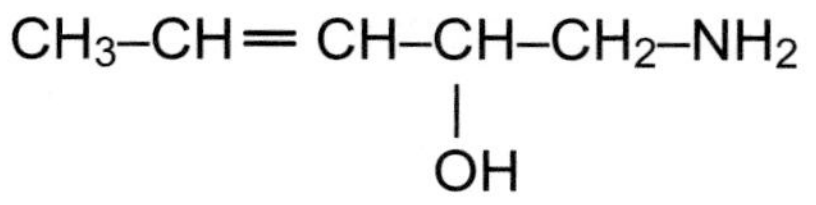

10. List the functional groups by name that are shown in the following molecules:

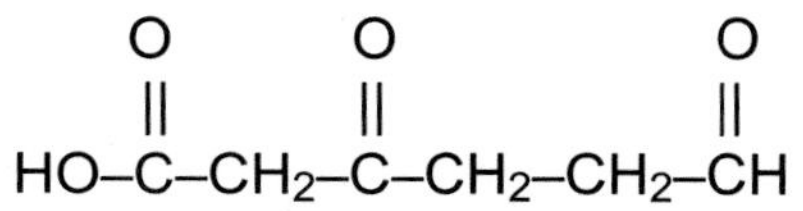

   $$\underset{HO-\overset{O}{\overset{||}{C}}-CH_2-\overset{O}{\overset{||}{C}}-CH_2-CH_2-\overset{O}{\overset{||}{CH}}}{}$$

11. Give the structure of the organic product of each of the following:

a.

b.
CH₃CH with oxidizing agent (CH₃CHO structure, with =O)

$$CH_3\overset{O}{\overset{||}{C}}H \xrightarrow[\text{agent}]{\text{oxidizing}}$$

12. Give the structure of the organic product of each of the following:

a. $CH_3C \equiv CH + Cl_2 \rightarrow$

b. $C_3H_8 + Cl_2 \xrightarrow{\text{light}}$

c. (cyclohexene) $+ H_2 \xrightarrow[\text{catalyst}]{\text{Pt}}$

13. Give the structure of the organic product of each of the following:

a.
$$CH_3CH_2CH=CHCH_2CH_3 + H_2O \xrightarrow[\text{dil}]{H_2SO_4}$$

b. $CH_3\underset{\underset{OH}{|}}{C}HCH_3 + Cr_2O_7^{2-} \rightarrow$

14. Give the structure of the organic product of each of the following:

a.
$$CH_3CH_2\overset{O}{\overset{||}{C}}CH_3 + Cr_2O_7^{2-} \rightarrow$$

b. (cyclohexene) $+ H_2O \xrightarrow[\text{dil}]{H_2SO_4}$

15. Give the structure of the organic product of each of the following:

a. $CH_3\overset{O}{\overset{||}{C}}OH + CH_3OH \xrightarrow{H^+}$

b. $CH_3CH_2CH = CHCH_3 + H_2 \xrightarrow{Pt}$

16. a. Sketch structural formulas of the hydrolysis products of the following ester:

$$CH_3-\underset{\underset{CH_3}{|}}{C}H-\overset{O}{\overset{||}{C}}-O-CH_3 + H_2O \xrightarrow{H^+}$$

b. Sketch structures of the products of the following saponification reaction:

$CH_3(CH_2)_{12}COO(CH_2)_4CH_3 + NaOH \rightarrow$

17. Define the following terms:
   a. an alkene
   b. homologous series
   c. an addition reaction
   d. functional group

18. How many structural isomers are there of $C_4H_{10}O$?

19. Show how the following chemical change can be carried out in two steps.

$$CH_3CH_2CH_2OH \rightarrow CH_3\underset{\underset{OH}{|}}{C}HCH_3$$

24

24

24

24

24

24

24

24

24

# Chapter 25.  Synthetic and Natural Polymers

**Synthetic Organic Polymers (Sections 25.1 – 25.2)**
**Proteins (Section 25.3)**
**Nucleic Acids (Section 25.4)**


## SUMMARY

### Synthetic Organic Polymers (Sections 25.1 – 25.2)

*Polymers.*  The word polymer means "many parts."  A polymer is a compound with an unusually high molecular mass, consisting of a large number of small molecular units that are linked together. The small unit that is repeated many times is called a monomer. A typical polymer molecule contains a chain of monomers several thousand units long. Polymers are often called macromolecules.

    Proteins, nucleic acids, carbohydrates, and rubber are natural polymers. Synthetic polymers such as nylon, polyester, and polyethylene are organic compounds.

*Addition Polymers.*  Addition polymers are made by adding monomer to monomer until a long chain is produced. Ethylene and its derivatives are excellent monomers for addition polymers. When ethylene is heated to 250 °C under high pressure (1000–3000 atm) in the presence of a little oxygen or benzoyl peroxide (the initiator), addition polymers with molecular masses of about 30,000 amu are obtained. This reaction is represented by:

$$n \; \underset{\text{ethylene}}{C=C} \;\rightarrow\; \underset{\text{a segment of polyethylene}}{-C-C-C-C-C-C-C-C-}$$

The general equation for addition polymerization is:

$$n \; \underset{\text{monomer}}{C=C} \;\rightarrow\; \underset{\text{repeating unit}}{\left[ -C-C- \right]_n}$$

    Polyethylene is an example of a homopolymer, which is a polymer made up of only one type of monomer. Substitution of one or more hydrogen atoms in ethylene with Cl atoms, phenyl groups, acetate, cyano groups, and F atoms provides a wide selection of monomers from which to make various homopolymers. For example, substitution of a chlorine atom, Cl for a hydrogen atom, H, ethylene gives the monomer called vinyl chloride, $CH_2 = CHCl$. Polymerization of vinyl chloride yields the polymer, polyvinyl chloride.

$$n \ \underset{\underset{\displaystyle H \quad H}{|} \quad |}{\overset{\overset{\displaystyle H \quad H}{|} \quad |}{C=C}} \ \rightarrow \ \left[ \underset{\underset{\displaystyle H \quad Cl}{|} \quad |}{\overset{\overset{\displaystyle H \quad H}{|} \quad |}{-C-C-}} \right]_n$$

vinyl chloride    polyvinyl chloride
monomer        repeating unit

Table 25.1 in the text gives the names, structures, and uses of a number of monomers and addition polymers.

***Addition Polymerization.*** The addition polymerization process occurs by several steps which are summarized below. The steps are: initiation, chain growth process, and finally a termination reaction.

1. *Initiation.* Polymerization is initiated by a radical. A radical (also called a free radical) is a species which has an unpaired electron. The symbol for a radical contains the usual letters for the elements followed by a dot to represent the unpaired electron. $CH_3\cdot$ is the symbol for a methyl radical. The letter R is the general symbol for an alkyl group and $R^\bullet$ for an alkyl radical.
   In a polymerization process, first the initiator molecule is dissociated by heating to yield two radicals. Then, the radical adds to an ethylene molecule, which generates a new radical.

   $$R{-}R \rightarrow 2R\cdot$$
   $$R^\bullet + CH_2{=}CH_2 \rightarrow R{-}CH_2{-}CH_2\cdot$$

2. *Chain Growth.* Next, the free radical formed above can add to another molecule of ethylene.

   $$R{-}CH_2{-}CH_2^\bullet + CH_2{=}CH_2 \rightarrow R{-}CH_2{-}CH_2{-}CH_2{-}CH_2^\bullet$$

   The length of the carbon chain grows rapidly as the last free radical formed reacts with yet another ethylene molecule, and so on.

   $$R{-}(CH_2{-}CH_2)_n{-}CH_2{-}CH_2^\bullet$$

3. *Termination.* The polymerization process continues until a termination reaction occurs. When the radical ends of two chains meet, they may combine. When this happens, there is no new radical formed and the chain lengthening process ceases. Of course, the result of this termination is the formation of a molecule of polyethylene which may contain up to 800 carbon atoms.

   $$R{-}(CH_2{-}CH_2)_n{-}CH_2{-}CH_2^\bullet + {}^\bullet CH_2{-}CH_2{-}(CH_2{-}CH_2)_n{-}R \rightarrow$$
   $$R{-}(CH_2{-}CH_2)_n{-}CH_2{-}CH_2{-}CH_2{-}CH_2{-}(CH_2{-}CH_2)_n{-}R$$

***Condensation Polymers.*** Copolymers are polymers that contain two or more different monomers. When monomers A and B are linked by condensation reactions, a uniform copolymer, —ABABAB— can be formed. Nylon and polyesters, such as the well-known Dacron, are copolymers.

Polyester is made by an esterification reaction. When one monomer is an alcohol and the other is a carboxylic acid, they can be joined by an esterification reaction. The alcohol and the acid both must contain two functional groups. The monomers in polyester are the dicarboxylic acid called phthalic acid, and the dialcohol, ethylene glycol.

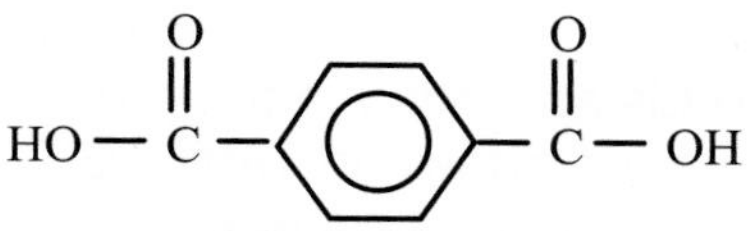

phthalic acid

HO–CH₂CH₂–OH
ethylene glycol

Condensation reactions differ from addition reactions in that the former always results in the formation of a small molecule such as water. Polyesters are produced from the esterification reaction between an alcohol and an acid. When phthalic acid and ethylene glycol react to form an ester, the first products are:

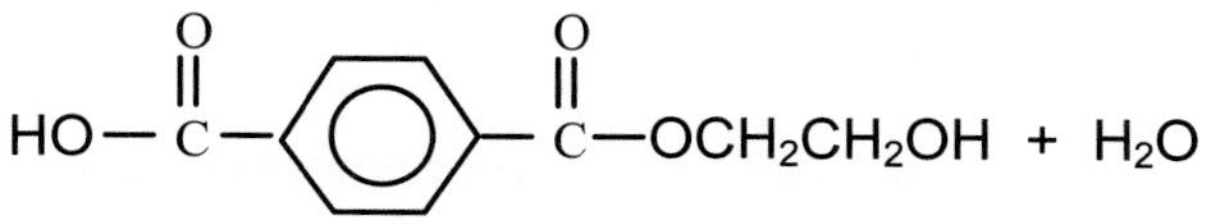

This molecule can react further from both ends. When this product reacts with another molecule of the diacid, the polymer chain grows longer.

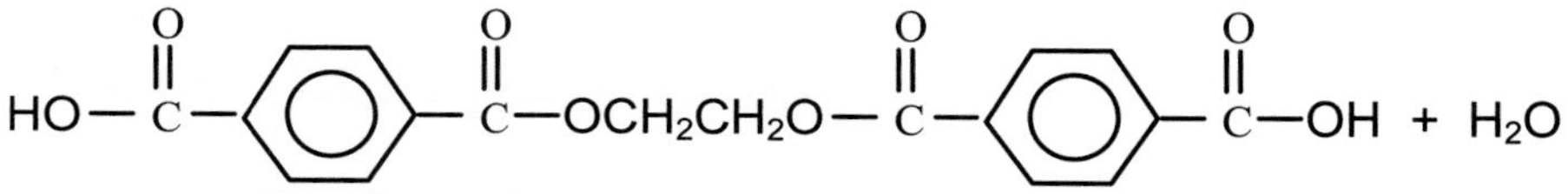

Segment of a condensation polymer chain

The general formula for this polyester is:

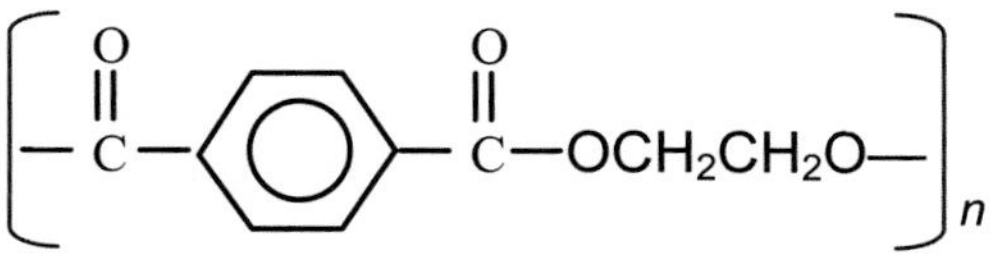

> Example
> 25.1
>
> Exercises
> 1 – 3

## Proteins (Section 25.3)

**Proteins.**  Proteins are truly giant molecules having molecular masses that range from about 10,000 to several million amu. Proteins play many roles in living organisms where they function as catalysts (enzymes), transport molecules (hemoglobin), contractile fibers (muscle), protective agents (blood clots), hormones (chemical messengers), and structural members (feathers, horn, nails). The word *protein* comes from the Greek word *proteios,* meaning "first."  From this partial list of functions, it is easy to see why proteins occupy "first place" among biomolecules in their importance to life.

**Amino Acids.**  Even though each protein is unique, all proteins are built from the same set of amino acids. An amino acid consists of an amino group, a carboxylic acid group, a hydrogen atom, and a distinctive R-group, all bonded to the same carbon atom.

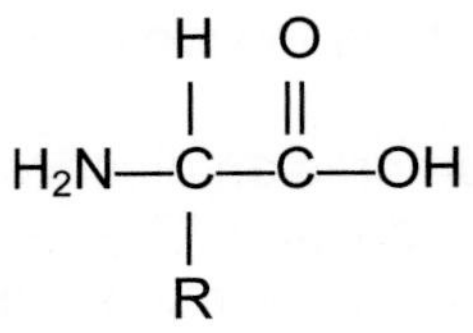

$$\begin{array}{c} \quad\quad\; H \quad O \\ \quad\quad\; | \quad\; || \\ H_2N\!-\!C\!-\!C\!-\!OH \\ \quad\quad\; | \\ \quad\quad\; R \end{array}$$

All amino acids in proteins have a common structural feature. This is the attachment of the amino group to the carbon atom adjacent to the carboxylic acid group. The R-group is different in each amino acid, and some 20 different R-groups are found in the proteins from natural sources. The structural formulas of the amino acids are shown in Table 25.2 (text).

Amino acids in solution at near neutral pH, exist as dipolar ions. This form results from the transfer of a proton from the carboxylic acid group to the basic amine group

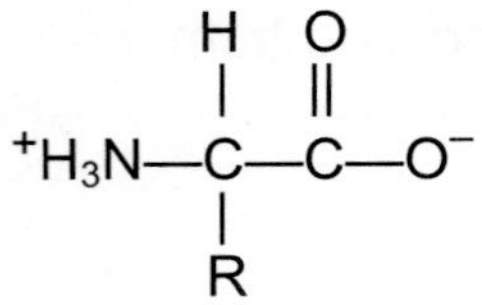

$$\begin{array}{c} \quad\quad\quad H \quad O \\ \quad\quad\quad | \quad\; || \\ {}^{+}H_3N\!-\!C\!-\!C\!-\!O^{-} \\ \quad\quad\quad | \\ \quad\quad\quad R \end{array}$$

In proteins, the amino acid units are hooked together to form a polypeptide chain. The carboxyl group of one amino acid is joined to the amino group of another amino acid by the formation of a peptide bond.

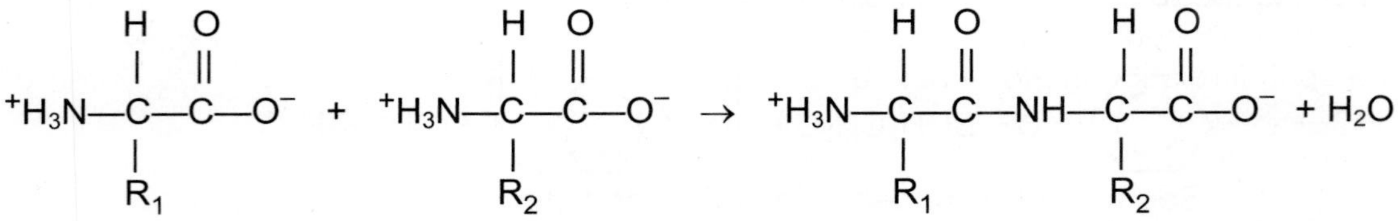

$$\begin{array}{c} H \quad O \\ | \quad\; || \\ {}^{+}H_3N\!-\!C\!-\!C\!-\!O^{-} \\ | \\ R_1 \end{array} \;+\; \begin{array}{c} H \quad O \\ | \quad\; || \\ {}^{+}H_3N\!-\!C\!-\!C\!-\!O^{-} \\ | \\ R_2 \end{array} \;\rightarrow\; \begin{array}{c} H \quad O \\ | \quad\; || \\ {}^{+}H_3N\!-\!C\!-\!C\!-\!NH \\ | \\ R_1 \end{array}\begin{array}{c} H \quad O \\ | \quad\; || \\ \!-\!C\!-\!C\!-\!O^{-} \\ | \\ R_2 \end{array} \;+\; H_2O$$

This type of reaction is another example of a condensation reaction. The new C—N covalent bond is called a peptide bond or an amide bond. The amide functional group is present in all proteins:

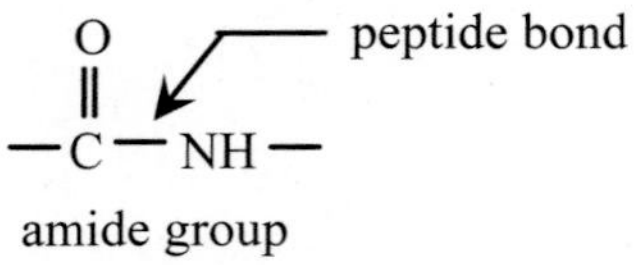

$$\begin{array}{c} O \\ || \\ -\!C\!-\!NH\!- \end{array} \quad \text{peptide bond}$$

amide group

The molecule above, in which two amino acids are joined, is called a dipeptide. Peptides are structures intermediate in size between amino acids and proteins. The term polypeptide refers to long molecular chains containing many amino acid units. An amino acid unit in a polypeptide chain is called a residue.

**Structure of Proteins.** Proteins are so complex that four levels of structural features have been identified. The structure of proteins is extremely important in determining just how efficiently and effectively a protein will function. A stretched-out or unfolded polypeptide chain does not exhibit biological activity, and is said to be denatured. The four levels of protein structure are summarized. See Figure 25.13 of the text.

1. *The Primary Structure.* Each protein has a unique amino acid sequence of its polypeptide chain. It is the amino acid sequence that distinguishes one protein from another. Proteins differ in the numbers and kinds of amino acids, but especially in the sequence of amino acid residues.

2. *The Secondary Structure.* This refers to the spatial relationship of amino acid units that are close to one another in the sequence. A configuration that appears in many proteins is the $\alpha$-helix shown in Figure 25.11 (text). In this configuration the polypeptide is coiled much like the arrangement of stairs in a spiral staircase. The tightly coiled polypeptide chain forms the inner part of the spiral stairway, and the R-groups extend outward forming a helical pattern. The figure shows that the $\alpha$-helix is stabilized by the presence of hydrogen bonds (dashed lines) between the $C=O$ and NH groups in the peptide chains. The CO group of each amino acid is hydrogen bonded to the NH group of the amino acid that is located four amino acids ahead in the sequence. Thus amino acids spaced four apart in the sequence are actually close to one another due to the coiled or spiral arrangement.

   The $\alpha$-helix is the main structural feature of the oxygen-storage protein, myoglobin, and of many other proteins. The presence of this feature in a polypeptide chain is shown in Figure 25.11 (text). The $\beta$-pleated sheet is another common protein structure. In this structure a polypeptide chain interacts strongly with adjacent chains by forming many hydrogen bonds (Figure 25.12 in text).

3. *The Tertiary Structure.* Myoglobin like many other proteins is a globular protein. The polypeptide chain is folded into a compact globular shape. The folding of the chain results in some amino acid units being in very close proximity to each other even though they are widely separated in the amino acid sequence. The term tertiary structure refers to the spatial relationship of amino acid units that are far apart in the sequence, or more simply, to just its three-dimensional structure. In myoglobin, for example, the tertiary structure is a unique three-dimensional shape resulting from the folding of the polypeptide chain. The folding takes place quite naturally in aqueous medium due to the interactions of hydrophobic and hydrophilic R-groups with water. The tertiary structure is stabilized by hydrogen bonding, dispersion forces, and ionic forces.

4. *The Quaternary Structure.* Proteins that consist of more than one polypeptide chain exhibit an additional level of structural organization called the quaternary structure. This structural feature involves the way in which separate chains fit together. Hemoglobin, the oxygen-carrying molecule of the blood, exhibits a quaternary structure (Chemistry in Action). It consists of four separate polypeptide chains or subunits (labeled $\alpha_1$, $\alpha_2$, $\beta_1$, $\beta_2$). The quaternary structure results from interaction between chains. Ionic forces and hydrogen bonds are important in holding the subunits together.

> Examples
> 25.2 – 25.4
>
> Exercises
> 4 – 8

## Nucleic Acids (Section 25.4)

***Components of DNA and RNA.*** The chemical composition of the cell nucleus was first studied in the 1860s by Friedrich Miescher. He found the major components to be protein and a new material not previously isolated. This material was found to be acidic, and so it was referred to as nucleic acid.

Nucleic acids are now known to be giant molecules with molecular masses in the range 1 to 10 billion amu. These molecules carry information in the form of the genetic code. Enough information is stored in nucleic acid molecules to allow the complete assembly of an entire organism. The nucleic acid molecule called DNA carries this information from generation to generation.

Hydrolysis of nucleic acid showed that it is composed of one part phosphoric acid, one part sugar, and one part nitrogen base. One of two sugars are present—ribose or deoxyribose. Five different nitrogen bases are found in nucleic acids. Their names are adenine, thymine, guanine, cytosine, and uracil (Figure 25.17 text).

Two types of nucleic acids are recognized. These are called DNA and RNA. The compositions of DNA and RNA are quite similar, but they differ in two significant ways. DNA contains the sugar deoxyribose and is called deoxyribonucleic acid. RNA contains the sugar ribose, hence the name ribonucleic acid. The second difference is that DNA contains the four bases adenine, thymine, guanine, and cytosine, while RNA contains adenine, uracil, guanine, and cytosine. DNA contains thymine but no uracil, whereas RNA contains uracil but no thymine. Table 25.1 summarizes the building blocks of DNA and RNA and gives useful abbreviations.

Ribose            Deoxyribose

**Table 25.1** The Component Parts of DNA and RNA

|          | DNA | RNA |
|----------|-----|-----|
| Acid:    | phosphoric acid (P) | phosphoric acid (P) |
| Sugar:   | deoxyribose (D) | ribose (R) |
| N bases: | adenine (A) | adenine (A) |
|          | thymine (T) | uracil (U) |
|          | guanine (G) | guanine (G) |
|          | cytosine (C) | cytosine (C) |

**Nucleotides.** The repeating unit in nucleic acids is called a nucleotide. It is a combination of a phosphate group, a five-carbon sugar, and a nitrogen base. See Figure 25.18 (text) for the full structure of the nucleotide, deoxyadenosine monophosphate (dAMP), which contains the sugar deoxyribose, and the base adenine. Figure 25.1 below shows the structure of the nucleotide containing the base cytosine, the sugar ribose and the phosphate group. It is named cytosine monophosphate (CMP).

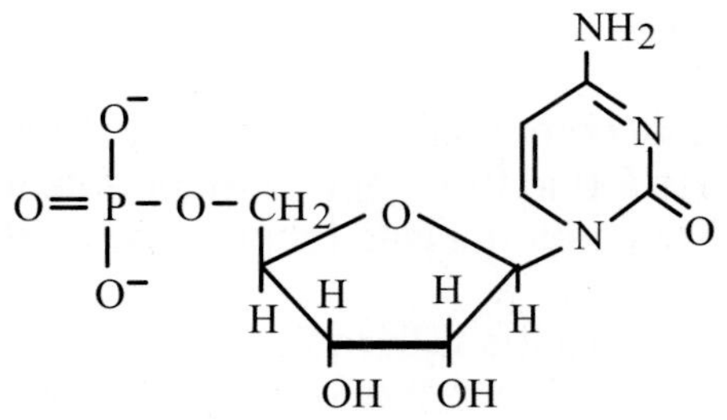

**Figure 25.1** The molecular structure of the nucleotide cytosine monophosphate (CMP). The phosphoric acid groups of nucleotides are quite acidic, and at pH 7, these groups exist mainly as the dianions as shown.

***Nucleic Acids.*** A strand of DNA or RNA is constructed by linking the nucleotides together with a bond from the sugar unit of one nucleotide to the phosphate unit of another nucleotide. Such a strand of DNA is represented in abbreviated form below. To simplfy DNA and RNA structures, we will use the abbreviations in Table 25.1: P = phosphate, D = deoxyribose, and A, G, T, and C for the bases.

```
   /
  P
   \
    D—A
   /
  P
   \
    D—G
   /
  P
   \
    D—T
   /
  P
   \
    D—C
   /
  P
   \
```

In the 1940s, Edwin Chargaff studied the composition of DNA. His analysis of the base composition of DNA showed that the amount of adenine (A) always equaled that of thymine (T), and that the amount of guanine (G) equaled that of cytosine (C). These relations became known as Chargaff's rules. Watson and Crick in 1953 proposed a two stranded structure for DNA which provided an explanation of Chargaff's rules. The two-stranded structure can be shown as follows:

```
   /              \
  P                P
   \              /
    D—C · · · G—D
   /              \
  P                P
   \              /
    D—A · · · T—D
   /              \
  P                P
   \              /
    D—G · · · C—D
   /              \
  P                P
   \              /
    D—T · · · A—D
   /              \
  P                P
   \              /
```

Adenine in one strand is always paired with thymine in the other strand. Guanine is always paired with cytosine. The two strands are not identical, rather they are complementary. Base pairing and the resulting association of the two strands are the result of hydrogen bonding. Hydrogen atoms in a base in one strand form hydrogen bonds to oxygen and nitrogen atoms of a base attached to the other strand, see Figure 25.19(a) (text). X-ray data suggested that DNA had the helical structure as shown in Figure 25.19(b) of the text.

> Examples
> 25.5, 25.6
>
> Exercises
> 9 – 12

RNA, on the other hand, does not follow the base-pairing rules. X-ray data and other evidence ruled out a double-helical structure for RNA. RNA is single stranded.

## GLOSSARY LIST

| | | |
|---|---|---|
| polymer | amino acid | nucleic acid |
| monomer | protein | nucleotide |
| homopolymer | denatured protein | deoxyribonucleic acid (DNA) |
| copolymers | | ribonucleic acid (RNA) |

## WORKED EXAMPLES

**EXAMPLE 25.1  Monomers and Polymers**

Write the formulas of the monomers used to prepare the following polymers:
a. Teflon     b. polystyrene     c. PVC

• **Solution**

Refer to Table 25.1 of the text.
a.   Teflon is an addition polymer with the formula $-(CF_2-CF_2)_n$. It is prepared from the monomer tetrafluoroethylene ($CF_2=CF_2$).
b.   The monomer used to prepare polystyrene is styrene.

polystyrene                     styrene

c.   Polyvinylchloride is prepared by the successive addition of vinyl chloride molecules, $CH_2=CHCl$.

**Work EXERCISES & PROBLEMS: 1 – 3**

## EXAMPLE 25.2  Amino Acids

Find and name five amino acids in Table 25.2 (textbook) that have polar R-groups.

### • Solution

Recall that polar groups are those that have relatively large electronegativity differences between bonded atoms. Thus, —COOH, —OH, and —SH are polar R groups. Answer: The R groups in aspartic acid and glutamic acid contain carboxylic acid groups. The R-groups in serine and threonine contain hydroxyl groups. And the R-group in cysteine contains a polar —SH group.

## EXAMPLE 25.3  A Polypeptide Chain

Sketch a portion of a polypeptide chain consisting of the amino acids; cysteine, glycine, valine, and phenylalanine. Point out the peptide bonds and amide groups.

### • Solution

The main backbone of a polypeptide chain is made up of carbon atoms, and the amide group repeating alternately along the chain. The abbreviation for the chain is cys—gly—val—phe.

Table 25.2 (text) gives the structures of the amino acids. Substitute for each R-group shown previously, the distinctive R-groups of the four amino acids. The chain structure will be:

There are four peptide bonds. They are the C—N bonds. Each –CO—NH– is an amide group.

**Work EXERCISES & PROBLEMS: 4 – 8**

**EXAMPLE 25.4  Tertiary Structure of Proteins**

What kinds of forces stabilize the tertiary structures of proteins?

**• Solution**

The polypeptide chain folds into a specific three-dimensional structure under the influence of hydrophobic and hydrophilic interactions. The hydrophilic R-groups (polar or ionic groups attracted to water) form the outside surface of the folded polypeptide chain where they interact with water via hydrogen bonding and ion-dipolar interactions. The hydrophobic R-groups (nonpolar groups repelled by water) form the interior of the molecule shielded from water molecules.

**EXAMPLE 25.5  Nucleotide Structures**

Sketch the structure of the nucleotide containing uracil that appears in RNA.

**• Solution**

Nucleotides from RNA contain ribose as the sugar and a phosphate dianion. The phosphate group is bonded to the sugar, which in turn, is bonded to the base, uracil.

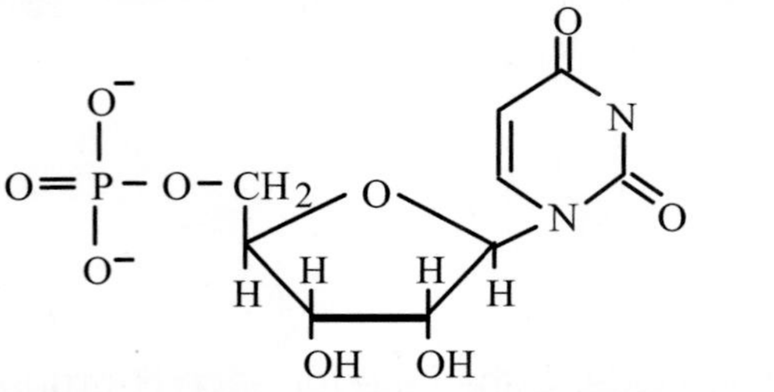

Phosphate unit   Ribose unit   Uracil unit

**EXAMPLE 25.6  Base Pairing**

What types of forces cause base pairing in the double-stranded helical DNA molecule?

**• Solution**

The N-bases thymine (T) and cytosine (C) in one strand of DNA, form hydrogen bonds to the N-bases adenine (A) and guanine (G), in the other DNA strand. Hydrogen atoms that are covalently bonded to nitrogen carry a partial positive charge. These H atoms are attracted to lone electron pairs on oxygen and nitrogen atoms of another base. Since two complementary bases are attached to different strands, the hydrogen bonds hold the strand together.

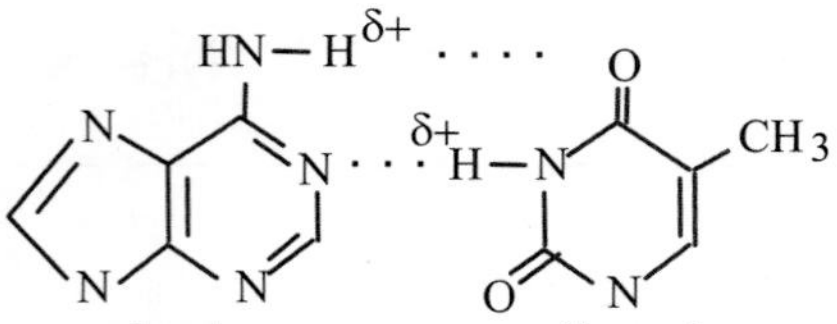

adenine        thymine

## EXERCISES & PROBLEMS

1. List three steps occurring during addition polymerization.

2. What monomer is used to make Teflon $-(CF_2-CF_2)_n$ ?

3. Is Teflon a homopolymer or a copolymer?

4. What is an amino acid?

5. Sketch the dipolar ion of alanine.

6. What is an amide group and a peptide bond?

7. What is the primary structure of a protein?

8. What is a denatured protein?

9. What is the 5-carbon sugar in DNA?  In RNA?

10. Name four nitrogen bases found in DNA.

11. What type of bond is responsible for base pairing in DNA?

12. What is the repeating unit in nucleic acids?

## PRACTICE TEST QUESTIONS
See notes on taking practice test in the Preface

1. Sketch the structures of the monomers from which these polymers are formed.

   a.

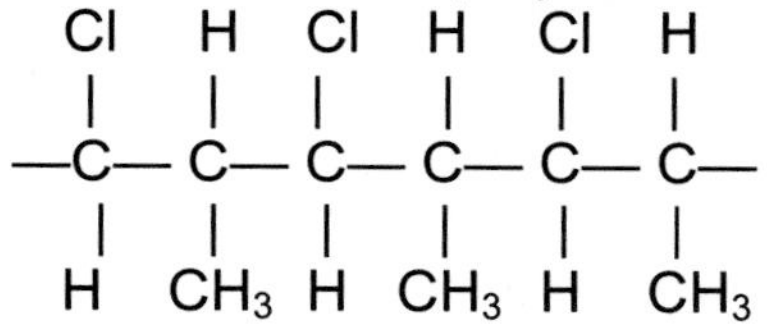

   $$
   \begin{array}{ccccccc}
   & Cl & H & Cl & H & Cl & H \\
   & | & | & | & | & | & | \\
   -C- & C- & C- & C- & C- & C- \\
   & | & | & | & | & | & | \\
   & H & CH_3 & H & CH_3 & H & CH_3
   \end{array}
   $$

   b. $-CH_2-CCl_2-CH_2-CCl_2-CH_2-$

2. Draw structures for the monomers used to make the following polyester.

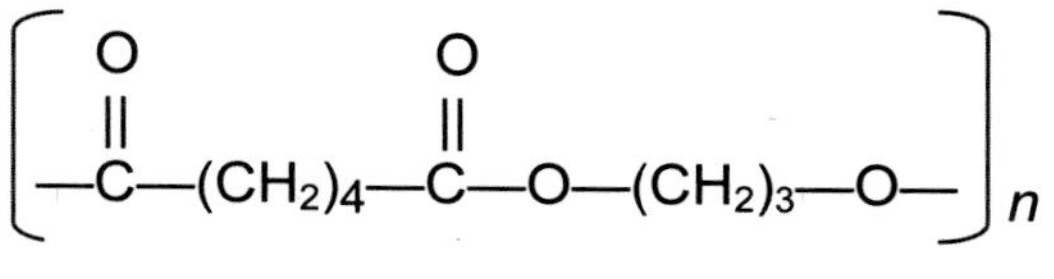

   $$
   \left[ \begin{array}{c} O \\ \| \\ -C-(CH_2)_4-\end{array} \begin{array}{c} O \\ \| \\ C-O-(CH_2)_3-O- \end{array} \right]_n
   $$

3. Sketch the structure of the monomer of natural rubber.

4. What chemical elements are found in proteins?

5. List several functions of proteins in living systems.

6. What is the role of enzymes in biochemical systems?

7. Describe the structure of an amino acid.

8. Describe a dipeptide. A polypeptide.

9. Draw the structural formula of the tripeptide formed from the sequence of amino acids below. (See Table 25.2 of the textbook.) Label the peptide bonds and R-groups.

   Ser—Cys—Lys

10. Consider two polypeptide chains
    1. Leu—Phe—Pro—Gly—Ala
    2. Gly—Ser—Lys—Asp—Tyr
    a. Which would be more soluble in water?
    b. Which would be more soluble in a nonpolar solvent?

11. Explain how the primary structure of a protein differs from the tertiary structure.

12. List four differences between DNA and RNA.

13. Sketch the structure of the nucleotide consisting of phosphate, deoxyribose, and adenine.

14. What is the biological role of DNA?

15. If the base sequence in one strand of DNA is A, T, G, C, T, then the base sequence in the complementary strand is __, __, __, __, __.

25   25

25   25

25

# APPENDIX: Answers to Exercises & Problems and Practice Tests

## Chapter 1. Chemistry: The Study of Change

### EXERCISES and PROBLEMS
1. a. compound   b. element   c. mixture   d. mixture   e. mixture
2. a. d. e. are heterogeneous      b. c. are homogeneous
3. a. b. physical properties   c. d. chemical properties
4. a. c. physical change      b. d. chemical change
5. 3300 cm$^3$
6. 11 g/cm$^3$
7. a. $2.4 \times 10^4$  b. $1.4 \times 10^{-4}$   c. $7.4 \times 10^8$   d. $9.06 \times 10^{-2}$
8. a. 0.0052   b. 1400   c. 7,500,000   d. 0.0000706
9. a. 3  b. 2  c. 3  d. 4  e. 4
10. a. 0.609  b. $1.0 \times 10^3$   c. 0.000222   d. 238.0   e. 1.3
11. a. $2.6 \times 10^4$  b. 0.4  c. $3.1 \times 10^3$   d. 13.92  e. $4.80 \times 10^3$
12. a. 200. g   b. 12.0 cm   c. $2 \times 10^1$ g/mL
13. a. $7 \times 10^{-9}$ kg (remember, the SI unit for mass is kg)   b. $8.0 \times 10^{-9}$ m   c. $1.4 \times 10^5$ L
    d. $1.0 \times 10^3$ s
14. a. 0.125 m  b. 80 nm   c. $4.45 \times 10^{-3}$ km   d. $3.25 \times 10^4$ μm   e. $5.73 \times 10^{-3}$ mm
15. 13.5 g/cm$^3$
16. 22.9 mi per hr
17. 16.7 mL
18. 5.74 L
19. 83.1 °F
20  53.3 °C

### PRACTICE TEST

1. a. physical  b. chemical  c. physical  d. physical  e. physical  f. chemical
2. a. chemical  b. chemical  c. physical  d. physical  e. chemical
3. a. heterogeneous  b. homogeneous  c. pure substance
   d. pure substance  e. homogeneous
4. a. mixture  b. compound  c. element  d. compound  e. mixture
   f. mixture
5. a. 98 mm  b. 17.0 cm  c. 0.325 L  d. 22 m/s  e. 20,000,000 kJ
6. 21.14
7. 2.37 g/mL
8. 0.93 kg
9. 23.5 cm$^3$
10. SUV: $3000 per 10,000 miles; compact car: $1300 per 10,000 miles
11. $1.0 0\times 10^3$ kg/m$^3$
12. $8 \times 10^7$ atoms (80 million)
13. $1.03 \times 10^{-3}$ Mg/cm$^2$
14. 14.3 g solder
15. 530 mL
16. 11 g gasoline
17. $5.0 \times 10^5$ kg
18. 36 mph
19. 2804 °F

20. 305 K

# Chapter 2.  Atoms, Molecules & Ions

## EXERCISES and PROBLEMS

1. 24
2. 62
3. a. $^{67}_{30}Zn$  b. $^{120}_{51}Sb$
4. Protons, electrons, and neutrons. Protons have one unit of positive charge, and electrons have one unit of negative charge. Neutrons have no charge. Protons and neutrons are found in the nucleus.
5. a. 3p, 2e    b. 38p, 36e    c. 26p, 23e    d. 7p, 10e    e. 34p, 36e    f. 17p, 18e
6. CH
7. a. $NO_2$    b. $CH_2$    c. $AlCl_3$    d. $Fe_2O_3$    e. $SF_5$
8. a, c, e.
9. a. $BaCl_2$    b. $Mg_3N_2$    c. $Fe_2O_3$    d. $FeF_2$
10. a. $NH_4Cl$    b. $Na_3PO_4$    c. $K_2SO_4$    d. $CaCO_3$    e. $KHCO_3$
    f. $Mg(NO_2)_2$    g. $NaNO_3$    h. $NH_4ClO_4$    i. $Sr(OH)_2$    j. $Cu(CN)_2$
11. a. potassium nitride  b. silver carbonate  c. magnesium hydroxide  d. sodium cyanide
    e. ammonium iodide  f. iron(II) nitrate    g. calcium sulfate dihydrate
12. a. phosphorus pentachloride    b. sulfur trioxide    c. tetraphosphorus decaoxide
    d. dinitrogen monoxide        e. nitrogen dioxide
13. a. nitric acid    b. nitrous acid    c. hydrobromic acid    d. hydrocyanic acid
    e. chlorous acid
14. a.                 b. 
15. See the box on page 23.
16. The evidence was related to the direction of scattering of the positively charged alpha particles by gold atoms. The fact that 1 in 10,000 alpha particles was deflected "backward" meant that all the positive charge of the gold atom must be concentrated into a very small part of the atom.
17. An atom is the smallest particle of an element.  A molecule is a particle that is made of two or more atoms.  When the atoms in a molecule are identical we have a molecular form of an element.  When the atoms are different we have the smallest particle of a compound.
18. a. $Ca^{2+}$ is an example. b. $SO_4^{2-}$ is an example. See Table 2.5 for other examples.
19. a. compound    b. compound  c. element   d. element   e. element  f. compound
20. The symbol $O_3$ represents a molecule of ozone, a molecular form of oxygen containing 3 oxygen atoms. Whereas, 3O represents 3 separate oxygen atoms.
21. The empirical formula of a compound gives the simplest whole-number ratio of the atoms of the elements making up the compound. Ionic compounds consist of positive and negative ions in the ratio needed to give an electrically neutral substance. It is this ratio of positive to negative ions that is essential to the formation of a neutral ionic compound.  Any crystal of the compound having the proper ratio of positive and negative ions will always be electrically neutral, no matter what its size.

**PRACTICE TEST**

1.

| Name | Symbol | Number of Protons | Number of Electrons | Number of Neutrons | Mass Number |
|------|--------|-------------------|---------------------|--------------------|-------------|
| Sodium | $^{23}Na$ | 11 | 11 | 12 | 23 |
| Argon | $^{40}Ar$ | 18 | 18 | 22 | 40 |
| Arsenic | $^{75}As$ | 33 | 33 | 42 | 75 |
| Lead | $^{208}Pb$ | 82 | 82 | 126 | 208 |
| Potassium | $^{39}K$ | 19 | 19 | 20 | 39 |

2. a. 26 p, 30 n, 26 e   b. 64
3. a. 8 p, 9 n, 10 e   b. 47 p, 60 n, 46 e   c. 86 p, 136 n, 86 e
4. $^{46}_{20}X$ and $^{43}_{20}X$ are isotopes.
5. p: 49.952%   n: 50.021%   e: 0.027%
6. $P_4$, He, $N_2$, $O_3$
7. a. $C_3H_4O_3$   b. CH   c. HgCl   d. HO   e. $CHO_2$   f. $MgCl_2$
8. a. $Ca(ClO)_2$   b. $HgSO_4$   c. $BaSO_3$   d. ZnO   e. $N_2O$   f. $Na_2CO_3$   g. CuS   h. $PbO_2$
9. a. sodium hydrogen phosphate   b. hydrogen iodine   c. tetraphosphorus hexoxide
   d. lithium nitrate   e. hydroiodic acid   f. strontium nitrite   g. sodium hydrogen
   carbonate   or sodium bicarbonate   h. potassium sulfite   i. sodium phosphate   j.
   aluminum hydroxide
10. a. sulfurous acid   b. hypochlorous acid   c. perchloric acid   d. phosphoric acid
    e. hydrocyanic acid

# Chapter 3.    Mass Relationships in Chemical Reactions

**EXERCISES and PROBLEMS**
1. 17.9990 amu
2. 107.87 amu
3. a. 153.81 amu   b. 30.03 amu   c. 169.3 amu
4. a. 142.05 g   b. 162.20 g   c. 171.3 g   d. 80.07 g
5. a. 63.55 g   b. 79.55 g   c. 159.62 g   d. 249.7 g
6. a. 0.0463 mol Ag   b. 0.217 mol Na
7. a. $2.8 \times 10^{22}$ atoms   b. $1.2 \times 10^{24}$ molecules   c. $9.38 \times 10^{23}$ molecules
8. a. 4.0 g   b. 5.00 g   c. $4.48 \times 10^{-23}$ g
9. a. 0.103 mol $NaNO_3$   b. 0.308 mol O
10. a. 27.29% C, 72.71% O
    b. 2.43% H,   52.8% As,   45.1% O
    c. 10.06% C, 0.844% H,   89.09% Cl
    d. 33.32% Na, 20.30% N, 46.38% O
    e. 2.06% H, 32.70% S, 65.24% O
11. a. HO   b. $CaF_2$   c. $CH_2O$   d. $BH_3$
12. a. NO   b. $N_2O$   c. $K_3PO_4$   d. $K_2Cr_2O_7$
13. $Fe_2S_3$
14. $C_3H_4O_3$
15. $C_6H_{12}$
16. $4Fe + 3O_2 \rightarrow 2Fe_2O_3$

17. $2H_2 + CO \rightarrow CH_3OH$
18. a. $CH_4 + 2H_2O \rightarrow CO_2 + 4H_2$
    b. $H_2SO_4 + 2NaOH \rightarrow Na_2SO_4 + 2H_2O$
    c. $4NH_3 + 5O_2 \rightarrow 4NO + 6H_2O$
19. a. 2.12 mol    b. 360 g
20. 9.58 g $H_2$
21. 1.79 g
22. 1.09 kg
23. 1.67 mol
24. a. $Fe_2O_3$ is the limiting reactant; 140 g Fe    b. 78.6% yield
25. a. 21.5 g    b. 66.0%
26. $3 \times 10^{22}$ Ag atoms
27. a. 330.37 g/mol    b. male, $1.6 \times 10^{23}$ molecules;  female, $1.3 \times 10^{23}$ molecules
28. 178 amu
29. Atoms cannot be weighed directly. However, how many times heavier an atom of one element is compared to an atom of another element can be measured. These relative masses allow one to order the elements by their increasing atomic mass. Assigning a mass to one of the atoms makes it the standard for the scale. The individual masses of atoms of all the elements are assigned by comparison to the standard. The present standard of the atomic mass unit scale is the carbon-12 atom which is assigned a mass of exactly 12 amu.
30. The average atomic mass of sulfur is 32.07 amu and that of oxygen is 16.00 amu. Because of the presence of isotopes all sulfur atoms do not weigh 32.07 amu, and all oxygen atoms do not weigh 16.00 amu. So only on average do S atoms weigh 2.0 times as much as the average O atom.

## PRACTICE TEST

1. oxygen would be 8.4, and argon would be 21.0
2. 3.96 times
3. 25.62 amu
4. a. 5.2 g    b. $2 \times 10^{-11}$ g
5. $3.16 \times 10^{-23}$ g
6. a. $2 \times 10^{24}$ atoms  b. $6 \times 10^{23}$ atoms  c. $4.3 \times 10^{23}$ atoms
7. 51.7 g
8. a. 4.41 mol $CaSO_4$    b. 17.6 mol O   c. $1.06 \times 10^{25}$ O atoms
9. $6.02 \times 10^{24}$ Sb atoms
10. a. $1.40 \times 10^3$ g    b. 0.10 g
11. a. $C_3H_4O_3$        b. $P_2O_5$    c. $MgCl_2$
12. a. 52.91% Al, 47.08% O    b. 1.60% H, 22.23% N, 76.17% O
    c. 9.93% C, 58.63% Cl, 31.43% F
13. 74.5% F
14. $C_8H_8$
15. $CrO_3$
16. $C_3H_8O_3$
17. a. $P_4O_{10} + 6H_2O \rightarrow 4H_3PO_4$
    b. $2Ga + 3H_2SO_4 \rightarrow Ga_2(SO_4)_3 + 3H_2$
    c. $2C_4H_{10} + 13O_2 \rightarrow 8CO_2 + 10H_2O$

18.

| | mol $SO_2$ | g $O_2$ | mol $SO_3$ | g $SO_3$ |
|---|---|---|---|---|
| a | 1.50 | 24.0 | 1.50 | 120 |
| b | 1.25 | 20.0 | 1.25 | 100 |
| c | 5.21 | 83.4 | 5.21 | 417 |

19. a. 71.8 g $Na_2CO_3$    b. 36.8 g $Na_2S$
20. a. $H_2SO_4$ is the limiting reactant; 30.6 g HF
    b. 3.27 g $CaF_2$    c. 85.6% yield

# Chapter 4.    Reactions in Aqueous Solutions

## EXERCISES and PROBLEMS
1.    a, c, d, and e are ionic compounds.
2.    a, c, e, and f.
3.    a. $Mg^{2+}$ and $NO_3^-$    b. $K^+$ and $OH^-$    c. $Ca^{2+}$ and $F^-$
4.    a, and e.
5.    b, c, and e.
6.    a. yes, $Mg(OH)_2$    b. yes, AgI
7.    a. $Na^+$ and $Br^-$    b. $Na^+$ and $NO_3^-$
8.    a. $Pb^{2+}(aq) + 2Br^-(aq) \rightarrow PbBr_2(s)$
     b. $Ag^+(aq) + Br^-(aq) \rightarrow AgBr(s)$
9.    a. $HBr \rightarrow H^+ + Br^-$    b. $HClO_4 \rightarrow H^+ + ClO_4^-$    c. $H_2SO_4 \rightarrow H^+ + HSO_4^-$
10.   a. $Ca(OH)_2 \rightarrow Ca^{2+} + 2OH^-$    b. $HN(CH_3)_2 + H_2O \rightarrow H_2N(CH_3)_2^+ + OH^-$
     c. $CsOH \rightarrow Cs^+ + OH^-$
11.   a. $NaOH(aq) + HClO_4(aq) \rightarrow H_2O(\ell) + NaClO_4(aq)$
     b. $2NaOH(aq) + H_2SO_4(aq) \rightarrow 2H_2O(\ell) + Na_2SO_4(aq)$
12.   a. $HNO_3(aq) + LiOH(aq) \rightarrow H_2O(\ell) + LiNO_3(aq)$
     b. $Ca(OH)_2(aq) + 2HCl(aq) \rightarrow 2H_2O(\ell) + CaCl_2(aq)$
     c. $3KOH(aq) + H_3PO_4(aq) \rightarrow 3H_2O + K_3PO_4(aq)$
13.   a. +4    b. +6    c. –2    d. +1    e. 0    f. –1    g. +4    h. +3    i. +6    j. +7
14.   +7 → +2
15    a. Ag is oxidized and N is reduced.    b. $Fe^{2+}$ is oxidized and Cr is reduced
     c. Zn is oxidized and $Cu^{2+}$ is reduced
16.   a. Ag is the reducing agent.    HNO3 is the oxidizing agent.
     b. $Fe^{2+}$ is the reducing agent. $Cr_2O_7^{2-}$ is the oxidizing agent.
     c. Zn is the reducing agent.  $Cu^{2+}$ is the oxidizing agent.
17.   a. and c.
18.   0.983 *M*
19.   0.169 mol
20.   4.56 g
21.   a. 0.0138 mol    b. 0.110 *M* $Mg^{2+}$    c. 0.220 *M* $Cl^-$ ion
22.   1.25 *M*
23.   Take 33.3 mL of 5.0 *M* KOH and dilute to a final volume of 250 mL.
24.   0.113 mol
25.   0.236 *M*
26.   20.7%
27.   18.7 mL
28.   9.8 mL
29.   50 mL

30.  0.157 $M$
31.  53 mL
32.  0.20 $M$
33.  A strong electrolyte is a compound that is 100% dissociated in solution. Its solubility may be low or its solubility may be high.  The term strong does not relate to the solubility. It only relates to the degree of dissociation.  So a low solubilty compound that is 100% dissociated is still a strong electrolyte.
34.  Molarity is the moles of solute per liter of solution, and is useful because for a solution of known molarity and volume the product of $M$ x $V$ is equal to the number of moles of solute. Measuring liters of solution is a way to count moles of solute.
35.  $Ba(NO_3)_2(aq) + Na_2SO_4(aq) \rightarrow BaSO_4(s) + 2NaNO_3(aq)$
     $Ba(s) + H_2SO_4(aq) \rightarrow BaSO_4(s) + H_2(g)$

## PRACTICE TEST

1.  a. and b. are strong electrolytes     c. and d. are weak electrolytes     e. is a nonelectrolyte
2.  a. $H^+(aq) + OH^-(aq) \rightarrow H_2O(\ell)$
    b. $CH_3COOH(aq) + OH^-(aq) \rightarrow CH_3COO^-(aq) + H_2O(\ell)$
    c. $H^+(aq) + OH^-(aq) \rightarrow H_2O(\ell)$
3.  a. $Ba^{2+}(aq) + CO_3^{2-}(aq) \rightarrow BaCO_3(s)$

    b. $Ag^+(aq) + Cl^-(aq) \rightarrow AgCl(s)$
    c. $Pb^{2+}(aq) + S^{2-}(aq) \rightarrow PbS(s)$
4.  a. b. c. and d. are insoluble     e. and f. are soluble
5.  a. $H_2SO_4$ and $Cu(OH)_2$     b. HBr and KOH     c. $H_3PO_4$ and $Ca(OH)_2$
6.  a. $Na^+(aq) + Br^-(aq) + Ag^+(aq) + NO_3^-(aq) \rightarrow AgBr(s) + Na^+(aq) + NO_3^-(aq)$
       $Ag^+(aq) + Br^-(aq) \rightarrow AgBr(s)$
    b. $Mg^{2+}(aq) + 2Br^-(aq) + Pb^{2+}(aq) + 2NO_3^-(aq) \rightarrow PbBr_2(s) + Mg^{2+}(aq) + 2NO_3^-(aq)$
       $Pb^{2+}(aq) + 2Br^-(aq) \rightarrow PbBr_2(s)$
7.  a. +3          b. +3          c. +&          d. +4
    e. +2          f. +1
8.  a and c.
9.  a. $O_2$ is the oxidizng agent and S is the reducing agent.
    b. $BrO_3^-$ is the oxidizng agent and $I^-$ is the reducing agent.
    c. $NO_3^-$ is the oxidizng agent and As is the reducing agent.
10.  0.983 $M$ NaOH
11.  3.29 g NaCl
12.  1.1 L
13.  0.060 $M$ KCl
14.  12.0 g $PbSO_4$
15.  $7.06 \times 10^{-3}$ $M$ $Pb^{2+}$
16.  13 mL of 0.10 $M$ $H_2SO_4$
17.  1.19 $M$ NaOH
18.  5.0 mL
19.  54.6 mL $Ce^{4+}$ soln
20.  14.6% Fe
21.  0.0282 $M$ KMnO$_4$

# Chapter 5. Gases

## EXERCISES and PROBLEMS

1.  a. 0.849 atm  b. 86.0 kPa
2.  a. 125 torr  b. 96.6 kPa
3.  500 cm$^3$
4.  25.0 °C
5.  16.9 mL
6.  2.3 L
7.  3.6 atm
8.  84.5 mmHg
9.  96.9 atm
10. 1.34 atm
11. 0.230 mol
12. 2.69 g
13. 2.15 × 10$^{22}$ molecules
14. 0.0682 g/L
15. $CF_4$
16. 279 g/mol,  $HgCl_2$
17. 137 g/mol
18. a. 1.30 L    b. 191 g
19. 2.8 L
20. yes
21. 0.552 g $H_2$ and 0.329 g He
22. 314 m/s
23. 155 K (−118 °C)
24. 2.8 × 10$^2$ m/s
25. 0.0138
26. carbon tetrachloride according to van der Waals constant a.
27. Because of the large distances between molecules in the gaseous state as compared to the liquid state.
28. Equal volumes of two gases at the same temperature and pressure contain the same number of molecules. There are more grams of $SF_4$ because $SF_4$ molecules have a higher molecular mass than that of $O_2$ molecules.
29. Hint. See Application to the Gas Laws in Section 5.7 of the text.
30. 0.0169 mol $H_2$, 40.1 g/mol
31. $KrF_2$

## PRACTICE TEST

1.  a. 0.914 atm  b. 695 torr  c. 9.26 × 10$^4$ Pa
2.  1.2 L
3.  17.3 L
4.  5240 L
5.  790 mmHg
6.  0.0161 g/L
7.  22.3 g
8.  a. 336 K        b. 434 K        c. 6.39 mol    d. 7.36 × 10$^{-3}$ mol
    e. 197 L        f. 4.0 × 10$^3$ L  g. 1.38 atm    h. 153 atm

9. 11.4 L
10. 15.7 g/L
11. 279 g/mol, $HgCl_2$
12. 120 g/mol
13. 0.136 L
14. 0.0289 mol $N_2$
15. CO: 728 mmHg; X = 0.968
16. 0.782 $N_2$, 0.210 $O_2$, 0.008 Ar
17. $P_{O2}$ = 0.0999 atm   $P_{He}$ = 0.800 atm
18. a. 5.84 L  b. 3.44 L
19. 13 L
20. 62 L
21. $C_6H_{14}$ + 19/2 $O_2$ → 6$CO_2$ + 7$H_2O$;  178 L $O_2$
22. 1260 m/s
23. a. $SF_6$  b. $O_2$  c. $SO_2$
24. a. 1270 atm   b. 3250 atm

# Chapter 6. Thermochemistry

## EXERCISES and PROBLEMS

1. w = +202 L•atm
2. $\Delta E$ = −350 J
3. q = 625 J
4. 116 kJ
5. 162 g
6. 37.8 kJ
7. 3340 kJ
8. 489 °C
9. 5.77 kJ
10. 1150 J
11. 2.91 kJ/°C
12. 941 J/°C
13. −111 kJ/g
14. −862.7 kJ/mol
15. $\Delta H_f^\circ$ = −75 kJ/mol
16. −37 kJ/mol
17. $\Delta H_{hydr}$ = −974 kJ/mol
18. According to Hess's law of heat summation, if a reaction is the sum of several reaction steps, then the enthalpy change for the overall reaction is equal to the sum of the enthalpy changes of the intermediate steps. The law works because enthalpy is a state function. Starting with the reactants and forming the products will give the same enthalpy change whether we go by a set of intermediate steps or by a direct path. Only the difference in initial and final states determines the change in a state function such as enthalpy.

19. The lattice energy is the energy required to completely separate one mole of a solid ionic compound into gaseous ions. This energy depends on the strength of attraction between the cations and anions. The lattice energy is a positive quantity because oppositely charged particles attract one another, therefore an input of energy is required to separate them. The hydration energy is a negative quantity because there

are strong attractive forces between ions and polar water molecules that lead to ion hydration.

20. The battery contains energy which is the capacity to do work. Work is energy lost or gained by mechanical means, and heat is energy transferred because of a temperature difference. A toy truck or car can be moved (displaced) when energy in the battery is used to do work. A flashlight uses energy in the battery to produce a hot glowing filament in the bulb which transfers heat to the surroundings.

## PRACTICE TEST

1. A state function is a property of a system that depends only on its present condition (state) and not on how it got that way. Energy, enthalpy, and temperature are state functions.
2. a. 50 kJ is liberated  b. $2.0 \times 10^2$ kJ is liberated
3. $\Delta H_{rxn}^{\circ} = -26$ kJ/mol
4. $\Delta H_{rxn}^{\circ} = 366$ kJ/mol
5. 321 kJ
6. 61.6 kJ
7. a. $C_2H_5OH + \dfrac{7}{2} O_2 \rightarrow 2CO_2 + 3H_2O$

   b. $q = 2000(4.18)(1.6) + 950(1.6) = 14{,}900$ J

   c. $\Delta H_{rxn}^{\circ} = -1370$ kJ

8. $\Delta H_f^{\circ} (SO_2) = -296$ kJ/mol

9. $\Delta H_f^{\circ} (CH_4) = -75$ kJ/mol
10. $\Delta E = -1623$ J
11. $w = +322$ J, work done on system by surroundings

# Chapter 7.  Quantum Theory and the Electronic Structure of Atoms

## EXERCISES and PROBLEMS

1. 600 nm
2. $4.0 \times 10^{14}$ Hz
3. $4.11 \times 10^{-19}$ J
4. $2.97 \times 10^{-19}$ J/photon, 179 kJ/mol
5. $9.70 \times 10^{-7}$ m
6. $2.6 \times 10^{-19}$ J
7. a. 6   b. $n = 4$ to $n = 1$   c. $n = 4$ to $n = 3$
8. 95.0 nm, ultraviolet
9. $n = 5$
10. $4.09 \times 10^{-19}$ J, 4
11. $1.32 \times 10^{-5}$ nm
12. a, b are not allowed
13. b

14.   a, c
15.   a. 1   b. 5   c. 3   d. 9
16.   a. 3d, $m_l = -2, -1, 0, 1, 2$.  5 orbitals
      b. 4f, $m_l = -3, -2, -1, 0, 1, 2, 3$.  7 orbitals
      c. 5p, $m_l = -1, 0, 1$.  3 orbitals
17.   a. n = 2  $\ell = 0$, 1 orbital    b. n = 3  $\ell = 1$, 3 orbitals    c. n = 4  $\ell = 3$, 7 orbitals
18.   Pb
19.   P
20.   Sb $[Kr]5s^2 4d^{10} 5p^3$;   V $[Ar]4s^2 3d^3$;   Pb $[Xe]6s^2 4f^{14} 5d^{10} 6p^2$
21.   2
22.   Essentially each orbit in the H atom has a certain energy associated with it.  The emission of radiation by an excited H atom is explained by the electron dropping from a higher to a lower energy orbit.  Since only a few orbits are permitted, then only a few transitions are possible.  Each transition produces one line in the spectrum.  With the number of transitions restricted, one observes a line spectrum rather than a continuous spectrum.
23.   Only b. The wavelength of light emitted by H atoms can be observed.
24.   No, in the 1s orbital the electron density is more dense near the nucleus.
25.   In the hydrogen atom the 2s and 2p orbitals have the same energy. The energy of an electron in a many-electron atom depends not only on its principle quantum number (n) but also on its angular momentum quantum number ($\ell$). Electrons are assigned to subshells in order of increasing "n + $\ell$" value. For the 2s subshell n + $\ell$ = 2, while for the 2p subshell n + $\ell$ = 3. If two subshells have the same value of n + $\ell$, then electrons are assigned to the one with the lower n value first.
26.   6280 g
27.   $1.4 \times 10^{19}$ photons/s
28.   $1.20 \times 10^3$ nm


## PRACTICE TEST

1.    500 m
2.    150 s
3.    $2.18 \times 10^{-18}$ J
4.    179 kJ/mol
5.    n=4
6.    $\lambda = 0.145$ nm; ultraviolet
7.    

| n | $l$ | $m_l$ | $m_s$ |
|---|---|---|---|
| 2 | 0 | 0 | −1/2 |
| 2 | 0 | 0 | +1/2 |
| 2 | 1 | −1 | −1/2 |
| 2 | 1 | −1 | +1/2 |
| 2 | 1 | 0 | −1/2 |
| 2 | 1 | 0 | +1/2 |
| 2 | 1 | +1 | −1/2 |
| 2 | 1 | +1 | +1/2 |

8.    18, 32
9.    d. 3d
10.   2
11.   d.

12. a. $n = 2, l = 0, m_l = 0, m_s = ½$ (or $- ½$ )
    b. $n = 4, l = 0, m_l = 0, m_s = 1/2$
       $n = 4, l = 0, m_l = 0, m_s = -1/2$
       $n = 4, l = 1, m_l = -1, m_s = 1/2$
       $n = 4, l = 1, m_l = 0, m_s = 1/2$
13. Ar: $1s^2 2s^2 2p^6 3s^2 3p^6$ or $[Ne]3s^2 3p^6$
    Se: $[Ar]4s^2 3d^{10}4p^4$
    Ag: $[Kr]5s^1 4d^{10}$
14. 36
15. d
16 d. P

# Chapter 8. Periodic Relationships Among the Elements

## EXERCISES and PROBLEMS
1. Elements that have an incompletely filled set of s or p orbitals. Pick any from Groups 1A through 7A.
2. Transition metals have incompletely filled d subshells or readily form cations having incompletely filled d subshells. Pick any from Groups 2B through 8B in the periodic table.
3. These elements have an incompletely filled inner f subshell. Pick any between atomic numbers 58 and 71.
4. $6.5\,°C$
5. $ns^1(n-1)d^{10}$
6. a. $ns^2 np^1$  b. $ns^2 np^4$
7. The halogen elements. Group 7A
8. Group 4B
9. $[Ar]\,4s^2 3d^1$  Sc is a transition element and a metal.
10. a. Group 2A  b. Group 5A  c. Group 8B
11. a, c and f have 36 electrons. b and d have 18 electrons.
12. a. $K^+$ [Ar]  b. $O^{2-}$ [Ne]  c. $Sr^{2+}$ [Kr]  d. $I^-$ [Xe]  e. $Li^+$ [He]  f. $Mg^{2+}$ [Ne]
13. a. $[Ar]\,3d^6$  b. $[Ar]\,3d^5$  c. $[Ar]\,3d^9$  d. $[Ar]\,3d^{10}$  e. $[Ar]\,3d^2$  f. [Ar]
14. a. [Ar]  b. [Ar]  c. $[Xe]\,6s^2 4f^{14}5d^{10}$  d. $[Xe]\,4f^{14}5d^{10}$
15. Ar and $Cl^-$ are one isoelectronic pair, and $Na^+$ and Ne are another.
16. a. Na  b. Pb
17. $Se^{2-}$
18. Ne
19. a. $Co^{2+}$  b. $S^{2-}$
20. a. noble gases  b. alkali metals
21. K
22. Cl
23. $I_5$
24. Electron affinity is the negative of the energy change that occurs when an electron is accepted by an atom in the gaseous state to form an ion.
25. K
26. Br
27. V

28. As
29. diagonal relationship
30. +1 and +2, respectively
31. −2 and −1, respectively
32. As the principal quantum number n increases, the size of the orbital increases, and the farther, on average, electrons will be from the nucleus. When reading down a group, the outermost electrons have an increasingly higher n value, and so the outermost electrons are farther from the nucleus and the atomic radius increases.
33. When going across a period from left to right protons are being added one at a time to the nucleus and the corresponding electrons are added simultaneously to the outermost principle energy level. The first ionization energy increases across the period because the electrons in the same principle energy level do not completely shield the increasing nuclear charge due to the added protons. As the effective nuclear charge increases from left to right so will the ionization energy.
34. The number of outermost electrons and the type of orbital occupied is what determines the chemical behavior of an element.

## PRACTICE TEST

1. $1^{st}$ IE $\approx$ 350 kJ; density $\approx$ 2.0 g/cm$^3$
2. Density $\approx$ 11.3 g/cm$^3$
3. Ti, Zr, and Hf
4. d. $ns^2np^2$
5. a. Group 8A, noble gases　　b. Group 1A, alkali metals
   c. A transition metal　　　　d. Group 7A, halogens
6. Cr $[Ar]4s^23d^4$; the observed configuration is Cr $[Ar]4s^13d^5$, which is an exception to the rules we have developed.
   Sb $[Kr]5s^24d^{10}5p^3$
   Pb $[Xe]6s^24f^{14}5d^{10}6p^2$
7. 5
8. c. 4th
9. a. $Ca^{2+}$ [Ar]　　　b. $Se^{2-}$ [Kr]　　　　c. $Cl^-$ [Ar]
   d. $Mn^{2+}$ $[Ar]3d^5$　　e. $Co^{3+}$ $[Ar]3d^6$　　f. $Sc^{3+}$ [Ar]
10. c. $Sc^{3+}$
11. d. Ge
12. a. $I^-$　　　　　b. $K^+$　　c. Mg
13. e. S
14. The effective nuclear charge increases from left to right because the nuclear charge increases by one proton at a time, and as electrons are added to the atom they go into the same principal energy level where they do not shield each other effectively. As $Z_{eff}$ increases, the electrons are pulled closer to the nucleus and the atomic radius decreases correspondingly.
15. B(I = 801 kJ) and Si(I = 786 kJ) diagonal relationship.

# Chapter 9.　Chemical Bonding I:　Basic Concepts

## EXERCISES and PROBLEMS

1. 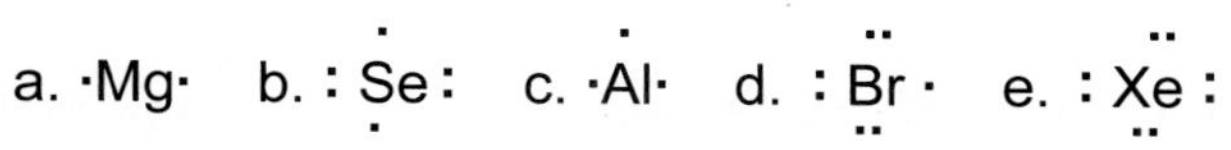  a. ·Mg·  b. : S̈e :  c. ·Al·  d. : B̈r ·  e. : Ẍe :

2. 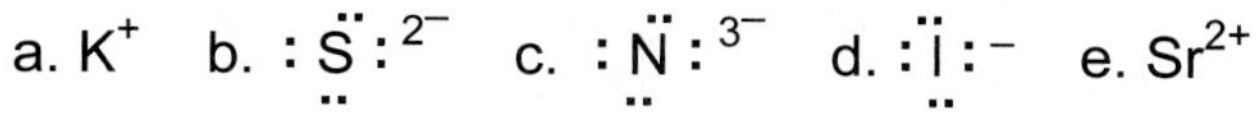  a. K⁺  b. : S̈ : ²⁻  c. : N̈ : ³⁻  d. : Ï : ⁻  e. Sr²⁺

3.  $MgO(s) \rightarrow Mg^{2+}(g) + O^{2-}(g)$   $\Delta H$ = lattice energy

4.  a. $CaCl_2$  b. NaI
5.  a. CaO  b. KCl
6.  736 kJ/mol
7.  a. 1  b. 2  c. 2

8. 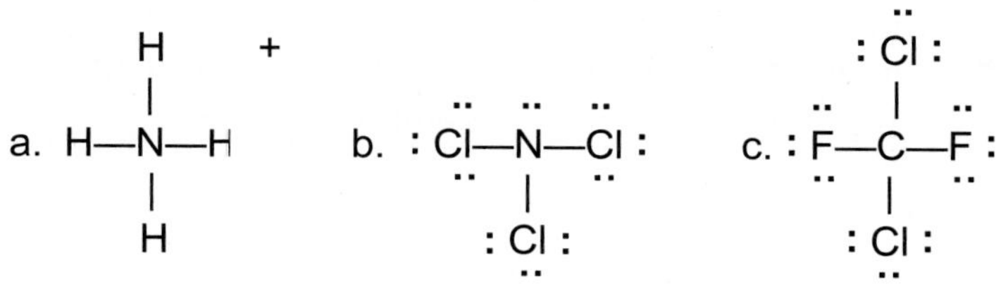
a.
```
    H   +
    |
 H—N—H
    |
    H
```
b.
```
       :Cl:
        |
 :Cl—N—Cl:
        |
       :Cl:
```
c.
```
       :Cl:
        |
 :F—C—F:
        |
       :Cl:
```

9.  P
10. O—O < S—O < N—O < Na—O
11. polar covalent
12. Br—Br
13. polar covalent, polar covalent (using electronegativity difference > 2.0), ionic, nonpolar covalent
14. a. The formal charges in $NH_3$ are:   The N atom: 0.
    The H atoms: 0.
    b. The formal charges in $SO_2$ are:   The S atom: 0.
    The O atoms: 0.

15. 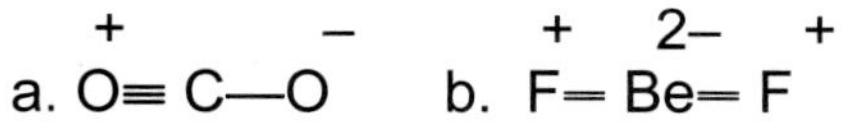 a. ⁺O≡ C—O⁻   b. ⁺F= Be²⁻ = F⁺

16. 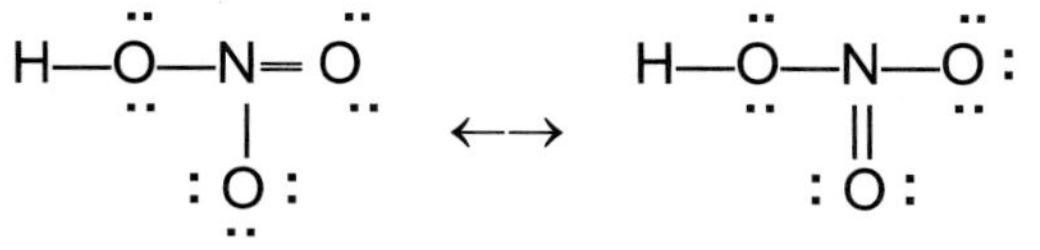
```
H—Ö—N= Ö          H—Ö—N—Ö:
     |      ←→          ||
    :O:               :O:
```

17. a. 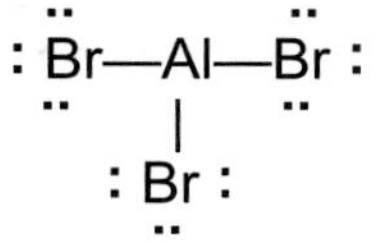 : B̈r—Al—B̈r :   c. · N= O
```
         |
        :Br:
```
b.
```
         :S:
          |
 :S —F—S:
          |
         :S:
```

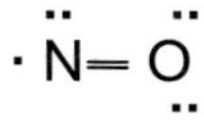

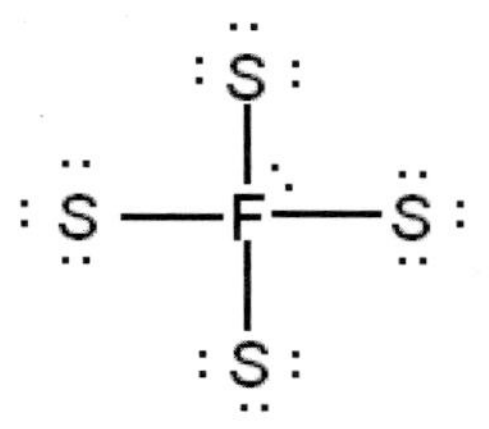

18. 390 kJ/mol

19. −54.5 kJ/mol
20. Electron affinity is a property of a gaseous atom, while electronegativity is a property of an atom within a molecule. Both have to do with attracting electrons but they are very different properties. Electron affinity is related to the energy released by an uncombined atom in the gaseous state when it adds an additional electron. Electronegativity is a measure of the ability of an atom within a molecule to attract an electron pair in a bond it has formed.
21. Formal charges do not represent actual charges of the atoms in a molecule. However they do give an indication of the sign (+ or −), if not the magnitude, of the charge and we use formal charges to choose from several possible Lewis structures.
22. When a molecule cannot be accurately represented by a single Lewis structure we resort to resonance structures. These structures have the same placement of atoms but different positions of electron pairs. The electron pairs do not actually move from position to position within the molecule. Therefore none of the resonance structures actually exist. Rather the real molecule is said to be a composite of the resonance structures.

## PRACTICE TEST

1. $RbI > CaBr_2 > MgO$
2. a. KI        b. $MgCl_2$

3. a.        b.

   c.

   d.

   e.        f.

   g.        h.

4.

$$: \ddot{O} : \qquad : O : \qquad : \ddot{O} :$$

$$\ddot{O} = \overset{\displaystyle |}{S} - \ddot{O} : \longleftrightarrow \quad : \ddot{O} - \overset{\displaystyle \|}{S} - \ddot{O} : \longleftrightarrow \quad : \ddot{O} - \overset{\displaystyle |}{S} = \ddot{O}$$

5.   b.

6.   b.

7.   d.

8.   a.

9.   c.

10.   a. $PCl_5$   b. $BCl_3$   c. $CH_3$   d. $SbCl_5$

11.   b.  Because fewer formal charges are present.  Particularly the formal charge on sulfur is zero compared to +2 in a.

12.

$$\text{a. } : \overset{-}{\ddot{N}} = \overset{+}{N} = \overset{O}{O} : \qquad \text{b. } \left[ : \overset{-}{\ddot{O}} - H \right]^{-} \qquad \text{c. } \left[ : \overset{-}{\ddot{O}} - \ddot{Cl} : \right]^{-}$$

13.   Lattice energy $= -\Delta H_f^\circ (CaCl_2) + \Delta H_{subl}(Ca) + I_1 (Ca) + I_2(Ca) + BE (Cl_2) + 2 (-EA (Cl))$

14.   946 kJ

15.   180 kJ

# Chapter 10.   Chemical Bonding II: Molecular Geometry and Hybridization of Atomic Orbitals

## EXERCISES and PROBLEMS

1.   a. linear   b. trigonal pyramidal   c. distorted tetrahedron (seesaw)   d. T-shaped
     e. square planar

2.   a. bent   b. trigonal planar   c. tetrahedral

3.   a. and c.

4.   a. 180°   b. 109.5°   c. less than 109.5°

5.   The two requirements are polar bonds and a molecular geometry in bonds dipoles do not cancel.

6.   c. $PCl_3$

7.   c. $SO_2$ and d. $NO_2$

8.   Each fluorine atom has the same electron configuration: $1s^2 2s^2 2p_x^2 2p_y^2 2p_z^1$. The $2p_z$ orbitals of each F atom are only half full. Therefore, a sigma bond can be formed by overlap of the $2p_z$ orbital from each atom. In HBr the 1s orbital of the H atom overlaps with the $4p_z$ orbital of Br to form a sigma bond.

9.   a. 180°   b. 120°   c. 109.5°

10.   a. sp   b. $sp^2$   c. $sp^2$   d. $sp^3 d$   e. $sp^2$

11.   a. $sp^3$   b. $sp^3 d$   c. $sp^3 d^2$

12.   octahedral

13.   2 $\sigma$ and 2 $\pi$

14.   a. The antibonding MO.   b. The atomic orbitals from which the bonding molecular orbital was created.

15.   The bond order for $Be_2$ is zero, while for $Be_2^+$ it is 0.5.

16.  $Li_2$ $(\sigma_{1s})^2(\sigma^*_{1s})^2(\sigma_{2s})^2$    $Li_2^+$ $(\sigma_{1s})^2(\sigma^*_{1s})^2(\sigma_{2s})^1$    $Li_2$ has the stronger bond.

17.  a. $O_2$    b. $O_2^-$

18.  The central oxygen atom is $sp^2$ hybridized and forms $\sigma$ bonds to the two terminal oxygen atoms. The central O atom has an unhybridized $2p_z$ orbital that overlaps "sideways" with the $2p_z$ orbitals of both terminal O atoms.

19.  Recall that lone-pair electrons take up more space than electrons in a bond. A lone pair in an axial position has three close neighbors in the bonding electron pairs in the equatorial positions. A lone pair in an equatorial position has two close neighbors in the bonding electron pairs in the axial positions. (The neighboring equatorial electron pairs are too far away). This means that a lone pair will have more room and have less repulsion when it is in an equatorial position.

20.  We know that the C—O bonds are polar. For $CO_2$ to be nonpolar, the bond dipoles must be pointing in opposite directions so that they cancel. The molecule must be linear.

21.  Compare the electron configurations.

$$N_2 \ (\sigma_{1s})^2(\sigma^*_{1s})^2(\sigma_{2s})^2(\sigma^*_{2s})^2(\pi_{2p_y})^2(\pi_{2p_z})^2(\sigma_{2p_x})^2$$

$$N_2^+ \ (\sigma_{1s})^2(\sigma^*_{1s})^2(\sigma_{2s})^2(\sigma^*_{2s})^2(\pi_{2p_y})^2(\pi_{2p_z})^2(\sigma_{2p_x})^1$$

$$O_2 \ (\sigma_{1s})^2(\sigma^*_{1s})^2(\sigma_{2s})^2(\sigma^*_{2s})^2(\pi_{2p_y})^2(\pi_{2p_z})^2(\sigma_{2p_x})^2(\pi_{2p_y}^*)^1(\pi_{2p_z}^*)^1$$

$$O_2^+ \ (\sigma_{1s})^2(\sigma^*_{1s})^2(\sigma_{2s})^2(\sigma^*_{2s})^2(\pi_{2p_y})^2(\pi_{2p_z})^2(\sigma_{2p_x})^2(\pi^*_{2p_y})^1$$

The bond order in $N_2$ is 3. Loss of 1 electron gives $N_2^+$. The electron came out of a bonding orbital and so the bond order is less than in $N_2$, b.o. = 2.5. The bond order in $O_2$ is 2. Loss of 1 electron gives $O_2^+$. The electron came out of an antibonding orbital and so the bond order is greater than in $O_2$, b.o. = 2.5.

## PRACTICE TEST

1.  a. square pyramid   b. linear   c. trigonal planar   d. bent   e. bent
2.  a. $sp^3d^2$  b. sp   c. $sp^2$  d. $sp^2$  e. $sp^3$
3.  a. trigonal planar   b. distorted tetrahedral (seesaw)  c. square planar
    d. trigonal planar   e. tetrahedral
4.  a. $sp^2$   b. $sp^3d$   c. $sp^3d^2$   d. $sp^2$   e. $sp^3$
5.  b. $SiH_4$
6.  a. < 120°          b. 90°          c. < 90° and 120°
7.  $CCl_4$ has four electron pairs about the central carbon atom, while $AsCl_4^-$ has 5 pairs about the As atom with only 4 Cl atoms attached. The extra pair of unshared electrons prevents the formation of a tetrahedron by $AsCl_4^-$.
8.  There are two electron pairs in the valence shell of Be in $BeCl_2$, but there are four pairs in the valence shell of Te in $TeCl_2$. These four pairs are at the corners of a tetrahedron. When two chlorine atoms bond to a Te atom via two of the pairs, the Cl—Te—Cl bond angle is 109.5°.
9.  $sp^3d$, $sp^3d^2$
10. No
11. HBr
12. b. $H_2S$
13. b. $BCl_3$   c. $BeCl_2$
14. c.

15. The O—H bonds are polar. Nonlinear.

16.

| | $H_2^+$ | $He_2$ | $He_2^+$ |
|---|---|---|---|
| $\sigma_{1s}^*$ | ___ | ↑↓ | ↑ |
| $\sigma_{1s}$ | ↑ | ↑↓ | ↑↓ |
| Bond order | 0.5 | 0 | 0.5 |

17. a. $C_2$ $(\sigma_{1s})^2(\sigma_{1s}^*)^2(\sigma_{2s})^2(\sigma_{2s}^*)^2(\pi_{2p_y})^2(\pi_{2p_z})^2$

$C_2^{2-}$ $(\sigma_{1s})^2(\sigma_{1s}^*)^2(\sigma_{2s})^2(\sigma_{2s}^*)^2(\pi_{2p_y})^2(\pi_{2p_z})^2(\sigma_{2p_x})^2$

b. For $C_2$ the bond order = 2. For $C_2^{2-}$ the bond order = 3.

18. The bond order in $N_2$ is 3, while in $N_2^+$ it is only 2.5. The greater the bond order, the stronger the bond and the shorter the bond distance.

# Chapter 11.   Intermolecular Forces, Liquids & Solids

## EXERCISES and PROBLEMS

1. a. dipole-dipole forces and dispersion forces   b. dispersion forces   c. dispersion forces   d. dipole-dipole forces and dispersion forces   e. ionic bonds

2. $CS_2$ because of its much larger number of electrons and therefore polarizability.

3. b, c, and d.

4. a. CO   b. ICl   c. $H_2O$   d. $AsH_3$

5. a. hydrogen bonds and dispersion forces   b. covalent bonds   c. dispersion forces

6. surface tension

7. $C_2H_5OH$

8. 8.94 g/cm$^3$

9. $6.02 \times 10^{23}$

10. 4 atoms, face-centered cubic

11. 361 pm

12. 204 pm

13. 0.154 nm

14. 0.188 kJ/g

15. 3.29 kJ

16. 9.18 g

17. 82.5 kJ/mol

18. $\Delta H_{vap}$ = 41.7 kJ/mol

19. 768 °C

20. 100 mmHg

21. a. $CH_3OH$   b. $Cl_2$   c. $CH_3Cl$

22. 3017 kJ

23. 81.5 kJ/mol

24. a. endothermic   b. endothermic   c. exothermic   d. exothermic   e. endothermic

25.   a. The flow is faster at first because the vapor pressure of $CO_2$ is higher at first. The temperature of the $CO_2$ is higher at first. Therefore more of the $CO_2$ is in the gas phase and the vapor pressure is higher at first. The liquid $CO_2$ cools down because it supplies some of the heat of vaporization.
b. The flow almost stops before the cylinder is empty because as the liqiud cools off the vapor pressure of $CO_2$ drops so low that very little of the liquid $CO_2$ can vaporize.
c. As the liquid $CO_2$ evaporates the heat of vaporization must be continually supplied. Heat is taken from the steel cylinder and then from the surrounding air. As the temperature of the steel drops below the freezing point of water the water vapor in the air condenses (deposition) as ice on the cold cylinder.

26.   The atmospheric pressure decreases as altitude increases. Therefore the higher one goes the lower the vapor pressure needed to form the bubbles observed during boiling. The lower the vapor pressure needed for boiling, the lower the temperature.

27.   Viscosity is the resistance to flow and is caused by intermolecular forces. As the temperature of water is raised, more and more H bonds are broken, and the viscosity decreases.

28.   Heat is required for water to evaporate. When water evaporates from your skin, the heat of vaporization is absorbed from the skin, and the skin cools. You feel warmer on a humid day because the condensation of water vapor on your skin slows the net rate of evaporation (cooling).

## PRACTICE TEST

1.   a. hydrogen bonding and dispersion forces.  b. dipole-dipole and dispersion forces c. dispersion forces  d. dipole-dipole and dispersion forces  e. hydrogen bonding and dispersion forces.

2.   $H_2O$ molecules are attracted to each other by hydrogen bonding, whereas $H_2S$ and $H_2Se$ are not.

3.   a. $SO_2$   b. $Br_2$   c. $H_2O$   d. HCl

4.   Grease and water do not mix, thus if grease remains on the glass surface, water will bead up. When the surface is clean, the cohesive forces between water and glass cause water to spread out in a thin film and "wet" the glass.

5.   The cohesive forces between Hg atoms are greater than the adhesives forces between glass and mercury.

6.   $\theta = 5.2°$

7.   a. 4       b. 478 pm      c. 3.41 g/cm$^3$

8.   $6.1 \times 10^{23}$

9.   573 pm

10.   a. KCl   b. Fe   c. Fe   d. $SiO_2$ , covalent crystal   e. $SiO_2$

11.   The bubble formation that signifies boiling cannot occur until the vapor pressure is great enough to push back the atmosphere. Thus the boiling point varies with atmospheric pressure. The normal boiling point of a liquid is the temperature at which its vapor pressure is 1.0 atm.

12.   237 mmHg

13.   a. liquid  b. gas, then liquid

14.   436 kJ

15.   a. Kr   b. $CH_3OH$

16.   c. $CH_3OH$

# Chapter 12. Physical Properties of Solutions

## EXERCISES and PROBLEMS

1. miscible
2. Hexane is a nonpolar liquid and water is a polar liquid.
3. a. soluble   b. soluble   c. insoluble   d. insoluble   e. soluble   f. soluble
4. Ammonia $NH_3$ can form hydrogen bonds to water, but hexane cannot form hydrogen bonds.
5. Solvation involves surrounding a solute molecule or ion with solvent molecules. The process is called hydration when the solvent is water. Solvation is the more general term.
6. a. NaCl(s)      b. $NH_3$(g)      c. $CH_3OH$($\ell$)
7. a. $I_2$(s)      b. $CS_2$($\ell$)      c. $CO_2$(g)
8. a. 7.96%      b. 10.9%      c. 1.6%
9. a. 1.48 *m*      b. 0.359 *m*
10. 2.33 *M* NaCl
11. 9.95% $AgNO_3$
12. 0.34 *M* $AgNO_3$
13. 48.5 g $KNO_3$
14. decreases
15. $1.5 \times 10^{-5}$ mol/L
16. 31.4 mmHg
17. −0.103 °C
18. −1.8 °C
19. 20. atm
20. 127 g/mol
21. 181 g/mol
22. a. 1   b. 1   c. 2   d. 4   e. 2   f. 3
23. 0.20 *m* $C_6H_{12}O_6$ > 0.15 *m* KI > 0.120 *m* NaCl > 0.15 *m* $MgCl_2$
24. The $T_f$ of 0.10 m HCl (a strong acid) is lower than that of 0.10 m acetic acid (a weak acid).
25. First, the water molecules in the solvent must spread apart a bit to make room for the solute particle. This requires a positive ΔH. Energy must be added in order to break hydrogen bonds between water molecules. Second, the ions in the lattice of the solid salt must be separated from all their neighbors. The energy for this step is related to the lattice energy. The forces holding the crystal together are ionic bonds. ΔH is positive. Next, water molecules and the anions and cations from the salt are mixed. Hydration occurs and heat is released (ΔH is negative) as ion-dipole forces bring water molecules and ions together.
26. Energy is only one factor affecting the solution process. A second factor which affects all natural events is a tendency to produce more disordered systems. In the pure state the solute and solvent molecules are arranged in a fairly ordered manner. Much of this order disappears when the molecules become mixed together. It is the increase in disorder of the system that favors the solubility of any substance.
27. saturated
28. Ion pairs (+/−) can form in solutions of electrolytes. The formation of an ion pair will reduce the number of solute particles in the solution. Therefore the osmotic pressure would decrease compared to a solution in which ions remained separated and there were no ion pairs.

## PRACTICE TEST

1.  a. HCl    b. $MgCl_2$        c. $CH_3Cl$        d. $CH_3OH$
2.  a. $CH_3CH_2CH_2CH_2CH_2CH_3$       b. $C_6H_6$          c. $I_2$
3.  1.04 *m* urea
4.  24.9 g KI
5.  0.88 *M* $H_2O_2$
6.  18 *M*
7.  $X_{CH_3OH} = 0.360$, $X_{H_2O} = 0.640$
8.  a. 1.28 *M*
    b. 17.0% $CuSO_4$
    c. $X_{CuSO_4} = 0.0227$
    d. 1.29 *m*
9.  $1.9 \times 10^{-8}$ *M* CO
10. 520 g
11. 100.84 °C
12. 180 g/mol
13. 63.8 g/mol
14. 650 g/mol
15. 0.20 *M* NaCl
16. 0.154 *M* NaCl
17. NaBr; $Ca(NO_3)_2$; HCl = ethanol
18. *i* = 1.76
19. $C_6H_6O_4$ when the solute dissolves in the solvent.


# Chapter 13.   Chemical Kinetics

## EXERCISES and PROBLEMS

1.  rate $= \dfrac{1}{2}\dfrac{\Delta[NH_3]}{\Delta t} = -\dfrac{\Delta[N_2]}{\Delta t} = -\dfrac{1}{3}\dfrac{\Delta[H_2]}{\Delta t}$

2.  a. $\dfrac{-\Delta[S_2O_3^{2-}]}{\Delta t} = 8.3 \times 10^{-4}$ *M*/s      b. $\dfrac{-\Delta[I_2]}{\Delta t} = 4.2 \times 10^{-4}$ *M*/s

3.  a. $5.0 \times 10^{-4}$ *M*/s     b. $3.5 \times 10^{-4}$ *M*/s    c. $1.8 \times 10^{-4}$ *M*/s
4.  9 times
5.  a. rate = k $[A][B]^2$    k = $4.8 / M^2 s$
6.  a. k = $6.8 \times 10^{-3}$ /s    b. 102 s

7.  a. ln $\dfrac{[A]_t}{[A]_0}$ = −kt    b. Plot ln[A] vs. t   where x = t, y = [A], and m = −k

    c. The initial concentration $[A]_0$ and $[A]_t$ again at a known time t.
8.  58%
9.  a. $6.0 \times 10^{-4}$/s  b. 2300 s   c. 1200 s
10. first-order: 1/s;   second-order:  L/mol•s or 1/*M*·s
11. a. 5.8 y    b. 0.28 y
12. 92.2 kJ/mol
13. $k_2 = 3.4 \times 10^{-3}$/s

14. a. According to Arrhenius's interpretation the slope is:  m $= -\dfrac{E_a}{R}$

b. Calculate the slope from m = $\dfrac{\ln k_2 - \ln k_1}{1/T_2 - 1/T_1}$,     $m = -1.0 \times 10^4$ K

c. 83.2 kJ/mol

15. $E_a(rev) = 26.5$ kJ/mol
16. products and intermediates
17. a. $rate_1 = k_1 [NO_2]^2$;     $rate_2 = k_2 [NO_3] [CO]$

b. when step 1 is rate determining:  rate = $k_1 [NO_2]^2$

when step 2 is rate determining:  rate = $\dfrac{k_2 k_1}{k_{-1}} [NO_2]^2 [CO][NO]^{-1}$

18. yes
19. A catalyst is a substance that increases the rate of a chemical reaction without itself being consumed. It may react to form an intermediary, but it is regenerated in a subsequent step of the reaction.
20. NO is the catalyst. $NO_2$ is the intermediate. The overall equation is:
$O_3 + O \rightarrow 2O_2$
21. Normally only a small fraction of the colliding molecules have enough energy to exceed the activation energy. As temperature increases the average kinetic energy of molecules increases. Therefore, the fraction of the colliding molecules having energies greater than the activation energy increases. More molecules are capable of reacting per second and the rate of reaction increases as temperature increases.
22. First of all cooking an egg is a chemical reaction and as such its rate will increase with temperature in accordance with the Arrhenius equation. At sea level water (used to cook the egg) boils at 100 °C, but at higher elevations such as 4452 m water boils at a temperature lower than 100 °C (86 °C on top of Pikes Peak). Therefore the rate of cooking an egg is much slower on mountain tops than at sea level.
23. An elementary reaction represents a reaction between individual molecules. The reaction occurs when the molecules collide. An overall reaction is the net change due to the sum of all the elementary reactions. It does not represent the collisions and reactions of any particular molecules. Indeed any intermediates have canceled out of the net reaction. The balanced equation shows only the reactants and products in the proper molar amounts.

## PRACTICE TEST

1. Rate = $k[B]^2$

2. Rate = $-\dfrac{1}{30}\dfrac{\Delta[CH_3OH]}{\Delta t} = -\dfrac{\Delta[B_{10}H_{14}]}{\Delta t} = \dfrac{1}{22}\dfrac{\Delta[H_2]}{\Delta t} = \dfrac{1}{10}\dfrac{\Delta[B(OCH_3)_3]}{\Delta t}$

3. a. [sucrose] = 0.083 $M$;  b. $t_{1/2} = 3.3 \times 10^5$ s
4. a. k = 0.106/h     b. 0.251     c. 6.54 h     d. 0.0174 $M$
5. a. first order    b. $4.6 \times 10^{-3}$ per day
6. a. $2.8 \times 10^{-2}$ /$M \cdot$ min    b. 89 min
7. Twelve-fold
8. 182 kJ/mol
9. k = $4.8 \times 10^{-5}$/$M$ s;  $k_2/k_1 = 3.2$
10. 16.8
11. 180 kJ/mol
12. Yes

13. The rate law for the rate determining step is
rate = $k[NO_2^-][O_2]$
The concentration of dissolved $O_2$ is a constant according to Henry's law (Chapter 12).
Therefore $k[O_2] = K$ (a constant), and rate = $K[NO_2^-]$.

14. The rate law for the rate determining step is
rate = $k[Ni(H_2O)_6^{2+}]$
This rate law matches the experimental rate law.

15. The rate law for the rate determining step is
rate = $k_2[I]^2[H_2]$
Obtaining a mathematical substitution for [I] from
$k_1[I_2] = k_{-1}[I]^2$

yields    rate = $\dfrac{k_2 k_1}{k_{-1}}[H_2][I_2]$

This predicted rate law matches the experimental rate law given in the problem.

16. a. The catalyst reacts in one of the first few steps, but then is regenerated in the last step.
b. If $E_a$ is lowered for the forward reaction it must also be lowered for the reverse.

17. a. $Br_2$    b. $2H_2O_2(aq) \rightarrow 2H_2O(\ell) + O_2(g)$    c. homogeneous

# Chapter 14.  Chemical Equilibrium

## EXERCISES and PROBLEMS

1. a. $K_c = [O_2]$        b. $K_c = \dfrac{[Ni(CO)_4]}{[CO]^4}$        c. $K_c = \dfrac{[NOBr]^2}{[NO]^2[Br_2]}$

2. a. $K_p = 1.4 \times 10^{-5}$        b. $K_p = 270$
3. $K_c = 1.5 \times 10^{-3}$
4. $K_c = \dfrac{[COCl_2]^2}{[CO_2][Cl_2]^2} = 4.2 \times 10^{11}$

5. $K_c = 0.169$
6. $K_c = 5.6 \times 10^{-4}$
7. $K_c = 73.6$
8. $K_c = 17$
9. a. no    b. reverse (left)
10. a. no    b. forward (right)
11. $[NOBr] = 5.85 \times 10^{-3}$ *M*
12. $[HI] = 0.53$ *M*,    $[H_2] = [I_2] = 0.06$ *M*
13. $[NO_2] = 0.17$ *M*,    $[N_2O_4] = 0.080$ *M*
14. a. right    b. no shift    c. left    d. left    e. no shift    f. no shift
15. right
16. decrease
17. increase
18. Yes some radioactive $^{131}I$ atoms will become part of $I_2$ molecules, and form of $I^{131}I$. Some will remain as $H^{131}I$. First of all $^{131}I$ atoms will react just like any other iodine atom. The reaction system is at equilibrium which is a dynamic equilibrium. This is the key. The forward and reverse reactions are occurring simultaneously and at the same rate. Any $H^{131}I$ will eventually dissociate by the reverse reaction and produce $I^{131}I$

molecules and $H_2$ molecules. All of the $^{131}I$ will not end up in the $I^{131}I$ molecules because the forward reaction converts $I^{131}I$ back into $H^{131}I$. Therefore the $^{131}I$ will be distributed between the two I-containing molecules.

19. The technician should have known that a catalyst cannot change the position of equilibrium. A catalyst lowers the activation energies of *both* the forward and reverse reactions, and so speeds up the rates of both the forward and reverse reactions. This does not change the equilibrium constant.

**PRACTICE TEST**

1.  a. $K_c = \dfrac{[N_2]^2[H_2O]^6}{[NH_3]^4[O_2]^3}$  b. $K_c = \dfrac{[N_2O_4]^2}{[N_2O]^2[O_2]^3}$  c. $K_c = \dfrac{[FClO_2]^2}{[ClO_2]^2[F_2]}$

    d. $K_c = \dfrac{[HBr]^2}{[H_2]}$  e. $K_c = \dfrac{[CO]^2}{[CO_2]}$  f. $K_c = \dfrac{[H_2O]}{[H_2]}$

2.  $2.0 \times 10^{-5}$
3.  $[HI] = 1.1 \times 10^{-4}$ *M*
4.  $K_p = 2.2 \times 10^{-5}$, $K_c = 5.40 \times 10^{-4}$ ($\Delta n = -1$)
5.  $K_c = 0.0314$
6.  $K_c = 34$
7.  net reverse reaction.
8.  $[HI] = 0.22$ *M*
9.  $K_c = 0.012$
10. $P_{NO} = 0.017$ atm; $P_{N_2} = 1.99$ atm; $P_{O_2} = 0.39$ atm
11. a. 0.36 mol $NO_2$  b. 56% $N_2O_4$ dissociated
12. a. Least extent of reaction  b. Greatest extent of reaction
    c. Intermediate of the three examples
13. a. The partial pressures of $SO_2$ and $O_2$ will increase.
    b. No changes in equilibrium partial pressures.
    c. The partial pressures of $SO_2$ and $O_2$ decrease while that of $SO_3$ increases.
    d. The partial pressures of $SO_2$ and $O_2$ decrease while that of $SO_3$ increases.
    e. The partial pressure of $SO_2$ increases and that of $SO_3$ decreases.
14. a. $K_c$ will decrease  b. $[PCl_3]$ decreases  c. $[PCl_5]$ increases; $[PCl_5]$ and $[Cl_2]$ decrease  d. $Cl_2$ pressure increases
15. $[SO_2] = [NO_2] = 0.0118$ *M*; $[NO] = [SO_3] = 0.108$ *M*
16. a. decreases  b. no change  c. decreases  d. increases
17. $K_p = 3.0 \times 10^{-8}$  $K_c = 4.7 \times 10^{-6}$

# Chapter 15.  Acids & Bases

## EXERCISES and PROBLEMS

1.  $CO_3^{2-}$
2.  a. $NH_4^+$ is the acid on the left and $NH_3$ is its conjugate base, $H_2O$ is the base on the left and $H_3O^+$ is its conjugate acid.
    b. $HNO_2$ is the acid on the left and $NO_2^-$ is its conjugate base, $CN^-$ is the base on the left and HCN is its conjugate acid.
3.  $1.0 \times 10^{-11}$ *M*

4. $1.3 \times 10^{-12}$ M
5. $[OH^-] = 0.0033$ M and $[H^+] = 3.0 \times 10^{-12}$ M
6. pH = 3.29
7. pH = 13.097
8. $1 \times 10^{-4}$ mol
9. $5.1 \times 10^{-3}$ M
10. $1.3 \times 10^{-5}$ M
11. 100
12. pH = 11.70
13. pH = 1.56
14. HCN
15. Left. $HNO_2(aq) + ClO_4^-$ (aq) predominate at equilibrium.
16. 3.4%
17. pH = 2.74
18. $1.3 \times 10^{-5}$
19. $1.4 \times 10^{-3}$
20. HF
21. $HNO_2$
22. 17%
23. a. $[OH^-] = 1.3 \times 10^{-4}$ M $[H^+] = 7.9 \times 10^{-11}$ M $\quad$ b. weak $\quad$ c. $1.4 \times 10^{-6}$
24. $5.9 \times 10^{-11}$
25. $K_a = 5.6 \times 10^{-10}$
26. 1. $H_3PO_4 \rightleftharpoons H^+ + H_2PO_4^-$ $\quad$ 2. $H_2PO_4^- \rightleftharpoons H^+ + HPO_4^{2-}$ $\quad$ 3. $HPO_4^{2-} \rightleftharpoons H^+ + PO_4^{3-}$
27. 1.21
28. a. $HNO_3$ $\quad$ b. $SiH_4$ $\quad$ c. HOBr $\quad$ d. $NH_3$
29. a. pH > 7 $\quad$ b. pH = 7 $\quad$ c. pH < 7
30. 10.82
31. Because N atoms in most compounds have an unshared electron pair.
32. Because they have empty orbitals available to accept electron pairs.
33. Any acid stronger than $H_3O^+$ ion reacts with water to produce $H_3O^+$. Hydronium ion is the strongest acid existing in water at equilibrium.
$HA + H_2O \rightarrow H_3O^+ + A^-$
Any base stronger than $OH^-$ ion accepts a proton from water yielding a $OH^-$ ion. Hydroxide ion is the strongest base that exists in water at equilibrium.
$B + H_2O \rightarrow HB + OH^-$
34. $O^{2-} + H_2O \rightarrow OH^- + OH^-$
$NH_2^- + H_2O \rightarrow NH_3 + OH^-$
35. We usually want to know the pH of an acid solution. For a weak acid we use $K_a$ because each weak acid dissociates to a different extent. The $K_a$ relates all the concentrations at equilibrium and allows us to find $[H^+]$. We know that all strong acids are 100% ionized. Therefore, we can calculate $[H^+]$ based on complete reaction.
36. a. and d.
37. NaF(*aq*) must contain a stronger base than NaCl(aq). Fluoride ion is a stronger base than chloride ion because $F^-$ ion is the conjugate base of a weak acid while $Cl^-$ ion is the conjugate base of a strong acid.

## PRACTICE TEST

1. a. $CH_3COO^-$ $\quad$ b. $HS^-$ $\quad$ c. $SO_3^{2-}$ $\quad$ d. $ClO^-$

2. a. $NH_3 + H_2CO_3 \rightleftharpoons NH_4^+ + HCO_3^-$
   base$_1$  acid$_2$      acid$_1$  base$_2$

   b. $CO_3^{2-} + HSO_3^- \rightleftharpoons HCO_3^- + SO_3^{2-}$
   base$_1$  acid$_2$      acid$_1$   base$_2$

   c. $HCl + HSO_3^- \rightleftharpoons Cl^- + H_2SO_3$
   acid$_1$  base$_2$    base$_1$   acid$_2$

   d. $HCO_3^- + HPO_4^{2-} \rightleftharpoons H_2CO_3 + PO_4^{3-}$
   base$_1$   acid$_2$      acid$_1$    base$_2$

3. a. $1 \times 10^{-7}$ $M$   b. $1$ $M$    c. $0.01$ $M$
4. a. $1 \times 10^{-7}$ $M$   b. $0.01$ $M$   c. $0.02$ $M$
5. $[H^+] = 1.0 \times 10^{-11}$ $M$
6. $[OH^-] = 5 \times 10^{-15}$ $M$
7. a. pH = 1.82     b. pH = 12.18
8. a. pOH = 0.22    b. pOH = 11.30
9. 100
10. $[H^+] = 3.5 \times 10^{-5}$ $M$
11. a. Strong    b. Weak    c. Weak    d. Strong    e. Weak    f. Strong
12. $Cl^- < HSO_4^- < NO_2^- < CH_3COO^-$
13. a. $H_2CO_3$   b. $H_3AsO_4$   c. $H_3PO_3$   d. $HClO_3$   e. $H_3PO_4$
14. a. HI   b. $H_2S$   c. HI   d. $H_2Se$
15. a. $H^+ + Cl^- \rightarrow HCl$   b. $H^+ + H_2O \rightarrow H_3O^+$   c. $BCl_3 + NH_3 \rightarrow BCl_3NH_3$
    d. $Al^{3+} + 4OH^- \rightarrow Al(OH)_4^-$   e. $Mg^{2+} + 2F^- \rightarrow MgF_2$

16. 3.35
17. to the right
18. c.
19. $[H^+] = 1.6 \times 10^{-4}$ $M$
20. 2.1%
21. pH = 2.90
22. $K_a = 6.9 \times 10^{-4}$
23. $[H^+] = [F^-] = 2.3 \times 10^{-3}$ $M$, $[HF] = 8 \times 10^{-3}$ $M$, $[OH^-] = 4.3 \times 10^{-12}$ $M$
24. pH = 11.78
25. $[H^+] = [HCO_3^-] = 1.8 \times 10^{-3}$ $M$
    $[CO_3^{2-}] = 4.8 \times 10^{-11}$ $M$
    $[H_2CO_3] = 0.080$ $M$
26. b. basic
27. a. acidic   b. slightly basic ($SO_4^{2-}$ is a proton acceptor) h   c. neutral   d. basic   e. acidic   f. basic
28. pH = 9.11
29. $[H^+] = 2.2 \times 10^{-11}$ $M$
30. d. KF

# Chapter 16.   Acid-Base and Solubility Equilibria

## EXERCISES and PROBLEMS

1.   a. pH = 2.87    b. pH = 4.05

2.  a. pH = 4.66    b. no
3.  a. initial pH is 9.39 and final is 9.45    b. $NH_4^+ \rightleftharpoons NH_3 + H^+$
    c. $OH^- + NH_4^+ \rightarrow H_2O + NH_3$
4.  6.50
5.  pH = 2.12
6.  pH > 7. Usually between 8 – 11 depending on the strength of the conjugate base of the weak acid.
7.  A titration of a strong acid with a weak base or titration of a weak base with a strong acid.
8.  a. $HBr(aq) + NaOH(aq) \rightarrow H_2O(\ell) + NaBr(aq)$    b. pH = 1.39
9.  4.38
10. 7
11. a.$BaCO_3(s) \rightleftharpoons Ba^{2+}(aq) + CO_3^{2-}(aq)$ $\qquad K_{sp} = [Ba^{2+}][CO_3^{2-}]$
    b. $CaF_2(s) \rightleftharpoons Ca^{2+}(aq) + 2F^-(aq)$ $\qquad K_{sp} = [Ca^{2+}][F^-]^2$
    c. $Al(OH)_3(s) \rightleftharpoons Al^{3+}(aq) + 3OH^-(aq)$ $\qquad K_{sp} = [Al^{3+}][OH^-]^3$
    d. $Ag_3PO_4(s) \rightleftharpoons 3Ag^+(aq) + PO_4^{3-}(aq)$ $\qquad K_{sp} = [Ag^+]^3[PO_4^{3-}]$
12. $K_{sp} = 4.87 \times 10^{-9}$
13. $K_{sp} = 1.69 \times 10^{-6}$
14. a. $s = 2.2 \times 10^{-4}\ M$    b. $[F^-] = 4.4 \times 10^{-4}\ M$
15. no
16. yes, $MgF_2$
17. $[Ba^{2+}] = 3.2 \times 10^{-5}\ M$
18. a. $CuI(s)$    b. For CuI to ppt $[I^-]$ must be $5.1 \times 10^{-8}\ M$. For $PbI_2$ to ppt $[I^-]$ must be $3.7 \times 10^{-3}\ M$.
    c. 0.0014%
19. a. $s = 6.9 \times 10^{-7}\ M$    b. $Ag_3PO_4(s) \rightleftharpoons 3Ag^+ + PO_4^{3-}$
20. $s = 23.5 \times 10^{-4}\ M$
21. $s = 8.0 \times 10^{-9}\ M$
22. b, c, and d.
23. $s = 1.2 \times 10^{-7}\ M$
24. pH = 9.10
25. a. $Cu^{2+}(aq) + 4NH_3 \rightleftharpoons Cu(NH_3)_4^{2+}$
    b. $Cu^{2+}(aq) + 4CN^- \rightleftharpoons Cu(CN)_4^{2+}$
    c. $Ag^+(aq) + 2CN^- \rightleftharpoons Ag(CN)_2^-$
26. $1.0 \times 10^{-22}\ M$
27. $K = K_{sp} \times K_f = 7.7 \times 10^8$
28. $s = 3.5 \times 10^{-3}\ M$
29. A solution of HBr and NaBr cannot be a buffer because the HBr is a strong acid. Its conjugate base $Br^-$ ion has no base strength. This means that the buffer cannot neutralize added strong acid. A reaction such as $H^+ + Br^- \rightarrow HBr$ cannot occur.
30. a. pH increases    b. no change in pH    c. pH increases
31. First determine what ions are present in the solution at the equivalence point. Here the salt formed by neutralization is NaCN. Second calculate the pH at the equivalence point. At the equivalence point we have a 0.050 $M$ NaCN solution. Cyanide ion undergoes hydrolysis $CN^- + H_2O \rightleftharpoons HCN + OH^-$. This equilibrium determines the pH. Knowing the $K_b$ for $CN^-$ ion calculate the $OH^-$ ion concentration, and then the pH as in Example 15.17. Third, look at Table 16.1 in the text, and find an indicator that changes color over a pH range that includes the pH at the equivalence point.

32. The solubility reaction of $Zn(OH)_2$ is:

$$Zn(OH)_2 \rightleftharpoons Zn^{2+}(aq) + 2OH^-(aq)$$

The pH of a solution directly affects the $OH^-$ ion concentration. To adjust the pH upward we must add more $OH^-$ ion to a solution. In terms of the solubility equilibrium $OH^-$ ion is a common ion and so the solubility equilibrium is shifted to the left as pH increases. Increasing the pH lowers the solubility of $Zn(OH)_2$.

33. Indicators are weak acids that ionize according to the equation:

$$HIn \rightleftharpoons H^+ + In^-$$

When [HIn] / [In⁻] > 10 the solution has the color of HIn. When [HIn] / [In⁻] > 0.1 the solution has the color of HIn. The equilibrium has to shift a long way for the ratio to change from 10:1 and 1:10. You can see from the equation that to shift the equilibrium you must change the $H^+$ ion concentration. Therefore, the indicator can change from its acid color to its base color only over a range of pH values.

## PRACTICE TEST

1. pH = 4.96
2. a. pH = 3.77   b. pH = 3.70
3. pH = 9.31
4. b. $NH_4Cl$ and $NH_3$
5. a. pH = 7.0   b. 40.0 mL   c. pH = 2.00   d. pH = 10.89
6. pH = 8.72
7. pH = 4.92
8. b. $Zn(OH)_2$
9. $K_{sp} = 3.2 \times 10^{-10}$
10. $K_{sp} = 1.8 \times 10^{-15}$
11. $s = 1.6 \times 10^{-5} M$
12. d. pure water
13. yes
14. yes
15. no
16. a. $CaCO_3$
17. a. $Mg^{2+}$   b. $[OH^-] = 3.5 \times 10^{-5} M$   c. $[Mg^{2+}] = 1.5 \times 10^{-7} M$   d. $1.5 \times 10^{-3}$% of $Mg^{2+}$ remains
18. c. $Ag(CN)_2^-$
19. $[Cu^{2+}] = 1.9 \times 10^{-16} M$
20. $s = 0.13 M$
21. $s = 0.042 M$
22. a. yes   b. $[Ag^+] = 3.8 \times 10^{-6} M$
23. $[NH_3] = 0.91 M$

# Chapter 17.   Chemistry in the Atmosphere

## EXERCISES and PROBLEMS

1. forming compounds from the element; $N_2(g) + O_2(g) \rightarrow 2NO(g)$
2. troposphere, stratosphere, mesosphere, and ionosphere The troposphere has constantly changing weather.
3. a. O* atoms   b. $N_2^+$* ions

4.  a. the stratosphere   b. 15–35 km
5.  10 ppm
6.  $3O_2(g) \rightarrow 2O_3(g)$
7.  $O_2(g) + h\nu \rightarrow 2O(g)$   $O_3(g) + h\nu \rightarrow O_2(g) + O(g)$   $CCl_2F_2 + h\nu \rightarrow CClF_2 + Cl$
8.  CFC = chlorofluorocarbon.   $CF_2Cl_2$ and $CFCl_3$
9.  chlorine monoxide
10. an international agreement to limit CFC production
11. combustion of a carbon compound.  respiration by animals and plants.  decompositon of dead animals and plants   volcanoes
12. methane, nitrous oxide, and CFCs
13. pH = 5.6
14. $SO_2(g) + H_2O(\ell) \rightarrow H_2SO_3(aq)$        $SO_3(g) + H_2O(\ell) \rightarrow H_2SO_4(aq)$
15. carbon monoxide, nitrogen monoxide, and unburned hydrocarbons
16. carbon monoxide, nitrogen monoxide, and unburned hydrocarbons
17. secondary pollutant
18. Ra   noble gas elements
19. plonium-214 and polonium-218
20. lung cancer
21. The answer lies in just where the ozone is. Ozone occurs naturally in the stratosphere which is too far away for people to come into contact with ozone. However, in the troposphere ozone concentrations are significantly increased by automobile exhaust reacting with sunlight. Humans (and other living things) can inhale too much ozone. Health effects include severe respiratory and eye irritation. Ozone in cities is a secondary pollutant.
22. Liming is the process of adding quicklime to lakes and soils to reduce their acidity. This process is very expensive due to the enormous transportation costs involved. It is much less expensive to remove $SO_2$ at the power plant as it is formed. Think about how much easier it is to "collect" sulfur atoms when they are concentrated in one place, rather than when they are spread out over the landscape as sulfuric acid.

## PRACTICE TEST

1. In the troposphere, the temperature decreases as altitude increases. In the stratosphere temperature increases with altitude.
2. a. $O_2(g) + h\nu \rightarrow 2O(g)$
      $O(g) + O_2(g) \rightarrow O_3(g)$
   b. $O_2$
3. a. $Cl(g) + O_3(g) \rightarrow ClO(g) + O_2(g)$
      $ClO(g) + O(g) \rightarrow O_2(g) + Cl(g)$
   b. The catalyst is atomic Cl, and the intermediate is chlorine monoxide, ClO.
4. HCFC stands for hydrochlorofluorocarbon.  $CF_3CHCl_2$
5. The trapping of heat near the earth's surface by atmospheric gases, particularly carbon dioxide.
6. Combustion and decomposition of trees and plants produces $CO_2$. Removing trees reduces photosynthesis which would normally consume $CO_2$.
7. landfills and natural gas leaks
8. emission of IR to space
9. $Fe(s) + 2H^+(aq) \rightarrow Fe^{2+}(aq) + H_2(g)$
   $CaCO_3(s) + 2H^+(aq) \rightarrow Ca^{2+}(aq) + CO_2(g) + H_2O(\ell)$
10. a. carbon monoxide, NO, and unburned hydrocarbons

b. CO is toxic but may not be involved in reactions producing smog. NO is oxidized to $NO_2$. Photodecomposition of $NO_2$ leads to ozone and PAN. Oxidation of unburned hydrocarbons produces alcohols and organic acids that eventually form aerosols and haze.

11. The wavelength is 376 nm, which is in the UV region.

12. 1. kills fish, salamanders, and frogs    2. damage to leaves and needles of trees; toxic to vegetation    3. damage to stone buildings and monuments

# Chapter 18.   Entropy, Free Energy & Equilibrium

## EXERCISES and PROBLEMS

1.    a. $CS_2(\ell)$    b. $SO_2(g)$    c. $BaSO_4(aq)$
2.    a. +    b. −    c. cannot predict the sign, $\Delta S$ will be essentially zero    d. −
3.    −198.5 J/K·mol
4.    8.7 kJ
5.    $\Delta G^\circ_{rxn}$ = 68.6 kJ/mol
6.    2440 K
7.    $\Delta S_{vap}$ = 92.9 J/K•mol
8.    negative
9.    a. yes    b. $7.50 \times 10^{20}$
10.    $\Delta G^\circ_{rxn}$ = 79.9 kJ/mol
11.    $\Delta G^\circ$ = +69.1 kJ and $\Delta G$ = +40.6 kJ. Therefore the reaction is nonspontaneous.
12.    $\Delta G^\circ$ refers to the free energy change when the reactants and products are both present in their standard states. Their concentrations are all 1 atm or 1 molar. $\Delta G$ refers to the free energy change when the reactants and products are present at concentrations other than those for the standard state.
13.    $\Delta G$ = 112.9 kJ/mol
14.    Think about the system. Let it be the oil or toxic substance. When an oil spill spreads out it is becoming more disordered as its molecules have a much larger area to occupy. Any process that gives rise to a more random distribution of the particles of the substance in space gives rise to an increase in entropy of the substance. In the reverse process, that of cleaning up a dispersed substance, work must be done to bring the substance together into a more concentrated space. This reduces the disorder of the system. This is a nonspontaneous process requiring a continual work input. Keeping the substance contained in the first place takes less work.

## PRACTICE TEST

1.    b and c
2.    a. $H_2O(g)$   b. $CO_2(g)$   c. $Ag^+(g)$   d. $Cl_2(g)$   e. $2Cl(g)$
3.    a. essentially zero    b. positive    c. positive    d. negative    e. negative
4.    $\Delta S_{vap}$ = 81 J/mol•K
5.    $\Delta G^\circ_{rxn}$ = −24.7 kJ/mol
6.    $\Delta S^\circ$ = − 111 J/K·mol
7.    a. $\Delta G^\circ$ = 236 kJ/mol    b. $\Delta G^\circ$ = 318.2 kJ/mol
8.    b. (not: don't forget to double $\Delta G_f^\circ$ in part a)
9.    $\Delta G^\circ_{rxn}$  = −23.3 kJ/mol
10.    $\Delta G^\circ_{rxn}$ = −34.85 kJ/mol; $K_p$ = $1.28 \times 10^6$

11.   yes, $\Delta G° < 0$
12.   a. o, $\Delta G° > 0$   b. T = 7320 K
13.   $\Delta G = -17.8$ kJ/mol  spontaneous in the forward direction


# Chapter 19.   Electrochemistry

## EXERCISES and PROBLEMS

1.   a. $14H^+(aq) + Cr_2O_7^{2-}(aq) + 6Br^-(aq) \rightarrow 2Cr^{3+}(aq) + 3Br_2(g) + 7H_2O(\ell)$

   b. $2MnO_4^-(aq) + 6I^-(aq) + 4H_2O(\ell) \rightarrow 2MnO_2(s) + 3I_2(aq) + 8OH^-(aq)$

2.   a. $5H_2C_2O_4(aq) + 2MnO_4^-(aq) + 6H^+(aq) \rightarrow 2Mn^{2+}(aq) + 10CO_2(g) + 8H_2O(\ell)$

   b. $Zn(s) + ClO^-(aq) + H_2O(\ell) \rightarrow Zn(OH)_2(s) + Cl^-(aq)$

3.   a. Cu    b. $Cu^{2+}(aq) + Pb(s) \rightarrow Cu(s) + Pb^{2+}(aq)$    c. Cu
   d. from Pb to Cu   e. Cu

4.   a. Al(s)    b. $Fe^{3+}$

5.   no

6.   a. no    b. yes

7.   a. $E°_{cell} = -2.58$ V    b. $E°_{cell} = 0.29$ V

8.   a. $\Delta G° = 498$ kJ/mol    b. $\Delta G° = -56.0$ kJ/mol

9.   $K_c = 3.1 \times 10^{-35}$

10.   E = 1.22 V

11.   $9.1 \times 10^{-4}$ M

12.   $E°_{cell} = 0.02$ V    $E_{cell} = 0.02$ V (note: Q=1, ln(1)=0, so (RT/nF)lnQ = 0)   $\Delta G = -11$ kJ

13.   a. $2Br^-(aq) \rightarrow Br_2(\ell) + 2e^-$
   b. $2H_2O(\ell) + 2e^- \rightarrow H_2(g) + 2OH^-(aq)$

14.   57,900 C

15.   1.9 mol $e^-$

16.   1.00 mol $e^-$

17.   63.3 min

18.   4.2 g Cd

19.   a. before: $[Cu^{2+}] = [SO_4^{2-}]$   b. after: $[Cu^{2+}] > [SO_4^{2-}]$   c. $Cl^-$ ions migrate into the solution
   from the salt bridge.  Two chloride ions enter the solution for each excess $Cu^{2+}$ ion.

20.   Avogadro's number is a ratio. Its units are "atoms per mole." We are given the mass of
   Ni so we can use it to find the number of moles. Now to count the atoms. Actually we
   will count the electrons and then divide by 2 to get the number of Ni atoms. As usual
   the total charge in coulombs flowing through the cell is equal to the current (A) times
   the time (t). Dividing the total charge by the charge on one electron gives the total
   number of electrons passing through the cell. The number of Ni atoms produced by the
   reduction of $Ni^{2+}$ ions is just half the number of electrons. Divide the number of Ni
   atoms by its corresponding number of moles in order to get Avogadro's number. The
   value obtained is $6.13 \times 10^{23}$.

21.   Molten $AlBr_3$ consists of $Al^{3+}$ ions and $Br^-$ ions. When it undergoes electrolysis the $Al^{3+}$
   ions are reduced to Al metal, and $Br^-$ ions are oxidized to $Br_2$. An aqueous solution of
   $AlBr_3$ consists of $H_2O$ molecules, as well as $Al^{3+}$ ions and $Br^-$ ions. As you can tell from
   reduction potentials, $Al^{3+}$ ions will not be reduced in aqueous solution. Rather the $H_2O$
   molecules are and the products are $H_2$ and $OH^-$ ions. In aqueous solution $Br^-$ ion will
   be oxidized just as they were in the molten state.

**PRACTICE TEST**

1. a. $Al(s) | Al^{3+}(aq) || H^+(aq) | H_2(g) | Pt(s)$
   b. $Pb(s) | Pb^{2+}(aq) || Fe^{3+}(aq) | Fe(s)$
   c. $Pt(s) | H_2(g) | H^+(aq) || Cu^{2+}(aq) | Cu(s)$
   d. $Pt(s) | Br^-(aq) | Br_2(\ell) || I_2(s) | I^-(aq) | Pt(s)$
   e. $Pt(s) | Fe^{2+}(aq), Fe^{3+}(aq) || Sn^{2+}(aq) | Sn(s)$
   f. $Pt(s) | Fe^{2+}(aq), Fe^{3+}(aq) || Sn^{4+}(aq), Sn^{2+}(aq) | Pt(s)$

2. a. Au   b. $2Au^{3+}(aq) + 3Fe(s) \rightarrow 2Au(s) + 3Fe^{2+}(aq)$
   c. 1.94 V   d. from Fe to Au   e. Au

3. $SO_4^{2-} < O_2 < Ce^{4+} < H_2O_2$

4. $F^- < H_2 < Ni < Zn$

5. a. yes   b. yes   c. no

6. $\Delta G° = 3$ kJ/mol;  $K = 0.3$

7. $[Fe^{2+}]/[Fe^{3+}] = 0.3$

8. $E = 1.21$ V

9. 0.73 V

10. The cell emf will increase due to a decrease in $[H^+]$. The cell reaction is $Cu^{2+} + H_2 \rightarrow Cu + 2H^+$.

11. $-1.84$ V

12. $[Cu^{2+}] = 3.3 \times 10^{-25}$ M

13. pH = 2.88

14. 0.45 mol $e^-$

15. 0.35 mol $e^-$

16. a. 5.7 g Ni   b. 3.8 g Co

17. 320 h

18. a. $H_2O_2 + 2I^- + 2H^+ \rightarrow I_2 + 2H_2O$
    b. $Cr_2O_7^{2-} + 3H_3AsO_3 + 8H^+ \rightarrow 2Cr^{3+} + 3H_3AsO_4 + 4H_2O$
    c. $Ag(s) + 2H^+(aq) + NO_3^-(aq) \rightarrow NO_2(g) + Ag^+(aq) + H_2O(\ell)$

19. $4Cl_2 + 8OH^- \rightarrow ClO_4^- + 7Cl^- + 4H_2O$

20. $K_w = 9.0 \times 10^{-15}$

21. pH = 3.38

22. 0.19 h


# Chapter 20.   Metallurgy and the Chemistry of Metals

**EXERCISES and PROBLEMS**

1. A mineral is a naturally occurring substance with a characteristic range of chemical composition. An ore is a mineral, or a mixture of minerals, from which a particular metal can be profitably extracted.

2. The electropositive metals can be reduced by a more electropositive metal such as Li, but there is no metal electropositive enough to reduce $Li^+$ to Li. Electrolytic reduction is the only way to prepare Li. It is possible to reduce other metals with Li. Rather than prepare Li by electrolysis and then use it to reduce aluminum, magnesium, or sodium, it is more convenient simply to prepare these metals in one step by electrolytic reduction.

3. Zone refining is a technique for purifying metals. Figure 20.8 in the text explains how it works.
4. Limestone is used to remove impurities from iron during production. Limestone decomposes to lime (CaO). CaO reacts with $SiO_2$ and $Al_2O_3$ impurities from the iron ore forming calcium silicate and calcium aluminate. The mixture of calcium silicate and calcium aluminate is known as slag.
5. $Cr_2O_3(s) + 2Al(s) \rightarrow 2Cr(\ell) + Al_2O_3(s)$
6. A band is a large number of molecular orbitals that are closely spaced in energy. The valence band is a set of closely spaced MOs that are filled with electrons. The conduction band is a set of closely spaced empty orbitals. An electron can travel freely through the metal since the conduction band is void of electrons.
7. A conductor is capable of conducting an electrical current. In a conductor there is essentially no energy gap between the valence band and the conduction band. Insulators are materials that do not conduct. In an insulator the energy gap between the valence band and the conduction band is much greater than the gap in a conductor. Semiconductors are normally not conductors, but will conduct electricity at higher temperatures, or when combined with small amounts of certain elements.
8. Good conductors of heat and electricity. Low density. Soft enough to cut with a knife. Low melting points.
9. Extremely reactive. React with water to form hydroxides. React with oxygen to form a variety of oxides, peroxides and superoxides.
10. Magnesium is precipitated from seawater. Calcium is obtained from limestone.
11. $CaCO_3$, $CaO$, and $Ca(OH)_2$, respectively
12. Molten cryolite is used as a solvent for alumina, $Al_2O_3$, in aluminum production. Cryolite melts at 1000 °C as compared to 2050 °C for alumina and therefore it lowers the energy consumption.
13. Reaction with acid:   $Al(OH)_3(s) + 3H^+(aq) \rightarrow Al^{3+}(aq) + 3H_2O(\ell)$
    Reaction with base:   $Al(OH)_3(s) + OH^-(aq) \rightarrow Al(OH)_4^-(aq)$
14. A fresh surface of aluminum does react readily with oxygen and forms a surface coating of $Al_2O_3$. This layer is very impenetrable and does not allow oxygen and water through to continue a reaction with the rest of the aluminum.
15. Alums have the general formula: $M^+M^{3+}(SO_4)_2 \cdot 12H_2O$ where $M^+ = K^+$, $Na^+$, $NH_4^+$ and $M^{3+} = Al^{3+}$, $Cr^{3+}$, $Fe^{3+}$. In baking powder, alum is used as a drying agent to keep it from caking.

## PRACTICE TEST

1. a. Leaching is the selective dissolution of a metal from an ore. Flotation is a technique used to separate mineral particles from waste clays and silicates called gangue.
   b. Roasting involves heating an ore in the presence of air. The idea is to convert metal carbonates and sulfides to oxides, which can be more conveniently reduced to yield pure metals. Reduction is the gain of electrons by a cation to yield the pure metal.
2. No. Use Appendix 3 of the text to calculate $\Delta G°$. The sign of $\Delta G°$ indicates whether the reaction will go or not.
3. $Al^{3+}(aq) + 3OH^-(aq) \rightarrow Al(OH)_3(s)$
   $Al(OH)_3(s) + OH^-(aq) \rightarrow Al(OH)_4^-(aq)$
4. $2Na + Cl_2 \rightarrow 2NaCl$
   $2Na + O_2 \rightarrow Na_2O_2$
   $2Na + \frac{1}{2}O_2 \rightarrow Na_2O$
5. Carbon monoxide. $Fe_2O_3 + 3CO \rightarrow 2Fe + 3CO_2$
6. An alloy is a mixture of two or more metals, or of metals and nonmetal elements that have metallic properties. An amalgam is an alloy containing mercury.

7. Copper atoms have a half-filled 4s orbital. This means that copper metal has a half-filled valence band. A partially filled valence band is a conduction band.
8. The increasingly random motion of electrons due to a temperature rise opposes motion in one unified direction.
9. As temperature increases more electrons acquire enough energy to "jump" the energy gap between the valence band and the conduction band.
10. The source of Al is bauxite. Potassium is obtained from sylvite.
11. Quicklime is CaO. It is made by the thermal decomposition of limestone.
12. $3K + AlCl_3 \rightarrow Al + 3KCl$
13. $Al_2O_3 \cdot 2H_2O$
14. The bauxite is pulverized and digested with sodium hydroxide solution. This converts the aluminum oxide to the aluminate ion $AlO_2^-$ which remains in solution. However treatment with base has no effect on the iron oxide which remains as insoluble $Fe_2O_3(s)$, and is removed by filtration. Aluminum hydroxide is then precipitated by acidification to about pH 6.
15. $Al_2(SO_4)_3$ is used to clarify water through its reaction with $Ca(OH)_2$. The precipitation of $Al(OH)_3$ traps dirt and dust particles.
16. pH = 2.00

# Chapter 21.   Non-Metallic Elements and their Compounds

## EXERCISES and PROBLEMS

1. $2Al(s) + 3HCl(aq) \rightarrow AlCl_3(aq) + \frac{3}{2}H_2(g)$
2. −1, +1
3. NaH
4. Brønsted base  $H^-(aq) + H_2O(\ell) \rightarrow OH^-(aq) + H_2(g)$
5. The melting point of diamond is related to the energy required to break C—C bonds. In diamond each carbon atom is covalently linked to four other carbon atoms. To turn diamond into a liquid about half of these bonds must be broken.  This requires extremely high temperatures.
6. $CN^-$, a $C_2^{2-}$ and a $C^{4-}$ ions
7. CO
8. −5 and −3
9. $4NH_3(g) + 5O_2(g) \rightarrow 4NO(g) + 6H_2O(g)$
   $2NO(g) + O_2(s) \rightarrow 2NO_2(s)$
   $2NO_2(g) + H_2O(\ell) \rightarrow HNO_2(aq) + HNO_3(aq)$
10. White phosphorus is very unstable compared to red phosphorus.
11. ozone
12. steelmaking
13. −1, −2, 0
14. $2H_2O_2(aq) \rightarrow 2H_2O(\ell) + O_2(g)$
15. $K_2O$, $K_2O_2$, and $KO_2$
16. $S(s) + O_2(g) \rightarrow SO_2(g)$
    $2SO_2(g) + O_2(g) \rightarrow 2SO_3(g)$
    $SO_3(g) + H_2O(\ell) \rightarrow H_2SO_4(aq)$
17. $F_2$ and $Cl_2$
18. $Cl_2(g) + 2Br^-(aq) \rightarrow 2Cl^-(aq) + Br_2(\ell)$
19. HCl, HBr, and HI
20. $F_2$

21.  a. $Cl_2(g) + 2I^-(aq) \rightarrow 2Cl^-(aq) + I_2(aq)$
     $Br_2(g) + 2I^-(aq) \rightarrow 2Br^-(aq) + I_2(aq)$
     b. $Cl_2(g) + 2NaOH(aq) \rightarrow NaOCl(aq) + NaCl(aq) + H_2O(\ell)$
     $Br_2(g) + 2NaOH(aq) \rightarrow NaOBr(aq) + NaBr(aq) + H_2O(\ell)$

22.  Because hydrogen can be oxidized to the $H^+$ ion, it resembles the alkali metals. And because it can be reduced to the $H^-$ ion it resembles the halogen elements. Hydrogen has an electron configuration of $1s^1$ making it similar to the alkali metals. It also forms diatomic molecules just as the halogens do. $H_2$ is a gas at room temperature. As a result of these properties, hydrogen resembles the alkali metals and the halogens.

23.  $Br_2$ can act as both an oxidizing agent and a reducing agent because Br has three oxidation states, namely $-1$ in $Br^-$ ion, 0 in $Br_2$, and $+1$ in $OBr^-$. Therefore, $Br_2$ can be an oxidizing agent by accepting electrons and being reduced to the $Br^-$ ion. And $Br_2$ can also be a reducing agent by donating electrons and being oxidized to the $OBr^-$ ion.

## PRACTICE TEST

1.  $C_3H_8(g) + 3H_2O(g) \rightarrow 3CO(g) + 7H_2(g)$
    $C(s) + H_2O(g) \rightarrow CO(g) + H_2(g)$
2.  to make ammonia and partially hydrogenated food products
3.  $Na_2B_4O_7 \cdot 4H_2O + H_2SO_4 + H_2O \rightarrow 4H_3BO_3 + Na_2SO_4$
4.  When coal is distilled in the absence of oxygen to release volatile hydrocarbons, the high carbon residue is called coke. Coke is made into graphite by the Acheson process, in which an electric current is passed through coke for several days.
5.  a. create atmospheres devoid of $O_2$; a coolant
    b. ammonia fertilizer
    c. nitrate fertilizer and explosives
6.  A mixture of 3 volumes of concentrated HCl to 1 volume of concentrated $HNO_3$.
7.  $NO_2$, NO, $P_2O_5$, $P_2O_3$
8.  $12NH_3 + 21O_2 \rightarrow 8HNO_3 + 14H_2O + 4NO$
9.  It is insoluble.
10. diammonium phosphate, $(NH_4)_2HPO_4$; superphosphate $Ca(HPO_4)_2 \cdot H_2O$
11. elemental oxygen, $O_2$, ozone, $O_3$; rhombic $S_8$, monoclinic $S_8$
12. in the blast furnace for steel making, medical uses, welding
13. peroxide ion $O_2^{2-}$; superoxide $O_2^-$
14. $CaF_2 + H_2SO_4 \rightarrow 2HF + CaSO_4$
    $2HF \rightarrow H_2 + F_2$
    $2NaCl + 2H_2O \rightarrow 2NaOH + H_2 + Cl_2$
    $Cl_2 + 2I^- \rightarrow 2Cl^- + I_2$
15. $HIO_3$, $HBrO_3$, $HClO_3$. The comparable fluorine acid does not exist.
16. $I_2(s)$, black crystals; $Br_2(\ell)$, dark red liquid; $Cl_2(g)$, a yellow-green gas; $F_2(g)$, pale yellow gas
17. $Cl_2 + 3F_2 \rightarrow 2ClF_3$
    $KBr + 3F_2 \rightarrow KF + BrF_5$
    $KI + 4F_2 \rightarrow KF + IF_7$

# Chapter 22. Transition Metal Chemistry and Coordination Compounds

## EXERCISES and PROBLEMS

1.  a. $[Ar]4s^2 3d^5$   b. $[Ar]3d^5$   c. $[Ar]3d^4$
2.  a. V   b. Nb
3.  Fe
4.  $-2$
5.  4
6.  $+2$
7.  coordination number = 6,   oxidation number = $+2$
8.  tribromotrichlorocobaltate(II) ion
9.  diamminebis(ethylenediamine)nickel(II) ion
10. a. $[Cu(NH_3)_2(C_2O_4)]$   b. $K[PtNH_3Cl_3]$
11. Geometric isomers have the same numbers and kinds of atoms but the arrangement of atoms is different even though the same bonds are present. Optical isomers (enantiomers) are mirror images that cannot be superimposed on each other.
12. Chiral objects are mirror images that cannot be superimposed on one another no matter how they are rotated.
13. orange
14. ethylenediamine
15. The terms low-spin and high-spin refer to magnetic properties of complex ions. In a high-spin complex the electrons are arranged so that they are unpaired as much as possible. In a low-spin complex there may be more electrons that are paired. High-spin complexes are more paramagnetic than low-spin complexes.
16. 4
17. a. Six d-electrons total. In the low-spin $O_2$-containing complex there are no unpaired electrons. In the high-spin complex with $H_2O$ there are four unpaired electrons.
    b. The two forms of hemoglobin have different colors due to the differing amount of crystal field splitting ($\Delta$) in the iron complexes. $O_2$ must be a stronger-field ligand than $H_2O$. The splitting is much greater in the low-spin $O_2$-containing complex than in the high-spin $H_2O$-containing complex. Therefore the color of light absorbed will be different for the two complexes, leading to different observed colors. From the complementary colors you can tell that the $O_2$ complex absorbs green light and the $H_2O$ complex absorbs orange light.

## PRACTICE TEST

1.  Fe $[Ar]4s^2 3d^6$
    Fe $[Ar]3d^6$
    Fe $[Ar]3d^5$
2.  $[Pt(NH_3)_3Cl_3]NO_3$
3.  $+2$
4.  The coordination number is 6; the oxidation state is $+3$.
5.  The coordination number is 6; the oxidation state is $+2$.
6.  a. tetraamminedichloroplatinum(IV) chloride
    b. hexaaminenickel(II) ion
    c. tetrahydroxochromate(III) ion
    d. hexacyanocobaltate(III) ion
    e. potassium tetracyanocuprate(II)

7.    a. $[Al(OH)_4]^-$
      b. $[HgI_4]^{2-}$
      c. $K_3[Co(C_2O_4)_2Cl_2]$
      d. $[Ni(en)_3]SO_4$

8.

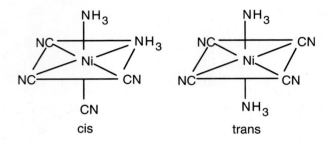

             cis                trans

9.    The two structures are not enantiomers.

10.  Because when plane-polarized light passes through a solution of a chiral substance, the plane of the polarized light is rotated.

11.  a.

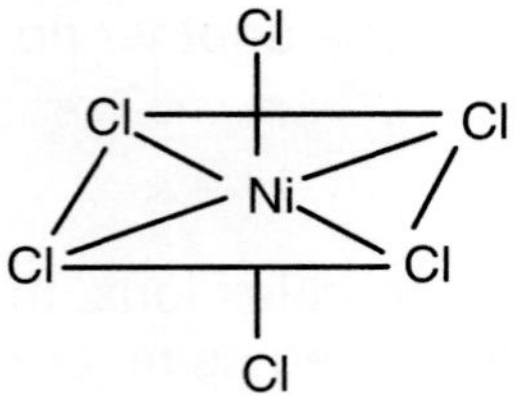

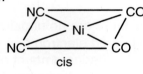

    b.

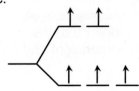

    c.

    d.    O  O
            ||  ||
       ⁻O—C—C—O⁻

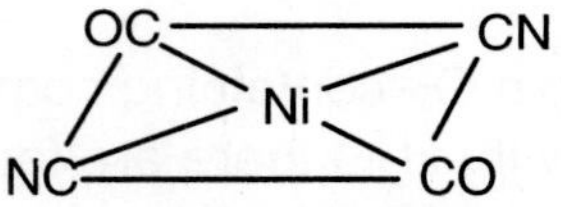

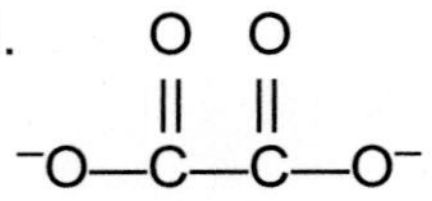

    e. $[H_3N{-}Ag{-}NH_3]^+$

12.  red

13.

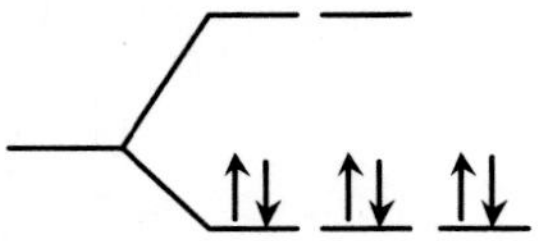

14.

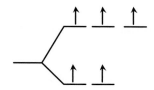

15. The terms inert and labile are kinetic terms, that is, they refer to relative rates of ligand substitution reaction. Complexes that undergo rapid ligand exchange are said to be labile complexes. In inert complexes the bonds between the central atom and the ligands are broken and re-formed relatively infrequently.
16. Zinc has ten d electrons that completely fill the five d orbitals.
17. 4
18. $[Ni(en)_3]^{2+}$

# Chapter 23. Nuclear Chemistry

## EXERCISES and PROBLEMS

1. $_2^4\alpha$ $\quad\quad\quad\quad\quad _{-1}^0\beta$ $\quad\quad\quad\quad\quad _0^0\gamma$
2. 15 protons, 17 neutrons, and 15 electrons
3. a. $_1^1H$ b. $_{82}^{206}Pb$ c. $_0^1n$ d. $_8^{15}O$
4. a. $_2^4He$ b. $_{13}^{27}Al$
5. lower
6. a. $_9^{18}F \rightarrow {}_8^{18}O + {}_{+1}^0\beta$

   b. $_9^{21}F \rightarrow {}_{10}^{21}Ne + {}_{-1}^0\beta$
7. $6.022 \times 10^{26}$ amu = 1.00 kg
8. 13 protons and 14 neutrons
9. $1.34 \times 10^{-12}$ J/nucleon
10. $1.45 \times 10^{-12}$ J
11. 8 alpha particles and 6 beta particles
12. the integrated first-order rate equation
13. a. 0.132/yr b. 6.3 mg
14. a. $4.89 \times 10^{-18}$ s$^{-1}$ b. $4.49 \times 10^9$ y
15. 3520 y
16. $5.5 \times 10^9$ y
17. $2.8 \times 10^9$ y
18. a. $_{12}^{27}Mg$ b. $_6^{14}C$
19. a. $_4^8Be$ b. $_4^{10}Be$

20. $_{42}^{98}Mo + {}_0^1n \rightarrow {}_{42}^{99}Mo$

    $_{42}^{99}Mo \rightarrow {}_{43}^{99}Tc + {}_{-1}^0\beta$
21. nuclear fusion
22. $_{36}^{93}Kr$
23. $_2^4He$

24. Both a rad and a rem refer to the absorption of the same amount of energy per kilogram of an object. However, for living matter the actual damage done varies for each type of particle absorbed. The rem is the dosage unit used for biological material. It is the absorbed dose in rads multiplied by the relative biological effectiveness factor, *RBE*. For beta and gamma rays *RBE* = 1.0; for fast neutrons and alpha particles *RBE* = 10.  dose (rem) = *RBE* × dose (rad)

## PRACTICE TEST

1. Beta particles and positrons have the same mass, 0.00055 amu. Beta particles have one unit of negative charge while positrons have one unit of positive charge.

2. a. $^{239}_{94}Pu \rightarrow \, ^{4}_{2}\alpha + \, ^{235}_{92}U$

   b. $^{2}_{1}H + \, ^{6}_{3}Li \rightarrow 2 \, ^{4}_{2}He$

   c. $^{90}_{38}Sr \rightarrow \, ^{90}_{39}Y + \, ^{0}_{-1}\beta$

   d. $^{10}_{5}B + \, ^{4}_{2}\alpha \rightarrow \, ^{13}_{7}N + \, ^{1}_{0}n$

   e. $^{56}_{26}Fe + \, ^{1}_{0}n \rightarrow \, ^{57}_{26}Fe$

3. $^{11}_{5}B < \, ^{39}_{20}Ca < \, ^{40}_{20}Ca$. B-11 has an odd number of protons and an odd number of neutrons (rule 2).
   Ca-39 has an even number of protons and an odd number of neutrons (rule 1). Ca-40 has an even number of protons and an even number of neutrons. It also has a "magic number" of protons and neutrons (rule 2).

4. a. positron emission   b. positron emission   c. beta decay

5. a. $3.6 \times 10^{-11}$ J/atom; $1.3 \times 10^{-12}$ J/nucleon
   b. $3.8 \times 10^{-11}$ J/atom; $1.4 \times 10^{-12}$ J/nucleon

6. a. $\lambda = 0.132$ /yr   b. $N/N_0 = 0.205$

7. $t_{1/2} = 12.4$ yr

8. $t = 5.1$ yr

9. $t = 1.5 \times 10^{9}$ yr

10. c. $\Delta m = -1.98 \times 10^{-2}$ amu

11. $^{209}_{83}Bi + \, ^{58}_{26}Fe \rightarrow \, ^{266}_{109}Mt + \, ^{1}_{0}n$

12. $^{238}_{92}U + \, ^{1}_{0}n \rightarrow \, ^{239}_{92}U$

    $^{239}_{92}U \rightarrow \, ^{239}_{93}Np + \, ^{0}_{-1}\beta$

    $^{239}_{93}Np \rightarrow \, ^{239}_{94}Pu + \, ^{0}_{-1}\beta$

13. $^{235}_{92}U + \, ^{1}_{0}n \rightarrow \, ^{137}_{55}Cs + \, ^{96}_{37}Rb + 3 \, ^{1}_{0}n$

14. positron emission

15. 8 alpha particles account for the mass change; 8 alpha and 6 beta particles account for the change in positive charge

16. 0.10 rad

17. 10 rem

18. $2.3 \times 10^{19}$ Ra atoms decayed in 10 years; $2.3 \times 10^{19}$ He atoms have a mass of $2 \times 10^{-4}$ g and occupy a volume of 1 cm$^3$ at STP.

# Chapter 24.   Organic Chemistry

## EXERCISES and PROBLEMS

1. $C_nH_{2n+2}$, $C_nH_{2n}$, and $C_nH_{2n-2}$
2. 3-methylhexane
3. 4-methyl-2-pentene
4. a. 2-fluoropropane     b. 2-chloro-4, 4-dimethylpentane
5.

$$CH_3-CH_2-\overset{\displaystyle CH_3}{\underset{\displaystyle CH_3}{\overset{|}{\underset{|}{C}}}}-CH_2-CH_3$$

6. a. 2-heptene     b. 1-hexyne
7. alkenes and alkynes
8. $CH_3CHCl-CH_2Cl$
9. $CH_3CHBr-CH_3$
10. Six isomers. 1–Butene, *cis*- and *trans*-2-butene, 2-methyl propene and 2 cycloalkanes
11. a. o-dinitrobenzene     b. p-dibromobenzene     c. m-bromonitrobenzene
12. Benzene is much less reactive than alkenes. Rather than addition reactions, benzene undergoes substitution reactions.
13.

$$\overset{\displaystyle O}{\overset{\|}{-C-H}} \qquad \overset{\displaystyle H}{\overset{|}{-N-H}} \qquad \overset{\displaystyle O}{\overset{\|}{-C-O-H}} \qquad \overset{\displaystyle O}{\overset{\|}{-C-O-R}}$$

14. nitrogen and hydrogen
15. a. alcohol   b. aldehyde   c. ketone   d. ester   e. alkane
16. Alcohols react with carboxylic acids to form esters. Also alcohols are easily oxidized.
17. a secondary alcohol
18. carboxylic acids and alcohols
19. an ester; $C_6H_{13}COOC_2H_5$ and NaOH

## PRACTICE TEST

1. a. *n*-nonane              b. 3-methyl-1-pentene
2. a. 2-ethyl-1-butanol   b. 5-methyl-3-octyne
3. 3,4-diethyl-2-methylheptane
4. 4-ethyl-3-methylheptane
5. *trans*-2-pentene
6. a. 2-butanol              b. methanol

7. a.

$$CH_3-CH_2-CH-\overset{\displaystyle C_2H_5}{\underset{\displaystyle CH_3-CH-CH_3}{\overset{|}{\underset{|}{C}}}}\overset{\displaystyle CH_3}{\overset{|}{}}-CH_2-CH_2-CH_3$$

b.

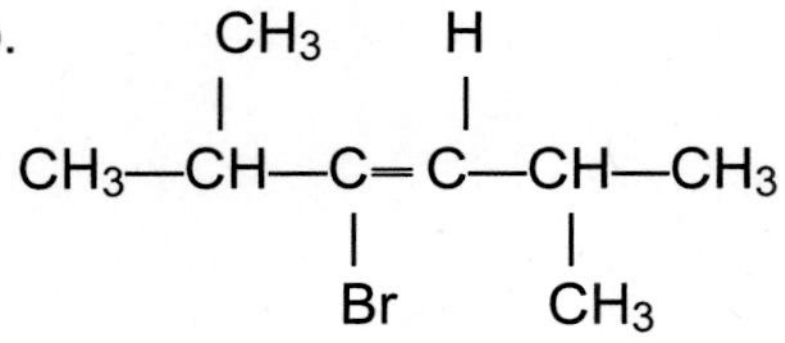

$$\underset{\underset{Br}{|}}{CH_3-CH-\underset{CH_3}{\overset{CH_3}{|}}C}=\overset{H}{\underset{\underset{CH_3}{|}}{C}}-CH-CH_3$$

c.

$$CH_3-CH_2-\underset{\underset{OH}{|}}{\overset{\overset{CH_3}{|}}{C}}-CH_2-CH_2-CH_3$$

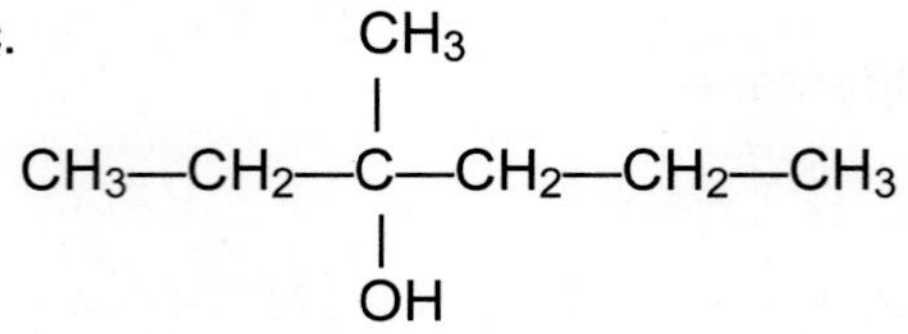

d.        CCl₃–COOH

8.   There are five alkenes:  1-pentene, 3-methyl-1-butene, 2-methyl-1-butene, and *cis*- and *trans*-2-pentene.  There are three cycloalkanes:  cyclopentane, methylcyclobutane, and dimethylcyclopropane.
9.   carbon-carbon double bond, hydroxyl, amine
10.  carboxyl, carbonyl (ketone), carbonyl (aldehyde)

11.  a.

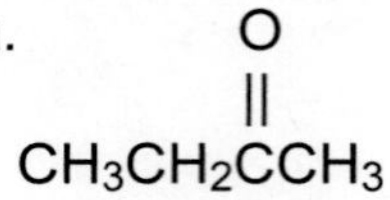

$$CH_3CH_2\overset{\overset{\displaystyle O}{\|}}{C}CH_3$$

     b. CH₃COOH
12.  a. CH₃CCl₂CHCl₂
     b. C₃H₇Cl,  C₃H₆Cl₂,  C₃H₅Cl₃,  etc.

c.

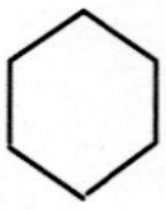

13.  a. 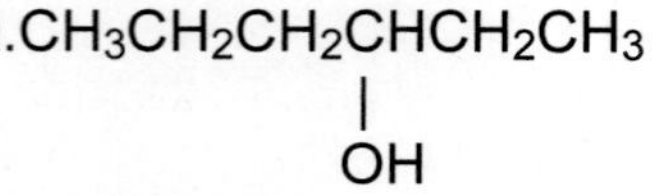 CH₃CH₂CH₂CHCH₂CH₃
                     |
                     OH
     b.

$$CH_3\overset{\overset{\displaystyle O}{\|}}{C}CH_3$$

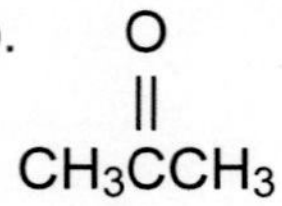

14.  a. no reaction

b.

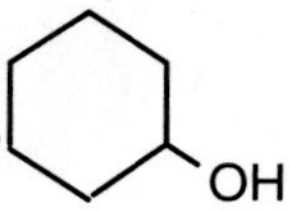

15. a.
$$\underset{\substack{|| \\ CH_3COCH_3}}{O}$$

   b. $CH_3CH_2CH_2CH_2CH_3$

16. a.
$$\underset{\substack{| \quad || \\ CH_3—CH—C—OH}}{\overset{CH_3 \quad O}{\phantom{x}}} + CH_3OH$$
   b. $CH_3(CH_2)_{12}COONa + CH_3(CH_2)_4OH$

17. a. A straight chain or branch chain hydrocarbon with one or more double bonds
    b. Organic compounds with the same functional group that differ only in the number of carbon atoms.
    c. A reaction in which a small molecule is added to an organic compound. The organic molecule must contain a C= C bond or C= O (carbonyl) group.
    d. A characteristic grouping of atoms that imparts certain chemical and physical properties when incorporate into organic molecules

18. Seven (four alcohols and three ethers)

19.
$$CH_3CH_2CH_2OH \xrightarrow[\text{conc}]{H_2SO_4} CH_3CH= CH_2 + H_2O$$

$$CH_3CH= CH_2 + H_2O \xrightarrow[\text{dil}]{H_2SO_4} \underset{\substack{| \\ OH}}{CH_3CHCH_3}$$

# Chapter 25.  Synthetic and Natural Polymers
## EXERCISES and PROBLEMS

1. initiation, chain growth, and termination
2. $CF_2= CF_2$
3. a homopolymer
4. An amino acid is a carboxylic acid that contains an amino group (—$NH_2$) attached to the carbon atom adjacent to the carboxylic acid functional group (—COOH).

5.
$$\underset{\substack{| \\ CH_3}}{\overset{\substack{H \quad O \\ | \quad ||}}{^+H_3N—C—C—O^-}}$$

6. An amide group is a –CO—NH– group.  The peptide bond is the C—N bond that links amino acids in a polypeptide or protein.
7. the sequence of amino acids in a peptide chain.
8. Denaturation is the loss of structure in a protein. The structure change may be a loss of secondary, tertiary, or quaternary structure. Sometimes the primary structure is lost.
9. deoxyribose.  Ribose
10. adenine, guanine, cytosine, and thymine
11. hydrogen bond
12. nucleotide

## PRACTICE TEST

1.  a. $CH_3CH= CHCl$          b. $CH_2= CCl_2$

2.
$$HO-\overset{\overset{\displaystyle O}{\|}}{C}-(CH_2)_4-\overset{\overset{\displaystyle O}{\|}}{C}-OH \qquad HOCH_2CH_2CH_2OH$$

3.
$$\underset{CH_2= C-CH= CH_2}{\overset{\overset{\displaystyle CH_3}{|}}{}}$$

4.  C, H, O, N, and S
5.  catalysts, transport, contractile, protection, hormones, structural elements
6.  Enzymes are catalysts.
7.  Amino acids consist of an amino group, a carboxylic acid group, a hydrogen atom, and R-roup all bonded to the same carbon atom.
8.  A dipeptide is two amino acids linked by a peptide bond. A polypeptide consists of many amino acid residues each linked to the next by peptide bonds.

9.

      ser               cys               lys

10.  a. 2    b. 1
11.  The primary structure of a protein refers to the number and the sequence of amino acids within the polypeptide chain. The chain twists and folds so that some amino acids far apart in sequence come into proximity with each other. The tertiary structure is the 3D structure that shows the folding of the polypeptide chain.
12.  (1) RNA contains uracil but no thymine; DNA contains thymine but no uracil. (2) RNA contains ribose as the 5-carbon sugar; DNA contains deoxyribose. (3) RNA is single stranded; DNA is double helical. (4) DNA is found only in the nucleus; RNA is found outside the nucleus.

13.

14. DNA is the molecule of heredity; it is responsible for passing genetic information from one generation to the next.
15. T, A, C, G, A

# Notes

# Notes

# Notes

# Notes

# Notes

# Notes

# Notes

# Notes

# Notes

# Notes

# Notes